Chemical Process Design, Simulation and Optimization

Chemical Process Design, Simulation and Optimization

Editors

Jean-Pierre Corriou
Jean-Claude Assaf

MDPI • Basel • Beijing • Wuhan • Barcelona • Belgrade • Manchester • Tokyo • Cluj • Tianjin

Editors

Jean-Pierre Corriou	Jean-Claude Assaf
University of Lorraine	Lebanese University (UL)
France	Lebanon

Editorial Office
MDPI
St. Alban-Anlage 66
4052 Basel, Switzerland

This is a reprint of articles from the Special Issue published online in the open access journal *Processes* (ISSN 2227-9717) (available at: https://www.mdpi.com/journal/processes/special_issues/chemical_design).

For citation purposes, cite each article independently as indicated on the article page online and as indicated below:

LastName, A.A.; LastName, B.B.; LastName, C.C. Article Title. *Journal Name* **Year**, *Volume Number*, Page Range.

ISBN 978-3-0365-0096-6 (Hbk)
ISBN 978-3-0365-0097-3 (PDF)

Cover image courtesy of Jean-Pierre Corriou.

Contents

About the Editors

Jean-Pierre Corriou is emeritus Professor at the National School of Chemical Industries, University of Lorraine, Nancy, France. Prof. Corriou is a specialist in the modeling, simulation, optimization, and control of chemical processes. He is the author of three books, including *Process Control–Theory and Applications* (Springer, 2nd edition, 2018), *Commande des Procédés* (in French), and *Méthodes Numériques et Optimization* (in French). He has published more than 100 articles in international journals. Since his retirement in 2014, Prof. Corriou has worked as a simulation director in a startup company specializing in the simulation of fine chemicals and biochemical processes. He has also been an invited professor in Canadian, Brazilian, and Chinese universities.

Jean-Claude Assaf (Associate Professor) is a lecturer with the chemical department of the Faculty of Engineering at Lebanese University and Saint Joseph University. Dr. Assaf has won several national and international awards including the Lebanese Industrial Research Achievements (LIRA) award. His expertise lies in chemical and petrochemical process design, simulation, and optimization. Dr. Assaf is the owner of several industrial patents and has published numerous articles in peer-reviewed journals. Since 2018, Dr. Assaf has been recognized as an outstanding reviewer for a number of high impact journals in which he conducted successful reviews.

Editorial

Special Issue on "Chemical Process Design, Simulation and Optimization"

Jean-Pierre Corriou [1,*] and **Jean-Claude Assaf** [2,3,4,*]

1   LRGP-CNRS-ENSIC, University of Lorraine, 1 rue Grandville BP 20451, 54001 Nancy, France
2   Laboratory of Microbiology, Department of Life and Earth Sciences, Faculty of Sciences I,
    Lebanese University, Hadat Campus, Beirut, Lebanon
3   Platform of Research and Analysis in Environmental Sciences (PRASE), Doctoral School of Sciences and
    Technologies, Lebanese University, Hadat Campus, Beirut, Lebanon
4   Ecole Doctorale 'Sciences et Santé', Université Saint-Joseph de Beyrouth, Campus des Sciences Médicales
    et Infirmières, Riad El Solh, Beyrouth, Liban
*   Correspondence: jean-pierre.corriou@univ-lorraine.fr (J.-P.C.); jeanclaude.assaf@net.usj.edu.lb (J.-C.A.)

Received: 30 November 2020; Accepted: 2 December 2020; Published: 4 December 2020

Since humanity has been able to transform materials, such as raw minerals, and produce food or beverages, a central question was the type of operation and how and where it should be performed. Thus, furnaces, ovens and simple distillation stills were operated, to cite a few. Thermodynamics already started as a new science with the development of vapor engines, from Denis Papin in the seventeenth century to James Watt in the eighteenth. The nineteenth century saw the huge development of industries devoted to large productions and the arrival of electricity. New methods started to be available to engineers for the design of workshops in factories. However, chemical engineering started as a real science with the concept of unit operations first taught by G. Davis at the University of Manchester in 1887 and quickly spread to MIT. At the beginning of the twentieth century, the development of the petroleum industry and the building of refineries strengthened the need for new and more efficient calculation procedures; thus, new concepts were invented. Transport phenomena, heat and mass transfer and chemical reaction engineering became new fields of research and found important applications. After the Second World War, the arrival of computers modified the landscape for engineers as, for the first time, it allowed them to perform extremely complex calculations and no frontiers seemed impossible to reach. First, custom computing codes were developed, then commercial codes, often created in research labs before industrial development, were largely available to design engineers for their simulation needs. Simultaneously, it seemed that chemical engineering was too restrictive and its methods could be applied in other domains, such as metallurgy and biotechnology, hence the broader denomination of chemical process engineering. Batch as well as continuous processes are considered, and dynamic simulation as well as steady-state simulation is a usual practice. Furthermore, with more and more powerful computers, it was possible to aggregate the optimization with respect to parameters and even unit operations together with the design. Nowadays, even process control seems possible at early stages of the design. It is also usual to use dedicated codes of computational fluid dynamics integrating heat transfer and possibly chemical reactions, in industries as well as in research departments.

In this context, it is natural that this Special Issue of "Chemical Process Design, Simulation and Optimization" was likely to attract a large number of proposals for authors of many different countries. It must be underlined that these proposals are very different from what they would have been forty or twenty years before with the availability of computing codes.

General Methods

Some articles deal with general methods which belong to the methodology of chemical process engineering. The article by Furda et al. [1] proposes a general approach, coupling two commercial codes, Aspen Plus® and MATLAB®, for optimal design, with a special focus on energy saving, particularly with regard to steam use, such as in turbines, compressors, pumps and fans. They describe a coupled steam- and process-side modeling approach, taking into account the varying inlet steam parameters or shaft work requirements and implementing fix turbine and driven equipment efficiencies. They even show the influence on the balances. They apply this method to an industrial case of the heat-pump-assisted distillation of a liquid propane–propylene mixture, where the energy for separation is provided by a condensing steam turbine. They propose a serious change due to the increase in high-pressure steam. The impact on the gas emissions is also studied and the economic gains are assessed. Zecevic et al. [2] are also concerned by energy savings applied to a real industrial steam methane reformer unit used for ammonia production. They develop an integrated detailed model which takes into account the production parameters. This enables them to exchange data between the distributed control system and the model to continuously monitor and optimize the steam methane reformer catalyst and tubes. Thus, the overall energy saving is 3% in the real plant. Dizabar et al. [3] compare the exergy and advanced exergy analysis in three different organic Rankine cycles. Exergy is the extent of energy to the second law of thermodynamics by considering entropy and it allows to take into account the irreversibilities in a process. Thus, again, this study focuses on energy saving and process optimization. By advanced exergy analysis, they can separate the exergy loss between unavoidable and avoidable and endogenous and exogenous contributions for each component. This allows them to improve the design by concentrating on the most important components from an energy point of view. Wang et al. [4] propose an efficient numerical method to solve the calculation of a set of nonlinear equations applied to a reactive distillation column and to a distillation column. Their method, called inside-out, divides the calculation into two loop iterations. Sun et al. [5] study the problem of model-based fault diagnosis for a distillation column and, again, this study is strongly related to a numerical and algorithmic point of view. Although it might seem a particular case, their approach is general and can be used for different applications. It consists in adopting a hybrid inverse problem approach using partial least squares to fit and forecast trajectories of the fault parameters generated by least squares. It is applied to the classical Tennessee Eastman process. Skorych et al. [6] study the solution of population balance equations. This type of equation occurs in the description of the distributions of sizes of granular materials, such as polymers and crystals. This is an important problem which often poses numerical difficulties, particularly when it is met in flowsheets and in dynamic problems, resulting in partial differential equations. The authors propose a new model based on transformation matrices. They also use finite volumes to describe the phenomena of agglomeration and breakage. They are able to implement their method in the specially designed Dyssol simulator, which uses a sequential-modular approach. Simulation examples are taken from the pharmaceutical industry. This sufficiently general approach allows the users to apply this method to their own population balance models. Mukhtar et al. [7] study a problem very close to the previous one as it deals with the numerical study of a batch crystallization. They include fines dissolution, which is critical in the industrial practice. Their numerical method is based on a quadrature method of moments. The application refers to test problems, not real problems.

Coupling with Fluid Mechanics

As previously mentioned, due to the progress of computer efficiency, memory and rapidity and the availability of both commercial and open-source codes of fluid mechanics, these advanced numerical techniques are more and more frequently applied to traditional processes, which are, thus, revisited and improved. Zeng et al. [8] study a particular unit, the turbo air classifier, which is used to separate powders into coarse and fine. Based on the governing equations and using a commercially available software, they study the influence of operation parameters, such as rotation speed and air volume. Studies are performed for multiple particles. Simulation and experimental results are favorably compared. Cheng et al. [9] also use a computational fluid dynamics code to better understand bubble columns, which are frequently used in process industries and whose behaviour is complex and not totally predictable. They use the Euler–Euler approach, consisting in obtaining mean local properties to describe the bubble plume, and compare their simulation results to laboratory experiments. They reveal the correlation of operating conditions with the gas mixing and plume oscillation period. The commercial code of CFD is the same as used by [8].

Studies of Given Processes

It is natural and usual that some given processes are studied by researchers who either try to promote specialized models or improve their behavior by examining operation parameters. Many studies can be gathered in this category. Yang et al. [10] study an in-situ combustion method that can be used for bitumen sands or heavy oil production. Their numerical study deals with a given block of a Chinese oil field. They analyze the influence of production (huff and puff rounds, air injection speed and air injection temperature) and geological parameters (bottom water thickness, stratigraphic layering, permeability ratio and formation thickness). They are thus able to extract the main operation parameters. Shakeel et al. [11] compare the industrial production of formaldehyde using two different catalysts. They use a commercial steady-state process simulator to simulate the process and deduce the advantages and drawbacks of these configurations. Chinh et al. [12] study a nitrogen gas separator. This is a fixed bed operated by pressure swing adsorption. Thus, it is an appropriate subject for modeling and simulation with its dynamic and sequential operation, but the authors also compare their results to a laboratory pilot plant. The model, composed of partial differential and algebraic equations, is fully described together with its operating parameters. Chen et al. [13] study the production of butyric anhydride by means of a single reactive distillation column. Although this process has already been described in the literature, its applications are not numerous due to many operation difficulties and lack of generality. Moreover, the authors have replaced a two-column process with a single one by using the internal circulation of acetic anhydride. They show that this can be extended to similar reactive distillation two-column processes. The simulation is performed by a commercially available steady-state process simulator by which they optimize the process. Finally, they perform an economic analysis of the novel process. Marecka-Migacz et al. [14] study a membrane process—more specifically, ceramic nanofiltration for the separation of succinic acid aqueous solutions. They provide a detailed model of nanofiltration for ion transport, taking into account convection, diffusion and electromigration. They compare the results obtained by their elaborate model to those issued from more standard approaches. In their case, all species are considered, as well as solutes, ions and solvent, pH-regulating solutions and water. The conditions of use of their more complete model are emphasized. Experiments are also performed for comparison. Nguyen et al. [15], again, study a membrane process of ultrafiltration of protein. The model is simpler than in [14]. Their objective is to perform an economic assessment of the process and, for this purpose, their focus is related to the design parameters. With the latter being numerous, they perform the optimization by means of a genetic algorithm in a black-box manner, although they rely on their analytical model.

Original Processes

By original process, we mean processes that are rarely studied from a process point of view and/or that recently emerged. Among these, the process synthesis of carbon dots studied by Pudza et al. [16] can be ranked. A part of the article deals with the chemical synthesis of these carbon dots and has little connection with process engineering. However, the latter is concerned when the authors apply a surface response methodology using design of experiments to optimize their process. This strategy is often used when researchers or engineers meet great difficulties in modeling the process. From that angle, it could also mean that such a process presents interest for modeling based on first principles.

Nevertheless, the authors are able to propose conditions for efficient and sustainable synthesis of carbon dots.

Safety Studies

Finally, one article deals with safety. This is an extremely important aspect of any practical operation, especially at the industrial level. Often, safety studies are published in dedicated journals and less in more general journals, such as *Processes*. Wang et al. [17] study the process of natural gas explosion in the case of linked vessels with three structures. They perform a numerical simulation by means of an available and efficient software designed for predicting the consequences of gas explosions in three dimensions. They also build an apparatus to infer some parameters. Simulations deal with the influence of parameters such as pipe length, ignition position, length of connection pipes and sizes of vessels. This article demonstrates how a very elaborate code can be useful in performing a safety study.

Conclusions

A large variety of manuscripts were proposed. Without aiming at a particular classification chosen a priori, it appears that the accepted articles are mainly divided in two categories. The first one deals with general methods applicable to many processes and pursuing the goal of improving the methods of process engineering, either from an analytical or from a numerical point of view. General simulation codes are regularly used and sometimes coupled between themselves, even with advanced computational fluid dynamics, to provide new insights. The second category deals with dedicated processes, often related to an experimental laboratory or industrial process unit. In this case, the objective is rather to show how a refined model can improve the simulation or simply how a process can be optimized. Economic aspects are treated in several articles, revealing an important purpose of process engineering. Even if only one article dealing with safety was retained, it shows how process studies are interlinked, as it also makes use of computational fluid dynamics. In general, when possible, the authors try to compare their simulation results with experimental results obtained at the laboratory or plant level.

Funding: This paper received no funding support.

Conflicts of Interest: The authors declare no conflict of interest.

References

1. Furda, P.; Variny, M.; Labovská, Z.; Cibulka, T. Process drive sizing methodology and multi-level modeling linking MATLAB®and Aspen Plus®environment. *Processes* **2020**, *8*, 1495. [CrossRef]
2. Zečević, N.; Bolf, N. Integrated method of monitoring and optimization of steam methane reformer process. *Processes* **2020**, *8*, 408. [CrossRef]
3. Yousefifard, M.; Salehi, G.; Davarpanah, A. Comparison of exergy and advanced exergy analysis in three different organic Rankine cycles. *Processes* **2020**, *8*, 586. [CrossRef]
4. Wang, L.; Sun, X.; Xia, L.; Wang, J.; Xiang, S. Inside-out method for simulating a reactive distillation process. *Processes* **2020**, *8*, 604. [CrossRef]
5. Sun, S.; Cui, Z.; Zhang, X.; Tian, W. A hybrid inverse problem approach to model-based fault diagnosis of a distillation column. *Processes* **2020**, *8*, 55. [CrossRef]
6. Skorych, V.; Das, N.; Dosta, M.; Kumar, J.; Heinrich, S. Application of transformation matrices to the solution of population balance equations. *Processes* **2019**, *7*, 535. [CrossRef]
7. Mukhtar, S.; Sohaib, M.; Ahmad, I. A numerical approach to solve volume-based batch crystallization model with fines dissolution unit. *Processes* **2019**, *7*, 453. [CrossRef]
8. Zeng, Y.; Si, Z.; Zhou, Y.; Li, M. Numerical simulation of a flow field in a turbo air classifier and optimization of the process parameters. *Processes* **2020**, *8*, 237. [CrossRef]
9. Cheng, Y.; Zhang, Q.; Jiang, P.; Zhang, K.; Wei, W. Investigation of plume offset characteristics in bubble columns by Euler-Euler simulation. *Processes* **2020**, *8*, 795. [CrossRef]

10. Yang, Z.; Han, S.; Liu, H. Numerical simulation study of heavy oil production by using in-situ combustion. *Processes* **2019**, *7*, 621. [CrossRef]

11. Shakeel, K.; Javaid, M.; Muazzam, Y.; Naqvi, S.R.; Taqvi, S.A.; Uddin, F.; Mehran, M.T.; Sikandar, U.; Niazi, M.B.K. Performance comparison of industrially produced formaldehyde using two different catalysts. *Processes* **2020**, *8*, 571. [CrossRef]

12. Van Chinh, P.; Hieu, N.T.; Tien, V.D.; Nguyen, T.-Y.; Nguyen, H.N.; Anh, N.T.; Van Do, T. Simulation and experimental study of a single fixed-bed model of nitrogen gas generator working by pressure swing adsorption. *Processes* **2019**, *7*, 654. [CrossRef]

13. Chen, G.; Jin, F.; Guo, X.; Xiang, S.; Tao, S. Production of butyric anhydride using single reactive distillation column with internal material circulation. *Processes* **2019**, *8*, 1. [CrossRef]

14. Marecka-Migacz, A.; Mitkowski, P.T.; Antczak, J.; Różański, J.; Prochaska, K. Assessment of the total volume membrane charge density through mathematical modeling for separation of succinic acid aqueous solutions on ceramic nanofiltration membrane. *Processes* **2019**, *7*, 559. [CrossRef]

15. Nguyen, T.-A.; Yoshikawa, S. Modelling and economic optimization of the membrane module for ultrafiltration of protein solution using a genetic algorithm. *Processes.* **2019**, *8*, 4. [CrossRef]

16. Pudza, M.Y.; Abidin, Z.Z.; Rashid, S.A.; Yasin, F.M.; Noor, A.; Issa, M.A. Sustainable synthesis processes for carbon dots through response surface methodology and artificial neural network. *Processes* **2019**, *7*, 704. [CrossRef]

17. Wang, Q.; Sun, Y.; Li, X.; Shu, C.-M.; Wang, Z.; Jiang, J.; Zhang, M.; Cheng, F. Process of natural gas explosion in linked vessels with three structures obtained using numerical simulation. *Processes* **2020**, *8*, 52. [CrossRef]

Publisher's Note: MDPI stays neutral with regard to jurisdictional claims in published maps and institutional affiliations.

Article

Process Drive Sizing Methodology and Multi-Level Modeling Linking MATLAB® and Aspen Plus® Environment

Patrik Furda [1,*]**, Miroslav Variny** [1,*]**, Zuzana Labovská** [1] **and Tomáš Cibulka** [2]

[1] Department of Chemical and Biochemical Engineering, Faculty of Chemical and Food Technology, Slovak University of Technology in Bratislava, Radlinského 9, 812 37 Bratislava, Slovakia; zuzana.labovska@stuba.sk

[2] SLOVNAFT, a.s., Vlčie hrdlo 1, 824 12 Bratislava, Slovakia; tomas.cibulka@slovnaft.sk

* Correspondence: patrik.furda@stuba.sk (P.F.); miroslav.variny@stuba.sk (M.V.)

Received: 27 October 2020; Accepted: 17 November 2020; Published: 19 November 2020

Abstract: Optimal steam process drive sizing is crucial for efficient and sustainable operation of energy-intense industries. Recent years have brought several methods assessing this problem, which differ in complexity and user-friendliness. In this paper, a novel complex method was developed and presented and its superiority over other approaches was documented on an industrial case study. Both the process-side and steam-side characteristics were analyzed to obtain correct model input data: Driven equipment performance and efficiency maps were considered, off-design and seasonal operation was studied, and steam network topology was included. Operational data processing and sizing calculations were performed in a linked MATLAB®–Aspen Plus® environment, exploiting the strong sides of both software tools. The case study aimed to replace a condensing steam turbine by a backpressure one, revealing that: 1. Simpler methods neglecting frictional pressure losses and off-design turbine operation efficiency loss undersized the drive and led to unacceptable loss of deliverable power to the process; 2. the associated process production loss amounted up to 20%; 3. existing bottlenecks in refinery steam pipelines operation were removed; however, new ones were created; and 4. the effect on the marginal steam source operation may vary seasonally. These findings accentuate the value and viability of the presented method.

Keywords: process steam drive; software linking; heat pump; propane–propylene separation; steam network; pressure and heat losses; energy efficiency

1. Introduction

Cogeneration and polygeneration systems are an essential part of industrial production of materials and energies, consuming or generating heat, electricity, mechanical, and chemical energy [1,2]. Ambitious efforts of national and European institutions to reduce greenhouse gases emissions and to ensure sustainable industrial production [3–9] at the same time cannot be successfully met without increasing material and energy efficiency of the industry [10–13] by simultaneous energy, economic, environmental, and risk and safety optimization [14–16] of existing industrial cogeneration and polygeneration systems [17,18].

Efficient steam production, and its transport and use for both process heating and polygeneration purposes, has been targeted on various complexity levels in numerous recent studies [19–38]. Starting with techno-economic studies optimizing the efficiency of a single equipment unit [24,27,30,34,35], through steam consumption optimization in a single production unit [25,32,38,39], and cogeneration potential exploitation [19,33,36,40,41] to total site heat and power integration [20–23,28,33,37,42], the goal is always to reduce operational expenses, improve steam system stability, and decrease fuel consumption in industrial plants. Steam system topology and the impact of pressure and heat losses from steam pipelines on optimal cogeneration system sizing and operation has been addressed in several papers [22,37,43–45], but most studies consider neither off-design operation of steam turbines nor variable steam pressure levels as important aspects in the optimization procedure. A systematic method comprising characteristics of a real steam system operation [35,46] (variable pipeline loads, steam pressures, and temperatures) as well as real process steam/work demands has the potential to fill the knowledge and experience the gap between the modeling approach in utility systems' optimization and real steam system operation.

Process steam drives are important steam consumers in heavy industry [20,31,34] and play a significant role in the design and operation of complex steam networks. The most common driven equipment includes compressors, pumps, and fans (blowers) [34,38,47]. The steam turbines used can be of simple condensing, backpressure, or of a combined extraction condensing type [36]. Steam consumption is influenced by several factors that include the actual steam inlet parameters, steam discharge pressure, actual turbine revolutions, as well as the shaft work needed, which varies according to the process requirements. Process compressors driven by steam turbines are standard equipment of ethylene production and gas processing and fractionation plants [38,47], and they are also frequently used in compression heat pump-assisted distillations [48–51]. Thus, they are deeply integrated in the process. The shaft work needed depends on several process parameters, including (but not limited to) the distillation feed amount and composition, desired product quality, and column and compressor design parameters. This highlights the pressing need to develop a robust method for process steam drive sizing which would incorporate not only the real steam-side condition variations but also the process-side shaft work variations. Improper sizing results either in limited shaft work delivery (undersizing), causing possible process throughput limitations, or inefficient steam use (oversizing). Moreover, the steam drive design and operation have to be optimized with respect to the whole steam system, always taking into account the marginal steam source, its seasonal operation variations [52], and the possible steam pipeline capacity constraints [26,43].

Given all the prerequisites, it is only natural that examination and precise evaluation of such complex process parameters poses a challenge which can hardly be faced successfully without employing robust simulation environment. For almost two decades, researchers have strived to combine the colossal computing capacity of the Aspen Plus® simulation engine with the exceptional data-processing capabilities of the MATLAB® software [53,54]. Several papers have been published, mostly focusing on multi-objective optimization [55–59] or automation problems solution [60,61]. Unfortunately, only scarce details regarding the chosen approach can be found. Hence, the perspective to close this gap in knowledge remains particularly attractive.

The contribution of this paper to the field of knowledge is twofold:

- First, it presents a robust method for optimal process drive sizing and integration considering all relevant factors affecting the design, while the method is suitable for operational optimization as well;
- Second, both the process-side and steam-side are modeled using Aspen Plus® and MATLAB® linking, which is a novel and promising approach for complex systems' operation analysis and optimization purposes.

Table 1 provides a comparison of the key parameters and characteristics of relevant recently published methods with the proposed method, all of them aiming at: 1. Maximization of the cogeneration potential exploitation; 2. optimal process steam drive sizing; and 3. optimal steam vs. electrodrive use. As is seen in Table 1, the relevant methods are focused mostly on steam-side modeling and optimization, while the proposed method presents a coupled steam- and process-side modeling approach. Moreover, several of the relevant methods do not incorporate such important aspects as the varying inlet steam parameters or shaft work requirements and implement fix turbine and driven equipment efficiencies instead of considering their variations in the real operation. As documented on an industrial case study, failure to implement the real operational parameters of a system leads to steam drive undersizing and, consequently, to limited process throughput.

The effect is that the more pronounced, the farther the process drive is located from the main steam pipeline. Further findings from the industrial case study include the fact that the incorporation of a steam drive into an existing steam network can significantly affect its balance (and, thus, its operation) and thus create a new operation bottleneck, or remedy its existing ones. Furthermore, the marginal steam source operation mode is also affected, which has to be considered when evaluating the economic feasibility of such an investment proposal.

Paper organization is as follows: Part 2 presents the proposed complex steam drive sizing method and is subdivided into process-side and steam-side model subparts. Following that, part 3 introduces an industrial case study with the description of the existing system layout, proposed change, available process data, and their processing, including initial analyses and their results serving as additional model input parameters. Part 4 presents the calculation results, including a comparison of the presented steam drive sizing method with several others (included in Table 1), and evaluating the economic feasibility of the proposed system change. Discussion is followed by a concise conclusion part summing up the novelty and significance of the presented method and the key findings extracted from the industrial case study results.

Table 1. Comparison of key parameters of the proposed method with recently published papers. Legend: BE = balance equations, Calc. = calculated, N/A = not applicable, NP = not provided, Reg. = calculated based on statistic regression, SA = sensitivity analysis, SS = saturated steam, WS = wet steam.

	Method								
Parameter/Feature	Wu et al. [36]	Ng et al. [31]	Frate et al. [24]	Marton et al. 2017 [29]	Sun et al. 2016 [34]	Bütün et al. [43]	Mrzljak et al. [30]	Tian et al. [35]	Proposed Method
Inlet steam temperature & pressure	Fixed	Fixed	SS; SA	NP	SA	Fixed	Varying; process data	SS; SA	Varying; process data
Discharge steam pressure	Fixed	Fixed	SA	NP	NP	Fixed	Varying; process data	SA	Varying; process data
Discharge steam temperature	Fixed	Fixed	WS	NP	NP	Fixed	WS	WS	Calc.; polytropic expansion
Frictional pressure losses	✗	✗	✗	✗	✗	✓	✗	✓	✓
Heat losses from pipelines	✗	✗	✗	✗	✗	✓	✗	✗	✓
Turbine revolutions	✗	✗	✗	✗	Varying; NP	✗	Varying	SA	Varying; process data
Turbine efficiency	Reg.	Fixed	SA	NP	Varying	NP	Calc.	Calc.	Calc.
Driven equipment efficiency	Fixed	N/A	N/A	N/A	Varying; NP	N/A	Efficiency map	N/A	Efficiency map
Shaft work required	Fixed	N/A	N/A	N/A	Calc.; process-dependent	N/A	Varying; process data	N/A	Calc.; process-dependent
Process modeled	✗	N/A	N/A	Aspen utilities planner + Excel	Process and steam BE	Steam and heat BE	✗	N/A	Aspen Plus® linked with MATLAB®

2. Process Drive Sizing Method

2.1. Equipment Operation Assessment and Modeling

2.1.1. Heat Pump-Assisted C3 Fraction Splitting

Heat pump-assisted distillation is a good example of incorporating a steam drive into an industrial process; thus, it was chosen for illustration. In such systems, overhead distillate vapors from a separation column are compressed and subsequently condensed in the column reboiler [51,62]. For this type of heat pump to be applicable, the separated components have to be of similar boiling points [63]. However, this leads to a relatively low driving force in the reboiler; thus, to deliver the required power input, the vapor throughput needs to be sufficiently high. Hence, stable operation of the heat pump compressor is of the uttermost importance. Amongst the compressor drives, steam drives (turbines) are the most usual. These are normally shaft-bound with the compressor and, therefore, their correct sizing is just as important as that of the compressor itself. In the case of steam turbines, however, the whole sizing process is more complicated as steam quality fluctuations and overall steam network properties have to be considered.

A typical arrangement of a heat pump-assisted distillation is provided in Figure 1. Here, a propane–propylene mixture is split into separate components of high purity (>99.6% vol.). The energy necessary for the separation is provided by a condensing steam turbine.

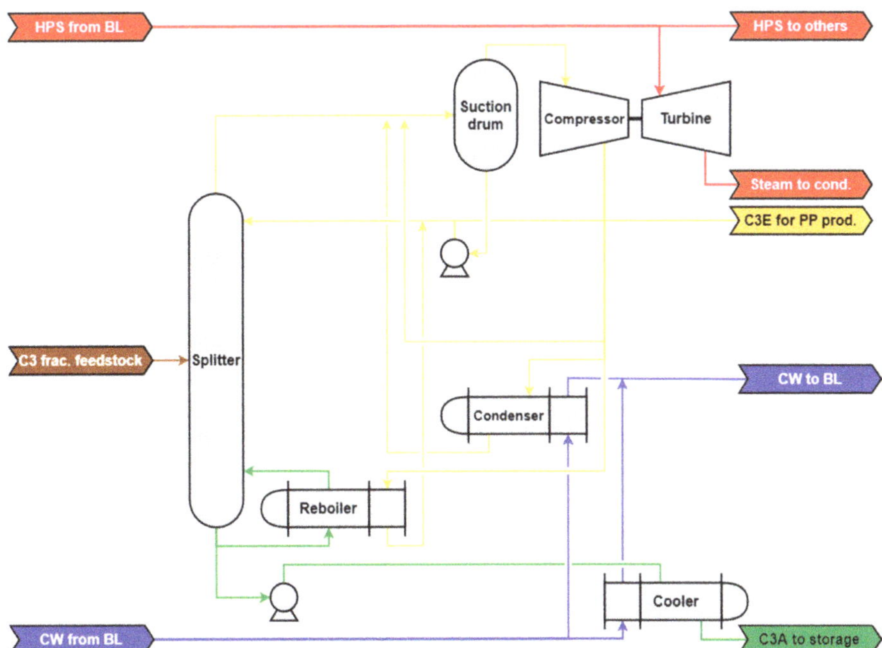

Figure 1. Process scheme. Legend: BL = battery limit, cond. = condenser, CW = cooling water, C3A = propane, C3E = propylene, frac. = fraction, HPS = high-pressure steam, PP = polypropylene, prod. = production.

To design (size) a process drive correctly, the process itself must be understood thoroughly. This encompasses not only the physical structure of the system but also the physicochemical and mechanical non-idealities and, most importantly, over-time variations in feedstock quality and mass flow [64]. To provide the most authentic results, the model was constructed as follows:

- Due to simplicity of the mixture being distilled, an equilibrium model using Murphree tray efficiency was chosen;
- To account for the vapor phase non-ideality, the Peng–Robinson equation of state [65] was applied, as it yielded the best results compared to the measured data [66];
- Because the operating temperature of the column (25–35 °C) was similar to ambient temperature, heat loss from the system was neglected;
- Technical details regarding the number of column trays and the position of the feed stage; exchanger areas, overall heat transfer coefficients, and logarithmic mean temperature differences; and compressor and condensing turbine characteristics were taken from technical documentation provided by the manufacturer.

The model briefly described above was, due to its complexity, constructed in the simulation environment of the Aspen Plus® software which provided fast-to-obtain and reliable results. Yadav et al. [67] provided a comprehensive tutorial on utilizing Aspen Plus® potential in distillation column operation simulation. For the separation column, the RadFrac model was chosen, which stands as a universal rigorous model for multi-stage component separation. Each of the three heat exchangers was modeled using a short-cut method (HeatX model) which proved satisfactory for the cause. Both the compressor and the turbine were modeled using the Compr model [68,69], though individual approaches differed significantly.

2.1.2. Compressor Operation

In general, there is a variety of approaches regarding compressor modeling—from the simplest calculations (e.g., polytropic work calculation with constant parameters and efficiencies) to the most complex ones (comprising efficiency and power maps, and shaft speed calculations) [70,71]. As shown later, the calculation should be carried out in the most precise way possible whenever the data is accessible as, for instance, compressor shaft revolutions affect the turbine efficiency drastically. Hence, performance and efficiency curves were included in the model via the Aspen Plus® interface. As a result, the model considered the mass flow through the compressor and the desired output pressure, and provided information on the required power input, polytropic efficiency, and shaft speed, as well as complete outlet stream results.

2.1.3. Process Drive Operation

Steam turbines, like compressors, can be modeled on many levels—from basic enthalpy balance with constant isentropic efficiency to more complex approaches comprising turbine characteristics and correction curves [20,30]. Aspen Plus® did not provide the option to model turbines with the same precision as the compressors. Therefore, the characteristics and correction curves had to be input "manually", via a calculator block. As only isentropic calculations were predefined for turbines, the model first calculated the turbine steam consumption based on the characteristics (required power output) and correction curves (live steam pressure, live steam temperature, exhaust pressure, and compressor revolutions). Then, the model iteratively calculated the isentropic efficiency of the turbine so that, based on the before-calculated steam consumption and enthalpy balance, the provided power output correlated with the required value.

2.1.4. Steam Network

While most industrial facilities have a major steam source (e.g., a combined heat and power unit (CHP)), the utilization of waste heat from production units in means of steam production is also a common practice [43]. Moreover, the overall steam supply of these local producers can even exceed the major steam source production. Thus, the quality of the supplied steam varies significantly depending not only on the CHP unit's operation but on the whole production network [22,72]. It is therefore vital for process drive sizing to account for steam quality fluctuations.

When analyzing steam networks, as well as any other pipeline systems, momentum and heat transfer must always be considered [73]. Pressure loss caused by friction and/or various installed armatures and potential heat loss to ambient space can affect the steam quality significantly. When analyzing the heat transfer, steam–ambient temperature difference plays the most significant role. The impact of heat loss on the medium quality is thus best observable during extreme ambient conditions, i.e., temperature peaks during winter. Even though most adverse effects of ambient temperature changes can be suppressed by effective insulation, Hanus et al. [26] showed that the effectivity of outdoor insulation decreases heavily after years of operation and it is in designer's best interest to investigate the true heat conductivity of the used insulation.

Effects of both pressure and heat losses vary greatly depending on pipeline geometry and mostly on the steam network topology and plant infrastructure, i.e., the resulting pipeline length. While the distance (pipe length) of the considered steam drive from the battery limit header can range in terms of tens of meters in some cases, there are units where the pipe length can reach hundreds of meters or even a kilometer. This is mostly the case of mid-twenty century refineries where individual units and their subunits were built far apart for safety reasons [26].

Lastly, it is necessary to assess the impact of steam drive alteration on the steam transport velocity. Naturally, high transport velocities may cause serious erosion of pipeline walls and damage installed armatures. However, low velocities in the pipelines may result in excessive heat loss and decrease in steam quality or even condensate production [26]. Hence, the velocity changes with the steam consumption of the drive, and pipeline suitability assessment is inevitable.

The process described in the previous text draws high-pressure steam from a steam network using a CHP unit as the main steam producer. As illustrated below (Figure 2), steam crossing the fluid catalytic cracking (FCC) unit battery limit is first supplied to the modeled system (propylene recovery unit) and then to other parts of the FCC unit. To account for all above-mentioned aspects, pipeline topology, geometry, and heat transfer properties were incorporated in the Aspen Plus® process model.

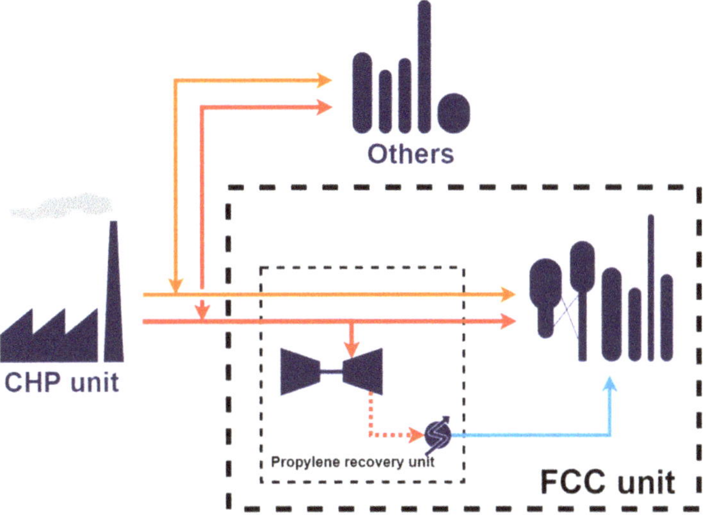

Figure 2. Simplified plant steam network. Legend: Red line = HPS pipelines, orange line = middle-pressure steam (MPS) pipelines, dotted red line = turbine exhaust; blue line = condensate.

2.1.5. Combined Heat and Power Unit (CHP) as Marginal Steam Source

Industrial combined heat and power plants (CHP) traditionally employ steam boilers and steam turbines as the cogeneration technology; some of them including gas turbines or combustion engines [40,74]. Their task is to cover the steam network imbalance on all pressure levels (i.e., to serve as a marginal steam source) while the cogenerated electric energy is utilized in the industrial facility or sold to an external grid. Seasonal steam demand variations influence their operation and the resulting backpressure power production decreases in summer below a pre-set acceptable level, and has to be compensated by other means (condensing power production). The reason for this specific system feature is explained in more detail in Section 3.2. Techno-economic assessment of a process steam drive sizing inevitably impacts the CHP operation and should be evaluated correctly. Last, but not least, the CHP operation is influenced not only by internal steam demand, but by external factors as well (energy management strategy of the refinery, changeable energy prices and their uncertainty, etc.) [75].

The considered CHP comprised high-pressure steam boilers (9 MPa, 530 °C) and a set of backpressure and extraction condensing turbines exporting high-pressure (HPS, 3.5 MPa), medium-pressure (MPS, 1.1 MPa), and low-pressure (LPS, 0.5 MPa) steam to the industrial facility. Constant marginal heat production efficiency of the CHP and constant marginal backpressure and condensing power production per one ton of steam were assumed. The effect of a change in the production unit steam balance on a certain pressure level was transposed to the corresponding change in backpressure electric energy production and fuel consumption in the CHP. A detailed description of the CHP unit operation and its further characteristics can be found elsewhere [26,47].

2.2. Data Processing

Data analyses are an inherent part of any technological proposal. To be able to size the desired equipment, it is necessary to understand the unit's operation, unearth any possible off-design performance and to map the system responses. In order to do so, measured data regarding the modeled unit and auxiliaries have to be studied.

To evaluate the maximal power requirements of the modeled unit and to assess seasonal performance variations, a year-long operation of the unit was monitored. To understand the effects of steam quality fluctuations on the proposed new process drive performance, daily measurements of the properties of steam in the network were taken. Finally, to examine possible bottlenecks linked to the CHP unit operation, its performance during the evaluated period was considered.

2.2.1. Measured Data

Table 2 summarizes all process data used in the simulation of the propylene recovery unit performance and serves as a guide for heat pump-assisted distillation simulation (and process drive sizing) data gathering. These data include feedstock mass flow and composition which are usually not measured continually. However, highly reliable records of products' mass flow and quality are generally recorded and they can be combined to obtain the desired information.

2.2.2. Aspen–MATLAB Linking

The proposed methodology exploits the full potential and relevance of multi-software modeling via linking Aspen Plus® software (Aspen Plus® V10, Aspen Technology Inc., Bedford, MA, USA) with MATLAB® software (MATLAB® 2020a, The MathWorks Inc., Natick, MA, USA). While the state-of-the-art simulation engine of Aspen Plus® provides swift and rigorous results due to the sequential modular algorithm, its use in evaluation of large datasets containing numerous different variables is somewhat laborious. On the other hand, the programming language of the MATLAB software is tailored for handling such tasks. Thus, an effort to link these two software environments persists. Fontalvo et al. [54] introduced the idea of software linking in the early 2000s, though no details were presented. Several years later, in 2014, Fontalvo described the linking principles [53];

however, these are of limited relevance today. Other authors have continued working on the idea with different aims or providing insufficient details. A MATLAB® sub-software, Simulink, was used by Dos Santos Vidal et al. [60] and Ryu et al. [61] to solve rather complicated automation problems. Muñoz et al. [57] used MATLAB®-to-Aspen linking in a gradient-based multi-objective optimization via ACSII file exchange with Aspen Plus® working in the equation-oriented mode. The published freeware by Abril [76] caused a breakthrough with instructions for component object model (COM) interface linking. Following this publication, several papers [58,59,77,78] were published providing scarce information about the interface build and utilization. Even though the works of Ramirez et al. [56] and Darkwah et al. [55] contain specific programming tips, these are rather unclear to a non-advanced user. Hence, to this day, a simple but complex interface linking methodology remains elusive. Details regarding the program capabilities and the linking procedure are presented in the following text.

Table 2. Considered measured data. Legend: BL = battery limit, CHP = combined heat and power plant, HPS = high-pressure steam, MPS = middle-pressure steam.

Equipment/Material Stream/Unit	Data	Purpose	Details
Propylene (product stream)	Total mass flow	Simulation	Together with propane stream represents the feed stream
	Propylene content		
Propane (component analysis)	Simulation	Feed stream composition	
Column	Head pressure	Simulation	
	Bottoms pressure	Simulation	
Suction drum	Pressure	Simulation	
Compressor	Exhaust pressure	Simulation	
	Exhaust temperature	Verification	Compressor exhaust temperature documents isentropic efficiency calculation accuracy
	Shaft speed	Verification	Shaft speed documents compressor performance calculation accuracy
Turbine	Steam consumption/condensate mass flow	Verification	For systems where steam consumption is not measured directly, it is possible to measure mass flow of turbine condensate
	Condensate pump by-pass valve position	Verification	When measuring condensate mass flow, it is sensible to check whether condensate pump by-pass is in operation and to what degree
	Live steam temperature	Simulation	
	Exhaust pressure/condenser temperature	Simulation	For systems where exhaust pressure is not measured directly, it is possible to estimate it based on the condenser temperature
HPS from BL (utility stream)	Mass flow	Simulation	
	Temperature	Simulation	
	Pressure	Simulation	
MPS from BL (utility stream)	Mass flow	Simulation	
	Temperature	Simulation	
	Pressure	Simulation	
CHP unit	HPS mass flow	Simulation	
	MPS mass flow	Simulation	

Aspen Plus® cooperation with external Windows applications is enabled via ActiveX Automation Server. In this way, the applications can interact with Aspen Plus® through a programming interface while the automation server exposes objects through the COM object model [79]. Through this interface it is possible:

- to connect inputs and results of Aspen Plus® simulations to other applications;
- to manipulate (create, reconnect, delete, etc.) Aspen Plus® objects;

- to control the Aspen Plus® user interface (handle events, suppress dialog boxes, disable user interface features, etc.);
- to control a simulation (run, stop, reinitialize, etc.).

As the original user guide for Aspen Plus® is written for the Visual Basic programming language, a simple step-by-step manual for Aspen–MATLAB linking is presented:

- First, a local ActiveX server is created where the component object model is situated using the inbuilt function "actxserver". The syntax is as follows: *var = actxserver(ProgID)*, where *var* is a structured variable used to access the server and *ProgID* is the program identifier. The program identifier for Aspen Plus® documents is "Apwn.Document.X" where X depends on the Aspen Plus® version: 34.0 for V8.8, 35.0 for V9.0, and 36.0 for V10.0 (e.g., "Apwn.Document.36.0" for Aspen Plus® V10);
- After the server creation, the whole system is initialized as shown in Figure 3. There are three initialization methods depending on the format of the simulation: "InitFromArchive2" (for use with .bkp and .apw archive files), "InitFromTemplate2" (for use with templates), and "InitFromFile2" (for use with .apwn compound files). No difference has been observed in their performance, though .bkp files are generally the smallest in size and thus recommended. As with other MATLAB® scripts, all files have to be located in the same folder;

```matlab
%% Aspen link-up
Aspen = actxserver('Apwn.Document.36.0'); % Creating a local COM server; creating a
structured variable "Aspen"
[~, mess] = fileattrib; % Accessing the folder
Simulation_Name = 'Turbine'; % Name of the desired simulation to run
Aspen.invoke('InitFromArchive2',[mess.Name '\' Simulation_Name '.apw']); % Linking Aspen
Plus simulation with MATLAB via created server environment
Aspen.Visible = 1; % Whether or not will Aspen Plus be physically opened
Aspen.SuppressDialogs = 1; % Whether or not will contextual windows be displayed
Aspen.Run2(); % Starting the initial simulation
```

. . .

```matlab
%% Excel input: steam-size data
first = '97'; % First row of the database to evaluate
last = '427'; % Last row of the database to evaluate
TK401.P = xlsread('Steam Properties.xlsx', 'TK401 steam consumption', ['B' first ':B'
last]);
```

. . .

```matlab
% Input adjustment
Aspen.Tree.FindNode("\Data\Blocks\10IN35\Input\FLUX").Value = - q.in10;
```

. . .

```matlab
% Starting the simulation:
Aspen.Run2();
```

. . .

```matlab
% Gathering results:
PA.m(i, 1) = Aspen.Tree.FindNode("\Data\Streams\PA10-II\Output\MASSFLMX\MIXED").Value;
```

Figure 3. Example of Aspen Plus®–MATLAB® link utilization (for details see Appendix A).

- From this point on, the Aspen–MATLAB® link is ready to use. To access results, manage inputs, and control the simulation and/or the user interface, dot notations are used. Examples of the syntax for various commands are given below:

    a. Simulation control Syntax: var.command (e.g., var.Run2, var.Reinit, …)
    b. User interface control Syntax: var.attribute = value (e.g., var.Visible = 1, …)
    c. Input alteration Syntax: var.Tree.Findnode(path).Value = value_a
    d. Results gathering Syntax: value_b = var.Tree.Findnode(path).Value

All commands can be found in Appendix A or in Figure 3. The path to every individual variable of the simulation in Aspen Plus® can be accessed directly in the program via: Customize→ Variable explorer.

- Prior to launching the simulation, it is sensible to also link MATLAB to Excel for more flexible operation via simple and useful inbuilt functions "xlsread" and "xlswrite" enabling reading and writing data from and to the Excel spreadsheet, respectively, without the need for opening the data file manually. An example can be seen in Figure 3.

3. Industrial Case Study

3.1. System Description

The aforementioned propylene recovery unit is a subunit (a section) of a fluid catalytic cracking (FCC) unit, splitting the liquid propylene–propane fraction from the FCC gas plant into individual products of high purity. Due to low relative volatility of the components (and therefore low temperature difference between the head and bottom of the distillation column), a compression heat pump system can be utilized. However, the small difference in boiling points demands a large reflux ratio (>15) and numerous (>150) separation stages. Hence, such a system (as described in Figure 1) poses not only a technological but also a computational challenge, which once again underlines the pressing need for use of robust simulation software.

Performance of the considered propylene recovery unit was evaluated during an approximately one-year period, from 1 April 2018 to 28 February 2019, and provided the following observations:

- The unit's feedstock flow rate was flexible, ranging from 6.5 to 9.7 t/h;
- Feedstock quality ranged from 81.7 to 86.5 mass % of propylene;
- Turbine condensate pump was by-pass protected.

Based on the technical documentation, we can state:

- Maximal unit throughput was 10 t/h of the propane–propylene fraction;
- Maximal compressor power at the coupling is 1250 kW.

3.2. Proposed Change in Steam Drive Type

Currently, a condensing steam turbine is used as the heat pump compressor drive. Such a situation makes sense if there is excess steam that cannot be utilized for other purposes (process heating, stripping agent, etc.). In the presented industrial study, HPS, used as driving steam, was imported from the CHP. At the same time, enough variability in both HPS and MPS production and transport capacities allowed us to consider condensing steam drive replacement by a backpressure one, targeting fuel savings in the CHP at the expense of a certain loss of CHP power generation. The replacement of condensing mechanical power production by a backpressure unit is economically feasible at most fuel-power price ratios. An additional tangible benefit was the resulting decrease in CO_2 and other pollutant emissions in the CHP.

Another steam pressure level, LPS, can be considered as an alternative backpressure steam sink to the MPS, but it is less suitable for this purpose. The reason is an occasional LPS excess in summer and the resulting decrease of LPS export from the CHP to very low values, which negatively affects the steam quality in the main LPS pipelines. Additional LPS from a new backpressure steam drive would thus have to be vented into the atmosphere. This situation is well-documented in Figure 4, where occasional drops of LPS export from the CHP can be seen. Sudden one-direction changes in LPS export within a few days or one to two weeks (decrease in April, increase in November) result from switching on/off the space heaters and steam tracing in the refinery. The existing LPS excess in the refinery during summer months is a serious issue that has two consequences: First, lower economic attractiveness of incentives for LPS consumption decrease in the refinery. Second, low LPS export from the CHP leads to reduced backpressure power production. Long-term energy policy of the refinery includes the demand to secure the power supply to critical production units during outer grid outages (severe weather, unexpected events) from the CHP. Thus, a certain minimal power production in the CHP is required regardless of the actual season. As the backpressure power production is insufficient to cover this demand, additional power is produced in condensing steam turbines during warmer months. The average duration of this period is 40% of the year. Any changes in CHP backpressure power production resulting from the proposed change in the steam drive are reflected in the change of condensing power production in the opposite way during this period of the year.

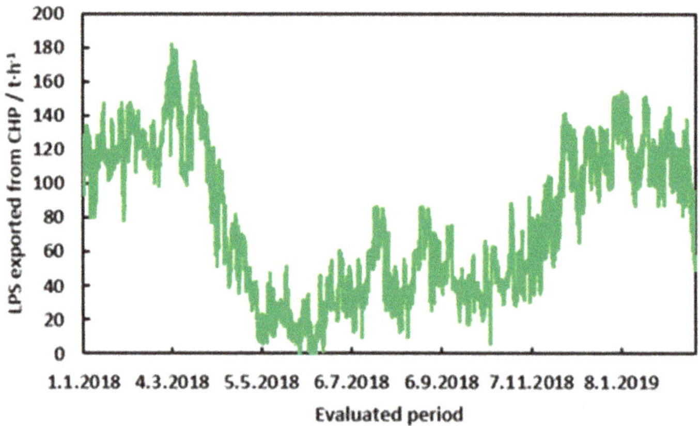

Figure 4. Seasonal fluctuations in low-pressure steam (LPS) export.

Apart from MPS and LPS, other steam pressure levels can be considered with sufficient capacity to absorb the steam produced from the new steam drive. A preliminary analysis of steam consumption in the FCC unit showed no such options, as their steam absorption capacity was only a small fraction of what was needed. Thus, only a backpressure steam drive with HPS as the driving steam and with MPS discharge was considered and further analyzed. A schematic of the analyzed proposal is provided in Figure 5. The FCC unit consumed between 25 to 40 t/h of MPS. During periods of lower MPS consumption, the excess of MPS produced in the new steam drive was exported to other refinery units. The CHP remained the marginal source of both HPS and MPS for the refinery. MPS export from the FCC unit was associated with increased MPS backpressure at the new steam drive discharge, which was considered in the proposed steam drive sizing method. Increase in the HPS export from the CHP increased the steam flow velocities in the HPS network, which was desirable following the outcomes of a study dedicated to HPS network operation [26]. However, transport capacity of both main HPS pipelines as well as those within the FCC unit has to be reviewed.

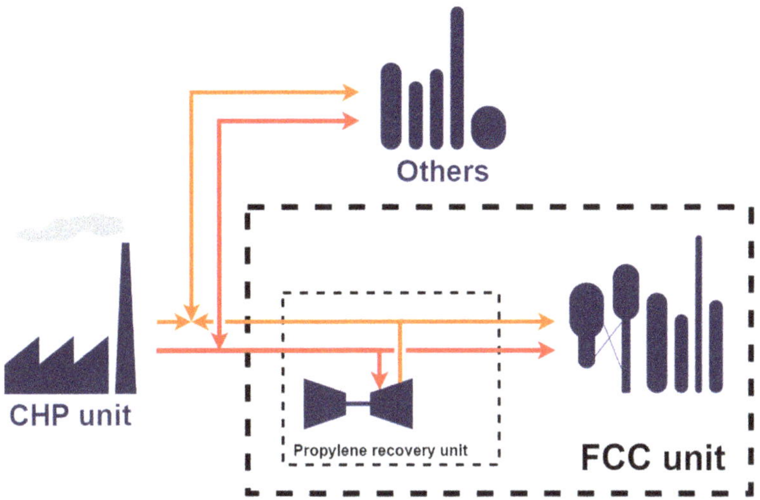

Figure 5. Simplified plant steam network alteration proposal. Legend: Red line = HPS pipelines; orange line = MPS pipelines.

3.3. Key System Analysis and Model Verification

Prior to designing a new process equipment, it is inevitable to confirm the relevance of the constructed model, based on which the proposal will be carried out. Designing a new process steam drive, the best practice is to prove that the model can predict actual steam consumption precisely. According to the enclosed graph (Figure 6), the model prediction seems to be incorrect though the trend is generally preserved. This was probably caused by the lack of live steam consumption measurement, except for the turbine condensate mass flow, which did not, however, totally correlate with the steam consumption because of the condensate pump by-pass increasing the condensate flow. Moreover, as shown in Figure 7, the calculated compressor shaft speed was in most cases slightly lower than the measured one due to the applied process control: For systems comprising such an enormous recycle stream flow rate (>150 t/h recycle vs. <10 t/h feedstock) it was near impossible to numerically obtain the exact same composition as analytically determined. Rather than that, an automation condition was set, forcing the process control to keep the product quality above 99.6 vol%. Even though real process control can occasionally achieve higher purity, there were numerous cases where the calculated purity was slightly below the measured. As a result, the compressor shaft speed may have, on a small scale, differed from the measured as well.

To account for both effects, the condensate mass flow rate with by-pass valve position was included in the calculation and the measured shaft speed was used for turbine performance assessment. The latter perfectly illustrated the impact of the shaft speed on the turbine efficiency (discussed in detail later). Based on the measured to calculated condensate mass flow rate comparison (Figures 8 and 9), it can be stated that the model provided reliable results and was thus verified.

Subsequently, to design a process drive correctly, it is necessary to:

- Obtain the process side characteristics—to evaluate the maximal power requirements and predict the process behavior in case of insufficient power supply as well as the daily performance;
- Map fluctuations in the steam network as to find the most appropriate design parameters for a new steam drive;
- Understand the effects of the process side (shaft speed and power requirements) and steam side (live steam pressure, live steam temperature, and exhaust pressure) parameters on turbine efficiency.

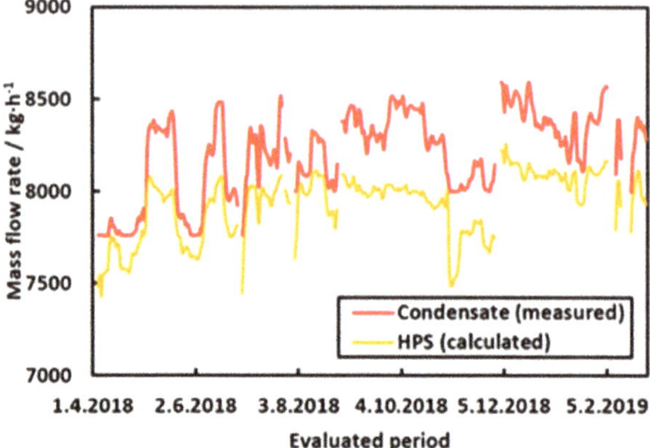

Figure 6. Condensate to steam mass flow comparison.

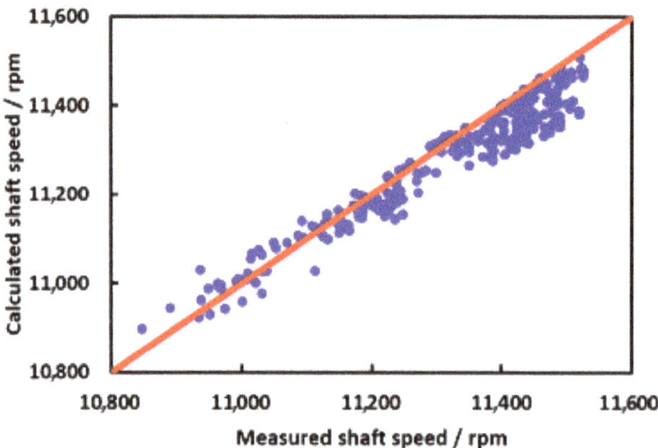

Figure 7. Measured to calculated shaft speed comparison.

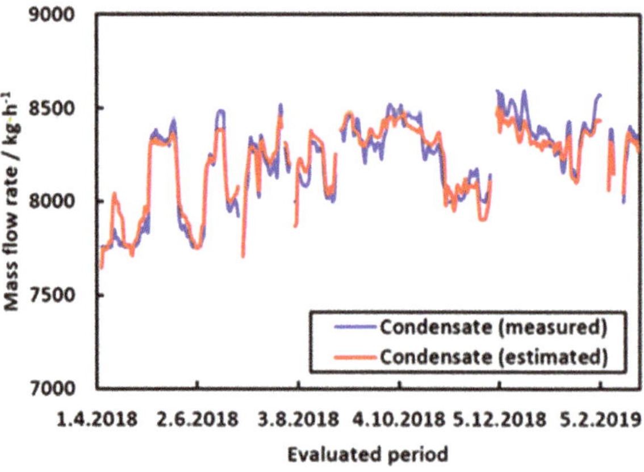

Figure 8. Comparison of measured and estimated condensate mass flow rate over the evaluated period.

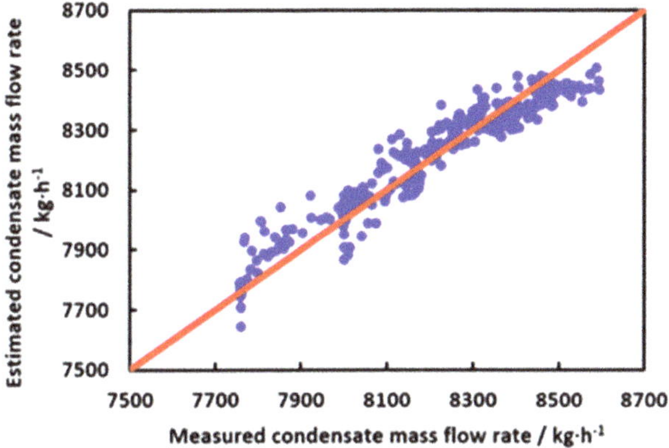

Figure 9. Measured to estimated condensate mass flow rate comparison.

Considering the above, process side characteristics (Figure 10) were constructed. Despite expectations, the primary assumption that the feedstock quality affects the process side power requirements was refuted. The best explanation was that the impact of large reflux efficiently suppressed the impact of the feedstock quality and so the characteristics were practically linear, with the feed flow rate as the independent and the power requirement as the dependent variable. The slope of the linear function used to fit the data was later used in results.

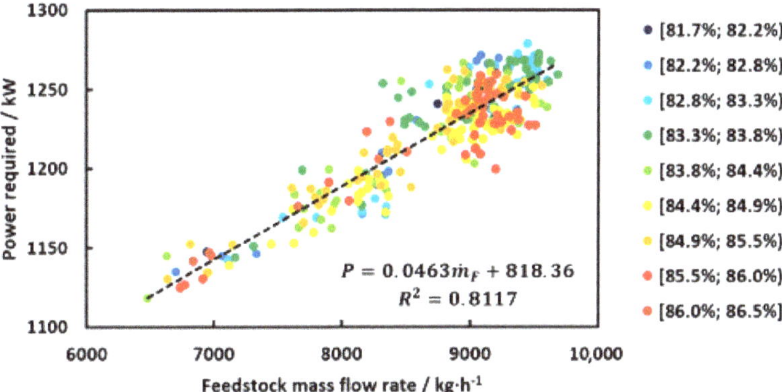

$$P = 0.0463\dot{m}_F + 818.36$$
$$R^2 = 0.8117$$

Figure 10. Process side characteristics.

The installed condensing steam turbine provided 1250 kW of power at nominal conditions, i.e., live steam pressure of 2.95 MPa, live steam temperature of 300 °C, exhaust pressure of 25 kPa, and shaft speed of 11,548 rpm, while consuming 2.1 kg/s of HPS. At these conditions, isentropic efficiency of 71.9% was declared. Based on the enclosed technical documentation (correction curves), the effect of alteration of these parameters was studied. The results can be observed in the figures below. It was proven that steam side parameters (Figure 11) affect the turbine isentropic efficiency insignificantly, as even the highest relative parameter deviations caused no more than a 2% difference in isentropic efficiency. Moreover, it is worth mentioning that no such drastic deviations occur in real steam networks, and thus the impact of steam side parameters on the isentropic efficiency can be neglected. However, as it is shown below, the shaft speed affected the isentropic efficiency to a high measure, while a change in shaft speed was a common phenomenon resulting from variations in the system performance and the feedstock throughput. Hence, the effect of shaft speed alteration on turbine isentropic efficiency was normalized (Figure 12) for designing the new turbine.

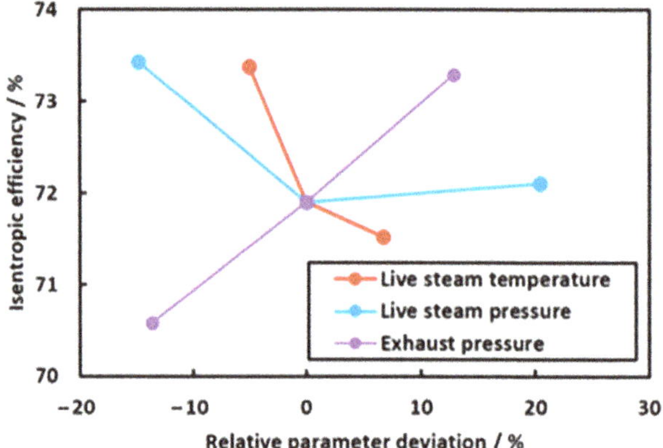

Figure 11. Effect of steam side parameters on isentropic efficiency.

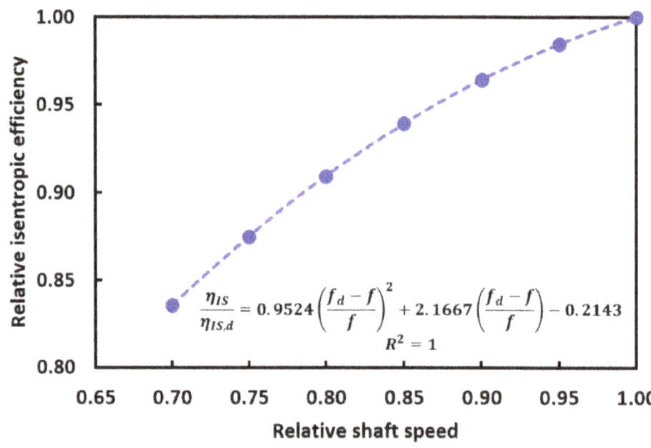

Figure 12. Relative isentropic efficiency as a function of relative shaft speed.

The study of actual steam side parameters, however, showed that the installed turbine operates off design as the real steam network measured values differ from the nominal ones. Hence, the parameters measured at the FCC unit battery limit were displayed in histograms (Figures 13–16) to obtain ideal design parameters for the new steam drive.

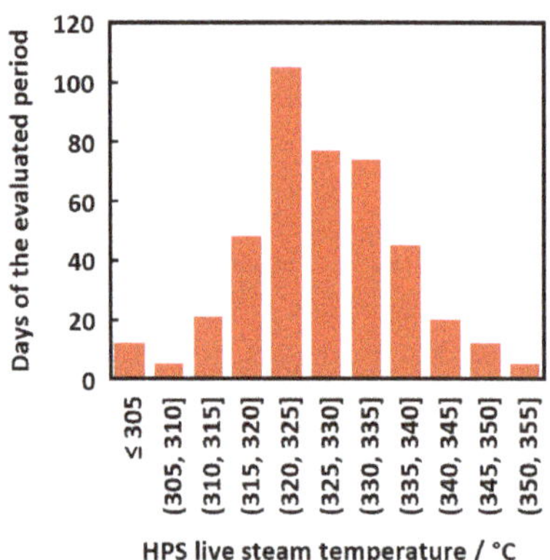

Figure 13. HPS temperature histogram.

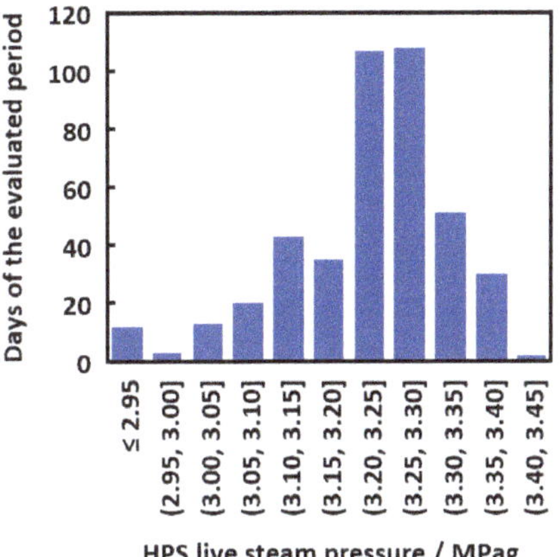

Figure 14. HPS pressure histogram.

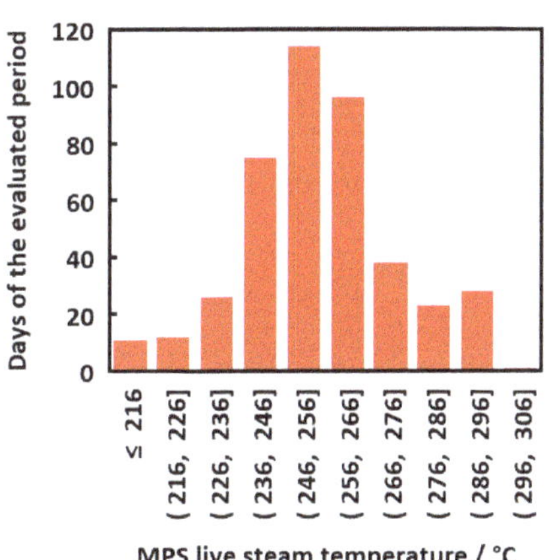

Figure 15. MPS temperature histogram.

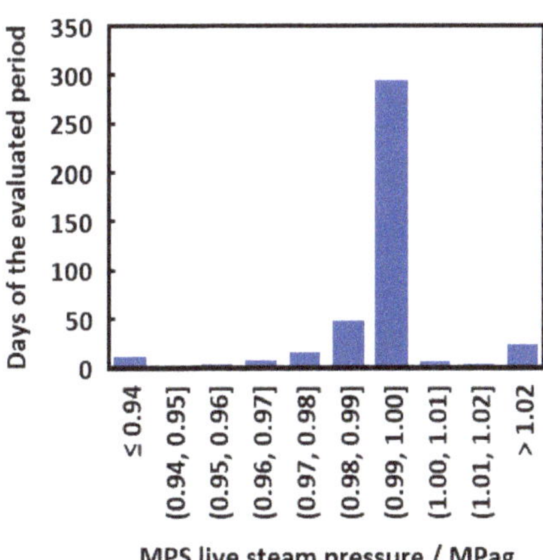

Figure 16. MPS pressure histogram.

Temperature and heat losses were studied on a plant-wide scale. Measured values of temperature and pressure of HPS exported from CHP were compared to those measured at the FCC battery limit. The results can be seen in Figure 17. Rather significant differences can be observed namely in temperatures where an almost 50 °C decrease was documented. These results illustrate the need for pressure and heat loss assessment prior to any proposals incorporating steam networks.

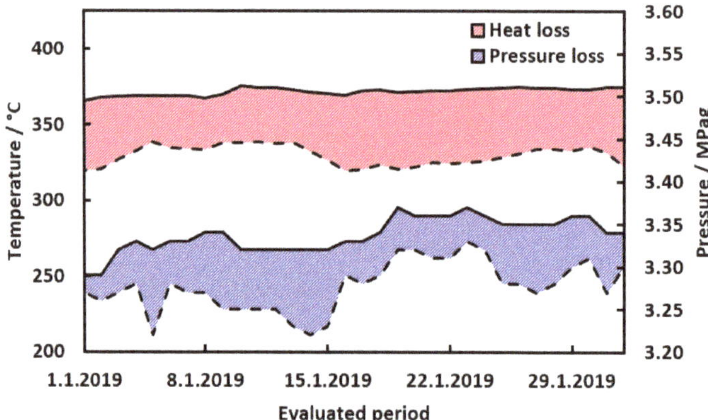

Figure 17. Heat and pressure losses in steam network. Legend: Solid line = conditions at the CHP; dashed line = conditions at the fluid catalytic cracking (FCC) battery limit.

3.4. Variable Approaches in Steam Drive Design

The steam turbine model was based on steam turbine characteristics well known as the Willan's line [80,81], i.e., linear dependency of the actual turbine's shaft output, P, on the actual steam mass flow, $\dot{m}$, Equation (1), applied to steam expansion between HPS and MPS, with I being the intercept of the linear relationship:

$$P = k\dot{m} - I \tag{1}$$

As reported by Mavromatis and Kokossis [80], the slope, k, in Equation (1) can be expressed as follows, in Equation (2):

$$k = \frac{1.2}{B}\left(\Delta h_{IS} - \frac{A}{\dot{m}_{max}}\right) \tag{2}$$

Parameters A and B can be correlated as a function of inlet steam saturation temperature (see [80] for further details); Δh_{IS} represents the isentropic enthalpy difference between inlet and discharge steam, depending on the inlet steam pressure and temperature as well as on the discharge pressure; $\dot{m}_{max}$ stands for the maximal (design) steam mass flow through the turbine. As verified in [47], values of k for common steam drive applications do not differ significantly from Δh_{IS} and, thus, Δh_{IS} is used as the slope of the steam drive characteristics in the steam turbine model. Once the nominal turbine steam consumption, the nominal obtained output, and the slope of the characteristics are known, the steam turbine characteristics can be constructed. Following the engineering practice of steam turbine vendors, when depicting the given relationship, the inlet steam mass flow is located on the y-axis and the obtained output on the x-axis. Thus, an inverse function to Equation (1) is depicted in Figure 18, with its slope being equal to Δh_{IS}^{-1}, as seen in Equation (3):

$$\dot{m}_s = \frac{1}{\Delta h_{IS}}P + \frac{I}{\Delta h_{IS}} = \frac{1}{\Delta h_{IS}}P + K \tag{3}$$

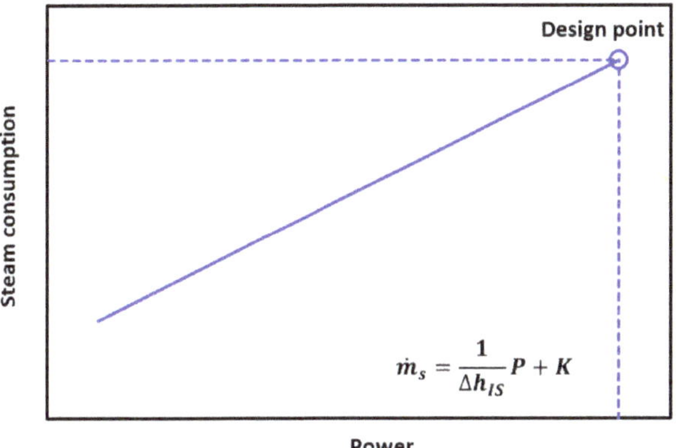

Steam consumption

$$\dot{m}_s = \frac{1}{\Delta h_{IS}} P + K$$

Power

Figure 18. Illustrative example of turbine characteristics.

Varying inlet steam conditions and discharge pressure impact the Δh_{IS} value and thereby affect the actual turbine's steam consumption necessary to achieve the desired output. The analysis of process side characteristics revealed the maximum power requirement of 1278 kW to be supplied by the turbine. For the new turbine, the design power output was thus set to 1300 kW. According to numerous publications on isentropic efficiency of industrial steam drives [30,82], an isentropic efficiency of 65% was assumed, which is typical for mid-size industrial steam drives operating at full load with a low-steam pressure ratio. Mechanical efficiency of such equipment can be estimated to be 85% [81]. While the mechanical efficiency changed in a very short interval and thus could be considered constant for a reasonable operational window [81], the isentropic efficiently changed significantly regarding the turbine load. The dependence between turbine power and isentropic efficiency has been well documented before [80]. It can be calculated directly from the Willan's line and described in the form of a parabolic function (Figure 19).

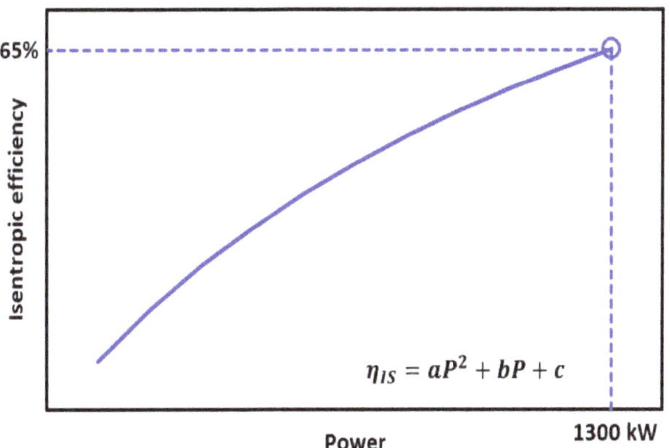

$$\eta_{IS} = aP^2 + bP + c$$

Figure 19. Isentropic efficiency as a function of power output.

To find the design point and subsequently construct the turbine characteristics, several approaches are available (Table 1). These vary depending on the number of variables considered, i.e., on the number of system properties that are neglected and/or simplified. To provide a comprehensive overview, a variety of possible approaches was exploited to illustrate the main differences in the resulting design (Table 3). These cases were subdivided into three groups: Cases 1–6 considered the properties and topology of the existing propylene recovery unit, with only cases 1–3 considering the features of the pipeline; cases 7–9 illustrated the changes in results for a steam drive located ten times farther (in means of pipeline length) from the battery limit; and case 10 was a standalone case considering only the basic enthalpy balance and constant values of all parameters.

Table 3. Steam drive design approach variations. Real insulation conductivity of 0.080 $W \cdot m^{-1} \cdot K^{-1}$, calculated previously by Hanus, et al. [26], was considered. Design insulation conductivity of 0.038 $W \cdot m^{-1} \cdot K^{-1}$ was considered as a common value for new steam pipeline insulations.

Approach	Pipe Length	Pipeline Heat Loss		Pressure Drop	Steam Quality Fluctuations	Varying Shaft Speed	Varying Isentropic Efficiency
		Real Insulation Conductivity	Design Insulation Conductivity				
Case 1	100%	✓	✗	✓	✓	✓	✓
Case 2	100%	✗	✓	✓	✓	✓	✓
Case 3	100%	✗	✗	✓	✓	✓	✓
Case 4	-	✗	✗	✗	✓	✓	✓
Case 5	-	✗	✗	✗	✗	✓	✓
Case 6	-	✗	✗	✗	✗	✗	✓
Case 7	1000%	✓	✗	✓	✓	✓	✓
Case 8	1000%	✗	✓	✓	✓	✓	✓
Case 9	1000%	✗	✗	✓	✓	✓	✓
Case 10	-	✗	✗	✗	✗	✗	✗

Pressure drop calculations for each segment of the pipeline came from the Bernoulli's equation, which can be generally written as Equation (4):

$$z_1 g + \frac{w_1^2}{2\alpha_1} + \frac{P_1}{\rho} = z_2 g + \frac{w_2^2}{2\alpha_2} + \frac{P_2}{\rho} + \varepsilon_{dis} \tag{4}$$

where z is the geodetic height, g the gravitational acceleration, w the mean steam transport velocity, P absolute pressure, ρ fluid density, ε_{dis} specific dissipated energy, and the dimensionless parameter, α, has the value of 0.5 or 1 for laminar or turbulent flow, respectively. Subscripts 1 and 2 refer to inlet and outlet of the pipe segment, respectively. For each segment, the inner pipe diameter remained constant. Thus, based on the continuity equation, the transport velocity remained constant as well. Furthermore, the difference in geodetic heights can be sensibly neglected and Equation (4) can be transformed into Equation (5):

$$\Delta P = \rho \varepsilon_{dis} \tag{5}$$

The overall specific dissipated energy comprises dissipation due to friction and local energy dissipation. Based on the Darcy's law, Equation (6) is deduced:

$$\Delta P = \rho \left(\lambda \frac{L w^2}{2D} + \sum \xi \frac{w^2}{2} \right) \tag{6}$$

where λ is the friction factor, L the length of the pipeline segment, D the inner diameter of the pipeline segment, and ξ is the coefficient of local dissipation. A generalized ready-to-use approach for friction factor estimation was proposed by Brkić and Praks [83].

Assuming that the only significant heat transfer resistances are that of the insulation and that of the ambient space, the following Equations (7) and (8) apply:

$$T_F \approx T_W \tag{7}$$

$$\dot{q}_L = \frac{\pi (T_F - T_I)}{\frac{1}{2\kappa} \ln \frac{D_I}{D_W}} = \frac{\pi (T_F - T_A)}{\frac{1}{2\kappa} \ln \frac{D_I}{D_W} + \frac{1}{\alpha_A D_I}} \tag{8}$$

where T_F is the temperature of transported fluid, T_W temperature of outer pipeline wall, T_I the outer temperature of insulation, T_A the ambient temperature, $\dot{q}_L$ the length-specific heat flux, κ the heat conductivity of the insulation, D_W the outer diameter of the pipe, D_I the outer diameter including insulation, and α_A is the combined radiation and convective heat transfer coefficient.

Therefore, the length-specific heat loss from a segment can be iteratively calculated for $f = 0$ from Equations (9)–(12) [84]:

$$\alpha_A = 9.74 + 0.07(T_{I,1} - T_A) \tag{9}$$

$$\dot{q}_L = \frac{\pi(T_F - T_A)}{\frac{1}{2\kappa}\ln\frac{D_I}{D_W} + \frac{1}{\alpha_A D_I}} \tag{10}$$

$$T_{I,2} = T_F - q\frac{\ln\frac{D_I}{D_W}}{2\pi\kappa} \tag{11}$$

$$f = T_{W,2} - T_{W,1} \tag{12}$$

where $T_{W,1}$ is the primary estimate of the temperature of the pipeline wall (under insulation).

4. Results and Discussion

The proposed change in the process steam drive from the actual condensing steam turbine to a new backpressure one anticipated a significant increase in HPS consumption. Whereas the actual consumption ranged roughly from 7.5 to 8.5 t/h (Figure 6), preliminary calculations for the turbine at nominal conditions (3.35 MPa, 325 °C HPS (BL), 1.0 MPa, 252 °C MPS (BL), 65% isentropic efficiency, 85% mechanical efficiency, 100% compressor shaft speed) revealed an increase in turbine HPS consumption to approx. 35 t/h. Hence, steam transport velocities in the interconnecting pipeline were studied (Figure 20).

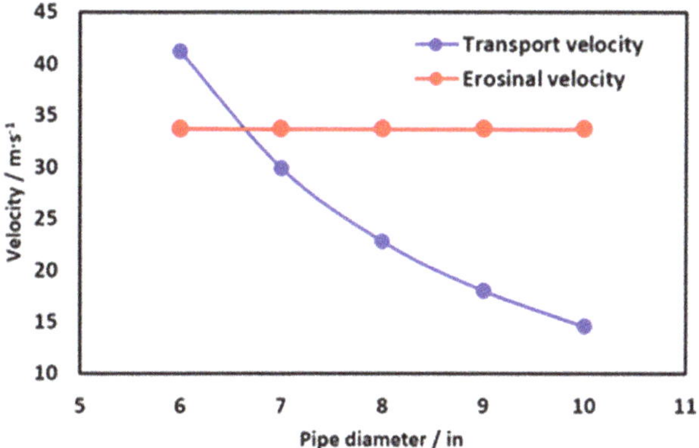

Figure 20. HPS transport velocity as a function of pipe diameter.

As the conducted case study showed (Figure 20), steam transported by a pipeline with a diameter below 7″ would breach the erosional velocity limit defined for HPS. Thus, a pipeline with such a diameter would encounter serious erosion. As the diameter of 7″ is not typical for industrial applications, an 8″ inlet pipeline was considered. A 10″ pipeline was chosen for the exhaust side (Figure 21).

The optimal inner pipe diameter was used to calculate nominal performance parameters. For each case, respective assumptions were incorporated in the simulation program to obtain information on nominal HPS consumption. The isentropic enthalpy difference was then determined from the actual enthalpy difference calculated by the model, and the characteristics' intercept was calculated directly, as seen in Equation (13):

$$K = \dot{m}_S - \frac{1}{\Delta h_{IS}}P \tag{13}$$

For any point of the characteristics (different from the nominal point), isentropic efficiency can be calculated from Equation (14):

$$\eta_{IS} = \frac{\Delta h}{\Delta h_{IS}} = \frac{\frac{W}{\dot{m}_S}}{\Delta h_{IS}} = \frac{\frac{\frac{P}{\eta_{mech}}}{\frac{1}{\Delta h_{IS}}P+K}}{\Delta h_{IS}} = \frac{P}{\eta_{mech}\left(P + \Delta h_{IS}K\right)} \tag{14}$$

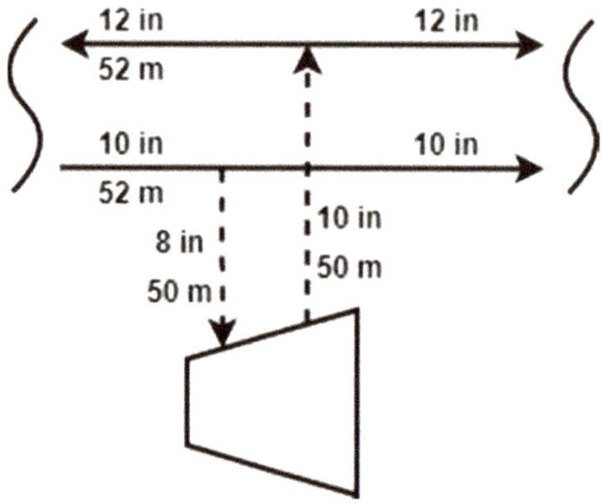

Figure 21. Pipeline inner diameter. Legend: Solid line = existing pipeline; dashed line = new pipelines.

Polynomial fitting of the isentropic efficiencies provided parameters for Figure 19, which, as well as other characteristic parameters of the turbine, are summed up in Table 4.

Table 4. Turbine characteristics.

	Ambient Temperature/°C	Nominal HPS Consumption/kg·h⁻¹	Nominal Δh/kJ·kg⁻¹	Δh_{IS}/kJ·kg⁻¹	K/kg·s⁻¹	a·10⁷	b·10⁴	c·10²
Case 1	10	34,703	158.6	244.0	4.312	−2.101	7.169	6.953
	35	34,703	158.6	244.0	4.312	−2.101	7.169	6.953
	−14	34,724	158.5	243.8	4.314	−2.101	7.170	6.954
Case 2	10	34,685	158.8	244.3	4.314	−2.098	7.164	6.933
	35	34,682	158.8	244.3	4.313	−2.099	7.164	6.935
	−14	34,688	158.8	244.3	4.314	−2.098	7.163	6.930
Case 3	N/A	34,623	159.0	244.6	4.303	−2.100	7.168	6.947
Cases 4–6	N/A	33,954	162.1	249.4	4.219	−2.101	7.169	6.953
Case 7	10	42,782	128.7	198.0	5.318	−2.100	7.166	6.942
	35	42,707	128.9	198.3	5.308	−2.100	7.168	6.948
	−14	42,855	128.5	197.7	5.328	−2.103	7.178	6.953
Case 8	10	42,172	130.5	200.8	5.239	−2.101	7.170	6.955
	35	42,137	130.7	201.1	5.240	−2.098	7.159	6.945
	−14	42,208	130.4	200.6	5.244	−2.103	7.176	6.961
Case 9	N/A	41,182	133.7	205.7	5.119	−2.100	7.166	6.942

To account for ambient temperature variations during the evaluated period, discrete temperature peaks were considered. These encompassed the highest, the lowest, and the average ambient temperature measured in 2018. Because cases 4–6 did not consider the pipeline properties, turbine characteristics for these cases were identical as they shared the same design point.

As can be seen in Table 4, temperature fluctuations affected the steam consumption minimally. Due to high temperature of the transported medium (HPS, ~325 °C), the driving force of heat transfer depended only insignificantly on the change of ambient temperature. Thus, the ambient temperature variations could be neglected, and were not taken into further consideration.

To visualize the differences between individual approaches, HPS consumption over the evaluated period using the methodology of each considered case was examined. For cases incorporating heat loss, the average ambient temperature (10 °C) was considered. In Figure 22, an evident discrepancy between the individual cases can be observed. Average deviations from the base cases (case 1 for actual pipeline length; case 7 for tenfold increase in pipeline length) are summarized in Table 5.

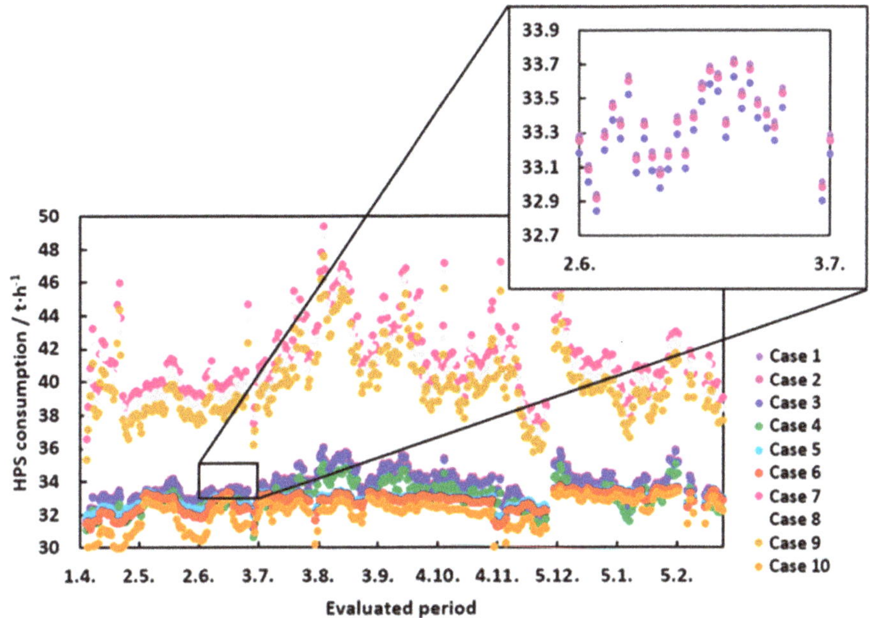

Figure 22. HPS consumption over the evaluated period.

Table 5. Comparison of different case results.

Base Case	Average Relative Deviation in HPS Consumption/%							
	Case 2	Case 3	Case 4	Case 5	Case 6	Case 8	Case 9	Case 10
Case 1	0.08	0.31	2.21	2.32	2.84	-	-	5.01
Case 7	-	-	20.35	20.09	20.52	1.61	3.74	22.28

While the effect of heat losses from pipelines to the ambient space were proven to be minimal for the actual length of piping (average deviation ≤ 0.31%), a tenfold increase in the pipe length increased the average deviation up to 3.74%. Hence, heat loss increased linearly with the distance. An almost identical trend could be observed in the pressure drop calculation, however, with dramatically different impact on the calculated HPS consumption. Thus, for calculations comprising long pipelines, severe errors are to be expected if pressure drop is neglected. Furthermore, models not considering steam quality fluctuations (case 5) and compressor shaft speed variations (case 6) are not capable of predicting peaks in HPS consumption (provide different trends) and are thus unsuitable even for systems comprising short pipelines, though their average deviation is only slightly different to case 4.

For a more comprehensive illustration, the yearly cumulative differences were displayed in Figures 23 and 24 for case 1 and case 7, respectively. For the actual-size pipeline, two most significant contributions were visible: The first caused by neglecting the pipeline pressure drop and the second by considering a constant isentropic efficiency. The latter has proven to be the most severe, resulting in an almost 13 kt/y difference.

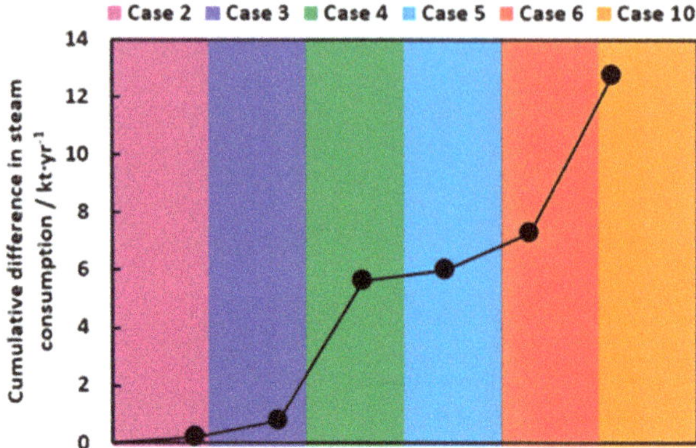

Figure 23. Cumulative difference in HPS consumption (base case 1).

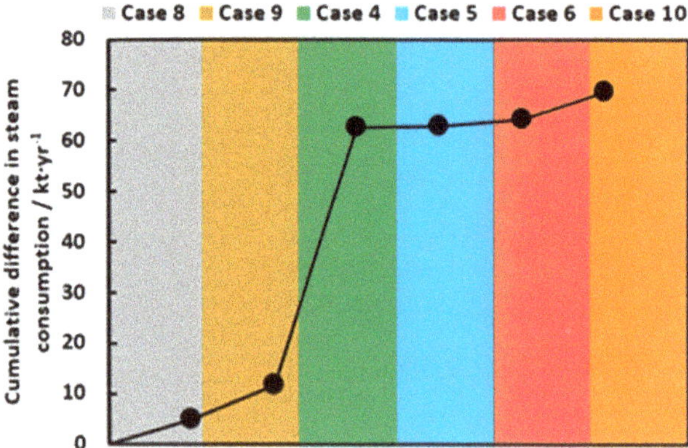

Figure 24. Cumulative difference in HPS consumption (base case 7).

When considering a pipeline ten times the length of the actual pipeline, the effect of every simplification in the process drive sizing becomes more evident. However, in contrast to calculations of the actual pipeline, the pressure drop is the most crucial. Moreover, cumulative differences in HPS consumption prove the overall trend to be monotonic as opposed to average deviation of case 5 from case 7 (Table 5). These findings underline the fact that even though average deviations for some cases showed only slight differences, significant cumulative differences can occur. This, above all, affected the final economic evaluation.

Practically every unit operates off-design for most of the operational time as the exact conditions defined as nominal are hardly ever met. However, the measure of deviation from the design parameters shows how well the equipment has been designed. To assess a possible negative impact of the steam drive replacement on the driven process, a study was conducted operating the turbine at full load for each case and each day of the evaluated period. The power output provided by the turbine consuming nominal amount of steam (Table 4) was compared to the actual power requirements of the process. A threshold of 5% (65 kW) was set, which is a typical design margin. In this study, again, results for the actual pipeline were compared to the ten-times-longer pipeline variant.

As shown in Figure 25, for a relatively short pipeline (≈100 m), a turbine designed with respect to all abovementioned aspects (Case 1) provided the process with the required power over the whole evaluated period. Simplified models, cases 4–6, also provided satisfactory results with hardly any threshold overshoots. However, the difference in the performance reserve was evident, pointing out nominal turbine inlet steam mass flow undersizing.

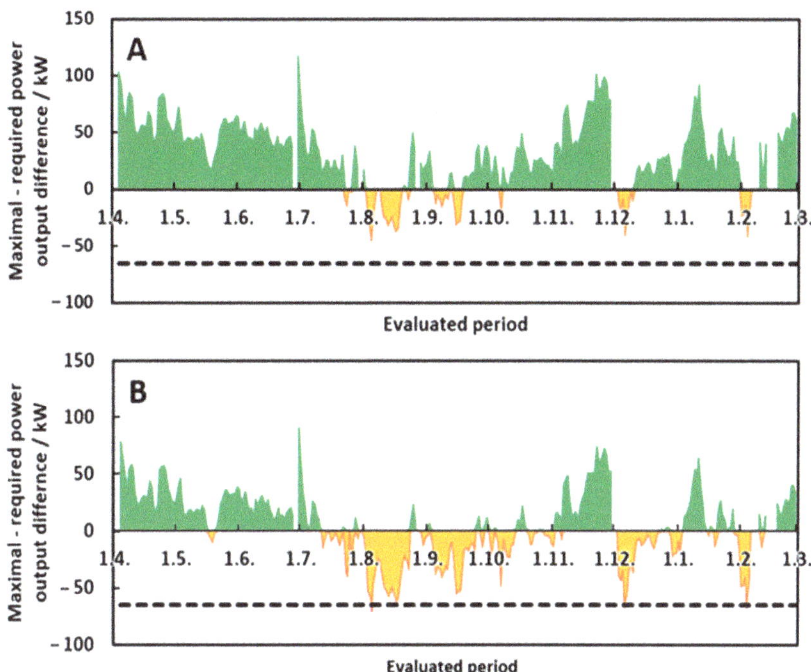

Figure 25. Required to maximal power output differences for case 1 (**A**) and cases 4–6 (**B**) considering 100% pipeline length. Dashed line represents a typical 5% design reserve.

The situation changed dramatically with the increase in pipeline length. As depicted in Figure 26, even when all the aspects were considered (case 7), there was a short period of time when the turbine was not capable of providing the required mechanical power. The more such periods appeared, the less system properties were considered (e.g., case 9, which did not take heat loss into account). Finally, the worst-case scenario did not take pressure drop into account. As proven below, a turbine designed without regard to pressure drop along the pipeline did not provide the process with the required mechanical power with its lack exceeding 100 kW (approximately 8% of nominal mechanical power requirement) very frequently. The inability of the steam drive designed without steam frictional pressure losses consideration to meet the process-side mechanical power demand inevitably resulted in a decrease of processed C3 fraction mass flow. The C3 fraction splitting process could thus become a bottleneck of the whole FCC reaction products separation section with serious consequences on its profitability.

Based on the process side characteristics (Figure 10), production loss due to insufficient power supply can be quantified. Each 100 kW of lacking power output (i.e., power output below the dashed line in Figure 26) represented 2.16 t/h of production loss (Figure 27). For the studied propylene recovery unit processing approx. 65.8 kt of feedstock yearly, the decrease of 13.2 kt/y (20%) was unacceptable.

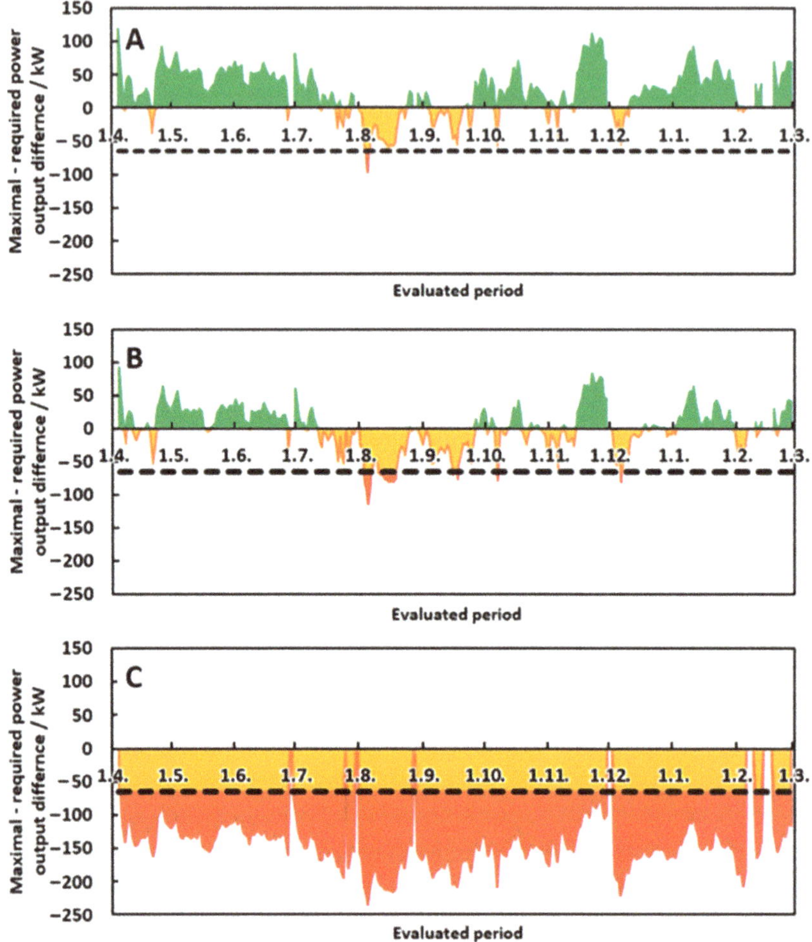

Figure 26. Required to maximal power output difference for case 7 (**A**), case 9 (**B**) and cases 4–6 (**C**) considering 1000% pipeline length. Dashed line represents a typical 5% design reserve.

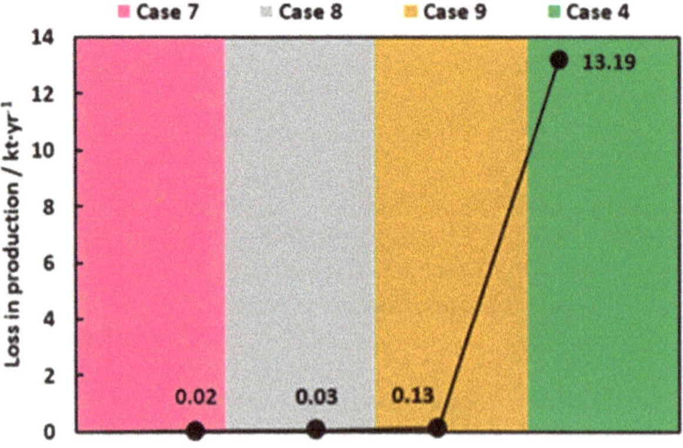

Figure 27. Loss in production for a system considering 1000% pipeline length.

Besides the impact of the proposed steam drive change on the propylene recovery process, its effect on the operation of the refinery's main HPS and MPS pipelines was assessed in terms of transported steam mass flow. A significant change in the steam mass flow may result in new bottlenecks or even an infeasible operation of the network [26]. Current HPS pipeline operation was analyzed in detail by Hanus et al. [26], defining the safe operation window between 20 and 60 t/h of exported HPS from CHP. Such conditions prevent excessive erosion and pressure losses on one hand and steam stagnation in pipelines on the other. An existing pipeline, returning HPS to the CHP, enables HPS network operation with zero or even negative net HPS export from the CHP. Furthermore, exporting more than 60 t/h of HPS from the CHP is possible by utilizing a second main HPS pipeline that is normally closed but can be activated if needed. However, even with both the main HPS pipelines active, exporting more than 80 t/h HPS from the CHP for longer periods is unwanted, though possible. As shown in Figure 28, the proposed steam drive change eliminated the occurrence of undesirably low HPS export from the CHP and simultaneously led to HPS export of over 80 t/h in certain periods, forcing the steam network operators to undertake additional measures to ensure the HPS network stability.

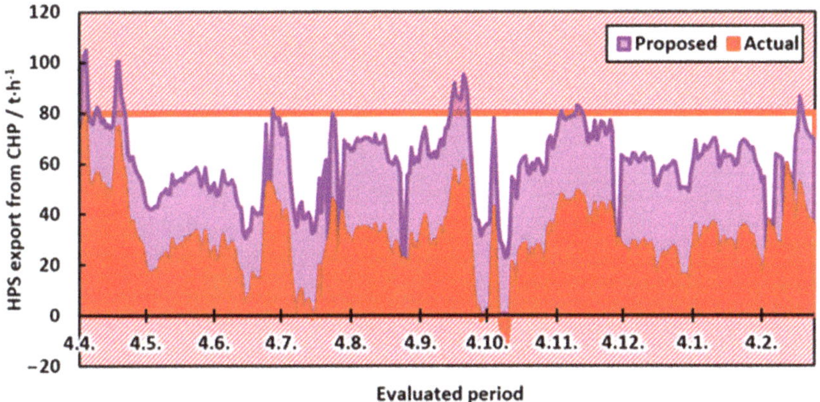

Figure 28. Impact of proposed steam drive change on the HPS network. Long-term operation of the HPS network outside the <0; 80> t/h HPS export interval is infeasible.

The MPS network operation was also investigated and the results are shown in Figure 29. Presently, the MPS demand of the refinery exceeds 80 or even 100 t/h, which strains the MPS production capacity of the CHP. MPS export of over 100 t/h cannot usually be met by steam extraction and the deficit has to be covered by 9 MPa steam throttling in the CHP. As presented in Figure 29, the occurrence of such unwanted states is strongly reduced by the new process drive in operation. A few periods appear, though, with MPS export below 20 t/h, which may affect the MPS network operation stability and have to be avoided by active MPS network management.

The change in exported steam mass flow at high-pressure and middle-pressure levels affects fuel consumption and carbon dioxide emissions of the CHP. The resulting effect differs according to the season of the year:

1. In colder months (October to April), fuel is saved, and CO_2 emissions are reduced. Backpressure power production in the CHP is also reduced;
2. In warmer months (May to September), the reduction in the CHP backpressure power production is compensated by an increase in the condensing production which keeps the total power output of the CHP unchanged. The resulting change in fuel consumption and in CO_2 emissions production is determined by the difference between: (a) Marginal condensing power production efficiency of the CHP and, (b) the condensing mechanical power production in the replaced condensing steam drive.

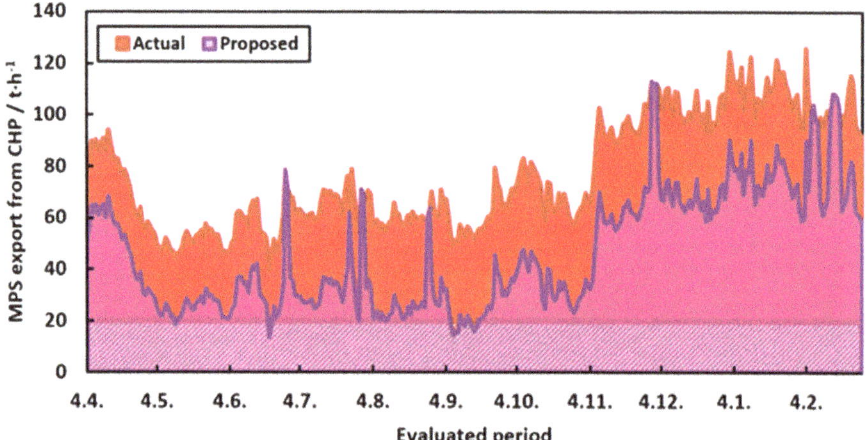

Figure 29. Impact of proposed steam drive change on the MPS network. Long-term operation of the MPS network below 20 t/h of exported MPS is infeasible.

Based on discussion with CHP managers and operators, the following input data were considered in calculations:

1. Average thermal efficiency of the CHP is 85%, determined as the ratio of the enthalpy in exported steam to the fuel lower heating value;
2. Heavy fuel oil combusted in the CHP produces 3.2 tons of CO_2 per 1 ton of oil;
3. Marginal efficiency of the condensing power production in the CHP is 3 MWh per ton of combusted fuel.

Trends of fuel consumption change and CHP electric power output decrease in the evaluated period are visualized in Figure 30.

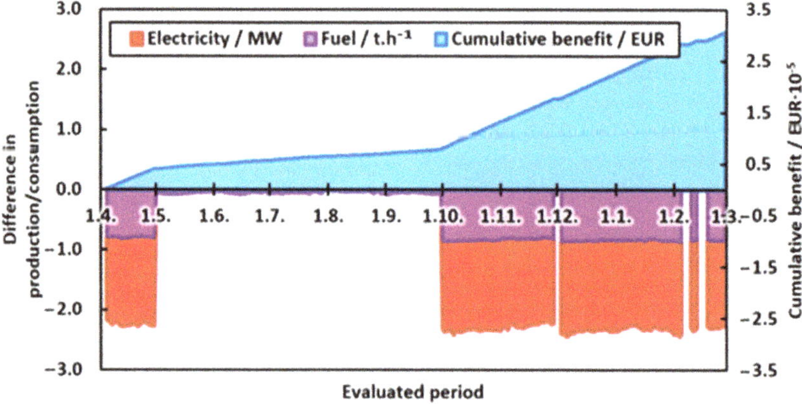

Figure 30. Economic evaluation.

Evaluation of the economic potential of the condensing steam drive replacement assumed the following prices: Electricity 45 €/MWh; heavy fuel oil 120 €/t; and CO_2 emissions 27 €/t. The hourly benefit results from (1) the achieved decrease in fuel consumption and in CO_2 emissions, and (2) the amount of additional power purchased from the outer grid. Additional power has to be purchased to balance the lower CHP output in colder months (Figure 30). The cumulative benefit curve (Figure 30) is obtained by summing up the hourly benefits over the evaluated period. As can be seen, most of the benefit is harvested during colder months. Overall, 3.72 kt/year of fuel can be saved and CO_2 emissions from the CHP can be reduced by 11.13 kt/year as a result of the process steam drive replacement. The resulting benefit is expected to be over 300,000 EUR/year. A preliminary estimate of the steam drive replacement cost including new pipelines is around 450,000 EUR, leading to a preliminary simple payback period of 1.5 years. Due to the envisioned CO_2 emission cost increase, economic feasibility of this proposal should be retained even if the fuel price decreases or that of purchased electricity increases.

Considering the results presented in Figure 25, it can be concluded that the C3 fraction feed processing capacity is met even with less rigorous steam drive sizing method. However, for the alternative with ten-times-longer pipelines, only the rigorous method presented in this paper ensures the retained process throughput capacity. Less complex sizing methods led to undersizing of the steam drive and the resulting processing capacity limitation (see Figure 27) outweighed the economic potential of the steam drive replacement (Figure 30) by far.

The presented technical and economic results varied significantly depending on the type and operation of the CHP that served as a marginal steam source. Modern industrial CHP units usually comprise a gas turbine-based combined cycle. If such a unit operates in a purely cogeneration mode (no or minimal condensing power production), the resulting change in fuel consumption and power production is different from that in Figure 30. Apart from the electricity production change due to the steam balance shift on HPS and MPS levels, either (a) an increase in condensing power production due to utilization of the saved HPS, or, (b) a decrease in the power output of gas turbine(s) and the associated fuel consumption reduction can be observed. The way is chosen by the CHP managers based on fuel costs compared to marginal power production cost in the combined cycle. In any case, since the combined cycle is a more efficient marginal cogeneration steam source than a traditional steam CHP, the resulting fuel consumption and CO_2 emissions release is lower. This further accentuates the need to analyze the marginal steam source operation carefully prior to steam drive sizing.

5. Conclusions

The presented method for process steam drive sizing comprises both process and steam network including the marginal steam source operation assessment. A proper investigation of both process-side and steam-side design and operation parameters is necessary and should precede the drive sizing itself. Attention should be paid to key process parameters including the required power input and its hourly and seasonal variability as well as the variability of the driven equipment frequency. Driving steam quality and discharge steam pressure variations should be evaluated carefully, and their change due to frictional pressure and heat losses from steam pipelines should also be assessed. Driving and discharge steam pressure levels should be chosen according to the analysis of individual steam pressure levels' operation range and their anticipated change resulting from steam drive implementation. Their effect on the marginal steam source operation should be evaluated considering all previous findings.

The proposed method comprises process data evaluation and calculations in a linked MATLAB® and Aspen Plus® environment. Thereby, the potential of MATLAB in process data analysis and results evaluation and visualization is coupled with the ability of rigorous process simulation by Aspen Plus®.

The industrial case study comprised the replacement of the existing condensing steam drive by a new backpressure one, driving the 1.25 MW compressor of the heat pump-assisted C3 fraction splitting process. Application of the proposed method and comparison with the results obtained using methods proposed by other authors revealed that:

1. Steam drive undersizing resulted from lower complexity of the sizing methods;
2. Neglecting the variable frequency of the driven equipment, frictional pressure losses, and the steam drive efficiency loss at partial load operation could decrease the ability of the steam drive to provide the power required for the process;

3. The simplest sizing method combined with the ten-times-longer steam pipeline led to a C3 fraction splitting capacity decrease of around 20%, which was unacceptable.

Examination of the changes in the HPS and MPS network operation revealed that several capacity bottlenecks could be removed by the steam drive replacement, but new ones could arise, which requires active steam network management. The impact on the CHP operation included fuel savings of up to 3.72 kt/year, and a CO_2 emissions reduction of 11.13 kt/year at the expense of an additional 17.29 GWh/year of power purchased from outer grid compensating for the power production decrease at the CHP. Warmer months' contribution to this cost was negligible as the CHP compensated for the lowered backpressure power production by the expensive condensing one. Despite this fact, the proposed steam drive replacement exhibited a simple payback period shorter than two years.

Further method improvement will be aimed in future work with the focus on implementing multi-objective optimization. The effects of steam drive implementation or replacement of the main steam network operation should be examined and assessed more closely.

Author Contributions: Conceptualization, P.F. and M.V.; data curation, P.F. and T.C.; funding acquisition, Z.L.; investigation, P.F. and T.C.; methodology, M.V. and P.F.; resources, T.C.; software, P.F; supervision, M.V. and Z.L.; validation, T.C.; visualization, P.F.; writing—original draft, P.F. and M.V.; writing—review and editing, Z.L. All authors have read and agreed to the published version of the manuscript.

Funding: This work was financially supported by the Slovak Research and Development Agency, Grant Nos. APVV-19-0170 and APVV-18-0134, and by the Slovak Scientific Agency, Grant No. VEGA 1/0659/18.

Conflicts of Interest: The authors declare no conflict of interest. The funders had no role in the design of the study; in the collection, analyses, or interpretation of data; in the writing of the manuscript; or in the decision to publish the results.

Nomenclature

Abbreviations

BE	balance equations
BL	battery limit
C3	propane–propylene mixture
C3A	propane
C3E	propylene
Calc.	calculated
CHP	combined heat and power unit
COM	component object model
Cond.	condenser
CW	cooling water
FCC	fluid catalytic cracking
frac.	fraction
HPS	high-pressure steam
MPS	middle-pressure steam
LPS	low-pressure steam
N/A	not applicable
NP	not provided
PP	polypropylene
prod.	production
Reg.	calculated based on statistic regression
SA	sensitivity analysis
SS	saturated steam conditions
WS	wet steam conditions

Symbols

a	first parameter of polynomial regression, Figure 19 (kW^{-2})
A	parameter, Equation (2) (kW)
b	second parameter of polynomial regression, Figure 19 (kW^{-1})
B	parameter, Equation (2)
c	third parameter of polynomial regression, Figure 19
D	diameter (m)
f	shaft speed (rpm)
g	gravitational acceleration, $g = 9.81\ m{\cdot}s^{-2}$
h	specific enthalpy ($kJ{\cdot}kg^{-1}$)
I	intercept, Equation (1) (kW)
k	slope, Equation (1) ($kJ{\cdot}kg^{-1}$)
K	intercept, Equations (3), (13) and (14) ($kg{\cdot}s^{-1}$)
L	length (m)
$\dot{m}$	mass flow rate ($kg{\cdot}s^{-1}$)
P	power (kW)
	pressure, Equations (4)–(6) (Pa)
$\dot{q}_L$	length-specific heat flux ($W{\cdot}m^{-1}$)
w	fluid mean transport velocity ($m{\cdot}s^{-1}$)
W	net work (kW)
z	geographical height (m)

Greek symbols

$\propto$	heat transfer coefficient ($W{\cdot}m^{-2}{\cdot}K^{-1}$)
	dimensionless parameter, Equation (4)
Δ	difference
ε	specific mechanical energy ($kJ{\cdot}kg^{-1}$)
η	efficiency
κ	overall heat transfer coefficient ($W{\cdot}m^{-1}{\cdot}K^{-1}$)
λ	friction factor
ξ	coefficient of local dissipation
ρ	density ($kg{\cdot}m^{-3}$)

Subscripts

A	ambient
d	design
dis	dissipation
F	fluid
I	insulation
IS	isentropic
max	maximal
mech	mechanical
W	wall

Appendix A

```
%% Aspen link-up
Aspen = actxserver('Apwn.Document.36.0'); % Creating a local COM server; creating a
structured variable "Aspen"
[~, mess] = fileattrib; % Accessing the folder
Simulation_Name = 'C3_simulation_3'; % Name of the desired simulation to run
Aspen.invoke('InitFromArchive2',[mess.Name '\' Simulation_Name '.apw']); % Linking Aspen
Plus simulation with MATLAB via created server environment
Aspen.Visible = 1; % Whether or not will Aspen Plus be physically opened
Aspen.SuppressDialogs = 1; % Whether or not will contextual windows be displayed
Aspen.Run2(); % Starting the initial simulation
```

```
%% Excel input: process data
first = '97'; % First row of the database to evaluate
last = '427'; % Last row of the database to evaluate
m.C3A = xlsread('Process Data.xlsx', 'Hárok1', ['O' first ':O' last]);
m.C3E = xlsread('Process Data.xlsx', 'Hárok1', ['N' first ':N' last]);
K401.P = xlsread('Process Data.xlsx', 'Hárok1', ['S' first ':S' last]);
steam.T = xlsread('Process Data.xlsx', 'Hárok1', ['Y' first ':Y' last]);
steam.kP = xlsread('Process Data.xlsx', 'Hárok1', ['AA' first ':AA' last]);
steam.P = xlsread('Process Data.xlsx', 'Hárok1', ['AC' first ':AC' last]);
```

```
%% Simulation
Aspen.Tree.FindNode("\Data\Blocks\K401\Input\MEFF").Value = 0.85; % Mechanic efficiency

% Preallocation of variables:
x.BTM = zeros(length(m.C3A), 1);
x.OVD = zeros(length(m.C3A), 1);
steam.m = zeros(length(m.C3A), 1);
K401.eff = zeros(length(m.C3A), 1);
K401.W = zeros(length(m.C3A), 1);
K401.f = zeros(length(m.C3A), 1);
m.W402 = zeros(length(m.C3A), 1);
m.E401 = zeros(length(m.C3A), 1);
Q.E401 = zeros(length(m.C3A), 1);
Convergence = zeros(length(m.C3A), 1);

for i = 1 : length(m.C3A)
    if isnan(m.C3A(i)) % If dataset incomplete, simulation proceeds with another day
        x.BTM(i, 1) = NaN;
        x.OVD(i, 1) = NaN;
```

Figure A1. Process-side calculation (part 1).

```matlab
        steam.m(i, 1) = NaN;
        K401.eff(i, 1) = NaN;
        K401.W(i, 1) = NaN;
        K401.f(i, 1) = NaN;
        m.W402(i, 1) = NaN;
        m.E401(i, 1) = NaN;
        Q.E401(i, 1) = NaN;
        Convergence(i, 1) = NaN;
    else
        % Input adjustment:
        Aspen.Tree.FindNode("\Data\Streams\C3F-C401\Input\FLOW\MIXED\C3").Value = m.C3A(i);
        Aspen.Tree.FindNode("\Data\Streams\C3F-C401\Input\FLOW\MIXED\C3=").Value =
m.C3E(i);
        Aspen.Tree.FindNode("\Data\Blocks\K401\Input\PRES").Value = K401.P(i);
        Aspen.Tree.FindNode("\Data\Streams\PA-3,5\Input\TEMP\MIXED").Value = steam.T(i);
        Aspen.Tree.FindNode("\Data\Blocks\TK401\Input\PRES").Value = steam.kP(i);
        Aspen.Tree.FindNode("\Data\Streams\PA-3,5\Input\PRES\MIXED").Value = steam.P(i);


        % Calculation boundaries:
        Aspen.Tree.FindNode("\Data\Convergence\Conv-Options\Input\DIR_MAXIT").Value = 1500;
% Initial number of maximum tear iterations
        Aspen.Tree.FindNode("\Data\Convergence\Conv-Options\Input\SEC_MAXIT").Value = 300;
% Initial number of maximum design specifications iterations
        Aspen.Tree.FindNode("\Data\Flowsheeting Options\Design-Spec\BTM-
C401\Input\LOWER").Value = 0.012; % Initial lower bound for BTM-C401
        Aspen.Tree.FindNode("\Data\Flowsheeting Options\Design-Spec\BTM-
C401\Input\UPPER").Value = 0.015; % Initial upper bound for BTM-C401
        Aspen.Tree.FindNode("\Data\Flowsheeting Options\Design-Spec\OVD-
C401\Input\LOWER").Value = 4500; % Initial lower bound for OVD-C401
        Aspen.Tree.FindNode("\Data\Flowsheeting Options\Design-Spec\OVD-
C401\Input\UPPER").Value = 6500; % Initial upper bound for OVD-C401
        Aspen.Tree.FindNode("\Data\Convergence\Conv-Options\Input\SEC_XTOL").Value = 1e-6;
% Initial Secant X tolerance
        Aspen.Tree.FindNode("\Data\Convergence\Conv-Options\Input\TOL").Value = 1e-4; %
Initial tear tolerance
        Aspen.Tree.FindNode("\Data\Convergence\Conv-Options\Input\FLASH").Value = 'NO'; %
Whether or not to flash tear streams after update (saves computation time)


        % Starting the simulation:
        Aspen.Run2();
```

Figure A2. Process-side calculation (part 2).

```matlab
        % Gathering results:
        x.BTM(i, 1) = Aspen.Tree.FindNode("\Data\Streams\C3A-
P401\Output\MASSFRAC\MIXED\C3=").Value;
        x.OVD(i, 1) = Aspen.Tree.FindNode("\Data\Streams\C3E-
PP3\Output\MASSFRAC\MIXED\C3=").Value;
        steam.m(i, 1) = Aspen.Tree.FindNode("\Data\Streams\KOT-
3,5\Output\MASSFLMX\MIXED").Value;
        K401.eff(i, 1) = Aspen.Tree.FindNode("\Data\Blocks\K401\Output\EFF_POLY").Value;
        K401.W(i, 1) = Aspen.Tree.FindNode("\Data\Blocks\K401\Output\WNET").Value;
        K401.f(i, 1) = Aspen.Tree.FindNode("\Data\Blocks\K401\Output\SH_SPEED").Value;
        m.W402(i, 1) = Aspen.Tree.FindNode("\Data\Streams\VAP-
W402\Output\MASSFLMX\MIXED").Value;
        m.E401(i, 1) = Aspen.Tree.FindNode("\Data\Streams\VAP-
E401\Output\MASSFLMX\MIXED").Value;
        Q.E401(i, 1) = Aspen.Tree.FindNode("\Data\Blocks\E401\Output\HX_DUTY").Value;
        Convergence(i, 1) = Aspen.Tree.FindNode("\Data\Results Summary\Run-
Status\Output\CVSTAT").Value;

        % Troubleshooting:
        if Convergence(i, 1) ~= 0 % Resolves insufficient iterations count
            Aspen.Tree.FindNode("\Data\Convergence\Conv-Options\Input\DIR_MAXIT").Value =
1500; % Adjusted number of maximum tear iterations
            Aspen.Tree.FindNode("\Data\Convergence\Conv-Options\Input\SEC_MAXIT").Value =
300; % Adjusted number of maximum design specifications iterations
            Aspen.Run2();
            Convergence(i, 1) = Aspen.Tree.FindNode("\Data\Results Summary\Run-
Status\Output\CVSTAT").Value;
        end
        if Convergence(i, 1) ~= 0 % Resolves design specifications problems
            if x.BTM(i, 1) < 0.02
                Aspen.Tree.FindNode("\Data\Flowsheeting Options\Design-Spec\BTM-
C401\Input\UPPER").Value = 0.03;
            elseif x.BTM(i, 1) > 0.06
                Aspen.Tree.FindNode("\Data\Flowsheeting Options\Design-Spec\BTM-
C401\Input\LOWER").Value = 0.005;
            elseif x.OVD(i, 1) < 0.996
                Aspen.Tree.FindNode("\Data\Flowsheeting Options\Design-Spec\OVD-
C401\Input\UPPER").Value = 7000;
            elseif x.OVD(i, 1) > 0.998
                Aspen.Tree.FindNode("\Data\Flowsheeting Options\Design-Spec\OVD-
C401\Input\LOWER").Value = 4000;
            end
```

Figure A3. Process-side calculation (part 3).

```matlab
        Aspen.Run2();
        Convergence(i, 1) = Aspen.Tree.FindNode("\Data\Results Summary\Run-
Status\Output\CVSTAT").Value;
    end
    if Convergence(i, 1) ~= 0
        Aspen.Reinit();
    end
  end
end
```

```matlab
%% Results
% Overal results variable:
X = [steam.m x.BTM x.OVD K401.eff K401.W K401.f m.W402 m.E401 Q.E401 Convergence];

% Saving the results:
save results.dat X -ascii

% Closing the simulation:
Aspen.Close
Aspen.Quit
```

Figure A4. Process-side calculation (part 4).

```matlab
%% Aspen link-up
Aspen = actxserver('Apwn.Document.36.0'); % Creating a local COM server; creating a
structured variable "Aspen"
[~, mess] = fileattrib; % Accessing the folder
Simulation_Name = 'Turbine'; % Name of the desired simulation to run
Aspen.invoke('InitFromArchive2',[mess.Name '\' Simulation_Name '.apw']); % Linking Aspen
Plus simulation with MATLAB via created server environment
Aspen.Visible = 1; % Whether or not will Aspen Plus be physically opened
Aspen.SuppressDialogs = 1; % Whether or not will contextual windows be displayed
Aspen.Run2(); % Starting the initial simulation
```

```matlab
%% Excel input: steam-side data
first = '97'; % First row of the database to evaluate
last = '427'; % Last row of the database to evaluate
TK401.P = xlsread('Steam Properties.xlsx', 'TK401 steam consumption', ['B' first ':B'
last]);
TK401.f = xlsread('Steam Properties.xlsx', 'TK401 steam consumption', ['C' first ':C'
last]);
FCC35.m = xlsread('Steam Properties.xlsx', 'TK401 steam consumption', ['D' first ':D'
last]);
BL35.T = xlsread('Steam Properties.xlsx', 'TK401 steam consumption', ['E' first ':E'
last]);
BL35.P = xlsread('Steam Properties.xlsx', 'TK401 steam consumption', ['F' first ':F'
last]);
FCC10.m = xlsread('Steam Properties.xlsx', 'TK401 steam consumption', ['G' first ':G'
last]);
BL10.T = xlsread('Steam Properties.xlsx', 'TK401 steam consumption', ['H' first ':H'
last]);
BL10.P = xlsread('Steam Properties.xlsx', 'TK401 steam consumption', ['I' first ':I'
last]);
```

```matlab
%% Simulation
% Prealocation of variables:
PA.m = zeros(length(TK401.P), 1);
PA.T_in = zeros(length(TK401.P), 1);
PA.T_out  = zeros(length(TK401.P), 1);
PA.P_in = zeros(length(TK401.P), 1);
PA.P_out  = zeros(length(TK401.P), 1);
TK401.eff = zeros(length(TK401.P), 1);
BL10.m_import = zeros(length(TK401.P), 1);
BL10.m_export = zeros(length(TK401.P), 1);
```

Figure A5. Steam-side calculation (part 1).

```matlab
BL35.m = zeros(length(TK401.P), 1);

for i = 1 : length(TK401.P)
    if isnan(TK401.P(i)) % For an incomplete dataset, simulation proceeds with another day
        PA.m(i, 1) = NaN;
        PA.T_in(i, 1) = NaN;
        PA.T_out(i, 1) = NaN;
        PA.P_in(i, 1) = NaN;
        PA.P_out(i, 1) = NaN;
        TK401.eff(i, 1) = NaN;
        BL10.m_import(i, 1) = NaN;
        BL10.m_export(i, 1) = NaN;
        BL35.m(i, 1) = NaN;
    else
        % Heat loss calculation
        lambda = 0.038;
        d.pipe_8in = 0.219075;
        d.pipe_10in = 0.27305;
        d.insul_8in = d.pipe_8in + 0.18;
        d.insul_10in = d.pipe_10in + 0.18;
        t_ambient = 10;

        [q.in8, t_w.in8] = heat_flux(BL35.T(i), t_ambient, lambda, d.pipe_8in,
d.insul_8in);
        [q.in10, t_w.in10] = heat_flux(BL35.T(i), t_ambient, lambda, d.pipe_10in,
d.insul_10in);

        % Input adjustment
        Aspen.Tree.FindNode("\Data\Blocks\10IN35\Input\FLUX").Value = - q.in10;
        Aspen.Tree.FindNode("\Data\Blocks\8IN35\Input\FLUX").Value = - q.in8;
        Aspen.Tree.FindNode("\Data\Streams\BL-PA10\Input\TEMP\MIXED").Value = BL10.T(i);
        Aspen.Tree.FindNode("\Data\Streams\BL-PA10\Input\PRES\MIXED").Value = BL10.P(i);
        Aspen.Tree.FindNode("\Data\Streams\BL-PA35\Input\TEMP\MIXED").Value = BL35.T(i);
        Aspen.Tree.FindNode("\Data\Streams\BL-PA35\Input\PRES\MIXED").Value = BL35.P(i);
        Aspen.Tree.FindNode("\Data\Streams\FCC-CONS\Input\TOTFLOW\MIXED").Value =
FCC10.m(i);
        Aspen.Tree.FindNode("\Data\Blocks\S1\Input\BASIS_FLOW\PA35-FCC").Value =
FCC35.m(i);
        Aspen.Tree.FindNode("\Data\Flowsheeting Options\Design-Spec\PA-
TK401\Input\EXPR2").Value = - TK401.P(i);
        Aspen.Tree.FindNode("\Data\Flowsheeting
Options\Calculator\RPM\Input\FVN_INIT_VAL\RPM").Value = TK401.f(i);
```

Figure A6. Steam-side calculation (part 2).

```matlab
        Aspen.Tree.FindNode("\Data\Flowsheeting
Options\Calculator\RPM\Input\FVN_INIT_VAL\POWER").Value = TK401.P(i);
        Aspen.Tree.FindNode("\Data\Flowsheeting
Options\Calculator\RPM\Input\FVN_INIT_VAL\A").Value = -0.0000209951e-2;
        Aspen.Tree.FindNode("\Data\Flowsheeting
Options\Calculator\RPM\Input\FVN_INIT_VAL\B").Value = 0.0716640927e-2;
        Aspen.Tree.FindNode("\Data\Flowsheeting
Options\Calculator\RPM\Input\FVN_INIT_VAL\C").Value = 6.9424077737e-2;


        % Starting the simulation:
        Aspen.Run2();


        % Gathering results:
        PA.m(i, 1) = Aspen.Tree.FindNode("\Data\Streams\PA10-
II\Output\MASSFLMX\MIXED").Value;
        PA.T_in(i, 1) = Aspen.Tree.FindNode("\Data\Streams\PA35-
IV\Output\TEMP_OUT\MIXED").Value;
        PA.T_out(i, 1)  = Aspen.Tree.FindNode("\Data\Streams\PA10-
II\Output\TEMP_OUT\MIXED").Value;
        PA.P_in(i, 1) = Aspen.Tree.FindNode("\Data\Streams\PA35-
IV\Output\PRES_OUT\MIXED").Value;
        PA.P_out(i, 1) = Aspen.Tree.FindNode("\Data\Streams\PA10-
II\Output\PRES_OUT\MIXED").Value;
        TK401.eff(i, 1) = Aspen.Tree.FindNode("\Data\Blocks\TK401\Output\EFF_ISEN").Value;
        BL10.m_import(i, 1) = Aspen.Tree.FindNode("\Data\Streams\BL-
PA10\Output\MASSFLMX\MIXED").Value;
        BL10.m_export(i, 1) = Aspen.Tree.FindNode("\Data\Streams\PA10-
BL\Output\MASSFLMX\MIXED").Value;
        BL35.m(i, 1) = Aspen.Tree.FindNode("\Data\Streams\BL-
PA35\Output\MASSFLMX\MIXED").Value;
    end
end

%% Results
% Overal results variable:
X = [PA.m PA.T_in PA.T_out PA.P_in PA.P_out TK401.eff BL10.m_import BL10.m_export BL35.m];
% Saving the results:
save turbine_results.dat X -ascii
% Closing the simulation:
Aspen.Close
Aspen.Quit
```

Figure A7. Steam-side calculation (part 3).

References

1. Murugan, S.; Horák, B. Tri and polygeneration systems—A review. *Renew. Sustain. Energy Rev.* **2016**, *60*, 1032–1051. [CrossRef]
2. Jana, K.; Ray, A.; Majoumerd, M.M.; Assadi, M.; De, S. Polygeneration as a future sustainable energy solution—A comprehensive review. *Appl. Energy* **2017**, *202*, 88–111. [CrossRef]
3. Baláž, V.; Nežinský, E.; Jeck, T.; Filčák, R. Energy and Emission Efficiency of the Slovak Regions. *Sustainability* **2020**, *12*, 2611. [CrossRef]
4. Brożyna, J.; Strielkowski, W.; Fomina, A.; Nikitina, N. Renewable Energy and EU 2020 Target for Energy Efficiency in the Czech Republic and Slovakia. *Energies* **2020**, *13*, 965. [CrossRef]
5. Rehfeldt, M.; Worrell, E.; Eichhammer, W.; Fleiter, T. A review of the emission reduction potential of fuel switch towards biomass and electricity in European basic materials industry until 2030. *Renew. Sustain. Energy Rev.* **2020**, *120*, 109672. [CrossRef]
6. Korkmaz, P.; Gardumi, F.; Avgerinopoulos, G.; Blesl, M.; Fahl, U. A comparison of three transformation pathways towards a sustainable European society—An integrated analysis from an energy system perspective. *Energy Strategy Rev.* **2020**, *28*, 100461. [CrossRef]
7. Martins, F.; Felgueiras, C.; Smitkova, M.; Caetano, N. Analysis of Fossil Fuel Energy Consumption and Environmental Impacts in European Countries. *Energies* **2019**, *12*, 964. [CrossRef]
8. Malinauskaite, J.; Jouhara, H.; Ahmad, L.; Milani, M.; Montorsi, L.; Venturelli, M. Energy efficiency in industry: EU and national policies in Italy and the UK. *Energy* **2019**, *172*, 255–269. [CrossRef]
9. Linares, P.; Pintos, P.; Würzburg, K. Assessing the potential and costs of reducing energy demand. *Energy Transit.* **2017**, *1*, 4. [CrossRef]
10. Maciková, L.; Smorada, M.; Dorčák, P.; Beug, B.; Markovič, P. Financial Aspects of Sustainability: An Evidence from Slovak Companies. *Sustainability* **2018**, *10*, 2274. [CrossRef]
11. Lieskovský, M.; Trenčiansky, M.; Majlingová, A.; Jankovský, J. Energy Resources, Load Coverage of the Electricity System and Environmental Consequences of the Energy Sources Operation in the Slovak Republic—An Overview. *Energies* **2019**, *12*, 1701. [CrossRef]
12. Ghoniem, A.F. Needs, resources and climate change: Clean and efficient conversion technologies. *Prog. Energy Combust. Sci.* **2011**, *37*, 15–51. [CrossRef]
13. Sutherland, B.R. Sustainably Heating Heavy Industry. *Joule* **2020**, *4*, 14–16. [CrossRef]
14. Rivas, D.F.; Castro-Hernández, E.; Villanueva Perales, A.L.; van der Meer, W. Evaluation method for process intensification alternatives. *Chem. Eng. Process.-Process Intensif.* **2018**, *123*, 221–232. [CrossRef]
15. Ifaei, P.; Safder, U.; Yoo, C. Multi-scale smart management of integrated energy systems, Part 1: Energy, economic, environmental, exergy, risk (4ER) and water-exergy nexus analyses. *Energy Convers. Manag.* **2019**, *197*, 111851. [CrossRef]
16. Safder, U.; Ifaei, P.; Yoo, C. Multi-scale smart management of integrated energy systems, Part 2: Weighted multi-objective optimization, multi-criteria decision making, and multi-scale management (3M) methodology. *Energy Convers. Manag.* **2019**, *198*, 111830. [CrossRef]
17. Rong, A.; Lahdelma, R. Role of polygeneration in sustainable energy system development challenges and opportunities from optimization viewpoints. *Renew. Sustain. Energy Rev.* **2016**, *53*, 363–372. [CrossRef]
18. Pintarič, Z.N.; Varbanov, P.S.; Klemeš, J.J.; Kravanja, Z. Multi-objective multi-period synthesis of energy efficient processes under variable environmental taxes. *Energy* **2019**, *189*, 116182. [CrossRef]
19. Al Moussawi, H.; Fardoun, F.; Louahlia, H. Selection based on differences between cogeneration and trigeneration in various prime mover technologies. *Renew. Sustain. Energy Rev.* **2017**, *74*, 491–511. [CrossRef]
20. Bamufleh, H.S.; Ponce-Ortega, J.M.; El-Halwagi, M.M. Multi-objective optimization of process cogeneration systems with economic, environmental, and social tradeoffs. *Clean Technol. Environ. Policy* **2012**, *15*, 185–197. [CrossRef]
21. Chen, Z.; Wang, J. Heat, mass, and work exchange networks. *Front. Chem. Sci. Eng.* **2012**, *6*, 484–502. [CrossRef]
22. Chew, K.; Klemeš, J.; Alwi, S.; Manan, Z.; Reverberi, A. Total Site Heat Integration Considering Pressure Drops. *Energies* **2015**, *8*, 1114–1137. [CrossRef]
23. Fan, Y.V.; Chin, H.H.; Klemeš, J.J.; Varbanov, P.S.; Liu, X. Optimisation and process design tools for cleaner production. *J. Clean. Prod.* **2020**, *247*, 119181. [CrossRef]

24. Frate, G.F.; Ferrari, L.; Lensi, R.; Desideri, U. Steam expander as a throttling valve replacement in industrial plants: A techno-economic feasibility analysis. *Appl. Energy* **2019**, *238*, 11–21. [CrossRef]
25. Ge, Z.; Zhang, F.; Sun, S.; He, J.; Du, X. Energy Analysis of Cascade Heating with High Back-Pressure Large-Scale Steam Turbine. *Energies* **2018**, *11*, 119. [CrossRef]
26. Hanus, K.; Variny, M.; Illés, P. Assessment and Prediction of Complex Industrial Steam Network Operation by Combined Thermo-Hydrodynamic Modeling. *Processes* **2020**, *8*, 622. [CrossRef]
27. Kler, A.M.; Stepanova, E.L.; Maksimov, A.S. Investigating the efficiency of a steam-turbine heating plant with a back-pressure steam turbine and waste-heat recovery. *Thermophys. Aeromech.* **2019**, *25*, 929–938. [CrossRef]
28. Liew, P.Y.; Wan Alwi, S.R.; Varbanov, P.S.; Manan, Z.A.; Klemeš, J.J. Centralised utility system planning for a Total Site Heat Integration network. *Comput. Chem. Eng.* **2013**, *57*, 104–111. [CrossRef]
29. Marton, S.; Svensson, E.; Subiaco, R.; Bengtsson, F.; Harvey, S. A Steam Utility Network Model for the Evaluation of Heat Integration Retrofits—A Case Study of an Oil Refinery. *J. Sustain. Dev. Energy Water Environ. Syst.* **2017**, *5*, 560–578. [CrossRef]
30. Mrzljak, V.; Poljak, I.; Mrakovčić, T. Energy and exergy analysis of the turbo-generators and steam turbine for the main feed water pump drive on LNG carrier. *Energy Convers. Manag.* **2017**, *140*, 307–323. [CrossRef]
31. Ng, R.T.L.; Loo, J.S.W.; Ng, D.K.S.; Foo, D.C.Y.; Kim, J.-K.; Tan, R.R. Targeting for cogeneration potential and steam allocation for steam distribution network. *Appl. Therm. Eng.* **2017**, *113*, 1610–1621. [CrossRef]
32. Sanaye, S.; Khakpaay, N.; Chitsaz, A. Thermo-economic and environmental multi-objective optimization of a novel arranged biomass-fueled gas engine and backpressure steam turbine combined system for pulp and paper mills. *Sustain. Energy Technol. Assess.* **2020**, *40*, 100778. [CrossRef]
33. Sun, L.; Doyle, S.; Smith, R. Heat recovery and power targeting in utility systems. *Energy* **2015**, *84*, 196–206. [CrossRef]
34. Sun, W.; Zhao, Y.; Wang, Y. Electro- or Turbo-Driven?—Analysis of Different Blast Processes of Blast Furnace. *Processes* **2016**, *4*, 28. [CrossRef]
35. Tian, Y.; Xing, Z.; He, Z.; Wu, H. Modeling and performance analysis of twin-screw steam expander under fluctuating operating conditions in steam pipeline pressure energy recovery applications. *Energy* **2017**, *141*, 692–701. [CrossRef]
36. Wu, L.; Liu, Y.; Liang, X.; Kang, L. Multi-objective optimization for design of a steam system with drivers option in process industries. *J. Clean. Prod.* **2016**, *136*, 89–98. [CrossRef]
37. Wu, Y.; Wang, R.; Wang, Y.; Feng, X. An area-wide layout design method considering piecewise steam piping and energy loss. *Chem. Eng. Res. Des.* **2018**, *138*, 405–417. [CrossRef]
38. Zhao, L.; Zhong, W.; Du, W. Data-Driven Robust Optimization for Steam Systems in Ethylene Plants under Uncertainty. *Processes* **2019**, *7*, 744. [CrossRef]
39. Huang, Y.; Hou, W.; Huang, Y.; Li, J.; Li, Q.; Wang, D.; Zhang, Y. Multi-Objective Optimal Operation for Steam Power Scheduling Based on Economic and Exergetic Analysis. *Energies* **2020**, *13*, 1886. [CrossRef]
40. Marton, S.; Svensson, E.; Harvey, S. Operability and Technical Implementation Issues Related to Heat Integration Measures—Interview Study at an Oil Refinery in Sweden. *Energies* **2020**, *13*, 3478. [CrossRef]
41. Beangstrom, S.G.; Majozi, T. Steam system network synthesis with hot liquid reuse: II. Incorporating shaft work and optimum steam levels. *Comput. Chem. Eng.* **2016**, *85*, 202–209. [CrossRef]
42. Min, K.-J.; Binns, M.; Oh, S.-Y.; Cha, H.-Y.; Kim, J.-K.; Yeo, Y.-K. Screening of site-wide retrofit options for the minimization of CO_2 emissions in process industries. *Appl. Therm. Eng* **2015**, *90*, 335–344. [CrossRef]
43. Bütün, H.; Kantor, I.; Maréchal, F. Incorporating Location Aspects in Process Integration Methodology. *Energies* **2019**, *12*, 3338. [CrossRef]
44. Wu, Y.; Wang, Y.; Feng, X. A heuristic approach for petrochemical plant layout considering steam pipeline length. *Chin. J. Chem. Eng.* **2016**, *24*, 1032–1037. [CrossRef]
45. Svensson, E.; Morandin, M.; Harvey, S.; Papadokonstantakis, S. Studying the Role of System Aggregation in Energy Targeting: A Case Study of a Swedish Oil Refinery. *Energies* **2020**, *13*, 958. [CrossRef]
46. Chowdhury, J.I.; Hu, Y.; Haltas, I.; Balta-Ozkan, N.; Matthew, G., Jr.; Varga, L. Reducing industrial energy demand in the UK: A review of energy efficiency technologies and energy saving potential in selected sectors. *Renew. Sustain. Energy Rev.* **2018**, *94*, 1153–1178. [CrossRef]
47. Variny, M.; Blahušiak, M.; Mierka, O.; Godó, Š.; Margetíny, T. Energy saving measures from their cradle to full adoption with verified, monitored, and targeted performance: A look back at energy audit at Catalytic Naphtha Reforming Unit (CCR). *Energy Effic.* **2019**, *12*, 1771–1793. [CrossRef]

48. Variny, M.; Furda, P.; Švistun, L.; Rimár, M.; Kizek, J.; Kováč, N.; Illés, P.; Janošovský, J.; Váhovský, J.; Mierka, O. Novel Concept of Cogeneration-Integrated Heat Pump-Assisted Fractionation of Alkylation Reactor Effluent for Increased Power Production and Overall CO_2 Emissions Decrease. *Processes* **2020**, *8*, 183. [CrossRef]

49. van de Bor, D.M.; Infante Ferreira, C.A.; Kiss, A.A. Low grade waste heat recovery using heat pumps and power cycles. *Energy* **2015**, *89*, 864–873. [CrossRef]

50. Gangar, N.; Macchietto, S.; Markides, C.N. Recovery and Utilization of Low-Grade Waste Heat in the Oil-Refining Industry Using Heat Engines and Heat Pumps: An International Technoeconomic Comparison. *Energies* **2020**, *13*, 2560. [CrossRef]

51. Kazemi, A.; Hosseini, M.; Mehrabani-Zeinabad, A.; Faizi, V. Evaluation of different vapor recompression distillation configurations based on energy requirements and associated costs. *Appl. Therm. Eng.* **2016**, *94*, 305–313. [CrossRef]

52. Holmberg, H.; Ruohonen, P.; Ahtila, P. Determination of the Real Loss of Power for a Condensing and a Backpressure Turbine by Means of Second Law Analysis. *Entropy* **2009**, *11*, 702–712. [CrossRef]

53. Fontalvo, J. Using user models in Matlab® within the Aspen Plus® interface with an Excel® link. *Ing. Investig.* **2014**, *34*, 39–43. [CrossRef]

54. Fontalvo, J.; Cuellar, P.; Timmer, J.M.K.; Vorstman, M.A.G.; Wijers, J.G.; Keurentjes, J.T.F. Comparing Pervaporation and Vapor Permeation Hybrid Distillation Processes. *Ind. Eng. Chem. Res.* **2005**, *44*, 5259–5266. [CrossRef]

55. Darkwah, K.; Knutson, B.L.; Seay, J.R. Multi-objective versus single-objective optimization of batch bioethanol production based on a time-dependent fermentation model. *Clean Technol. Environ. Policy* **2018**, *20*, 1271–1285. [CrossRef]

56. Briones Ramírez, A.; Gutiérrez Antonio, C. Multiobjective Optimization of Chemical Processes with Complete Models using MATLAB and Aspen Plus. *Comput. Sist.* **2018**, *22*, 1157–1170. [CrossRef]

57. Muñoz, C.A.; Telen, D.; Nimmegeers, P.; Cabianca, L.; Logist, F.; Van Impe, J. Investigating practical aspects of the exergy based multi-objective optimization of chemical processes. *Comput. Aided Chem. Eng.* **2017**, *40*, 2173–2178. [CrossRef]

58. Cui, C.; Zhang, X.; Sun, J. Design and optimization of energy-efficient liquid-only side-stream distillation configurations using a stochastic algorithm. *Chem. Eng. Res. Des.* **2019**, *145*, 48–52. [CrossRef]

59. Capra, F.; Magli, F.; Gatti, M. Biomethane liquefaction: A systematic comparative analysis of refrigeration technologies. *Appl. Therm. Eng.* **2019**, *158*, 113815. [CrossRef]

60. dos Santos Vidal, S.F.; Schmitz, J.E.; Franco, I.C.; Frattini Fileti, A.M.; Da Silva, F.V. Fuzzy Multivariable Control Strategy Applied to a Refrigeration System. *Chem. Prod. Process Modeling* **2017**, *12*, 20160033. [CrossRef]

61. Ryu, H.; Lee, J.M. Model Predictive Control (MPC)-Based Supervisory Control and Design of Off-Gas Recovery Plant with Periodic Disturbances from Parallel Batch Reactors. *Ind. Eng. Chem. Res.* **2016**, *55*, 3013–3025. [CrossRef]

62. Silva, W.C.; Araújo, E.C.C.; Calmanovici, C.E.; Bernardo, A.; Giulietti, M. Environmental assessment of a standard distillery using aspen plus®: Simulation and renewability analysis. *J. Clean. Prod.* **2017**, *162*, 1442–1454. [CrossRef]

63. Ping, W.; Changfang, X.; Shiming, X.; Yulin, G. Study of Direct Compression Heat Pump Energy-saving Technology. *Procedia Environ. Sci.* **2012**, *12*, 394–399. [CrossRef]

64. Gao, X.; Gu, Q.; Ma, J.; Zeng, Y. MVR heat pump distillation coupled with ORC process for separating a benzene-toluene mixture. *Energy* **2018**, *143*, 658–665. [CrossRef]

65. Peng, D.-Y.; Robinson, D.B. A New Two-Constant Equation of State. *Ind. Eng. Chem. Fundam.* **1976**, *15*, 59–64. [CrossRef]

66. Ho, Q.N.; Yoo, K.S.; Lee, B.G.; Lim, J.S. Measurement of vapor–liquid equilibria for the binary mixture of propylene (R-1270)+propane (R-290). *Fluid Phase Equilibria* **2006**, *245*, 63–70. [CrossRef]

67. Sarath Yadav, E.; Indiran, T.; Nayak, D.; Aditya Kumar, C.; Selvakumar, M. Simulation study of distillation column using Aspen plus. *Mater. Today Proc.* **2020**. in Press. [CrossRef]

68. Querol, E.; Gonzalez-Regueral, B.; Ramos, A.; Perez-Benedito, J.L. Novel application for exergy and thermoeconomic analysis of processes simulated with Aspen Plus®. *Energy* **2011**, *36*, 964–974. [CrossRef]

69. Lan, W.; Chen, G.; Zhu, X.; Wang, X.; Liu, C.; Xu, B. Biomass gasification-gas turbine combustion for power generation system model based on ASPEN PLUS. *Sci. Total Environ.* **2018**, *628–629*, 1278–1286. [CrossRef]

70. Li, S.; Li, F. Prediction of Cracking Gas Compressor Performance and Its Application in Process Optimization. *Chin. J. Chem. Eng.* **2012**, *20*, 1089–1093. [CrossRef]

71. Liu, Z.; Karimi, I.A. Simulating combined cycle gas turbine power plants in Aspen HYSYS. *Energy Convers. Manag.* **2018**, *171*, 1213–1225. [CrossRef]

72. Pouransari, N.; Bocquenet, G.; Maréchal, F. Site-scale process integration and utility optimization with multi-level energy requirement definition. *Energy Convers. Manag.* **2014**, *85*, 774–783. [CrossRef]

73. Zhu, Q.; Luo, X.; Zhang, B.; Chen, Y.; Mo, S. Mathematical modeling, validation, and operation optimization of an industrial complex steam turbine network-methodology and application. *Energy* **2016**, *97*, 191–213. [CrossRef]

74. Sun, L.; Doyle, S.; Smith, R. Understanding steam costs for energy conservation projects. *Appl. Energy* **2016**, *161*, 647–655. [CrossRef]

75. Golmohamadi, G.; Asadi, A. Integration of joint power-heat flexibility of oil refinery industries to uncertain energy markets. *Energies* **2020**, *13*, 4874. [CrossRef]

76. Abril, A.F. Aspen Plus—Matlab Link. Available online: https://www.mathworks.com/matlabcentral/fileexchange/69464-aspen-plus-matlab-link (accessed on 9 September 2020).

77. Lu, J.; Tang, J.; Chen, X.; Cui, M.; Fei, Z.; Zhang, Z.; Qiao, X. Global Optimization of Reactive Distillation Processes using Bat Algorithm. *Chem. Eng. Trans.* **2017**, *61*, 1279–1284. [CrossRef]

78. Gulied, M.; Al Nouss, A.; Khraisheh, M.; AlMomani, F. Modeling and simulation of fertilizer drawn forward osmosis process using Aspen Plus-MATLAB model. *Sci. Total Environ.* **2020**, *700*, 134461. [CrossRef]

79. Aspen Technology Inc. *Aspen Plus User Guide*; Version 10.2; Aspen Technology Inc.: Cambridge, MA, USA, 2000.

80. Mavromatis, S.P.; Kokossis, A.C. Conceptual optimisation of utility networks for operational variations—I. Targets and level optimisation. *Chem. Eng. Sci.* **1998**, *53*, 1585–1608. [CrossRef]

81. Varbanov, P.S.; Doyle, S.; Smith, R. Modelling and Optimization of Utility Systems. *Chem. Eng. Res. Des.* **2004**, *82*, 561–578. [CrossRef]

82. Mavromatis, S.P.; Kokossis, A.C. Conceptual optimisation of utility networks for operational variations—II. Network development and optimisation. *Chem. Eng. Sci.* **1998**, *53*, 1609–1630. [CrossRef]

83. Brkić, D.; Praks, P. Unified Friction Formulation from Laminar to Fully Rough Turbulent Flow. *Appl. Sci.* **2018**, *8*, 2036. [CrossRef]

84. Wang, H.; Wang, H.; Zhu, T.; Deng, W. A novel model for steam transportation considering drainage loss in pipeline networks. *Appl. Energy* **2017**, *188*, 178–189. [CrossRef]

Publisher's Note: MDPI stays neutral with regard to jurisdictional claims in published maps and institutional affiliations.

 *processes*

Article

Integrated Method of Monitoring and Optimization of Steam Methane Reformer Process

Nenad Zečević * and Nenad Bolf

Department of Measurements and Process Control, Faculty of Chemical Engineering and Technology, University of Zagreb, 10000 Zagreb, Croatia; bolf@fkit.hr
* Correspondence: nenad.zecevic@petrokemija.hr

Received: 14 March 2020; Accepted: 24 March 2020; Published: 31 March 2020

Abstract: Reforming of natural gas with steam represents the most energy-intensive part of ammonia production. An integrated numerical model for calculating composition of primary reforming products with cross-checking of outlet methane molar concentration, heat duty, maximum tube wall temperature, tube pressure drops, and approach to equilibrium was set up involving production parameters. In particular, the model was used for continuous monitoring and optimization of a steam methane reformer (SMR) catalyst in ammonia production. The calculations involve the solution of material and energy balance equations along with reaction kinetic expressions. Open source code based on Matlab file was used for modelling and calculation of various physical properties of the reacting gases. One of the main contributions is development of the rapid integrated method for data exchange between any distributed control system (DCS) and the model to accomplish continuous monitoring and optimization of SMR catalyst and reformer tubes. Integrated memory block was proposed for rapid synchronization between commercial DCS with the model solver. The developed model was verified with the industrial top-fired SMR unit in ammonia production located in Petrokemija, Croatia. Practical application of proposed solution can ensure overall energy savings of up to 3% in ammonia production.

Keywords: ammonia production; numerical modelling; steam methane reforming; simulation

---

1. Introduction

The importance of chemical process simulators is well-documented, as they are important tools for modelling chemical plants, while providing opportunities for optimization and debottlenecking of existing processes [1,2]. All commercially available process simulators follow the traditional development method, which in almost 100% of cases hides the source code from the end users thus relying on a closed, black-box approach. Open source code software can be a useful tool for development of an integrated process simulator and related simulation model that would be able to serve industrial users in improving their production operations. One of the interesting cases for application of the open source code software in fertilizer industry is ammonia production as the most energy-intensive process which gives the main raw material for different end products. The main energy consumer in ammonia plants is the steam methane reformer (SMR) furnace. In typical SMR furnace approximately 50% of the heat generated by combustion of natural gas in the burners is transferred through the reformer tubes and absorbed by the process gas. This unit presents the primary focus for the operators to minimize their costs in the whole ammonia plant.

The industrial users of any ammonia plant would like to have the ability to rapidly monitor and evaluate the performance of the SMR in its regular operation life in steady-state and dynamic mode. Many models have been created to describe SMR units with varying levels of details [3–5].

Sophisticated simulators have been used to describe the performance of the units with a high accuracy [6]. Nevertheless, many of these take a long time to converge, however, which is impractical for regular industrial uses and at the same time does not allow for the continuous prediction of process parameters with possibility for their adjustment and improvement to achieve the best available performance. Besides that, a series of catalyst beds within ammonia plants needs simple and practical on-line monitoring of their performance regarding the activity, selectivity, and lifetime.

The one-dimensional heterogeneous numerical model based on the open source code which replicates some of the work of Xu and Froment is proposed and extended to include main process control parameters in a form that is applicable to the vast majority of commercially available industrial SMR catalysts [7–10]. The developed model has been compared with a real top-fired SMR unit from a stand-alone ammonia plant in Petrokemija fertilizer production complex involving a well-defined SMR catalyst. The developed model takes into account reaction kinetic constants, thermodynamic equations of state, heat fluxes, pressure drop, temperature approach to equilibrium, and catalyst properties. The model can be used in continuous monitoring and optimization of the performance of many different SMR catalysts through the application of predictive simulation as shown in Figure 1.

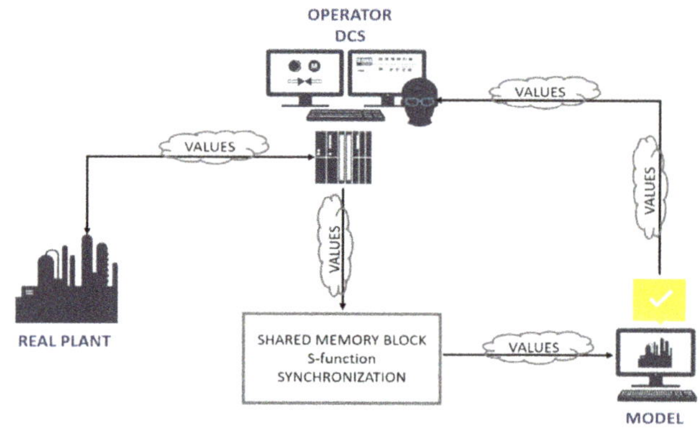

Figure 1. Integrated model for continuous monitoring and optimization of the steam methane reformer (SMR) catalyst.

The communication between a discrete model and any distributed control system (DCS) can be achieved by system-function which is compiled as MEX file using C++ language. This digital thread enables the model to access all specific information on each actual process parameter in the closed-loop system. In this way, the operators can directly and rapidly retrieve the necessary process data from the model at the end of each sampling period, which than can be immediately used as the new boundary condition in the optimization scheme.

The model can help operators to safely and securely optimize performance of the SMR catalyst, providing them with rapid, accurate, and predictive simulations. The simulation presents an exact and complete replica of the SMR catalyst, ensuring that users may interact with a control system interface to adjust better performance. It can be also easily used as a basis for implementation of an automatic advanced process (APC) control tool.

Implementation of the predictive simulation model for continuous monitoring and optimization of main process variables during operation of SMR catalyst will have direct economic benefits to industrial users in the range from 2% to 3% of overall energy consumption in running top-fired SMR. Besides that, the model will present a significant opportunity for digital transformation of existing industrial production units according to the goals of Industry 4.0.

2. Model Description

Steam reforming is the conversion of steam and hydrocarbons into a mixture of hydrogen, carbon monoxide, carbon dioxide, and unconverted reactants. Steam may be replaced either partially or completely by carbon dioxide as reactant [11]. The reforming of natural gas with steam utilizes reforming reactions and water–gas shift reaction. Besides these two main reactions there are nine other possible reactions in the SMR process. With respect to simplifying the model for industrial users, only first three reactions presented in Table 1 were used for description of industrial SMR [9–11]. According to literature data only those reactions are critical for thermodynamic and kinetic consideration [10,11].

Table 1. Steam methane reformer (SMR) reactions used in the model.

Reaction Description	Reaction	ΔH^o_{298} (kJ/mol)
Steam-methane reforming 1, SR1	$CH_4 + H_2O \leftrightarrow CO + 3H_2$	206.10
Water–gas shift, WGS	$CO + H_2O \leftrightarrow CO_2 + H_2$	−41.15
Steam-methane reforming 2, SR2	$CH_4 + 2H_2O \leftrightarrow CO_2 + 4H_2$	165.00
Carbon dioxide-methane reforming	$CH_4 + CO_2 \leftrightarrow 2CO + 2H_2$	247.30

The net reforming reaction is strongly endothermic and occurs by increasing the number of moles. The overall effect of increasing reforming temperature on the effluent gas composition is to reduce the methane and carbon dioxide content and increase carbon monoxide and hydrogen content. On decreasing reforming temperatures, the effects are reversed. The pressure of the system is so nearly fixed that reforming pressure should be considered invariable. However, increasing pressure has an effect similar to reducing temperatures and reformer designs always represent a compromise in economics. The water gas shift reaction is exothermic with shifting of the carbon monoxide favored by low temperature. However, the rate of reaction is favored by high temperature. It can be seen that with methane the stoichiometric requirement for steam per carbon atom is 1.0. However, it has been demonstrated that this is not practical because all primary reformer catalysts developed so far tend to promote carbon (char) forming reactions under such SMR conditions. Those reactions can only be suppressed by using an excess of steam, with the result that the minimum steam per carbon atom ratio is approximately 1.7 [12]. The reforming reaction itself is also promoted by an excess of steam and hence the ratios of 3.0 to 3.5 are commonly used in practice [12]. However, lower steam-to-carbon ratios are attractive from the economic point of view.

Important considerations during operation of SMR are the type of furnace used to transfer heat to the reactants, the catalyst properties such as activity, lifetime, size, and strength, operating conditions such as feed composition, pressure, temperature, and desired product composition. The catalyst in all of the SMR furnaces is contained in heat resistant alloy tubes that typically have outside diameters from 80 to 180 mm, wall thickness of 9 to 20 mm and overall length of 10,000 to 13,000 mm [12]. Fired lengths in commercial SMR furnaces vary from approximately 200 to as much as 3500 mm [12]. Firing is usually controlled such that tube wall temperatures are maintained at values that will give a reasonable tube lifetime. By design and industry practice, maximum allowable tube wall temperatures are set to give an in-service lifetime of 100,000 h when taking into account the stress-to-rupture properties of the particular alloy used in manufacturing the tube [13]. As noted by Schillmoller et al. [13] an increase in temperature of only 38 °C above design with the common HK-40 material (25% chrome, 20% nickel alloy) can shorten the tube lifetime from 10 to 1.4 years. The general SMR furnace classifications according to firing pattern are the following: top fired, side fired, and bottom fired [12]. All of the SMR furnaces have combustion heat recovery sections used for preheating feed streams, boiler feed water, etc. The schematic diagram of the top-fired SMR based on Kellogg Inc. design located in Petrokemija Plc., Kutina, used in the model is given in Figure 2.

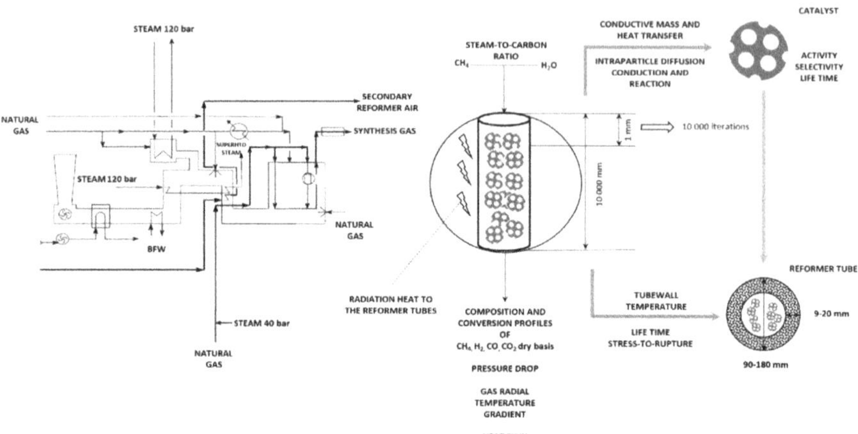

Figure 2. The schematic diagram of the top-fired SMR based on Kellogg Inc. design used in the model.

Catalyst activity has an important effect on the tube wall temperature in all SMR furnaces and is usually monitored by the plant operators by portable measurement devices. The heat transfer through the walls of the reformer tubes influences the catalyst activity. Of the total heat being transferred, the high activity catalyst uses more for the endothermic reforming reaction and less for raising the gas temperature. Besides that, feed gases can vary in composition from being nearly pure methane to the heavy hydrocarbons such as naphtha. As a result, specific catalyst formulations have been developed so far for different feed gas compositions. Xu and Froment [11] used in their work catalyst which contained 15.2% nickel, supported on magnesium spinel with a BET surface area of 58 m^2, nickel surface area of 9.3 m^2/g$_{cat}$, and with the void fraction of 0.528. The catalyst used in simulation contained 14.5% nickel, supported on calcium aluminate with BET surface area of 57 m^2, nickel surface area of 12.0 m^2/g$_{cat}$, and with the void fraction of 0.51963. The mentioned catalyst characteristics are in good alignment with the work of Xu and Froment [11] and therefore literature kinetics data are applicable for the model in spite of the fact that different catalyst support was used in the simulation procedure.

With respect to achieve an energy savings and to ensure mild temperature conditions against tubes during operation of SMR it is of utmost importance to maintain the catalyst performance at the best possible level. Therefore, a reliable and rapid mathematical model was important to be designed for continuous monitoring and evaluation of the steam methane reformer catalyst in an existing installation. Taking into account all theoretical explanations and literature review, model assumptions are given as follows:

1. The reaction mixture follows ideal gas law;
2. All reformer tubes within the furnace are identical; the overall performance of the reformer tubes is achieved by multiplication of one tube by the number of all other tubes;
3. The system operates in a dynamic state;
4. A 1-D analysis was used for mass flow, heat and momentum transfer. All important process variables are uniform at any cross section of the catalyst bed in the reformer tubes;
5. The model converts all higher hydrocarbons to an equivalent methane before any tube integrations take place. Kinetic expressions for SR1, SR2, and WGS reactions are only used;
6. There is no axial diffusion of mass and heat;
7. The model is heterogeneous;
8. There is no external mass and heat transfer resistance between the catalyst and the process gas;
9. Diffusion resistance between the catalyst is considerable and the same is considered;

SMR catalyst performance is usually evaluated in terms of its methane approach to equilibrium (ATE_{CH4}), furnace tube wall temperatures, heat flux, pressure drop, and presence or absence of hot bands on the tubes. The ATE_{CH4} is the difference between the actual gas temperature exiting the catalyst (T_1) and the temperature at which the measured outlet gas composition would be at equilibrium (T_2). By definition, the actual ATE_{CH4} cannot be less than zero. The temperature at which the outlet gas composition would be at equilibrium is determined in the model by calculating from the material balance the equilibrium constant for the SR1 reaction defined by [9,10]:

$$K_{SR1} = \frac{p_{CO} \cdot p_{H_2}^3}{p_{CH_4} \cdot p_{H_2O}} = 1.198 \times 10^{17} \exp\left(\frac{-26830}{T}\right) \tag{1}$$

and determining the corresponding temperature from a T_2-K_{SR1} correlating equation in the model, where K represents chemical equilibrium constant, and p is partial pressure of corresponding component (in bar), while T is the temperature (in K). Due to elevated temperature and pressure, the gaseous reaction mixture in the SMR process is not ideal and does not obey Mendeleev–Clapeyron ideal gas law. However, the Dalton law of partial pressure can be used in the model because the main assumption is that the reaction mixture behaves as an ideal gas.

Properly designed SMR should, with new catalyst, have ATE_{CH4} much lower than 5–10 °C. Those plants having effective desulphurization systems often have furnaces operating with ATE_{CH4} in the 0–5 °C range. When evaluations give these type approaches, it can be said the catalyst is giving satisfactory performance. Calculated ATE_{CH4} of 5–10 °C could be considered in the intermediate range and performance classified as acceptable, depending, of course, on the specific design in question. The ATE_{CH4} above 10 °C would correspond to marginal performance and would generally become a factor in discharging the catalyst. According to this it appears that continuous monitoring and evaluation procedure of ATE_{CH4} is a feasible perspective to keep catalyst performance at the best available level.

In many SMR furnace designs there can be a significant difference between measured outlet gas temperature and the actual outlet gas temperature. The measuring point is often in an outlet gas header, and because of heat losses from the bottom of the SMR furnace and exit lines, the resulting temperature is lower than the true catalyst exit value. For these cases, the ATE_{CH4} may well turn out to be a negative number. If the measured outlet gas composition is sufficiently accurate, the actual gas temperature exiting the catalyst can be calculated by assuming the gas is at equilibrium with respect to the WGS reaction [9,10]:

$$K_{WGS} = \frac{p_{CO_2} \cdot p_{H_2}}{p_{CO} \cdot p_{H_2O}} = 1.767 \times 10^{-2} \exp\left(\frac{4400}{T}\right) \tag{2}$$

The model assumes that WGS reaction reaches equilibrium in a SMR but the SR1 reaction does not. However, a clearance can be made for the deviation from equilibrium by an ATE_{CH4} and thus at a given exit temperature both K_{SR1} and K_{WGS} are known [14]. From Equations (1) and (2), the equilibrium constant for the SR1 reaction and WGS reaction at the respective temperature can be calculated. It is considered that equilibrium conversion and chemical equilibrium of the reactions depends on the thermodynamic properties of the gaseous mixture and catalyst characteristics.

To properly monitor and evaluate the catalyst performance it is necessary to compare DCS readings of the gas chromatography analysis data with predicted gas composition and how the pressure drop DCS readings data compare to the theoretically predicted values. Besides that, the temperature rise across the tube is required to indicate the kinds of adjustment that can be justified. The key components in the exit dry gas analyses are methane and, in addition, the carbon oxides (CO and CO_2), while hydrogen composition is then adjusted to give a total dry gas composition of 100%.

The accuracy of the pressure drop calculation is such that the actual value should be within 10% of the theoretical value. Actual pressure drops greatly exceeding the theoretical value would indicate some problem has or is occurring. Examples of such problems would be carbon formation catalyst breakage, faulty loading such as over-vibrating the tubes and/or failure to clean the tubes, pigtails, support plates, etc.

The calculations inside the developed model involve the solution of material and energy balance equations along with reaction kinetic expressions for the catalysts according to the literature findings [15–25]. The model calculates the material balance, ATE_{CH4}, gas radial temperature gradient, pressure drop, heat duty, molar fluxes, and reaction rates. The analytical expressions used in the one-dimensional heterogeneous model are shown in Figure 3. The calculation procedure is performed in Matlab (Version 2019b, The MathWorks, Inc., Natick, US, MA, 2019) file which is fairly easy to use.

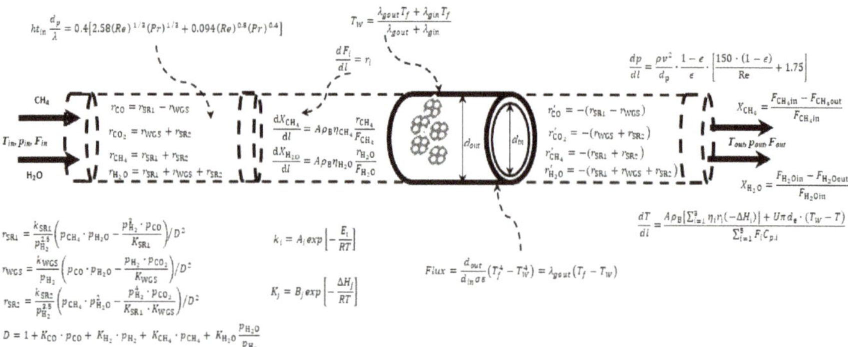

Figure 3. The analytical expressions used in the one-dimensional heterogeneous model according to literature findings [9,10,15–25].

The basic input requirements consist of two parts, one which must be manually filled in by the operator and another one which is automatically exchanged between DCS data and memory block by compiling with system-function. In Table 2 a list of all necessary variables needed for monitoring and evaluation of the SMR catalyst is shown.

The model converts the total wet feed flow to mole fractions of carbon, hydrogen, and oxygen with a base of the inlet dry gas as 100%. By using the mass balance and the equilibrium Equations (1) and (2), one can derive a quadratic equation:

$$a^2(1 - K_{WGS}) + \phi a - K_{WGS}(\phi - 4d)(B - 2d) = 0 \tag{3}$$

Solving for a give following expression:

$$a = \frac{-\phi \pm \left[\phi^2 + 4(1 - K_{WGS})(\theta - 4d)(B - 2d)K_{WGS}\right]^{1/2}}{2(1 - K_{WGS})} \tag{4}$$

where Φ and θ present intermediate calculation parameters as follows:

$$\phi = C + HK_{WGS} + 3d(1 - 2K_{WGS}) - \theta(1 - K_{WGS}) \tag{5}$$

$$\theta = 2C + H - 2O \ or \ a + b + 4d \tag{6}$$

where C, H, and O denote total molar quantities of carbon, hydrogen, and oxygen in compounds in the system (in kmol), while a, b, c, and d represent molar quantities of H_2, CO, CO_2, and CH_4 in reformed gas (in kmol), respectively.

Pressure (in bar or atm) in the tubes is calculated according to

$$p = \frac{C + H - 2d + i}{a}\left(K_{SR1}\frac{d(H - a - 2d)}{a(\theta - a - 4d)}\right)^{1/2} \tag{7}$$

where i denotes molar quantity of inert gases (Ar + He) in reformed gas (in kmol).

By using Equations (4) and (7) the model solves reforming calculations by assuming a methane content of the reformed gas (d), solving the corresponding hydrogen content (a) and pressure (p). The Newton–Raphson technique was used to get a rapid convergence in order to obtain the reformed gas composition. The technique uses previous values of the methane outlet molar concentration (d_{n-2} and d_{n-1}) to predict the value for the next calculation:

$$d_{n-1} = d_n - \frac{(p_n - p)}{(p_n - p_{n-1})}(d_n - d_{n-1}) \tag{8}$$

where the nth and $(n-1)^{th}$ results are used to predict the value of d to use in the $(n+1)^{th}$ calculation.

The activity multiplier will default to the standard values if there is no input (preferred values are in range from 0.25 to 0.65). The model minimizes the function which is the sum of the appropriately weighted squares of the difference between DCS readings and corresponding values from the simulation. Using input data, the model will converge in 30 s or less to give final results. Logic flow diagram for this model is shown in Figure 4.

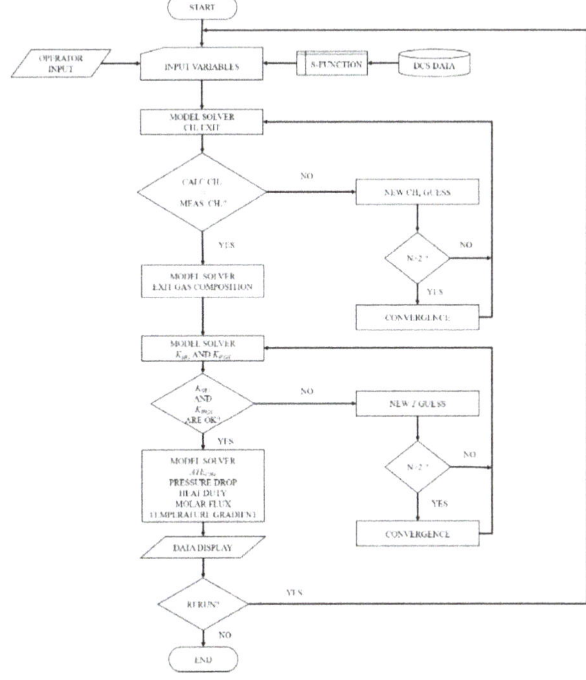

Figure 4. Model logic flow diagram.

For the purpose of synchronization between the model and any DCS system the C++ programming language was used for development of system-function and compiled with the MATLAB Compiler (MEX) (Version 2019b, The MathWorks, Inc. Natick, US, MA, 2019) [26]. The programming procedure which contained the basic functionality for SIMULINK shared memory block is shown as a logic flow diagram in Figure 5.

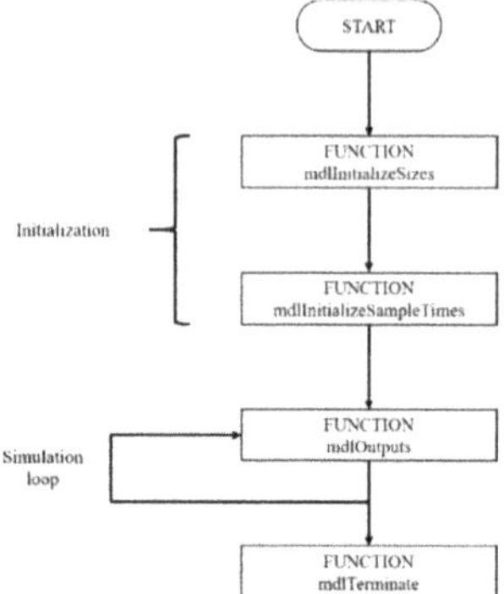

Figure 5. Logic flow diagram of developed system-function and shared memory block.

Characteristic process data of the top-fired SMR designed by Kellogg Inc. which is used in developed model for data reconciliation are given in Table 3.

Limitations of developed model are the following:

1. impossibility to evaluate carbon forming potential with higher hydrocarbon feed streams, due to the reason that the model converts all of these compounds to an equivalent methane before any tube integrations take place. However, it is believed that from about 30% tube level down, all higher hydrocarbons have been broken down. Carbon forming predictions from this point in the tube to the bottom are accurate;

2. the model does not provide any information about tube hot band problems. The model with the heat flux profile does show the region of the highest heat transfer, but nothing more as to hot band formation;

3. all of the reformer tubes within the furnace are assumed to be identical. Of course, this is not true in real conditions due to reason that tubes close to the refractory brick wall are under different heat fluxes in comparison with the tubes in the middle part of the furnace. In order to compensate this nonuniformity sets of simulations were performed and results are adjusted against real process data.

Table 2. List of variables for monitoring and evaluation of the SMR catalyst.

No.	Variable	Unit	Input by	Variable	Unit	Input by
1.	Reformer tube length	mm		Tube feed gas volume flow	m^3/h	
2.	ID of reformer tube	mm		Tube steam mass flow	t/h	
3.	OD of reformer tube	mm		Tube inlet temperature	°C	
4.	Number of reformer tubes	pcs		Tube outlet temperature	°C	
5.	Thermal conductivity of reformer tube material	$kJ/m^2 \cdot hr \cdot K$		Tube inlet pressure	bar	
6.	Fired length	mm		Tube outlet pressure	bar	
7.	Catalyst particle height	mm		Tube pressure drop	bar	
8.	Catalyst particle hole diameter	mm		CH_4 content in reformed gas	Mol %	
9.	No. of holes	-		Furnace fuel gas volume flow	m^3/h	DCS synchronization memory block by encoding S-function
10.	Catalyst bulk density	kg/m^3	Operator	Fuel gas temperature	°C	
11.	Activity multiplier	-		Furnace air volume flow	m^3/h	
12.	Tube wall temperature	°C		Furnace air temperature	°C	
13.	Heat transfer coefficient	$kJ/m^2 \cdot hr \cdot K$		CH_4 content in reformed gas	dry mol %	
14.	CH_4 content in feed gas	dry mol %		CO content in reformed gas	dry mol %	
15.	C_2H_6 content in feed gas	dry mol %		CO_2 content in reformed gas	dry mol %	
16.	C_3H_8 content in feed gas	dry mol %		H_2 content in reformed gas	dry mol %	
17.	C_4H_{10} content in feed gas	dry mol %		N_2 content in reformed gas	dry mol %	
18.	C_5H_{12} content in feed gas	dry mol %		Ar + He content in reformed gas	dry mol %	
19.	C_6H_{12} and C_6+ content in feed gas	dry mol %				
20.	CO content in feed gas	dry mol %				
21.	CO_2 content in feed gas	dry mol %				
22.	H_2 content in feed gas	dry mol %				
23.	N_2 content in feed gas	dry mol %				
24.	H_2S content in feed gas	dry mol %				
25.	Ar + He content in feed gas	dry mol %				

Table 3. Summary of process data of the top-fired SMR.

Reformer Tubes	Value/Unit	Catalyst Pellets	Value/Unit
Heated length of reformer tubes	10.00 m	Shape	Raschig rings with 12 holes
Inside diameter of reformer tubes	0.085 m	Dimensions	19×12 mm/30%; 19×16 mm/70%;
Outside diameter of reformer tubes	0.110 m	Bulk density	800 kg/m^3
Number of reformer tubes	520	Porosity	0.51963
Construction material of reformer tubes	MANAURITE XM and 900	Tortuosity	2.74
Number of rows	10	Mean pore radius	80 Å
Number of arch burners	198	Catalyst characteristic length	0.001948 cm
Heat duty of reformer	204.80 MWh	Catalyst material	$NiO + CaAl_{12}O_{19}$
Skin temperature	840 °C	Catalyst quantity	32.0 tons
Process Gas Flow Rate and Composition	**Value/Unit**	**Fuel Characteristics**	**Value/Unit**
Molar flow rate	1530 kmol/h	Molar flow rate	792 kmol/h; 3% of O_2 excess air at 503 K
Composition	mol %, dry basis	Composition	mol %, dry basis
C_2H_6	0.00	C_2H_6	1.11
C_3H_8	0.00	C_3H_8	0.00
i-C_4H_{10}	0.00	i-C_4H_{10}	0.00
H_2	1.43	H_2	1.43
CO_2	0.95	CO_2	0.95
N_2	0.95	N_2	0.95
Ar + He	0.01	Ar + He	0.01
Inlet Conditions	**Value/Unit**	**Outlet Conditions**	**Value/Unit**
Temperature	773 K	Temperature	800 °C
Pressure	30.5 bar	Pressure	30.0 bar
S/C ratio	3.60	S/C ratio	3.60
Maximum allowable outlet temperature		900 °C	
Ammonia production capacity		1360 MTPD	

3. Results and Discussion

The described model was integrated as an open source code software based on Matlab files, relying on the three reactions occurring in the SMR scheme which are given by Xu and Froment [9,10]. The kinetics are known for the catalyst used. According to three chemical equations given in Table 1 and using reaction rates a mathematical model was developed which is able to solve the molar flux profile of each species occurring in reaction scheme by iterative procedure. By using following equation, the calculation procedure is straightforward and converges readily to any desired degree of accuracy:

$$\frac{dF_i}{dl} = r_i \tag{9}$$

where F_i represents the molar flux of species i, r_i denotes the rate of formation or disappearance of species i given by Xu and Froment [9,10], and l is the distance along the reactor tube. The model is also able to calculate pressure drop and energy balance with related pressure and temperature profile along the reformer tube [9,10,23–27]. The Newton–Raphson iterative method of solving ordinary differential equations was used.

The developed model was tested against literature data for methane conversion in reformer tubes at 20 bar and at three different values for steam-to-carbon ratio, namely 1.0, 2.5, and 5.0 in the temperature range from 400 to 1000 °C [8]. The calculated data in comparison with the literature data are shown in Table 4 [8,11].

Table 4. Comparison between model and literature data for methane conversion at different steam-to-carbon ratios and temperature range from 400 to 1000 °C at 20 bar [8,11,27].

	CH$_4$ Conversion (%)								
	S/C = 1.0			S/C = 2.5			S/C = 5.0		
t (°C)	Literature	Model	Difference	Literature	Model	Difference	Literature	Model	Difference
400	2.19	0.50	1.69	5.01	3.22	1.79	7.20	5.45	1.75
450	3.88	2.10	1.78	8.05	6.25	1.80	12.23	10.52	1.71
500	6.89	4.95	1.94	11.97	10.20	1.77	17.87	16.04	1.83
550	9.74	7.88	1.86	17.25	15.51	1.74	26.24	24.23	2.01
600	13.83	12.08	1.75	24.17	22.46	1.71	36.56	34.53	2.03
650	18.54	16.77	1.77	32.00	30.24	1.76	47.77	45.84	1.93
700	24.49	22.48	2.01	41.23	39.22	2.01	59.84	57.97	1.87
750	31.00	29.31	1.69	51.38	49.41	1.97	72.20	70.23	1.97
800	39.43	37.44	1.99	63.46	61.41	2.05	84.35	82.32	2.03
850	47.66	45.77	1.89	73.40	71.37	2.03	93.75	91.83	1.92
900	56.28	54.22	2.06	83.16	81.17	1.99	96.85	95.14	1.71
950	69.05	67.32	1.73	90.84	88.81	2.03	99.15	97.11	2.04
1000	72.03	69.99	2.04	96.00	94.05	1.95	99.52	97.83	1.69

According to calculated data it can be concluded that the developed model fits almost perfectly against literature data in all tested temperature ranges from 400 to 1000 °C. The average deviation ranges from 1.69% to 2.06% for the three tested steam-to-carbon ratios.

Calculated data confirm the profile shapes regarding recently published work from Zecevic and Bolf [27]. The effectiveness factor of 0.001 was used as the typical value for industrial SMR catalyst installed in the reformer tubes [25]. The effectiveness factor is a function of Thiele Modulus which is related to the catalyst volume and external surface area of the catalyst. In the case of the large reforming catalyst pellet it is not possible to achieve equal accessibility implying that only a thin layer of catalyst close to the surface contributed to the reaction. Any differences may have contributed to the catalyst shape used, as a result of pellet geometry alteration. However, those alterations in pellet geometry of the commercially available reformer are not significant drawbacks to the developed model. All factors mentioned are more than acceptable for industrial practical usage.

The model was used to calculate the residence time required for a reasonable conversion in isothermal conditions and the results are compared with real process data from top-fired SMR unit. The feed composition, inlet molar flow rate, and mass of catalyst were taken from the real process conditions as given in Table 3. The resulting plot of outlet component molar flow rate as a function of residence time is shown in Figure 6. It can be seen that the reactor approaches equilibrium at residence times of 15–20 s. According to the results, the outlet component molar flow rates do not deviate more than 3% in comparison with the real process data taken from the real measurements during the operation of the top-fired SMR furnace at the same operating conditions. At the same inlet conditions given in Table 3 the measured process data are as follows: 10.31 mol % for CH_4, 72.23 mol % for H_2, 9.18 mol % for CO, and 10.63 mol % for CO_2. This also implies that the model is precise enough for industrial usage. The deviations of up to 3% could be explained with inaccurate plant measurements, as a result of temperature or pressure drops in the reactor outlet system or from non-uniform operation between different catalyst tubes inside the SMR. However, to get to closer alignment with real process equilibrium conversions in the future the model will be tested by varying the effectiveness factor values.

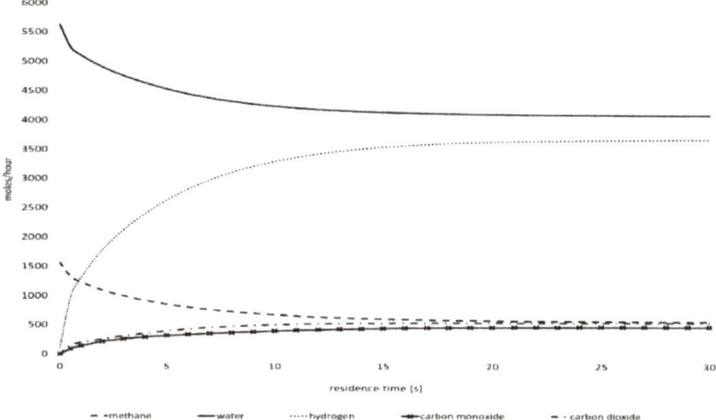

Figure 6. Dependence of the component molar flow as a function of residence time inside the reformer tubes according to the real process conditions given in Table 3.

The model is also able to calculate the necessary heat duty of the SMR box. The heat duty was calculated according to the inlet and outlet process data as given in Table 3, considering inlet and outlet temperatures of the reformer tubes of 500 and 800 °C. Additionally, the optimum working temperature for the reformer tubes made of MANAURITE XM and 900 of 820–840 °C was taken as the tube wall temperature of the reformer tubes. That tube wall temperature will guarantee the longer lifetime of the reformer tubes preventing their creep damage. The results are shown in Figure 7. It is evident that most of the heat duty is required in the first 3 m of the reformer tube, which is in accordance with the location of the top-fired burners. Namely, top-fired burners generate the 3–5 m long flame in the direction from the top to the bottom of the SMR furnace. This result is in complete accordance with the real process data given in Table 3, because the installed reformer tubes possess the overall length of 11.34 m, with heated length of 10.30 m and catalyst loaded length of 10.0 m.

With respect to inlet conditions given in Table 3 the molar flux of methane, carbon dioxide, carbon monoxide, hydrogen, and water as a function of the reformer tube length is shown in Figure 8. It can be seen that the production rate of hydrogen is increasing through the whole length of reformer tube. The model gives methane conversion of 64.64%, the methane molar outlet concentration of 10.54% per dry basis of process gas (water content is completely excluded), and temperature approach to equilibrium of 5 °C. The calculated data in comparison with actual process data at given inlet process conditions fits almost perfectly. It can be seen that reaction equilibrium is shifted in the direction of reactants in the approximately first 0.5 m of the reformer tube length. The reason for this is the strong endothermic behavior of the steam methane reforming reaction. Due to absorption of the heat from the burners the reforming reaction is going toward products until full equilibrium is achieved.

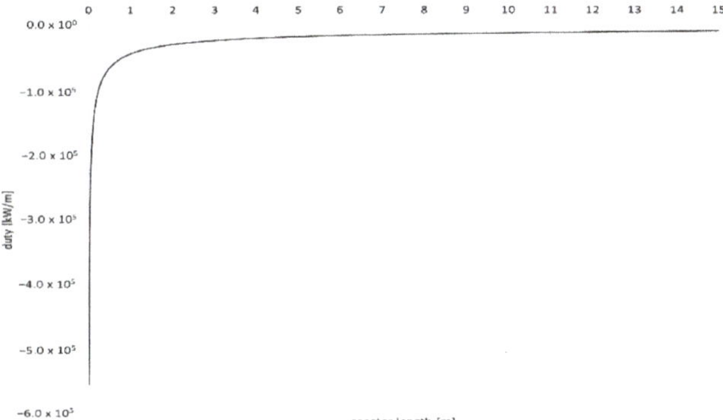

Figure 7. Dependence of heat duty as a function of reactor length.

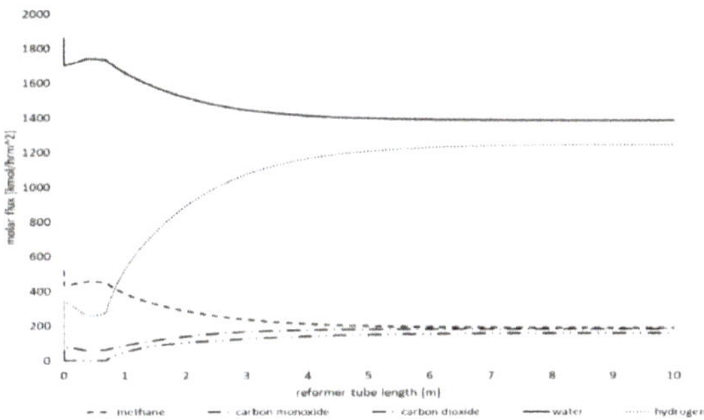

Figure 8. Molar flux profiles for the steam methane reforming reaction at given inlet process conditions from Table 3.

The reaction rate profile for three reactions is shown in Figure 9. Reaction rate profiles were tested according to the process data given in Table 3 and the same was compared with the literature data [9,10]. From the calculated data for the reaction rates at the given process data, it can be concluded that the model describes the general pattern of behavior very well. The reaction rate 2 reverses the direction between 8 and 9 m down the reformer tube as the temperature increases and the equilibrium point shifts. Regarding the calculated results of the reaction rates for SR1, SR2, and WGS it can be concluded that the model can be used reliably for the performance evaluation of the real process conditions.

The pressure drop profile was tested by application of the Ergun equation [20] taking into account all characteristics of the loaded catalyst. The characteristics of the catalyst are given in Table 3, while the inlet pressure and inlet temperature are 30.5 bar and 500 °C. The molar flow rate at the inlet of reformer tubes is 1530 kmol h^{-1}, which is equal to name plant capacity of 1360 MTPD. The pressure drop profile is shown in Figure 10. From the Figure 10 it is visible that the pressure drop along the reformer tubes is 1.21 bar. In real conditions the pressure drop was 1.58 bar. The difference of 0.37 bar can be attributed to uneven packaging and breaking of the SMR catalyst during charging of the reformer tubes and possible sulfur and carbon deposition. The model predicted the characteristics of the catalyst bed as at the start of the lifetime ideally packed with undamaged catalyst pellets and completely clean without any contamination at the available surface area.

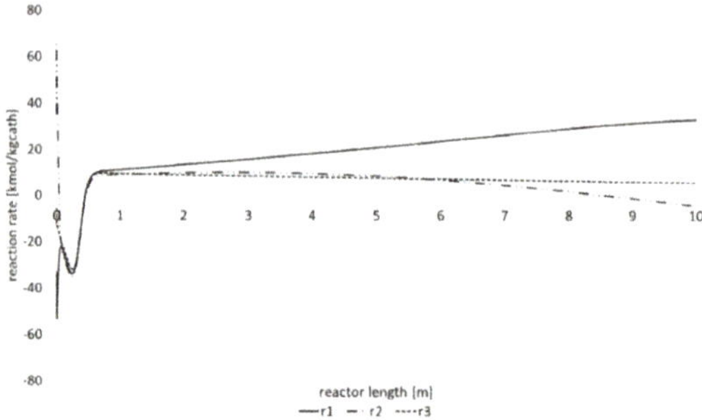

Figure 9. The intrinsic reaction rates for three reactions with the length of the reformer tubes at given inlet process conditions from Table 3.

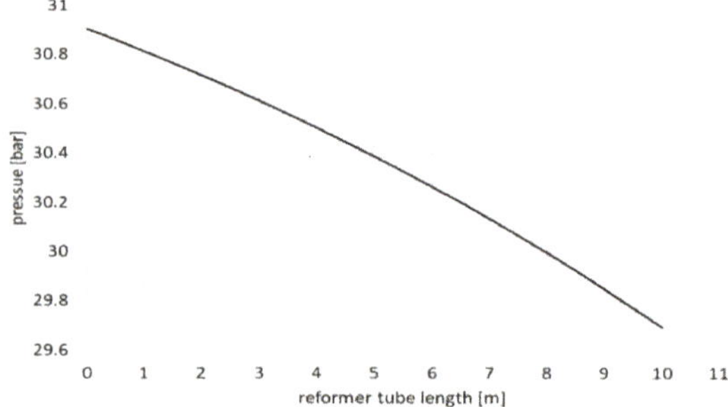

Figure 10. Calculated pressure drop profile along the reformer tubes at given process conditions.

The temperature inside the SMR furnace is one of the most important process parameters which influences the energy consumption, conversion of methane, kinetics of the steam methane reforming reaction, temperature approach to equilibrium, and lifetime of the reformer tubes due to creep damage. Optimization of this parameter can save a significant amount of energy during the operations and extend the reformer tube lifetime. In Figure 11 the temperature profile was calculated by the applied model at the inlet temperature and pressure of 500 °C and 30.5 bar, molar flow of 1530 kmolh^{-1}, skin temperature of the reformer tube of 840 °C, steam-to-carbon ratio of 3.60, and the exit temperature of 781 °C. In the real plant the average reformer tube outlet temperature at the same inlet conditions was 801 °C. The difference of only 20 °C points to the satisfactory agreement between model and real process conditions.

In order to give meaningful information about changes with time there are three general ways in which the performance of the SMR catalyst can be quantified. These are (a) measurement of composition at the outlet from the reformer tubes, (b) determination of the ATE_{CH4} at the outlet of reformer tubes, and (c) calculation of catalyst activity or active volume of catalyst. By calculating the ATE_{CH4}, it is possible to arrive at an expression for the trend in relative catalyst activity. The ATE_{CH4} can be used as a good measure of the catalyst performance when the operating temperature of the reactor is held constant and when the reaction is equilibrium limited, such as with SMR [12]. According to this, the model was used to evaluate the ATE_{CH4} values of the SMR catalyst in operation in comparison with ideal equilibrium curve at different outlet methane molar concentrations and temperatures. The relationship between calculated ATE_{CH4} values of the SMR catalyst in operation against the ideal equilibrium curve, is shown in Figure 12.

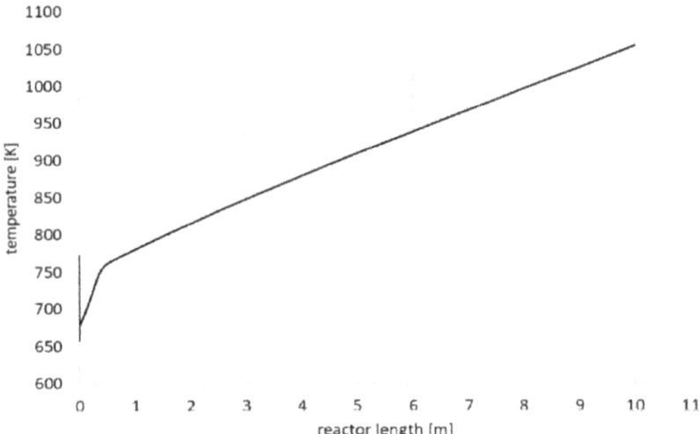

Figure 11. Temperature profile along the length of the reformer tubes at given process conditions.

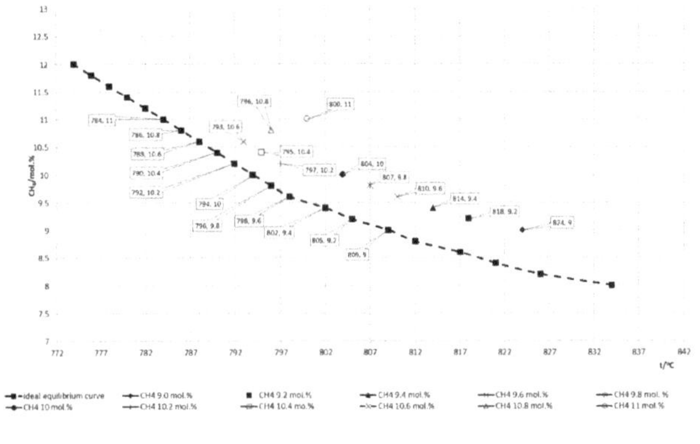

Figure 12. Calculated ATE_{CH4} values of the SMR catalyst in operation against the ideal equilibrium curve.

From the data given in Figure 12 it can be concluded that the catalyst achieves better ATE_{CH4} values in the outlet methane molar concentrations in the range of 10.2–10.6 mol. % and the same are at the level of 5 °C which reflects excellent performance of the catalyst activity. In the range of the higher outlet methane molar concentration the values of ATE_{CH4} are in the range of 5–10 °C which gives acceptable performance of the catalyst activity, while in the lower range the values of ATE_{CH4} are higher than 10 °C which correspond to marginal performance. This is the reason for appropriate adjustments of the process parameters to achieve better catalyst performance.

To test all features of developed model the same was connected with the DSC system in the existing ammonia plant through memory block and process data was retrieved for proper adjustments of parameters in order to improve catalyst performance and tube wall temperatures during ammonia operation. In Figure 13 catalyst outlet temperature readings regarding the equilibrium temperature at three different outlet methane molar concentrations are shown, namely 10.6, 10.4, and 10.2 mol. %, respectively. In Figure 14A,B the tube wall temperature readings are shown. In both cases the results are shown before and after implementation of recommendations given from the model in the period of 720 h.

From Figure 13 it can be seen that the ATE_{CH4} values before implementation of recommendations from the model are at the level of approximately 20 °C. After following model recommendations (adjustments of the steam-to-carbon ratio and firing in the reformer furnace), the ATE_{CH4} values are approaching the equilibrium temperature and are also in the range from 0 to 10 °C. The target value for outlet methane molar concentration was 10.4 mol. %.

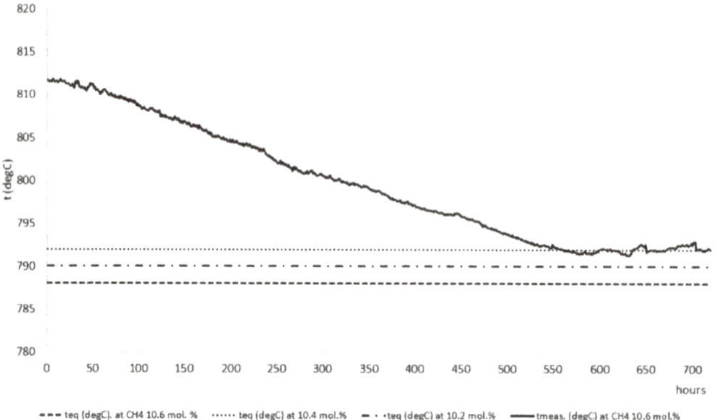

Figure 13. Catalyst outlet temperature readings regarding equilibrium temperature at outlet methane molar concentrations of 10.6, 10.4, and 10.2 mol. %.

From Figure 14A it can be seen that the tube wall temperatures are in the higher temperature range from 840 to 860 °C which reflects severe temperature conditions against tube metal alloy. At the same time, after following model recommendations by adjustment of the firing rate in the reformer furnace the tube wall temperature pattern is more uniform with the temperature range from 820 to 840 °C, which is shown in Figure 14B. On average this represents 20 °C lower temperature which is extremely favorable for tube performance, because a decrease of about 20 °C will significantly prolonged the tube lifetime.

During the test period the energy savings in terms of lower natural gas consumption were also measured. Before applying the recommendations from the model, the natural gas consumption was at the level of 1045 m^3 $tone^{-1}$ of ammonia. Following model recommendations, it was observed that the natural gas consumption was lower for 33 m^3 t^{-1} of ammonia which represents the savings in the amount of 3.15%. The achieved level of savings presents an extremely attractive savings scheme to be applied during operation of top-fired SMR in ammonia production.

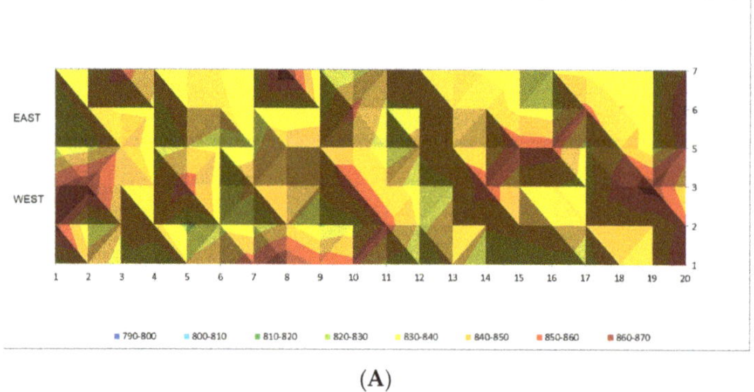

(**A**)

Figure 14. *Cont.*

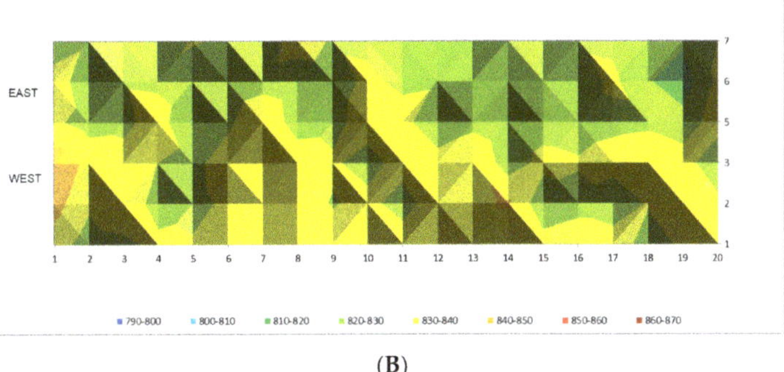

(B)

Figure 14. (**A**) Tube wall temperature readings before implementation of recommendations from the model. (**B**) Tube wall temperature readings after implementation of recommendations from the model.

4. Conclusions

A rapid integrated numerical model for industrial steam natural gas reformer within ammonia production based on m. files in an open source code software was developed. The developed model was tested, and calculated data was reconciled against real process data for the top-fired SMR designed by Kellogg Inc. within ammonia production. Applications of a series of differential equations were used for very close description of the reaction kinetics, molar flow, temperature, and pressure changes along the reformer tubes based on the previous literature models. Reaction rates follow the theoretical model from previous work very closely. Methane molar outlet concentration and temperature approach to equilibrium as two main process parameters for monitoring of the reformer operation has demonstrated satisfactory alignment with performance of the real plant. Pressure drop and temperature profile were taken, and the diffusional resistance of the catalyst bed at any point inside of the reformer tube and the real process conditions were reliably replicated. The computational speed is rapid enough for application in real conditions by the industrial users. The coupling between the discrete model and any DCS was achieved by compiling the model algorithm with standard system-function in an appropriately designed memory block. The model can be reliably used for evaluation of the performance of the vast majority commercially available SMR catalysts. The proposed model can be potentially used in application of an APC of any steam natural gas reformer in hydrogen production and for continuous on-line performance evaluation procedure of all relevant process parameters and catalyst performance through a predictive simulation model to clarify the cost and risk associated with making operational improvements of the plant, as well as support upskilling of the workforce. In the case of application of the developed model, energy savings up to 3% can be achieved and at the same time it can serve as the base for digital transformation of any syngas plant.

Author Contributions: N.Z. is responsible for writing the article and simulation calculations of the model. N.B. is responsible for the guidance of article writing, and for the language and format of the article. All authors have read and agreed to the published version of the manuscript.

Funding: This research received no external funding.

Conflicts of Interest: The authors declare no conflict of interest.

List of Symbols

A	reformer tube cross sectional area, m^2
A_i	pre-exponential factor, kmol bar^5 kg_{cat}^{-1} hr^{-1} or kmol kg_{cat}^{-1} bar^{-1}
a	molar quantity of H_2, kmol
ATE	approach to equilibrium, °C or K
b	molar quantity of CO, kmol

B_j	pre-exponential factor, bar^{-1}
C	molar quantity of carbon, kmol
c	molar quantity of CO_2, kmol
$c_{p,i}$	specific heat capacity of process gas i, $kJ\ kmol^{-1}\ K^{-1}$
D	denominator
d	molar quantity of CH_4, kmol
d_e	external reformer tube dimension, m
d_i	internal reformer tube dimension, m
d_p	catalyst particle diameter, m
E_i	activation energy, $kJ\ kmol^{-1}$
f	fluid
F_i	molar flow rate, $kmol\ hr^{-1}$
H	molar quantity of hydrogen, kmol
i	molar quantity of inert gases (Ar+He), kmol
K_i	adsorption constants for component i, bar^{-1}
K^{eq}_i	equilibrium constants for SR and WGS
k_i	rate coefficient for reaction, $kmol\ bar^5\ kg_{cat}^{-1}\ hr^{-1}$ or $kmol\ kg_{cat}^{-1}\ bar^{-1}$
l	reformer tube heated length, m
m	mass, kg, t
M	molecular weight, $kg\ kmol^{-1}$
MTPD	metric tons per day
O	molar quantity of oxygen, kmol
ODE	ordinary differential equation
p	pressure, Pa, bar
p_i	partial pressure of component i, bar
Pr	Prandtl number
R	gas constant, $kJ\ kmol^{-1}\ K^{-1}$
Re	Reynolds number
$r_{CO,\ CO2,\ CH4,\ H2}$	rate of formation and disappearance, $kmol\ kg_{cat}^{-1}\ hr^{-1}$
r_i	rate of chemical reaction, $kmol\ kg_{cat}^{-1}\ hr^{-1}$
S/C	steam-to-carbon ratio
SR	steam-reforming
T	temperature, ^oC, K
t	time, s, min, hr
U	heat transfer capacity, $kJ\ m^{-2}\ hr^{-1}\ K^{-1}$
v	velocity of the fluid, $m\ s^{-1}$
W	wall
$X_{CH4,\ H2O}$	molar rate of conversion, mol %
WGS	water–gas shift
ΔH^o_{298}	enthalpy change at 298 K, $kJ\ mol^{-1}$

List of Greek Letters

ϵ	catalyst bed voidage
λ_g	process gas thermal conductivity, $kJm^{-1}\ hr^{-1}\ K^{-1}$
λ_{st}	tube metal thermal conductivity, $kJm^{-2}\ hr^{-1}\ K^{-1}$
π	3.14
$\eta_{CH4,\ H2O}$	constant effectiveness factor
ρ	fluid density, $kg\ m^{-3}$
ρ_B	catalyst bed density, $kg\ m^{-1}$
σ_{ij}	collision diameter, Å

References

1. Grinthal, W. Chemputers, Process simulators shift into high gear. *Chem. Eng.* **1993**, *156*, 153.
2. Moneim, N.A.E.; Ismail, I.; Nasser, M.M. Simulation of Ammonia Production using HYSYS Software. *Chem. Proc. Eng. Res.* **2020**, *62*, 14.

3. Latham, D.A.; McAuley, K.B.; Peppley, B.A.; Raybold, T.M. Mathematical modeling of an industrial steam-methane reformer for on-line deployment. *Fuel Process. Technol.* **2011**, *92*, 8–1574. [CrossRef]

4. Lee, J.S.; Seo, J.; Kim, H.Y.; Chung, J.T.; Yoon, S.S. Effects of combustion parameters on reforming performance of a steam-methane reformer. *Fuel* **2013**, *111*, 461. [CrossRef]

5. Holt, J.E.; Kreusser, J.; Herritsch, A.; Watson, M. Numerical modelling of a steam methane reformer. *ANZIAM J.* **2018**, *59*, C112. [CrossRef]

6. Lao, L.; Aguirre, A.; Tran, A.; Wu, Z.; Durand, H.; Christofides, P.D. CFD modeling and control of a steam methane reforming reactor. *Chem. Eng. Sci.* **2016**, *148*, 78. [CrossRef]

7. Davis, B.; Okorafor, C. Hydrogen Fueling Station, Project Proposal, The Cooper Union for the Advancement of Science and Art Department of Chemical Engineering. *Process Eval. Des.* **2011**, *42*, 35–42.

8. Rostrup-Nielsen, J. Modelling of primary reformers. *Adv. Catal.* **2012**, *47*, 65.

9. Xu, J.; Froment, G.F. Methane steam reforming: Diffusional limitations and reactor simulation. *AIChE J.* **1989**, *35*, 97. [CrossRef]

10. Xu, J.; Froment, G.F. Methane steam reforming, methanation and water-gas shift: Intrinsic kinetics. *AIChE J.* **1989**, *35*, 88. [CrossRef]

11. Rostrup-Nielsen, J.; Christiansen, L.J. *Concepts in Syngas Manufacture, Catalytic Science Series*, 1st ed.; Imperial College Press: London, UK, 2011; pp. 14–21.

12. Twigg, M.W. *Catalyst Handbook*, 2nd ed.; Wolfe Publishing Ltd.: London, UK, 1989; pp. 230–241.

13. Schillmoller, C.M.; van den Bruck, U.W. Furnace Alloy Update. *Hydrocarb. Process.* **1984**, *63*, 55.

14. Hampson, G.M. Simple Solution to Steam Reforming Equations. *Chem. Eng.* **1979**, *1*, 523.

15. Elnashaie, S.; Uhlig, F. *Numerical Techniques for Chemical and Biological Engineers Using MATLAB, A Simple Bifurcation Approach*; Springer: New York, NY, USA, 2007; pp. 500–517.

16. Olivieri, A.; Veglio, F. Process simulation of natural gas steam reforming: Fuel distribution optimisation in the furnace. *Fuel Process. Technol.* **2008**, *89*, 622. [CrossRef]

17. Rennhack, R.; Heinisch, R. Kinetische Untersuchung der Reaktion Zwischen Methan und Wasserdampf an Nickel-Oberfächen. *Erdöl Kohle Erdgas Petrochem. Verinigt Brennst. Chem.* **1972**, *1*, 22.

18. Abas, S.Z.; Dupont, V.; Mahmud, T. Kinetics study and modelling of steam methane reforming process over a Ni/Al$_2$O$_3$ catalyst in an adiabatic packed bed reactor. *Int. J. Hydrog. Energ.* **2017**, *42*, 2889. [CrossRef]

19. Alhabadan, F.M.; Abashar, M.A.; Elnashaie, S.S. A Flexibile Computer Software Package for Industrial Steam Reformers and Methanators Based on Rigorous Heterogeneous Mathematical Model. *Math. Comput. Model.* **1992**, *16*, 77. [CrossRef]

20. Elnashaie, S.S.; Elshishini, S.S. *Modelling, Simulation and Optimization of Industrial Fixed Bed Catalytic Reactors*; Topics in Chemical Engineering; Gordon and Breach Science Publisher: Yverdon, Switzerland, 1993; pp. 69–76.

21. Leva, M.; Winstraub, M.; Grummer, M.; Pollchik, M.; Storch, H.H. *Fluid Flow Through Packed and Fluidized Systems*; US Government Printing Office: Washington, DC, USA, 1951; Volume 504, p. 45.

22. Cussler, E.L. *Diffusion and Mass Transfer in Fluid Systems*, 2nd ed.; Cambridge University Press: New York, NY, USA, 1997; pp. 189–197.

23. Nauman, E. *Chemical Reactor Design, Optimization, and Scaleup*; John Wiley and Sons Inc.: Hoboken, NJ, USA, 2008; pp. 144–153.

24. Froment, G.F.; Bischoff, K.B.; De Wilde, J. *Chemical Reactor Analysis and Design*, 3rd ed.; John Wiley & Sons: New York, NY, USA, 2010; pp. 98–103.

25. Elnashaie, S.; Adris, A.; Soliman, M.A.; Al-Ubaid, A.S. Digital simulation of industrial steam reformers. *Can. J. Chem. Eng.* **1992**, *70*, 786. [CrossRef]

26. Matlab (Matrix Laboratory), Version 2019b; Software for Numerical Computing; MathWorks: The MathWorks, Inc., 1994–2020 US. Available online: https://www.mathworks.com/products/matlab.html (accessed on 13 March 2020).

27. Zecevic, N.; Bolf, N. Advanced operation of the steam methane reformer by using gain-scheduled model predictive control. *Ind. Eng. Chem. Res.* **2020**, *59*, 3458–3474. [CrossRef]

Article

Comparison of Exergy and Advanced Exergy Analysis in Three Different Organic Rankine Cycles

Shahab Yousefizadeh Dibazar [1], Gholamreza Salehi [2] and Afshin Davarpanah [3,*]

[1] Energy System Engineering Department, Petroleum University of Technology, Abadan, Iran; shahab.yousefizadeh@gmail.com

[2] Department of Mechanical Engineering, Central Tehran Branch, Islamic Azad University, Tehran, Iran; gh.salehi@iauctb.ac.ir

[3] Department of Mathematics, Aberystwyth University, Aberystwyth SY23 3BZ, UK

* Correspondence: afd6@aber.ac.uk

Received: 17 April 2020; Accepted: 9 May 2020; Published: 14 May 2020

Abstract: Three types of organic Rankine cycles (ORCs): basic ORC (BORC), ORC with single regeneration (SRORC) and ORC with double regeneration (DRORC) under the same heat source have been simulated in this study. In the following, the energy and exergy analysis and the advanced exergy analysis of these three cycles have been performed and compared. With a conventional exergy analysis, researchers can just evaluate the performance of components separately to find the one with the highest amount of exergy destruction. Advanced analysis divides the exergy destruction rate into unavoidable and avoidable, as well as endogenous and exogenous, parts. This helps designers find more data about the effect of each component on other components and the real potential of each component to improve its efficiency. The results of the advanced exergy analysis illustrate that regenerative ORCs have high potential for reducing irreversibilities compared with BORC. Total exergy destruction rates of 4.13 kW (47%) and 5.25 kW (45%) happen in avoidable/endogenous parts for SRORC and DRORC, respectively. Additionally, from an advanced exergy analysis viewpoint, the priority of improvement for system components is given to turbines, evaporators, condensers and feed-water heaters, respectively.

Keywords: exergy; advanced exergy analysis; organic Rankine cycle; regenerative cycle

---

1. Introduction

Energy is known to be one of the most important elements in the development of any society. In recent years, many researchers have conducted many studies to discover ways to reduce energy consumption in different sections. Studies on industries have shown that a great amount of waste heat is generated during various processes that are placed in low-temperature ranges. This waste heat, which is released to the ambience directly, in many cases, causes a lot of problems for the environment, such as thermal pollution, ozone depletion, air pollution and so on [1]. There are some suggestions to use low-temperature waste heats, but recovering by organic Rankine cycles (ORCs) to produce power is offered as one of the best ways to increase process efficiency [2–4]. In the last two decades, different systems of ORCs have been studied widely by researchers, and their focus are more on working fluids of cycles and optimizing performance conditions [5–9]. Biomass, solar thermal energy and geothermal are some other heat sources for ORC applications to produce electricity [10]. The energy conservation law is not sufficient by itself, so the second low should be considered to have a wide view to design systems [11]. Huan et al. [12] selected a regenerative organic Rankine cycle (RORC) to analyze energy and exergy aspects of cycles with six different working fluids. In their study, they optimized the exergy efficiency of the cycles to find the best condition ranges for the inlet pressure and temperature of the turbines. R141b and R11 are suggested as better working fluids for systems, and the maximum exergy of 56.87% was obtained for ORC with double regeneration (DRORC). In another study by Roy et al. [13], inlet temperature in turbines and superheat in RORC were optimized. It was observed that R123, as working fluid of the system, showed its best performance at 2.5 MPa pressure at the evaporator compared with R123a. In 2011, Rashidi et al. [14] have investigated optimizing RORC using artificial bee colony-based neural network method. Results indicated that, for RORC, there is an optimum range for bleed pressure to reach the maximum thermal and exergy efficiency with the highest power output. If the pressure goes above or below this range, the thermal performance of the system will get worse.

As it is seen, most studies consider a conventional exergy analysis for ORCs. A conventional exergy analysis can just evaluate the performance of system components separately to find the component with the highest exergy destruction. This method does not show the share of each component of a system on other components' exergy loss. An advanced analysis divides exergy loss into unavoidable and avoidable and exogenous and endogenous for each component. By this analysis, the potential of each component is observed and discussed to improve its efficiency [15]. Conventional and advanced exergy analysis for gas turbines in different systems has been applied by Fallah et al. [16]. Among those systems, gas turbines with evaporative inlet air cooling have the best potential to reduce their destruction. It was also concluded that, when using an advanced exergy analysis, working conditions obtained by optimization for inlet cooling components were different from those obtained by a conventional exergy analysis. Galindo et al. [17] had discussed conventional and advanced exergy analysis in ORC as a bottoming cycle in an internal combustion (IC) engine in 2016. They indicated that boilers have the highest exergy destruction, but turbines have great potentials to improve their efficiency. Additionally, they suggested that about 36.5% of exergy loss in a system can be reduced by only the avoidable part of exergy loss in each component. Nami et al. [18], in 2017, worked on a binary fluid organic Rankine cycle with conventional and advanced exergy analysis. In this system, the low-pressure vapor generator (LPVG) has the highest exergy destruction among the components. Additionally, the advanced exergy analysis results showed that 15% of the condenser exergy destruction is placed in avoidable parts, which consists of 7% of the whole avoidable exergy destruction rate of the system. In addition, their study shows that, from an advanced exergy analysis view, more than 70% of total exergy destruction of the system is placed in endogenous exergy destruction parts, and among all the parts of the system, the endogenous exergy destruction is higher than the exogenous exergy destruction.

By studying many papers, it was observed that RORC has a really great efficiency and a potential to recover heat from low-temperature heat sources. RORC has been investigated by energy and exergy analysis, and the system was also studied from economics aspects [19–22]. To the best of the authors' knowledge, and by reviewing many papers, there is no study which has applied advanced exergy analysis to RORC for recovering low-temperature waste heat.

Therefore, the present study attempts to explain the system conditions, the first and the second law of thermodynamics and then to model our systems. Three cycles (BORC, single-regeneration ORC (SRORC) and DRORC) are selected to do the analysis. Then, the advanced exergy analysis aspects and their applications in the cycles are discussed. It is then followed by analyzing and discussing the results about the potential of each component to find and suggest some ways of decreasing the total exergy destruction rate in order to have the best design.

2. Description of Systems

For recovering heat with low-temperature sources, water is not a good option as a working fluid in cycles. ORCs are suitable cycles for the conversion of low-quality heat to electricity. ORCs are similar to the conventional steam Rankine cycles, but the working fluids in these cycles are organic fluids with low boiling points. Using this kind of fluid, it is possible to use heat from lower temperature heat sources. In ORC applications, organic fluids such as alkanes, aromates and siloxanes are used as working fluids instead of water. The reason to use these fluids is that they have lower boiling temperatures and higher molecular weights compared to water, which causes higher thermodynamic performances in lower temperatures. Since most of the organic fluids do not go to wet regions when they are expanded in turbines, they prevent erosion in turbine blades. Additionally, compared with the steam Rankine cycles, ORC applications work under lower working fluid pressures, which leads to lower turbine costs [23].

In ORC system designing, at least five components are needed for basic types: turbines, evaporators, condensers, pumps and working fluids. In the present study, R11 is selected as the working fluid for the analysis because of its high exergy efficiency [12]. Three different ORC systems (BORC, SRORC and DRORC) are chosen in this research. The schematic of these cycles with T-S (temperature-entropy) diagrams are shown respectively in Figures 1–3. In the basic ORC, first, the fluid is heated by the evaporator, and its temperature is increased. Then, the working fluid (R11) expands in the turbine from high pressure to low pressure to produce power. The outlet flow of the turbine goes to the condenser for cooling, and the saturated liquid enters the pump. The pump increases the flow pressure to reach the evaporator pressure.

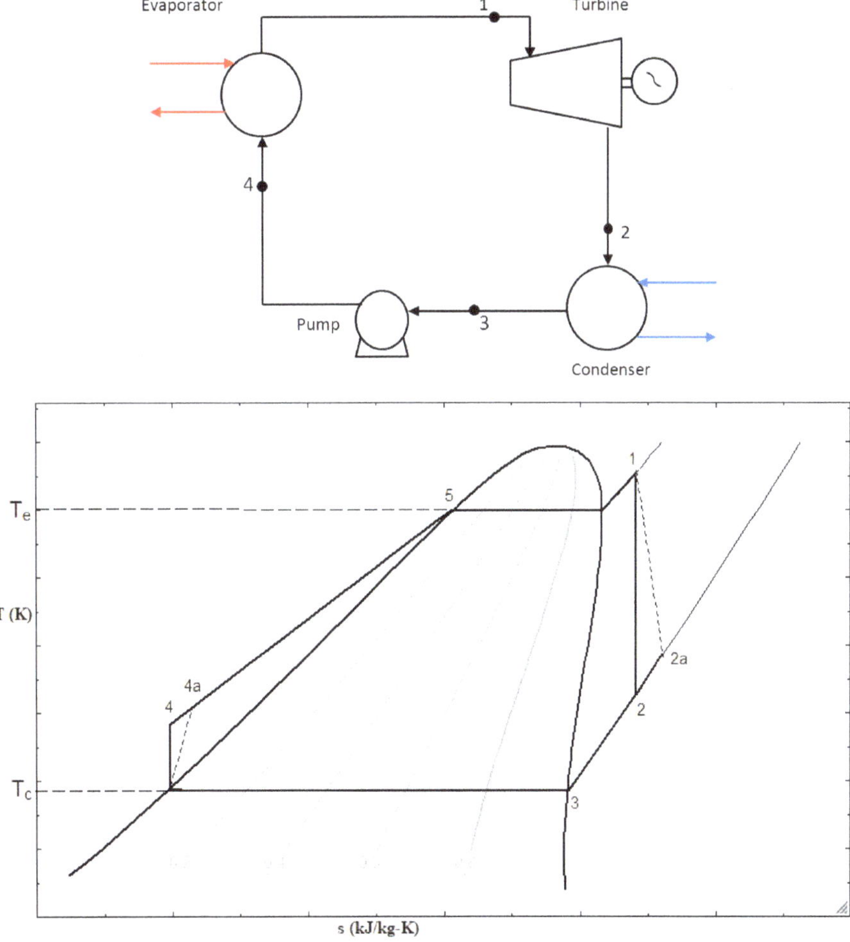

Figure 1. Schematic diagram of the cycle and T-S thermodynamic diagram of the basic organic Rankine cycle (BORC) system.

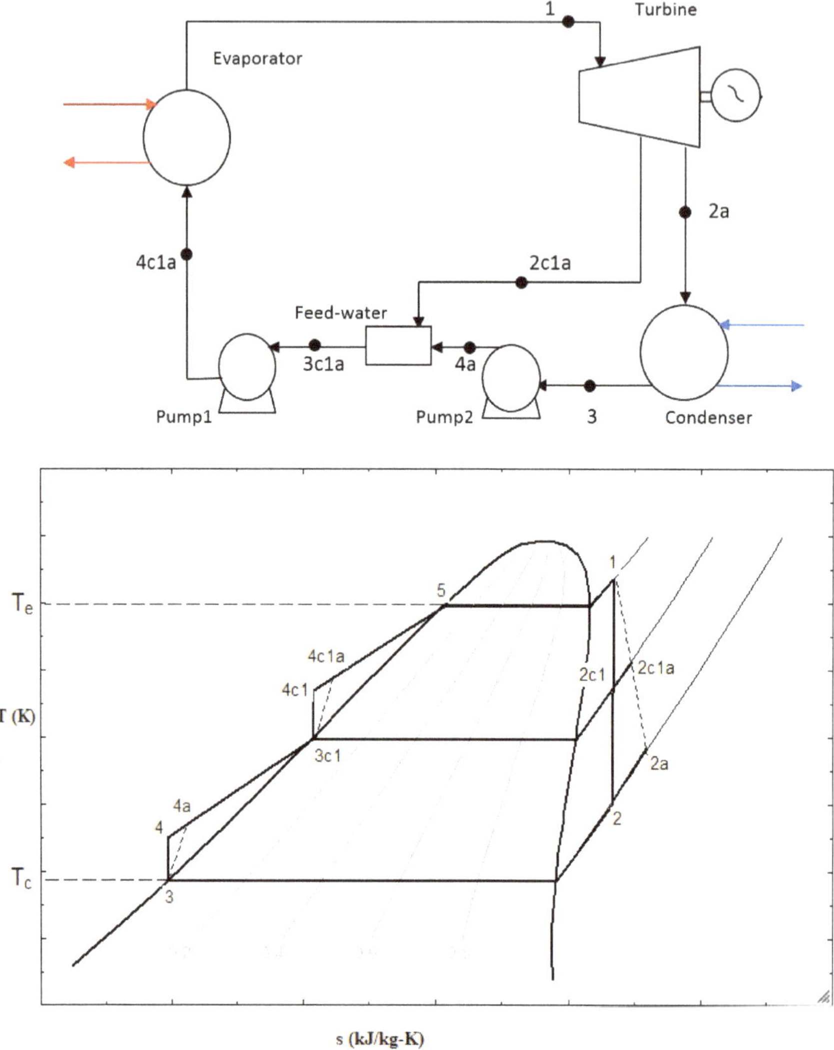

Figure 2. Schematic diagram of the cycle and T-S thermodynamic diagram of the regeneration ORC (RORC) system.

As it is seen in Figures 2 and 3, in regenerative cycles, the turbine outlet flow is split into two parts in single, and three parts in double, systems. These parts are in the vapor phase, which enters feed-water heaters to preheat the working fluid before going to the evaporator. By this method, more potential of the cycle energy is used, so a higher efficiency, compared with the basic type, is gained. In order to simulate the cycle performance in this study, the following assumptions are employed:

- All processes in cycles are assumed to be at a steady state and steady flow.
- There are no pressure drops in the pipes [24].
- Heat and friction losses, as well as the variation of potential and kinetic energies, are neglected [25].

Exiting fluid from the condenser is a saturated liquid.

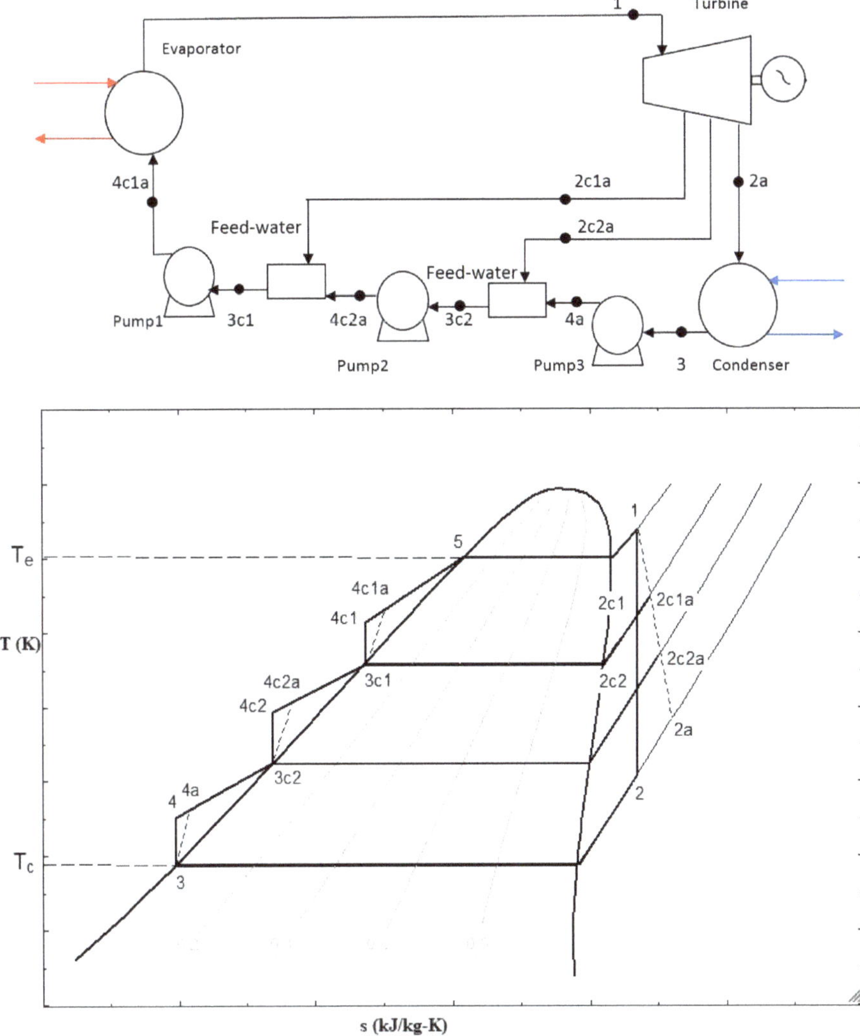

Figure 3. Schematic diagram of the cycle and T-S thermodynamic diagram of the double-regeneration ORC (DRORC) system.

3. Energy and Exergy Analysis

3.1. Energy and Conventional Exergy Analysis

The purpose of the energy analysis is to design thermal systems. However, the aim of the exergy analysis is to determine the thermodynamic efficiency of the cycles and compare them to find out the best system performance. There are two important basic laws in thermodynamics: the first and the second law. The first law analyzes the conservation of energy in processes, while the second law is used to discuss the quality of energy and materials. The main equations for energy and mass balance used in the first law analysis for a controlled volume of each component are written below:

Mass balance:

$$\sum \dot{m}_i = \sum \dot{m}_e \tag{1}$$

Energy balance:

$$\sum \dot{Q} - \sum \dot{W} + \sum \dot{m}_i h_i - \sum \dot{m}_e h_e = 0 \tag{2}$$

Exergy analysis has two advantages compared with the conventional energy analysis method. It provides a more precise measurement of the real inefficiencies in a system, as well as their exact places. The main goal of the second law is to recognize the place, type and quantity of energy loss resources in system processes and the factors affecting them. The exergy balance equation for a control volume of each component can be written as [24]:

$$\dot{I} = \dot{E}_{in} - \dot{E}_{out} + \dot{E}_Q - \dot{E}_W \tag{3}$$

where $\dot{I}$ is the total exergy rate of a destroyed system, and $\dot{E}_Q$ and $\dot{E}_w$ are the exergy transfer rates for heat and work, respectively. The equivalent work of any form of energy is defined as its exergy, so work is equivalent to exergy in every respect.

For a heat transfer rate ($\dot{Q}_r$) and system temperature where heat transfer happens (T_r), the maximum rate of conversion from thermal energy to work is:

$$\dot{w}_{max} = \dot{E}^Q = \dot{Q}^r \, \tau \tag{4}$$

$$\tau = 1 - T_0/T_r \tag{5}$$

τ is called the dimensionless exergetic temperature and is equal to the Carnot efficiency for special cases when the environment is at temperature T_0.

Exergy of each stream in a system, $\dot{E}$, is divided into four different parts:

$$\dot{E} = \dot{E}_k + \dot{E}_p + \dot{E}_{ph} + \dot{E}ch \tag{6}$$

where $\dot{E}_k$ is kinetic exergy, $\dot{E}_p$ is potential exergy, $\dot{E}_{ph}$ is physical exergy and $\dot{E}_{ch}$ is chemical exergy. $\dot{E}_k$ and $\dot{E}_p$ are associated with high-grade energies and $\dot{E}_{ph}$ and $\dot{E}_0$ with low-grade energies. In this research, kinetic and potential exergy is neglected.

Physical exergy of each stream can be calculated from the following Equation (0 refers to environmental state):

$$e_{ph} = (h - h_0) - T_0(S - S_0) \tag{7}$$

and the chemical exergy of the mixture for ideal gas can be expressed as below:

$$e_{ch} = \sum_{i=0}^{n} x_i ex_{chi} + RT_0 \sum_{i=0}^{n} x_i \ln(x_i) \tag{8}$$

Here, x_i is the mole fraction of the ith component in the mixture, and ex_{chi} in this relation is the chemical exergy of each component. As it is seen, exergy loss depends on the amount of heat that is transferred to ambient. So, convectional exergy analysis is only influenced by the entire condition of a system and does not deal with the effect of system components on other components' exergy destructions.

The basic equations used for the kth component in the conventional exergy analysis are shown in Equations (9)–(12).

$$\dot{E}_{D,k} = \dot{E}_{F,k} - \dot{E}_{P,k} \tag{9}$$

$$\varepsilon_k = \frac{\dot{E}_{P,k}}{\dot{E}_{D,k}} = 1 - \frac{\dot{E}_{D,k}}{\dot{E}_{F,k}} \tag{10}$$

$$y_k = \frac{\dot{E}_{D,k}}{\dot{E}_{F,k}} \tag{11}$$

$$y_k^* = \frac{\dot{E}_{D.k}}{\dot{E}_{D.tot}} \tag{12}$$

In these equations, $\dot{E}_{D.k}$, $\dot{E}_{F.k}$ and $\dot{E}_{P.k}$ represent the exergy destruction rate, exergy of fuel and exergy of product, respectively. Additionally, ε_k, y_k and y_k^* are the exergy efficiency, the exergy loss ratio and the exergy of fuel with the total exergy destruction in the kth component, respectively. The exergy equations for components of the three cycles are expressed in Table 1.

Table 1. Balance of exergy for any components of the systems. BORC: basic organic Rankine cycle, SRORC: single-regeneration ORC and DRORC: double-regeneration ORC.

Cycle Components	Exergy Balance Equations		
	BORC	**SRORC**	**DRORC**
Evaporator	$\dot{E}_4 + \dot{E}_{q,\,Eva} = \dot{E}_1 + \dot{E}_{D,\,Eva}$	$\dot{E}_{4c1a} + \dot{E}_{q,\,Eva} = \dot{E}_1 + \dot{E}_{D,\,Eva}$	$\dot{E}_{4c1a} + \dot{E}_{q,\,Eva} = \dot{E}_1 + \dot{E}_{D,\,Eva}$
Turbine	$\dot{E}_1 = \dot{E}_{w,\,tur} + \dot{E}_2 + \dot{E}_{D,\,Tur}$	$\dot{E}_1 = \dot{E}_{w,\,tur} + \dot{E}_{2a} + \dot{E}_{2c1a} + \dot{E}_{D,\,Tur}$	$\dot{E}_1 = \dot{E}_{w,\,tur} + \dot{E}_{2a} + \dot{E}_{2c1a} + \dot{E}_{2c2a} + \dot{E}_{D,\,Tur}$
Condenser	$\dot{E}_2 = \dot{E}_{q,\,con} + \dot{E}_3 + \dot{E}_{D,\,cond}$	$\dot{E}_{2a} = \dot{E}_{q,\,con} + \dot{E}_3 + \dot{E}_{D,\,cond}$	$\dot{E}_{2a} = \dot{E}_{q,\,con} + \dot{E}_3 + \dot{E}_{D,\,cond}$
Pump	$\dot{E}_3 + \dot{E}_{w,\,pump} = \dot{E}_4 + \dot{E}_{D,\,pump}$	$\dot{E}_{3c1} + \dot{E}_{w,\,pump1} = \dot{E}_{4c1a} + \dot{E}_{D,\,pump1}$ $\dot{E}_3 + \dot{E}_{w,\,pump2} = \dot{E}_{4a} + \dot{E}_{D,\,pump2}$	$\dot{E}_{3c1} + \dot{E}_{w,\,pump1} = \dot{E}_{4c1a} + \dot{E}_{D,\,pump1}$ $\dot{E}_{3c2} + \dot{E}_{w,\,pump2} = \dot{E}_{4c2a} + \dot{E}_{D,\,pump2}$ $\dot{E}_3 + \dot{E}_{w,\,pump3} = \dot{E}_{4a} + \dot{E}_{D,\,pump3}$
Feed-water heater	–	$\dot{E}_4 + \dot{E}_{2c1a} = \dot{E}_{3c1} + \dot{E}_{D,\,feed\text{-}water}$	$\dot{E}_{4c2a} + \dot{E}_{2c1a} = \dot{E}_{3c1} + \dot{E}_{D,\,feed\text{-}water1}$ $\dot{E}_{4a} + \dot{E}_{2c2a} = \dot{E}_{3c2} + \dot{E}_{D,\,feed\text{-}water2}$

In order to observe the system performance for the research, the first and the second law efficiencies are calculated from the equations below [12]:

$$\eta_t = \frac{W_{net}}{Q} \tag{13}$$

$$\eta_e = W_{net} / \left[Q\left(1 - \frac{T_0}{T_m}\right) \right] \tag{14}$$

where T_0 is environmental temperature, and T_m is source of heat; mean temperature can be calculated by the following equation:

$$T_m = (T_{in} - T_{out}) / \ln(T_{in}/T_{out}) \tag{15}$$

By the uses of Equations (13) and (14), the energy efficiency and exergy efficiency for the basic cycle can be obtained:

$$\eta_t = W_{net}/Q \tag{16}$$

$$\eta_e = \frac{W_{net}}{Q}\left(1 - \frac{T_0}{T_m}\right) \tag{17}$$

For SRORC, the first and the second law efficiencies are presented as:

$$\eta_t = \frac{W_{net}}{Q} = \frac{W_t - W_{p1} - W_{p2}}{Q} \tag{18}$$

$$\eta_e = \frac{W_{net}}{Q\left(1 - \frac{T_0}{T_m}\right)} = \frac{W_t - W_{p1} - W_{p2}}{\left[Q\left(1 - \frac{T_0}{T_m}\right)\right]} \tag{19}$$

For DRORC, efficiencies are calculated from the equations below:

$$\eta_t = \frac{W_{net}}{Q} = \frac{W_t - W_{p1} - W_{p2} - W_{p3}}{Q} \tag{20}$$

$$\eta_t = \frac{W_{net}}{Q\left(1 - \frac{T_0}{T_m}\right)} = \frac{W_t - W_{p1} - W_{p2} - W_{p3}}{Q(1 - T_0/T_m)} \tag{21}$$

3.2. Advanced Exergy Analysis

By a conventional exergy analysis, the only thing the researchers can discuss is the quantity of destruction rates in systems, and this method does not give any information about the type of these losses. Advanced exergy analysis is a new method that speaks about the details of the destructions to help researchers recognize different factors that affect component losses. Generally, advanced exergy analysis divides the loss rate of each component into two parts: endogenous/exogenous and avoidable/unavoidable.

3.2.1. Exogenous/Endogenous Exergy Destruction

To observe the effect of the parts on each other in a system, the kth component destruction rate is split into exogenous and endogenous parts:

$$\dot{E}_{D.k} = \dot{E}_{D.k}^{EN} + \dot{E}_{D.k}^{EX} \tag{22}$$

The endogenous exergy destruction ($\dot{E}_{D.k}^{EN}$) is the destruction rate associated with irreversibility due to inefficiency in the kth component, while all the residual components work preferably. The exogenous exergy destruction ($\dot{E}_{D.k}^{EX}$) is related to the destruction rate that happens in the remaining components. This dividing shows the real location of destructions in the systems, so that researchers can focus on the exact parts to improve system performance. There are two ways to compute the endogenous exergy destruction rate: engineering method and thermodynamic or hybrid cycle [26]. In this study, hybrid cycles of the systems are selected for the analysis. In this method, to calculate the rate of endogenous exergy destruction in the kth component, it is set on its real condition, and the other components work on their ideal conditions (with no irreversibilities). The components' ideal conditions selected for this study are shown in Table 2.

3.2.2. Avoidable/Unavoidable Exergy Destruction

Some values of exergy loss rates in each part, which are called unavoidable exergy destruction ($\dot{E}_{D.k}^{UN}$), cannot be reduced owing to technical limitations such as accessibility, cost of materials and making.

In order to calculate the unavoidable exergy destruction, some assumptions for the unavoidable conditions of the components are needed, which are usually obtained by researchers' experiences with component operations. In this study, these assumptions are selected from references [12,16,27–29], and these conditions for each component are written in Table 2. To calculate the unavoidable exergy destruction of the kth component, all components are set on their best possible conditions (unavoidable conditions), and the unavoidable loss rates for the kth component are obtained from the equation below [29]:

$$\dot{E}_{D.k}^{UN} = \dot{E}_{P.k}\left(\frac{\dot{E}_{D.k}}{\dot{E}_{P.k}}\right)^{UN} \tag{23}$$

Avoidable exergy loss in the kth part is obtained by the subtraction of the unavoidable portion from the exergy loss rate of that part (Equation (23)). The avoidable exergy loss rate is recoverable and can be decreased, so designers should localize this sector to improve system performance.

$$\dot{E}_{D.k} = \dot{E}_{D.k}^{UN} + \dot{E}_{D.k}^{AV} \tag{24}$$

Table 2. The assumed conditions for performing an advanced exergy analysis under real, unavoidable and ideal conditions [12,18,28,30].

Cycle Component	Real Conditions	Unavoidable Conditions	Ideal Conditions
Evaporator	$\Delta T_{pp} = 8\,^\circ\text{C}$ $\Delta p = 2\%$	$\Delta T_{pp} = 3\,^\circ\text{C}$ $\Delta p = 1\%$	$\Delta T_{pp} = 0\,^\circ\text{C}$ $\Delta p = 0\%$
Turbine	$\eta_{is.turbine} = 0.8\%$	$\eta_{is.turbine} = 0.95\%$	$\eta_{is.turbine} = 1\%$
Condenser	$\Delta T_{pp} = 8\,^\circ\text{C}$ $\Delta p = 2\%$	$\Delta T_{pp} = 3\,^\circ\text{C}$ $\Delta p = 1\%$	$\Delta T_{pp} = 0\,^\circ\text{C}$ $\Delta p = 0\%$
Pump	$\eta_{is.pump} = 0.7\%$	$\eta_{is.pump} = 0.9\%$	$\eta_{is.pump} = 1\%$
Feed-water heater	$\Delta p = 2\%$	$\Delta p = 1\%$	$\Delta p = 0\%$

3.2.3. Combination of Splitting

To have more information about the share of each component of a system in total exergy destruction, the two main parts discussed above can be combined. Researchers can use a combination of endogenous or exogenous parts with avoidable or unavoidable parts for a more detailed designing of the systems [30]. By combining the two main splitting parts, the exogenous/unavoidable and the endogenous/unavoidable, the exogenous/avoidable and the endogenous/avoidable rates are obtained. To reach the unavoidable endogenous rate of the kth component, the equation below is used [31,32]:

$$\dot{E}_{D.k}^{UN.EN} = \dot{E}_{P.k}^{EN}\left(\frac{\dot{E}_{D.k}}{\dot{E}_{P.k}}\right)^{UN} \tag{25}$$

In this equation, the ratio of exergy loss to exergy of product in this kth component is obtained in unavoidable conditions of the cycle, and the endogenous product exergy is obtained for the kth component when it is in hybrid cycle conditions. The unavoidable exogenous rate for the kth component is obtained from the following equation:

$$\dot{E}_{D.k}^{UN.EX} = \dot{E}_{D.k}^{UN} - \dot{E}_{D.k}^{UN.EN} \tag{26}$$

Then, the avoidable endogenous and exogenous parts of the exergy destruction for the kth component can be gained by reducing the unavoidable endogenous and exogenous exergy losses from the endogenous and exogenous exergy losses of that component, respectively.

$$\dot{E}_{D.k}^{AV.EN} = \dot{E}_{D.k}^{EN} - \dot{E}_{D.k}^{UN.EN} \tag{27}$$

$$\dot{E}_{D.k}^{AV.EX} = \dot{E}_{D.k}^{EX} - \dot{E}_{D.k}^{UN.EN} \tag{28}$$

The real potential of components in a system to improve their performances and to have the best efficiencies in a general system may be determined and analyzed by a mixture of the exergy loss sections mentioned above. There is no way to decrease the unavoidable/endogenous section of exergy loss because of technological limitations of the kth part, while avoidable/endogenous parts of the exergy destruction can be decreased by improving the efficiency of related components. The exogenous/unavoidable part of exergy loss for each component cannot be reduced because of technological constraints associated with other system components. However, the exogenous/avoidable parts may be decreased by the betterment of the total system performance, the efficiency of related components and, also, by improving the performance of other components in a system.

4. Simulation, Results and Discussion

The EES (engineering equation solver) is one of the most commonly used softwares to simulate power cycle operations. In this research, the EES was used to simulate cycle processes and to calculate properties of parameters in a system to reach cycle efficiencies. The input data and parameters for system components' working conditions of waste heat conditions used in this study are obtained from previous studies [12]. The conditions are shown in Table 3. These conditions are used for both conventional and advanced exergy analysis in this research.

Table 3. Heat source parameters and assumptions [12].

Parameter	Value
Temperature of heat source (K)	420
Mass flow of heat source (kg s^{-1})	14
Heat capacity of hot gases in constant pressure (kj kg^{-1} K^{-1})	1.1
Environmental temperature (K)	298.15
Environmental pressure (kPa)	101.35
Temperature of condensing (K)	303.15
Isentropic efficiency of turbine	0.8
Isentropic efficiency of pump	0.7
Minimum temperature difference in the evaporator (Pinch) (K)	8

4.1. Model Validation

For validation, the comparison between the present models and Reference [8] is considered in this section (Table 4). There might be some errors due to the basic equations that EES uses to derive the characteristics of fluids. As results show, the deviation is low, and the models can be used for this analysis.

Table 4. Model validation of simulations by R11 with Reference [8].

Parameter	BORC Results		SRORC Results		DRORC Results	
	This Study	Reference [12]	This Study	Reference [12]	This Study	Reference [12]
T_e (K)	401.263	401.6	407.412	407.7	407.682	408
P_3 (kPa)	125.961	125.3	125.961	125.3	125.961	125.3
T_{out} (K)	401.930	401.4	413.312	413.5	414.104	414.6
$e\eta$	50.61	58.4	55.00	63.19	56.87	65.2
η_t	13.89	15.99	15.64	17.98	16.21	18.62
W_{p1}(kW)	1.362	1.502	0.5967	0.6251	0.496	0.5175
W_{p2}(kW)			0.138	0.1457	0.128	0.1396
W_{p3}(kW)	-	-	-	-	0.074	0.07804
W_t (kW)	40.003	43.91	16.845	17.65	15.418	16.17
W_{net}(kW)	38.641	42.41	16.111	16.88	14.720	15.43
m (kg/s)	1.215	1.17	0.568	0.5212	0.536	0.4929
Q_e (kj/s)	278.267	265.1	102.999	96.63	90.800	91.32
Q_c (kj/s)	239.626	222.7	86.888	93.89	76.080	82.89
X_{c1}	-	-	0.20309	0.2031	0.132	0.1323
X_{c2}	-	-	-	-	0.128	0.129

As shown in Table 4, DRORC has a maximum exergy efficiency with 65.2% among the present systems. BORC produced more power compared with two other cycles, but as it is seen, this system has the lowest exergy efficiency because of its high exergy destruction rate in its components.

4.2. Conventional and Advanced Exergy Analysis

Considering conventional exergy analysis and by using the equations for each component of the cycles presented in Table 1, the exergy destruction rates are obtained. In the three cycles, the total exergy fuel rate is obtained from subtracting the exergy rate of the working fluid that leaves the evaporator from exhaust gases that enter the evaporator, and the total product exergy is the turbine power output. The main results of the conventional exergy analysis for BORC, SRORC and DRORC are presented in Tables 5–7, respectively. E_F, E_P, E_P, E_D, Y_K, Y_K^* in this tables are exergy of fuel, exergy of product, exergy destruction, exergy loss ratio and exergy of fuel with the total exergy destruction. It can be said that the components with higher exergy destruction rates have more effects on the efficiency of systems from an exergy point of view compared with other components. Referring to Table 5, in BORC, the maximum exergy destruction rate happens in the evaporator, followed by the turbine, the condenser and the pump. Table 6 shows the exergy destruction rates for the SRORC components. As it is seen, in single-regenerative systems, turbines show the highest exergy destruction rates among cycles components due to their design, and evaporators, condensers, feed-water heaters and pumps are in the next ranks, respectively. For DRORC, the exergy destructive rates are shown in Table 7. In these systems, as in single-regenerative cycles, turbines have the maximum exergy destructive rate, and they are followed by evaporators. However, unlike SRORC, pump1 and feed-water heater 2 are placed in the next ranks because of the second flow that extracts from the turbine. As a summary, it can be concluded that evaporators, turbines and feed-waters have more potential to reduce their destructive rates and to increase system efficiency. As it is seen, a conventional exergy analysis just focuses on components with high rates of exergy destruction, and it is not possible to specify whether these destructions occur in other components or in the component itself. These irreversibilities may only be specified by advanced exergy tools. As discussed above, an advanced exergy analysis evaluates the effects of component interactions and the real possibility of components to improve system efficiency.

Table 5. Results from exergy calculations for BORC.

Component	E_F (kW)	E_P (kW)	E_D (kW)	ε (%)	Y_K (%)	Y_K^* (%)
Evaporator	88.95	74.51	14.44	83.766	16.233	46.898
Turbine	53.1	43.68	9.42	82.259	17.740	30.594
Pump	1.538	1.538	0	100	0	0
Condenser	22.94	16.01	6.93	69.790	30.209	22.507

Table 6. Results from exergy calculations for SRORC.

Component	E_F (kW)	E_P (kW)	E_D (kW)	ε (%)	Y_K (%)	Y_K^* (%)
Evaporator	32.63	30.16	2.47	92.430	7.569	28.036
Turbine	21.5	17.64	3.86	82.046	17.953	43.813
Pump1	0.648	0.148	0.499	22.962	77.037	5.666
Pump2	0.151	0.015	0.135	10.257	89.742	1.541
Feed-water heater	4.778	3.84	0.938	80.368	19.631	10.647
Condenser	7.887	6.98	0.907	88.500	11.499	10.295

Table 7. Results from exergy calculations for DRORC.

Component	E_F (kW)	E_P (kW)	E_D (kW)	ε (%)	Y_K (%)	Y_K^* (%)
Evaporator	28.67	26.91	1.76	93.861	6.138	15.087
Turbine	24.39	16.06	8.33	65.846	34.153	71.408
Pump1	0.536	0.138	0.397	25.857	74.142	3.409
Pump2	0.145	0.028	0.116	19.545	80.454	1.001
Pump 3	0.079	0.008	0.0717	10.232	89.767	0.615
Feed-water heater1	3.274	3.044	0.23	92.974	7.025	1.971
Feed-water heater2	2.421	1.984	0.437	81.949	18.050	3.746
Condenser	6.867	6.545	0.322	95.310	4.689	2.760

By advanced exergy analysis, the exergy destructions of each component calculated in the previous section can be discussed in detail to find the sources of these destructions and real potentials of each component to amend the efficiency of the whole system. As mentioned above, these irreversibilities can be divided into exogenous, endogenous, unavoidable and avoidable parts to help researchers observe the effects of technological limitations and component interactions on the exergetic efficiency of a system for improvements [33]. In the advanced exergy analysis, the endogenous part of exergy loss for the kth component is calculated by defining real and ideal conditions for the cycles first. Then, the exogenous destruction rate is obtained by the difference of the total exergy and endogenous part (Equation (22)). To calculate the unavoidable exergy rate of loss in the kth part, instead of real conditions in cycles, unavoidable conditions are considered, and avoidable exergy is obtained from Equation (23). Furthermore, the values of exogenous/avoidable, endogenous/avoidable, exogenous/unavoidable and endogenous/unavoidable are determined using Equations (25) to (28). Advanced exergy analysis results for each part in three different cycles (BORC, SRORC and DRORC) are presented in Tables 8–10.

Table 8. Advanced exergy analysis results (kW) for BORC.

Component	$\dot{E}_D$	$\dot{E}_D^{UN}$	$\dot{E}_D^{AV}$	$\dot{E}_D^{EN}$	$\dot{E}_D^{EX}$	$\dot{E}_D^{AV.EX}$	$\dot{E}_D^{AV.EN}$	$\dot{E}_D^{UN.EX}$	$\dot{E}_D^{UN.EN}$
Evaporator	14.44	7.725	6.714	14.13	0.31	0.357	6.356	−0.047	7.773
Turbine	9.42	2.0194	7.400	3.135	6.285	4.935	2.464	1.349	0.670
Pump	0	0	0	0	0	0	0	0	0
Condenser	6.93	4.168	2.761	2.06	4.87	2.005	0.756	2.864	1.303

Table 9. Advanced exergy analysis results (kW) for SRORC.

Component	$\dot{E}_D$	$\dot{E}_D^{UN}$	$\dot{E}_D^{AV}$	$\dot{E}_D^{EN}$	$\dot{E}_D^{EX}$	$\dot{E}_D^{AV.EX}$	$\dot{E}_D^{AV.EN}$	$\dot{E}_D^{UN.EX}$	$\dot{E}_D^{UN.EN}$
Evaporator	2.47	1.027	1.442	2.561	−0.091	−0.084	1.527	−0.006	1.033
Turbine	3.86	0.824	3.035	2.083	1.777	1.397	1.637	0.379	0.445
Pump 1	0.499	0.067	0.431	0.263	0.235	0.203	0.228	0.032	0.034
Pump 2	0.135	0.014	0.121	0.067	0.067	0.060	0.060	0.007	0.007
Feed-water heater	0.938	0.299	0.638	0.479	0.458	0.313	0.324	0.144	0.155
Condenser	0.907	0.168	0.738	0.441	0.465	0.382	0.356	0.082	0.085

Table 10. The results of the advanced exergy analysis (per kW) for DRORC.

Component	$\dot{E}_D$	$\dot{E}_D^{UN}$	$\dot{E}_D^{AV}$	$\dot{E}_D^{EN}$	$\dot{E}_D^{EX}$	$\dot{E}_D^{AV.EX}$	$\dot{E}_D^{AV.EN}$	$\dot{E}_D^{UN.EX}$	$\dot{E}_D^{UN.EN}$
Evaporator	1.76	0.463	1.296	1.845	−0.085	−0.080	1.377	−0.004	0.467
Turbine	8.33	1.742	6.587	4.188	4.142	3.284	3.303	0.857	0.884
Pump 1	0.397	0.064	0.333	0.199	0.198	0.165	0.167	0.032	0.031
Pump 2	0.116	0.018	0.098	0.054	0.062	0.053	0.045	0.009	0.008
Pump 3	0.071	0.006	0.065	0.033	0.038	0.035	0.030	0.003	0.003
Feed-water heater 1	0.23	0.080	0.149	0.113	0.116	0.075	0.074	0.041	0.039
Feed-water heater 2	0.437	0.124	0.312	0.205	0.231	0.167	0.146	0.064	0.059
Condenser	0.322	0.072	0.249	0.138	0.183	0.145	0.104	0.038	0.034

As indicated in Table 9, for BORC, the exogenous exergy rate is greater than the endogenous exergy rate in system components, except the evaporator. So, system performance and modification of other components are necessary for turbines and condensers to decrease their exogenous destruction rates. The greater share of exergy destruction in the evaporator is caused by irreversibility of the component itself because of its high exergy loss rate in the endogenous part. As it was discussed before, the avoidable destruction rate in exergy can be controlled and reduced in practice. Table 8 shows that a turbine consists of a high value of avoidable destruction rates (7.4 kW) among the components of a system. Thus, the efficiency of this component can be improved using some technical modifications and new technologies or by replacing the component with the ones with higher efficiencies. It is important to note that, unlike the conventional analysis, a turbine is the most effective component due to its avoidable destruction rate to reduce irreversibilities. So, the main focus will be on the avoidable/endogenous parts of the exergy destruction, which can be decreased by improving the efficiency of the kth component. It follows by an investigation on the exogenous/avoidable exergy rates of loss, which can be reduced by improving the efficiency of other parts [34]. As it is stated in [17], Table 8 shows that the avoidable/endogenous exergy destruction rates in the turbine are greater than the unavoidable/endogenous destruction rates for BORC. This shows that the efficiency can be improved by technical modifications of this component. Table 8 also indicates that, except the condenser, the exogenous/avoidable exergy destruction rates are higher than the exogenous/unavoidable exergy rates of loss for the components. Results of the advanced exergy analysis for the division of the exergy rate of loss in the main parts in BORC are shown in Figure 4. As Figure 4 indicates, for the turbine, 79% of the total exergy rate of loss is avoidable, and from this rate, 53% can be reduced by amending other components efficiencies, and 26% of this rate depends on the performance of the component itself.

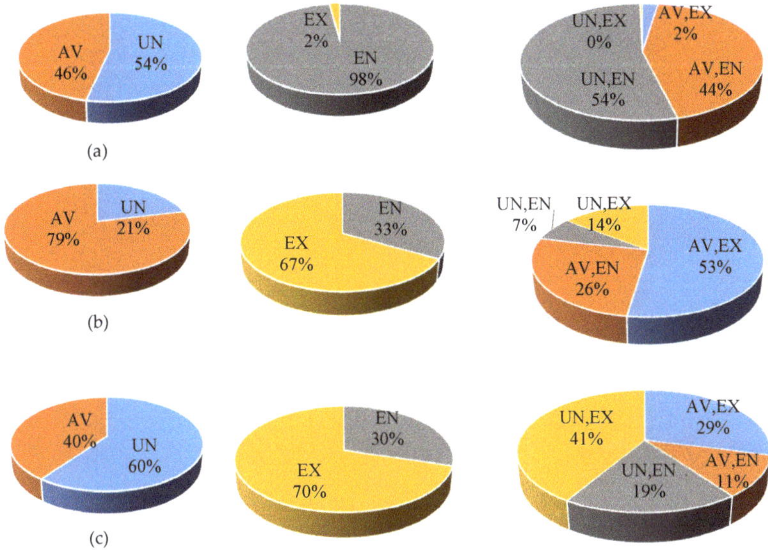

Figure 4. Splitting exergy destruction rates of components: (**a**) the evaporator, (**b**) turbine and (**c**) condenser in the BORC system.

In Table 9, for SRORC, except the condenser, the endogenous exergy destruction rate is higher than the exogenous exergy destruction rate in the system's components, which shows that, unlike BORC, the greater share of the destruction rate is because of the internal irreversibilities in the component itself. So, among all components in SRORC, the turbine consists of the highest destruction rate in the exogenous part (1.77 kW), and the evaporator has the maximum endogenous destruction rate (2.56 kW) among the system's components due to its irreversibilities. As it is seen in Table 9, the avoidable part of the exergy destruction rate is higher than the unavoidable part in all components of SRORC. This indicates that there is a great potential in a system to reduce its irreversibilities by using some efficient and new components. Splitting the exergy destruction rates into endogenous/avoidable and exogenous/avoidable parts provides some important information that helps researchers to optimize systems. As it is seen in Table 9, the avoidable/endogenous exergy destruction rates in all equipment are higher than the exogenous/avoidable rates, except for the condenser. Priority in the improvement process of a component should be given to the turbine and the evaporator because of their higher values in the endogenous/avoidable destruction rates. For a better analysis, results for splitting the exergy destruction rate for the components in SRORC are shown in Figure 5. The highest unavoidable/endogenous exergy destruction rates belong to the evaporator in the system, with about a 54% of the total exergy destruction rate, which shows low potential in reducing the irreversibility for this component.

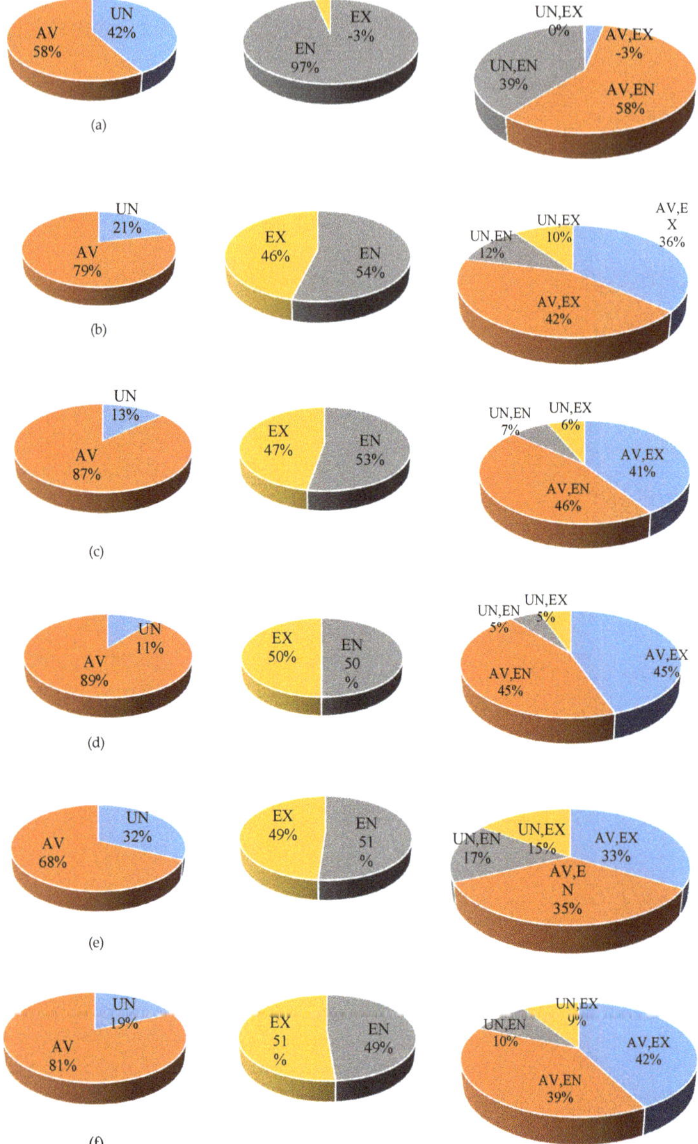

Figure 5. Splitting exergy destruction rates of components: (**a**) the evaporator, (**b**) turbine, (**c**) pump1, (**d**) pump2, (**e**) feed-water heater and (**f**) condenser in the SRORC system.

By referring to Table 10, it can be seen that, for DRORC, the endogenous exergy rate of loss is higher than the exogenous part for the evaporator and the turbine. This means that, in order to reduce the destruction rate, designers should focus on improving the efficiency of these components. Additionally, from these results, it is observed that there is a considerable contribution of the exergy rate of loss in pumps, feed-water heaters and condensers because of other components' performances in the systems. As it is mentioned above, the avoidable exergy destruction rate is an important part in reducing destruction. Table 10 shows that the avoidable exergy destruction rate is higher than the unavoidable part for all components in DRORC, which indicates that a system has a great potential for reducing its total exergy destruction by using modern technologies or replacing components with new, efficient ones. Among components, turbines with 6.587 kW consist of the highest avoidable exergy destruction rate, which should be paid more attention to in designing. As it is shown in Table 10, the endogenous/avoidable exergy destruction rates for the evaporator, turbine and pump1 is higher than the endogenous/unavoidable destruction rates in DRORC. Modifications in these components and improving working conditions should be noticed by designers to reduce this part of the exergy destruction. Priority in improving the performance of components in DRORC should be given to the turbine, the evaporator, pump 1 and feed-water heater 2. As for the previous cycles, the results of the advanced exergy analysis for DRORC components are shown in Figure 6.

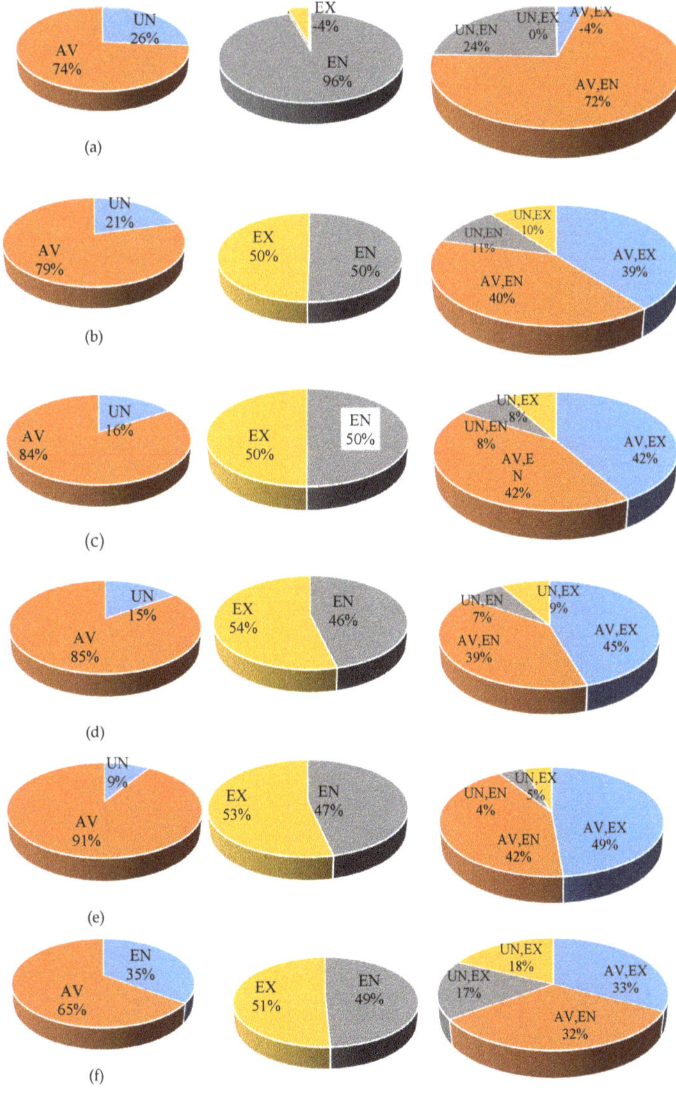

Figure 6. *Cont.*

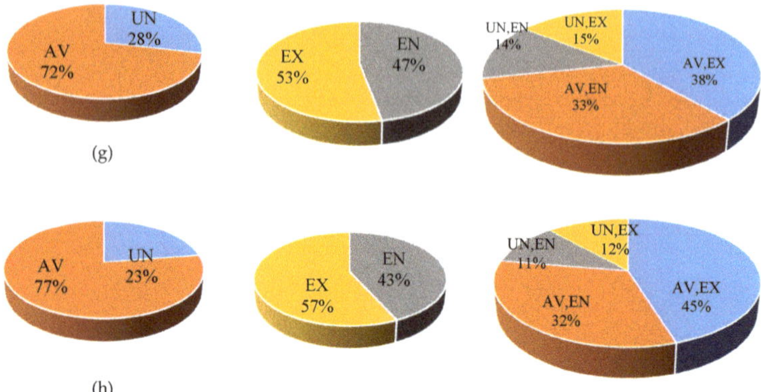

Figure 6. Splitting exergy destruction rates of components: (**a**) the evaporator, (**b**) turbine, (**c**) pump1, (**d**) pump2, (**e**) pump3, (**f**) feed-water heater1, (**g**) feed-water heater 2 and (**h**) condenser in the DRORC system.

In addition, the total exergy destruction rate for different parts obtained from an advanced exergy analysis for the three cycles are shown in Figures 7–9. As indicated, the total avoidable exergy rate of loss in these three systems is higher than unavoidable part, which shows a good potential to reduce irreversibilities. Avoidable exergy rate of loss with about 78% of the total rate of loss in DRORC is the first to be noticed by designers. SRORC and BORC with 74% and 55%, respectively, are given the next priorities. Additionally, in SRORC and DRORC, the total endogenous/avoidable exergy destruction rates are about 47% and 45% of the total exergy destruction rate, respectively, which are higher than other three combination parts.

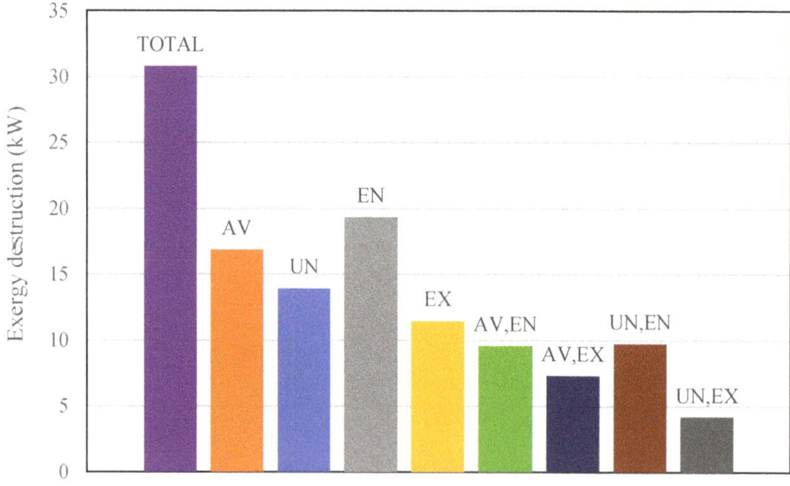

Figure 7. Total exergy destruction rate for the different parts in BORC.

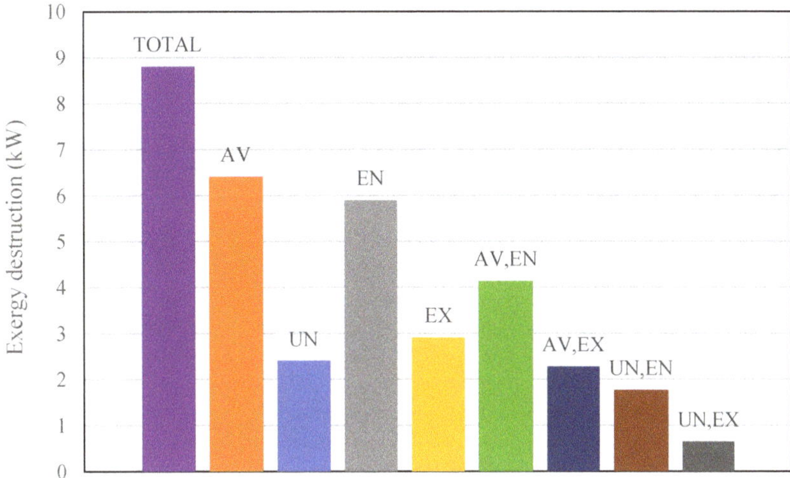

Figure 8. Total exergy destruction rate for the different parts in single-regeneration ORC (SRORC).

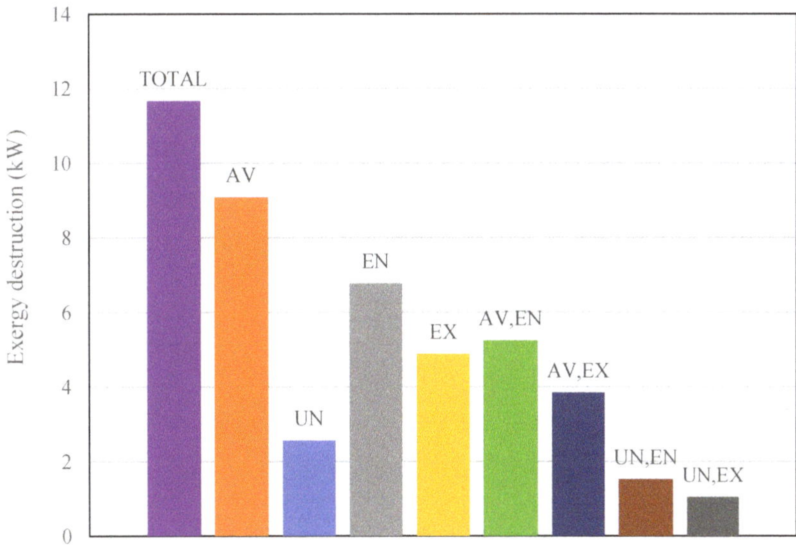

Figure 9. Total exergy destruction rate for the different parts in DRORC.

5. Conclusions

In this research, conventional and advanced exergy analysis are performed for three different models of organic Rankine cycles (BORC, SRORC and DRORC) at specified operating conditions by the use of a low-temperature heat source. The main results obtained from the present study are listed as below:

- According to the conventional exergy analysis, the highest exergy destruction was obtained in the evaporator for BORC, and for SRORC and DRORC, the maximum exergy destruction belongs to the turbine. Additionally, results of the conventional exergy analysis show that the condenser in three cycles is the third important component that should be noticed in designing. The exergy destruction rate for pumps may be neglected.

- The advanced exergy analysis showed that the exogenous exergy destruction rate is greater than the endogenous exergy destruction rate, except the evaporator, for BORC. However, in SRORC, the endogenous exergy rate of loss is higher than the exogenous part, except for the condenser. This indicates the greater share of the exergy destruction rate because of the internal irreversibilities in the component itself compared with BORC. Avoidable rate of exergy destruction in BORC, SRORC and DRORC are about 78%, 74% and 55% of the total destruction rate, respectively. Among the existing components, the condenser has the maximum value of exogenous exergy destruction rate, which can be reduced by the modification of other components.

- The endogenous/avoidable exergy destruction was identified as an important part of the destruction. Thirty-one percent, forty-seven percent and forty-five percent of the total exergy destruction rate in BORC, SRORC and DRORC placed in this part, respectively, can be reduced by improving the efficiency of the components.

- By the advanced exergy analysis, unlike the conventional exergy analysis, the preference of improvement should be given to turbines, evaporators, condensers and feed-water heaters, respectively. Additionally, regenerative cycles have high potentials to reduce their irreversibilities compared with basic systems.

Author Contributions: S.Y.D.; conceptualization; methodology; software; validation; formal analysis; writing—original draft preparation. G.S.; writing—original draft preparation; writing—review and editing; visualization; supervision. A.D.; writing—review and editing; supervision. All authors have read and agreed to the published version of the manuscript.

Funding: This research received no external funding.

Conflicts of Interest: The authors declare no conflicts of interest.

Nomenclature

parameter	define
ORCs	Organic Rankin cycles
BORC	Basic organic Rankin cycles
SRORC	Single regeneration organic Rankin cycles
DRORC	Double regeneration organic Rankin cycles
M	Mass flow rate
E	exergy
Eph	physical exergy
Ech	Chemical exergy
η	energy efficiency
ε	exergy efficiency
E_F	exergy of fuel
E_P	exergy of product
E_D	exergy destruction
Y_K	exergy loss ratio with the total exergy destruction
Y_K^*	exergy of fuel with the total exergy destruction
$\dot{E}_D^{AV}$	avoidable exergy destruction
$\dot{E}_D^{EN}$	endogenous exergy destruction
$\dot{E}_D^{EX}$	exogenous exergy destruction
$\dot{E}_D^{UN}$	unavoidable exergy destruction
$\dot{E}_D^{UN.EN}$	unavoidable endogenous exergy destruction
$\dot{E}_D^{AV.EN}$	avoidable endogenous exergy destruction
$\dot{E}_D^{AV.EX}$	avoidable exogenous exergy destruction

References

1. Yamamoto, T.; Furuhata, T.; Arai, N.; Mori, K. Design and testing of the organic Rankine cycle. *Energy* **2001**, *26*, 239–251. [CrossRef]
2. Bao, J.; Zhao, L. A review of working fluid and expander selections for Organic Rankine Cycle. *Renew. Sustain. Energy Rev.* **2013**, *24*, 325–342. [CrossRef]
3. Lecompte, S.; Huisseune, H.; Van Den Broek, M.; Vanslambrouck, B.; De Paepe, M. Review of organic Rankine cycle (ORC) architectures for waste heat recovery. *Renew. Sustain. Energy Rev.* **2015**, *47*, 448–461. [CrossRef]
4. Zhang, S.J.; Wang, H.X.; Guo, T. Performance comparison and parametric optimization of subcritical organic Rankine cycle (ORC) and transcritical power cycle system for low-temperature geothermal power generation. *Appl. Energy* **2011**, *88*, 2740–2754.
5. Li, J.; Pei, G.; Ji, J. Optimization of low temperature solar thermal electric generation with organic Rankine cycle in different areas. *Appl. Energy* **2010**, *87*, 3355–3365.
6. Davarpanah, A. Feasible analysis of reusing flowback produced water in the operational performances of oil reservoirs. *Environ. Sci. Pollut. Res.* **2018**, *25*, 35387–35395. [CrossRef]
7. Davarpanah, A.; Mirshekari, B. Experimental study of CO_2 solubility on the oil recovery enhancement of heavy oil reservoirs. *J. Therm. Anal. Calorim.* **2019**. [CrossRef]
8. Davarpanah, A.; Mirshekari, B. Experimental Investigation and Mathematical Modeling of Gas Diffusivity by Carbon Dioxide and Methane Kinetic Adsorption. *Ind. Eng. Chem. Res.* **2019**. [CrossRef]
9. Davarpanah, A.; Zarei, M.; Valizadeh, K.; Mirshekari, B. CFD design and simulation of ethylene dichloride (EDC) thermal cracking reactor. *Energy Sour. Part A Recovery Util. Environ. Eff.* **2018**, 1–15. [CrossRef]
10. Tchanche, B.F.; Lambrinos, G.; Frangoudakis, A.; Papadakis, G. Low-grade heat conversion into power using Organic Rankine Cycles—A review of various applications. *Renew. Sustai.n Energy Rev.* **2011**, *15*, 3963–3979. [CrossRef]
11. El-Emam, R.S.; Dincer, I. Exergy and exergoeconomic analyses and optimization of geothermal organic Rankine cycle. *Appl. Therm. Eng.* **2013**, *59*, 435–444. [CrossRef]
12. Xi, H.; Li, M.-J.; He, Y.-L. Parametric optimization of regenerative organic Rankine cycle (ORC) for low grade waste heat recovery using genetic algorithm. *Energy* **2013**, *58*, 473–482. [CrossRef]
13. Roy, J.P.; Misra, A. Parametric optimization and performance analysis of a regenerative Organic Rankine Cycle using R-123 for waste heat recovery. *Energy* **2012**, *39*, 227–235. [CrossRef]
14. Rashidi, M.M.; Galanis, N.; Nazari, F.; Basiri, P.; Shamekhi, L. Parametric analysis and optimization of regenerative Clausius and Organic Rankine Cycles with two feed-water heaters using artificial bees colony and artificial neural network. *Energy* **2011**, *36*, 5728–5740. [CrossRef]
15. Vatani, A.; Mehrpooya, M.; Palizdar, A. Advanced exergetic analysis of five natural gas liquefaction processes. *Energy Convers. Manag.* **2014**, *78*, 720–737. [CrossRef]
16. Fallah, M.; Siyahi, H.; Akbarpour Ghiasi, R.; Mahmoudi, S.M.S.; Yari, M.; Rosen, M.A. Comparison of different gas turbine cycles and advanced exergy analysis of the most effective. *Energy* **2016**, *116*, 701–715. [CrossRef]
17. Galindo, J.; Ruiz, S.; Dolz, V.; Royo-Pascual, L. Advanced exergy analysis for a bottoming organic rankine cycle coupled to an internal combustion engine. *Energy Convers. Manag.* **2016**, *126*, 217–227. [CrossRef]
18. Nami, H.; Nemati, A.; Jabbari Fard, F. Conventional and advanced exergy analyses of a geothermal driven dual fluid organic Rankine cycle (ORC). *Appl. Therm. Eng.* **2017**, *122*, 59–70. [CrossRef]
19. Imran, M.; Park, B.-S.; Kim, H.-J.; Lee, D.-H.; Usman, M.; Heo, M. Thermo-economic optimization of Regenerative Organic Rankine Cycle for waste heat recovery applications. *Energy Convers. Manag.* **2014**, *87*, 107–118. [CrossRef]
20. Anvari, S.; Taghavifar, H.; Parvishi, A. Thermo-economical consideration of Regenerative organic Rankine cycle coupling with the absorption chiller systems incorporated in the trigeneration system. *Energy Convers. Manag.* **2017**, *148*, 317–329. [CrossRef]
21. Hajabdollahi, H.; Ganjehkaviri, A.; Nazri Mohd Jaafar, M. Thermo-economic optimization of RSORC (regenerative solar organic Rankine cycle) considering hourly analysis. *Energy* **2015**, *87*, 369–380. [CrossRef]
22. Baccioli, A.; Antonelli, M.; Desideri, U. Technical and economic analysis of organic flash regenerative cycles (OFRCs) for low temperature waste heat recovery. *Appl. Energy* **2017**, *199*, 69–87. [CrossRef]

23. Chen, H.; Goswami, D.-Y.; Stefanakos, E.-K. A review of thermodynamic cycles and working fluids for the conversion of low-grade heat. *Renew. Sustain. Energy Rev.* **2010**, *14*, 3059–3067. [CrossRef]
24. Kotas, T.-J. *The Exergy Method of Thermal Plant*; Anchor Brendon: London, UK, 1985.
25. Wang, S.; Zhang, W.; Feng, Y.-Q.; Wang, X.; Wang, Q.; Liu, Y.-Z.; Wang, Y.; Yao, L. Entropy, Entransy and Exergy Analysis of a Dual-Loop Organic Rankine Cycle (DORC) Using Mixture Working Fluids for Engine Waste Heat Recovery. *Energies* **2020**, *13*, 1301. [CrossRef]
26. Khosravi, H.; Salehi, G.R.; Azad, M.T. Design of structure and optimization of organic Rankine cycle for heat recovery from gas turbine: The use of 4E, advanced exergy and advanced exergoeconomic analysis. *Appl. Therm. Eng.* **2019**, *147*, 272–290. [CrossRef]
27. Kelly, S.; Tsatsaronis, G.; Morosuk, T. Advanced exergetic analysis: Approaches for splitting the exergy destruction into endogenous and exogenous parts. *Energy* **2009**, *34*, 384–391. [CrossRef]
28. Rodríguez, C.E.; Palacio, J.C.; Venturini, O.J.; Lora, E.E.; Cobas, V.M.; Dos Santos, D.M.; Dotto, F.R.; Gialluca, V. Exergetic and economic comparison of ORC and Kalina cycle for low temperature enhanced geothermal system in Brazil. *Appl. Therm. Eng.* **2013**, *52*, 109–119. [CrossRef]
29. Petrakopoulou, F. Comparative Evaluation of Power Plants with CO_2 Capture: Thermodynamic, Economic and Environmental Performance. Ph.D. Thesis, Technische Universität Berlin, Berlin, Germany, 2011. [CrossRef]
30. Zare, V. A comparative exergoeconomic analysis of different ORC configurations for binary geothermal power plants. *Energy Convers. Manag.* **2015**, *105*, 127–138. [CrossRef]
31. Tsatsaronis, G.; Park, M. On avoidable and unavoidable exergy destructions and investment costs in thermal systems. *Energy Convers. Manag.* **2002**, *43*, 1259–1270. [CrossRef]
32. Fallah, M.; Mahmoudi, S.M.S.; Yari, M.; Ghiasi, R.A. Advanced exergy analysis of the Kalina cycle applied for low temperature enhanced geothermal system. *Energy Convers. Manag.* **2016**, *108*, 190–201. [CrossRef]
33. Petrakopoulou, F.; Tsatsaronis, G.; Morosuk, T.; Carassai, A. Conventional and advanced exergetic analyses applied to a combined cycle power plant. *Energy* **2012**, *41*, 146–152. [CrossRef]
34. Tsatsaronis, G.; Morosuk, T. Advanced exergetic analysis of a novel system for generating electricity and vaporizing liquefied natural gas. *Energy* **2010**, *35*, 820–829. [CrossRef]

Article

Inside–Out Method for Simulating a Reactive Distillation Process

Liang Wang [1], Xiaoyan Sun [1], Li Xia [1], Jianping Wang [2] and Shuguang Xiang [1,*]

[1] Institute for Process System Engineering, Qingdao University of Science and Technology, Qingdao 266042, China; wl60721102@126.com (L.W.); sun_xyan@163.com (X.S.); xiali@qust.edu.cn (L.X.)
[2] Petro-Cyber Works Information Technology Company Limited Shanghai Branch, Shanghai 200120, China; wangjianping178@163.com
* Correspondence: xsg@qust.edu.cn

Received: 30 March 2020; Accepted: 13 May 2020; Published: 19 May 2020

Abstract: Reactive distillation is a technical procedure that promotes material strengthening and its simulation plays an important role in the design, research, and optimization of reactive distillation. The solution to the equilibrium mathematical model of the reactive distillation process involves the calculation of a set of nonlinear equations. In view of the mutual influence between reaction and distillation, the nonlinear enhancement of the mathematical model and the iterative calculation process are prone to fluctuations. In this study, an improved Inside–Out method was proposed to solve the reaction distillation process. The improved Inside–Out methods mainly involved—(1) the derivation of a new calculation method for the K value of the approximate thermodynamic model from the molar fraction summation equation and simplifying the calculation process of the K value, as a result; and (2) proposal for an initial value estimation method suitable for the reactive distillation process. The algorithm was divided into two loop iterations—the outer loop updated the relevant parameters and the inside loop solved the equations, by taking the isopropyl acetate reactive distillation column as an example for verifying the improved algorithm. The simulation results presented a great agreement with the reference, and only the relative deviation of the reboiler heat duty reached 2.57%. The results showed that the calculation results were accurate and reliable, and the convergence process was more stable.

Keywords: reactive distillation; steady state simulation; numerical simulation; Inside–Out method

1. Introduction

Reactive distillation combines the chemical reaction and distillation separation process in a single unit operation [1]. As reactive distillation has the advantages of improving reaction conversion rate and selectivity and reducing costs, the research on design and optimization of the reactive distillation processes has increased rapidly in recent years [2]. Now reactive distillation technology has been successfully applied to the areas of esterification, etherification, hydrogenation, hydrolysis, isomerization, and saponification [3].

The production of methyl tert-butyl ether (MTBE), methyl tert-pentyl ether (TAME), and ethyl tert-butyl ether (ETBE), etc., adopts the reactive distillation technology, and the reaction conversion rate was significantly higher than the fixed-bed reactors [4]. Using the reactive distillation technology to produce methyl acetate can reduce investment and energy costs to one-fifth, and the reaction conversion rate is close to 100% [5]. Zhang et al. proposed a novel reaction and extractive distillation (RED) process and used it for the synthesis of isopropyl acetate (IPAc). The results showed that the purity of IPAc reached 99.5% [6].

As distillation and reaction take place simultaneously, there are complex interaction effects between them [7]. Therefore, it is very important to use mathematical methods to model and simulate the reactive distillation process, and optimize the operating conditions and equipment parameters [8]. The traditional algorithms for simulating reactive distillation include the tridiagonal matrix method, the relaxation method, the simultaneous correction method (Newton–Raphson), and the homotopy method.

Zhang et al. proposed a mathematical model of the heterogeneous catalytic distillation process. For the boundary value problem solved by the model, the multi-target shooting method and the Newton–Raphson iteration were used. The calculation results were consistent with the experimental data [9]. The partial Newton method proposed by Zhou et al. used the results from a repeated relaxation method as the initial value of the iteration variable, and then used the Newton–Raphson method to solve the MESHR equation (material balance, phase equilibrium, molar fraction summation, enthalpy balance, chemical reaction rate) [10]. Qi et al. aimed at the equilibrium reactive distillation process and transformed the physical variables such as the stage composition, flow, and enthalpy, such that the transformed mathematical model of the reactive distillation was completely consistent with the ordinary distillation and the modified Newton–Raphson method was used to solve it [11]. Juha introduced the homotopy parameters in the phase equilibrium, the enthalpy equilibrium, and the chemical reaction equilibrium equations, to establish a homotopy model of reactive distillation, and successfully simulated an MTBE reactive distillation column [12]. Wang et al. used the relaxation method to calculate the methyl formate hydrolysis catalytic distillation column. The simulation results were compared with the experimental ones and showed great agreement [13]. Steffen and Silva divided the reactive distillation model into several smaller sets of equations based on the equation tearing method, and obtained a set of linear and nonlinear equations each, which simulated chemical reactions [14].

Boston and Sullivan first proposed the Inside–Out method to simulate a multi-component ordinary distillation process [15,16]. Compared to other algorithms, two thermodynamic property models are used. The approximate thermodynamic models are used for frequent inside loop calculations to solve the MESH equation (material balance, phase equilibrium, molar fraction summation, enthalpy balance). The strict thermodynamic models are used for outer loop calculations of the phase equilibrium constant K and the approximate model parameters. This saves a lot of time when calculating thermodynamic properties. The inside loop uses a stripping factor (Sb) as an iterative variable, which integrates the effects of the stage temperature and the vapor and liquid flow rates, reduces the number of iterative variables, and improves the stability of convergence.

Russell further improved the Inside–Out method by solving the sparse tridiagonal matrix and applying the improved Thomas Chase method, and applied it to the calculation of crude oil distillation [17]. The modified Inside–Out method was proposed by Saeger, who added a two-parameter model for liquid composition, representing the effect on phase equilibrium constant, improving the convergence stability of highly non-ideal systems, and improving the convergence rate [18]. Jelinek simplified the Inside–Out method proposed by Russell; the simplified method could solve the distillation process specified by any operation and did not use an approximate enthalpy model in the inside loop, reducing the complexity of the calculation process [19].

Basing the Inside–Out method on simulating ordinary distillation process, increased the chemical reaction rate equation and established a mathematical model of the reactivedistillation. In view of the obvious enhancement of the nonlinearity of the mathematical model, it was more difficult to solve the problem than ordinary distillation [20]. This study used the Inside–Out method to solve the reactive distillation process, simplified the calculation of the K value of the approximate thermodynamic model, and proposed an initial value estimation method. This is an in-house model.

2. Mathematical Model

The schematic diagram of the reaction distillation column is shown in Figure 1. The whole column is composed of N stages. The condenser is regarded as the first stage and the reboiler is regarded as the last stage. There can be a liquid phase product, a vapor phase product, and a free water product at the top of the tower, and a liquid phase product at the bottom of the column. Chemical reactions can take place anywhere in the column. The general model of the jth theoretical stage is shown in Figure 2, where F_j is the feed flow rate of stage j, L_j is the liquid flow rate outputting stage j, and inputting stage j + 1, V_j is the vapor flow rate outputting stage j and inputting stage j − 1, U_j is the liquid side flow rate outputting stage j, W_j is the vapor side flow rate outputting stage j, Q_j is the heat duty from stage j, $R_{r,j}$ is the reaction extent of the rth reaction on stage j. The schematic representation from Figure 2 to represent all stages are shown in Figure 3.

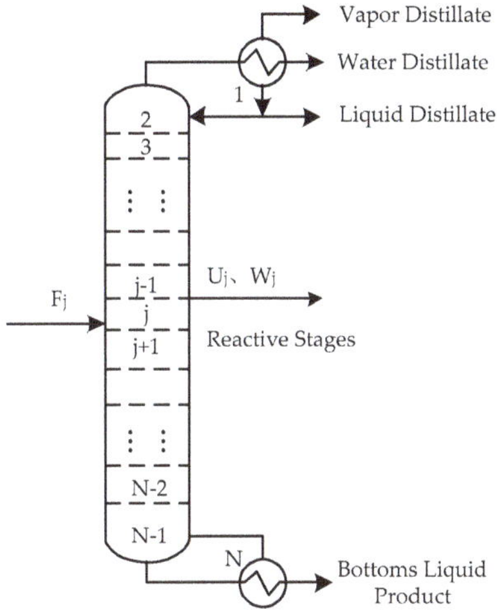

Figure 1. Schematic representation of the reactive distillation column.

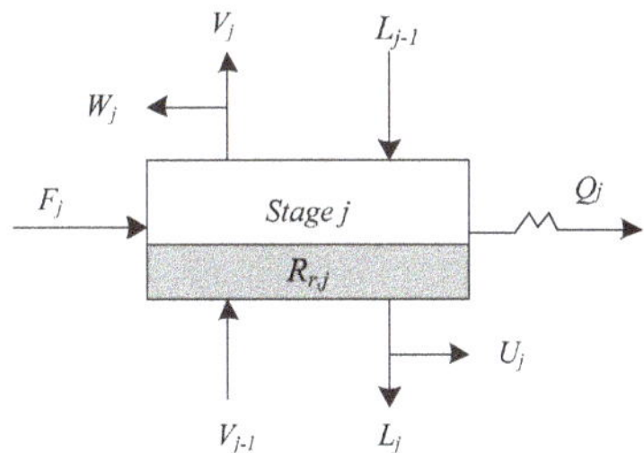

Figure 2. The jth equilibrium stage of the reactive distillation column.

The stages of the column are assumed to be in equilibrium conditions, the vapour and liquid phases leaving the stage are assumed to be in thermodynamic equilibrium, the stage pressure, temperature, flow, and composition are assumed to be constant at each stage. Five sets of equations are used to describe the equilibrium state of the stage—the material balances, the phase equilibrium, the molar fraction summation, the enthalpy balance, and the chemical equilibrium equations.

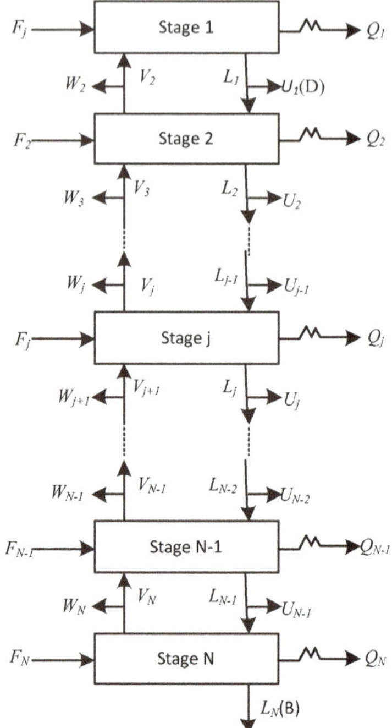

Figure 3. Schematic representation of input and output streams in all stages of a distillation column.

A set of highly nonlinear equations was obtained by modeling the steady state process of the reactive distillation column. The non-linearities were in the phase equilibrium, enthalpy balances, and chemical equilibrium equations. The solution of model equations is very difficult and the convergence depends strongly on the goodness of the initial guesses. After choosing appropriate iterative variables and providing the initial value of variables, the iterative calculation process can start. When the iterative calculation converges, the stage temperature, flow, and composition profiles can be obtained.

The mathematical model of reactive distillation [21]:

(1) Material balance Equation (M):

$$L_{j-1}x_{i,j-1} - (V_j + W_j)y_{i,j} - (L_j + U_j)x_{i,j} + V_{j+1}y_{i,j+1} + F_j z_{i,j} + \sum_r v_{r,i}R_{r,j} = 0, \tag{1}$$

(2) Phase equilibrium Equation (E):

$$y_{i,j} - K_{i,j}x_{i,j} = 0, \tag{2}$$

(3) Molar fraction summation Equation (S):

$$\sum_{i=1}^{c} y_{i,j} - 1 = 0 \quad \sum_{i=1}^{c} x_{i,j} - 1 = 0, \tag{3}$$

(4) Enthalpy balance Equation (H):

$$L_{j-1}h_{j-1} - (V_j + W_j)H_j - (L_j + U_j)h_j + V_{j+1}H_{j+1} + F_j H_{Fj} + Q_j - \Delta H_{r,j} = 0, \tag{4}$$

(5) Chemical reaction rate Equation (R):

$$r_{j,r} = f\Big(T_j, P_j, x_{i,j}, y_{i,j}\Big). \tag{5}$$

3. Algorithm

3.1. Inside–Out Method

The Inside–Out method divides the mathematical model into an inside and outer loop for iterative calculation, and uses two sets of thermodynamic models, a strict thermodynamic model and an approximate thermodynamic model (K value model and enthalpy value model). The simple approximate thermodynamic model was used for frequent inside loop calculations to reduce the time it takes to calculate the thermodynamic properties. The strict thermodynamic model was used for the outer loop calculations, and the parameters of the approximate thermodynamic model were corrected at the outer loop.

Defining the inside loop variables—relative volatility $\alpha_{i,j}$, stripping factors $S_{b,j}$ of the reference components, liquid phase stripping factors $R_{L,j}$, and the vapor phase stripping factors $R_{V,j}$.

$$\alpha_{i,j} = K_{i,j}/K_{b,j} \quad S_{b,j} = K_{b,j}V_j/L_j, \tag{6}$$

$$R_{L,j} = 1 + U_j/L_j \quad R_{V,j} = 1 + W_j/V_j. \tag{7}$$

The inside loop uses the vapor and liquid phase flow rates of each component instead of the composition, for the calculations. Their relationship with the vapor and liquid phase composition and flow is as follows:

$$y_{i,j} = v_{i,j}/V_j \quad x_{i,j} = l_{i,j}/L_j, \tag{8}$$

$$V_j = \sum_{i=1}^{c} v_{i,j} \quad L_j = \sum_{i=1}^{c} l_{i,j} \tag{9}$$

According to the inside loop variables defined by the mathematical model of the Inside–Out method, the material balance Equation (M) and the phase equilibrium Equation (E) can be rewritten into the following relationship.

Material Balance:

$$l_{i,j-1} - \big(R_{L,j} + \alpha_{i,j}S_{b,j}R_{V,j}\big)l_{i,j} + \big(\alpha_{i,j+1}S_{b,j+1}\big)l_{i,j+1} = -f_{i,j} + \sum_{r} v_{r,i}R_{r,j}. \tag{10}$$

Phase balance:

$$v_{i,j} = \alpha_{i,j}S_{b,j}l_{i,j}. \tag{11}$$

3.1.1. Strict Thermodynamic Model

The strict model is mainly used for the outer loop to correct the parameters in the approximate thermodynamic model, including the calculation of the phase equilibrium constant K value, and the vapor-liquid enthalpy difference, as follows:

$$K_{i,j} = K_{i,j}\Big(T_j, P_j, x_{i,j}, y_{i,j}\Big) \tag{12}$$

$$H_j = H_j\Big(T_j, P_j, y_{i,j}\Big) \tag{13}$$

$$h_j = h_j\Big(T_j, P_j, x_{i,j}\Big) \tag{14}$$

3.1.2. Approximate Thermodynamic Model

① *K*-value model

Define a new variable for the *K* value of the reference component, $K_{b,j}$, and the reference component b is an imaginary component. $K_{b,j}$ was calculated by the following formula, where $\omega_{i,j}$ is a weighting factor.

$$K_{b,j} = \exp\left(\sum_i \omega_{i,j} ln K_{i,j}\right),\tag{15}$$

$$\omega_{i,j} = \frac{y_{i,j}\left[\partial ln K_{i,j}/\partial(1/T)\right]}{\sum y_{i,j}\left[\partial ln K_{i,j}/\partial(1/T)\right]}.\tag{16}$$

The relationship between $K_{b,j}$ and the stage temperature T_j is related using Equation (17), T^* is the reference temperature.

$$ln K_{b,j} = A_j + B_j\left(\frac{1}{T} - \frac{1}{T^*}\right).\tag{17}$$

The value of the parameter B_j can be obtained by the differential calculation of $1/T$ by the K_b, and the temperatures T_{j-1} and T_{j+1} of the two adjacent stages are selected as T_1 and T_2.

$$B_j = \frac{\partial ln K_b}{\partial(1/T)} = \frac{ln\left(K_{b,T1}/K_{b,T2}\right)}{(1/T_1 - 1/T_2)},\tag{18}$$

$$A_j = ln K_{b,j} - B_j\left(\frac{1}{T} - \frac{1}{T^*}\right).\tag{19}$$

② Enthalpy model

The calculation of the enthalpy difference is the main time-consuming part of the entire enthalpy calculation, so the calculation of the enthalpy difference was simplified in the approximate enthalpy model, and a simple linear function was used to fit the calculation of the enthalpy difference.

$$\Delta H_j = c_j - d_j\left(T_j - T^*\right),\tag{20}$$

$$\Delta h_j = e_j - f_j\left(T_j - T^*\right).\tag{21}$$

3.1.3. Improved *K*-Value Model

The *K*-value model was improved by simplifying the calculation of the weighting factor. This was derived from the molar fraction summation equation. The derivation process is shown below.

The relationship between the *K* value of the reference component $K_{b,j}$ and the *K* value of the common component is:

$$ln K_{b,j} = \sum_i \omega_{i,j} ln K_{i,j}.\tag{22}$$

The weighting factors need to satisfy:

$$\sum \omega_{i,j} = 1.\tag{23}$$

The stage temperature needs to satisfy the following relationship:

$$\Phi\left(T_j\right) = \sum x_{i,j} - 1 = \sum \frac{y_{i,j}}{K_{i,j}} - 1 = 0.\tag{24}$$

The introduction of new variables $u_{i,j}$ was defined as:

$$u_{i,j} = ln\left(K_{i,j}/K_{b,j}\right). \tag{25}$$

The new relationship is:

$$\widetilde{\Phi}(T_j) = \sum \frac{y_{i,j}}{K_{b,j}e^{u_{i,j}}} - 1. \tag{26}$$

Since $K_{b,j}$ was calculated by $K_{i,j}$ through a weighting function, the trend with temperature changes was the same, so $u_{j,i}$ is a temperature-independent parameter, so at the stage temperature:

$$K_{b,j}(T_j)e^{u_{j,i}} = K_{j,i}(T_j), \tag{27}$$

$$\Phi(T_j) = \widetilde{\Phi}(T_j). \tag{28}$$

The derivative of $\widetilde{\Phi}(T_j)$ and $\widetilde{\Phi}(T_j)$ at the stage temperature was equal to:

$$\left. \frac{d\Phi}{dT} \right|_{T_j} = \left. \frac{d\widetilde{\Phi}}{dT} \right|_{T_j}, \tag{29}$$

$$\left. \sum \frac{y_{j,i}}{K_{j,i}} \left(\frac{\partial lnK_{j,i}}{\partial T} \right) \right|_{T_j} = \left(\sum \frac{y_{j,i}}{K_{j,i}} \right) \left. \frac{dlnK_{b,j}}{dT} \right|_{T_j}, \tag{30}$$

$$\left. \sum \frac{y_{j,i}}{K_{j,i}} \left(\frac{\partial lnK_{j,i}}{\partial T} \right) \right|_{T_j} = \left(\sum \frac{y_{j,i}}{K_{j,i}} \right) \sum \omega_{j,i} \left. \frac{\partial lnK_{j,i}}{\partial T} \right|_{T_j}. \tag{31}$$

The weighting function used the new calculation method:

$$\omega_{i,j} = \frac{y_{i,j}/K_{i,j}}{\sum y_{i,j}/K_{i,j}}. \tag{32}$$

3.2. Initial Value Estimation Method

The system of equations obtained from the modeling of the steady-state reactive distillation process was non-linear and was very difficult to solve. Therefore, it was important to provide good initial estimates, otherwise it would not be possible to reach a solution.

Before providing good initial estimates, all feed streams were mixed to perform chemical reaction equilibrium calculations to obtain a new set of compositions and flow rate, which were used to calculate the initial value of the variable.

Temperature: Calculate the dew and bubble point temperature by flash calculation, which was used as the initial temperature at the top and at the bottom of the tower. The initial temperature of the middle stages was obtained by linear interpolation.

Composition: The composition calculated by isothermal flash calculation was used as the initial value of the vapor-liquid composition of each stage.

Flow: According to the constant molar flow rate assumption and the total material balance calculation, the initial value of the vapor–liquid flow rate of each stage was obtained

Reaction extent: The reaction extent calculated by the chemical reaction equilibrium was taken as the maximum value of the reaction extent in the reaction section. According to the feeding condition of the reaction section, the position with the maximum reaction extent was selected. If there was only one feed in the reaction section, the position of the feed stage was selected. If there were multiple feeds in the reaction section, the position of the feed stage near the reboiler was chosen. If there was no feed in the reaction section, the position of the reaction stage closest to the feed was selected. The initial value of the reaction extent of the other stages were calculated according to the maximum value and a certain proportion of attenuation. The calculation is shown below.

Between the first reaction stage and the stage with the maximum reaction extent:

$$R_{j,r} = R_{\max} \times \frac{j - j_1 + 1}{(j_m - j_1 + 1)^2}.$$ (33)

Between the stage with the maximum reaction extent and the last reaction stage:

$$R_{j,r} = R_{\max} \times \frac{j_n - j + 1}{(j_n - j_m + 1)^2}.$$ (34)

where $R_{\max}$ is the maximum value of the reaction extent in the reaction section, j_m is the stage position where the maximum reaction extent occurs; j_1 is the first stage position in the reaction section; and j_n is the last stage position in the reaction section.

3.3. Calculation Steps and Block Diagram

The reactive distillation column adopts an improved Inside–Out method for calculation. The main working idea is that the the outer loop used a strict thermodynamic model to calculate the phase equilibrium constant and the vapor–liquid phase enthalpy difference, and the results were used to correct the approximate thermodynamic model parameters and update the reaction extents of each stage. The inside loop uses an approximate thermodynamic model to solve the MESHR equation to obtain the stage temperature, flow rate, and composition. After the inside loop calculation converges or reaches the number of iterations, the model returns to the outer loop to continue the calculation. When the inside and outer loop converges at the same time, the calculation ends.

The calculation steps are shown below:

1. Given initial value

(1) According to the feed, pressure, and operation specifications, the initial values of the temperature, flow rate, composition, and the reaction extent are provided (Section 3.2).

2. Outer loop iteration

(2) Calculate the phase equilibrium constant $K_{i,j}$ with a strict thermodynamic model, and then calculate parameters $K_{b,j}$, A, and B in the K-value model.

(3) Calculate the vapor–liquid phase enthalpy difference using a strict thermodynamic model, and fit the parameters c_j, d_j, e_j, and f_j in the approximate enthalpy model.

(4) Calculate the stripping factors $S_{b,j}$, relative volatility $\alpha_{i,j}$, liquid phase stripping factors $R_{L,j}$, and vapor phase stripping factors $R_{V,j}$.

(5) Calculate the reaction extent according to the reaction conditions. The chemical reaction can be a specified conversion rate or a kinetic reaction, and the kinetic reaction rate is calculated by a power law expression.

Power law expression:

T_0 is not specified:

$$r = kT^n e^{-E/RT} \prod (C_i)^{\alpha_i}.$$ (35)

T_0 is specified:

$$r = k(T/T_0)^n e^{-(E/R)(1/T - 1/T_0)} \prod (C_i)^{\alpha_i},$$ (36)

$$R_{j,r} = r_{j,r} \times \Omega_j.$$ (37)

(6) Outer convergence judgment conditions:

$$\sum_{j}^{n}\left(\frac{K_{b,j}^{n}-K_{b,j}^{n-1}}{K_{b,j}^{n-1}}\right)^{2}+\sum_{j}^{n}\sum_{i}^{c}\left(\frac{\alpha_{i,j}^{n}-\alpha_{i,j}^{n-1}}{\alpha_{i,j}^{n-1}}\right)^{2}+\sum_{j}^{n}\sum_{r}\left(\frac{R_{r,j}^{n}-R_{r,j}^{n-1}}{R_{r,j}^{n-1}}\right)^{2}<\varepsilon. \tag{38}$$

3. Inside loop iteration

(7) According to the material balance Equation (13), a tridiagonal matrix was constructed, and then the liquid flow rate $l_{i,j}$ of each component was obtained by solving the matrix. Using the phase equilibrium Equation (14), the vapor phase flow rate $v_{i,j}$ of each component was obtained.

(8) The vapor–liquid flow rates V_j and L_j can be calculated by Equation (9), the vapor–liquid phase composition $x_{i,j}$ and $y_{i,j}$ can be calculated by Equation (8).

(9) Combining the bubble point equation $\left(\sum_{i=1}^{c} K_i x_i = 1\right)$ with Equation (9) to obtain Equation (39), a new set of K values $K_{b,j}$ of the reference components can be calculated.

$$K_{b,j} = \frac{1}{\sum_{i=1}^{c}\left(\alpha_{i,j} x_{i,j}\right)}. \tag{39}$$

According to Equation (40), a new set of stage temperature can be calculated using the new $K_{b,j}$ values.

$$T_j = \frac{1}{\left(\ln K_{b,j} - A_j\right)/B_j + 1/T^*}. \tag{40}$$

Now, the correction values of the vapor–liquid flow rate, composition, and temperature are obtained. They satisfy the material balance and phase equilibrium equation, but does not satisfy the enthalpy equilibrium equation. In the following, the iterative variable of the inside loop should be modified according to the deviation of the enthalpy balance Equation (10). Select $\ln S_{b,j}$ as the inside loop iteration variable. If there are side products, the side stripping factors $\ln R_{L,j}$ and $\ln R_{V,j}$ should be added as the inside loop iteration variables. Taking the enthalpy balance equation as the objective function, the enthalpy balance equation of the first and last stage should be deleted and replaced with two operation equations of the reactive distillation column.

(11) Calculate the vapor–liquid phase enthalpy according to the simplified enthalpy model.

(12) Calculate the partial derivative of the objective function to the iterative variable and construct the Jacobian matrix. The Schubert method was used to calculate the modified value of the inside iteration variable [22,23]. Damping factors can be used if necessary.

(13) Using the new inside iteration variable value, repeat steps (7) to (9) to obtain the new stage flow, composition, and temperature, and calculate the error of the objective function. If the error was less than the convergence accuracy of the inside loop, return to the outer loop to continue the calculation, otherwise repeat the calculation from Step (11) to Step (13). The calculation block diagram of the improved Inside–Out method is shown in Figure 4.

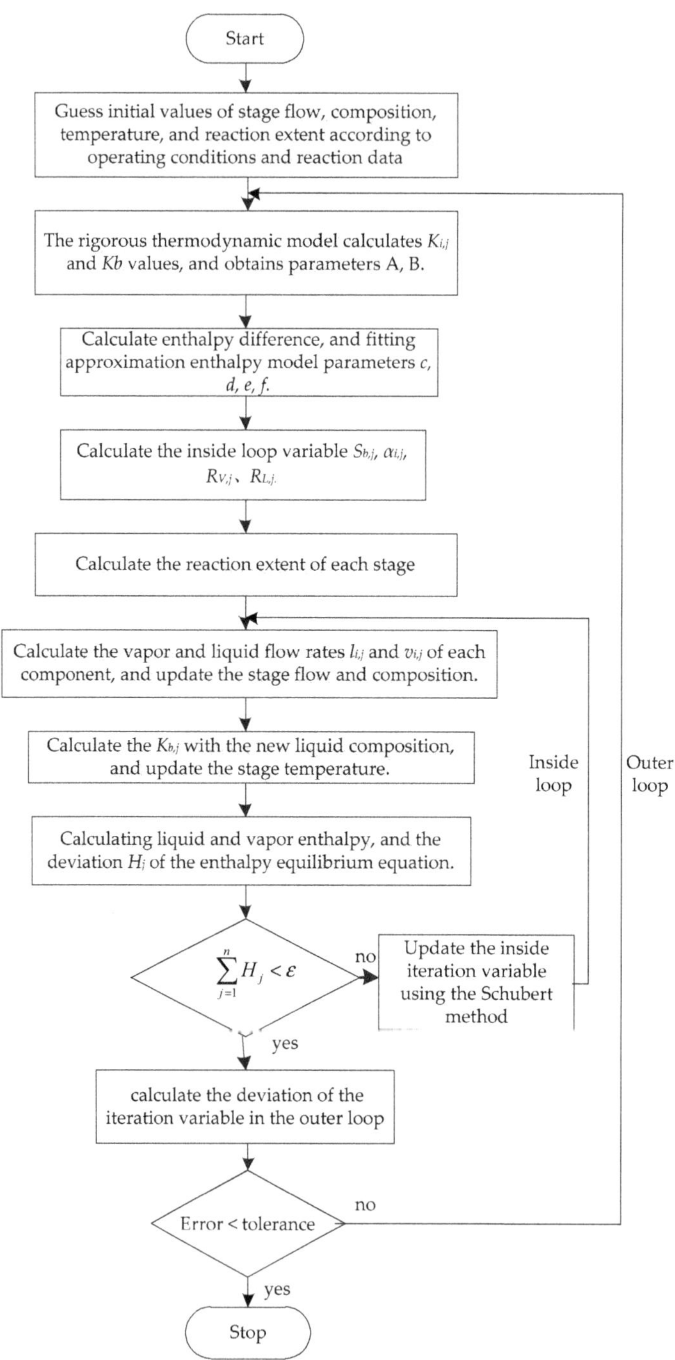

Figure 4. Calculation of the block diagram of the improved Inside–Out method.

4. Results

4.1. Example 1: Isopropyl Acetate

Taking the process of synthesizing isopropyl acetate (IPAc) with acetic acid (HAc) and isopropanol (IPOH) as an example, the esterification reaction is a reversible exothermic reaction, and the conversion rate is limited by the chemical equilibrium. Zhang et al. proposed reactive and extractive distillation technology for the synthesis of IPAc, and the extraction agent dimethyl sulfoxide (DMSO) was added in the tower to obtain high-purity IPAc [6]. Zhang et al. conducted a steady-state simulation of the reactive distillation column, and performed dynamic optimization control, based on the results. In this work, the improved Inside–Out method was used to solve the esterification process, and the steady-state simulation results were compared with that of Zhang et al. [6]. The chemical reaction equation of this esterification reaction was as follows:

$$CH_3COOH + (CH_3)_2CHOH \rightleftharpoons CH_3COOCH(CH_3)_2 + H_2O$$

Kong et al. determined the kinetic model of the esterification reaction through kinetic experiments [24]. The kinetic expressions and parameters are as follows:

$$r = k^+ C_{HAC} C_{IPOH} - k^- C_{IPAC} C_{H_2O}, \tag{41}$$

$$k^+ = 2589.1 \exp\left(-\frac{14109}{RT}\right) k^- = 2540.2 \exp\left(-\frac{18890}{RT}\right). \tag{42}$$

where r is the reaction rate, $mol \cdot L^{-1} \cdot min^{-1}$; C_i is the molar volume concentration of the i component, $mol \cdot L^{-1}$; k^+, and k^- are the forward and reverse reaction rate constants; R is the gas constant, with a value of 8.314 $J \cdot mol^{-1} \cdot K^{-1}$; and T is the temperature, in K.

The improved Inside–Out method was used to simulate the isopropyl acetate process. The property method was non-random two liquid (NRTL). The operating conditions of the isopropyl acetate reactive distillation column is shown in Table 1.

Table 1. Operational conditions of the reactive distillation column.

Variables	Specifications
Reaction section	1218– stages
Condenser pressure	0.4 atm
Stage pressure drop	0.689 kPa
Distillate rate	9.78 $kmol \cdot h^{-1}$
Reflux ratio	2.98
Liquid holdup	0.2 m^3
Condenser type	Total
Reboiler type	Kettle

The schematic diagram of the isopropyl acetate reactive distillation column and the parameters of the feed stream are shown in Figure 5.

Figure 6 is a comparison between the initial value of the reaction extent and the simulation result. It can be seen that the initial value and the simulation results were relatively close, which increased the possibility of convergence. The error behavior at each iteration, calculated by Equation (38), for the program using the improved Inside–Out method is shown in Figure 7. The outer loop calculation reached the convergence through 10 iterations, and the error decreased steadily, without oscillation.

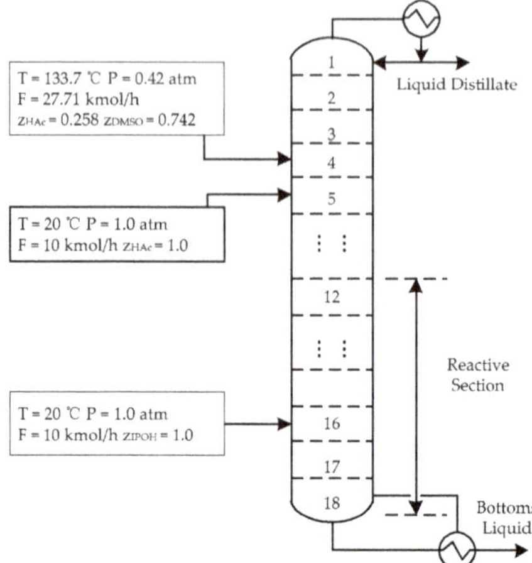

Figure 5. Schematic diagram of the isopropyl acetate reactive distillation column.

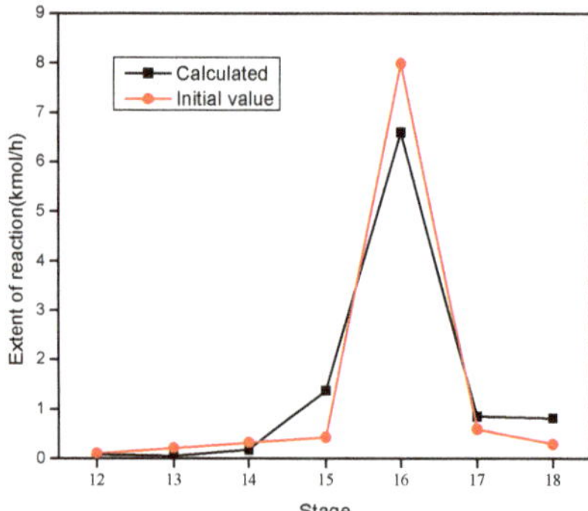

Figure 6. Comparison of the initial values of the extent of reaction and simulation results.

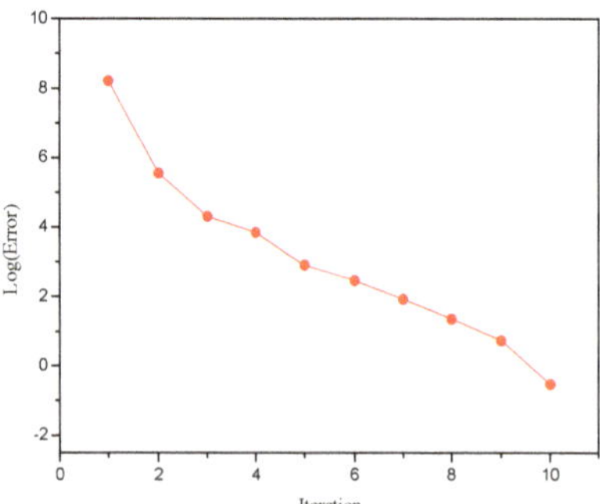

Figure 7. Error at each iteration for the simulation of isopropyl acetate reactive distillation column.

Table 2 shows the comparison between the simulation results of the isopropyl acetate reactive distillation column and the literature. It can be seen that the temperatures at the top and bottom of tower obtained in this work were similar to that in the literature, and the relative deviation was within

1%. The relative deviation of the condenser heat duty was 1.82% and the relative deviation of the reboiler heat duty was 2.47%.

Table 2. Comparison of the simulation results of Example 1.

Variables	Simulated	Zhang et al. [6]	δ1%
Temperature of top tower/°C	61.58	62.00	0.68
Temperature of bottom tower/°C	112.92	112.57	0.31
Condenser heat duty/kW	−382.71	−375.88	1.82
Reboiler heat duty/kW	312.11	304.58	2.47
Distillate of top tower/kmol·h^{-1}	9.78	9.78	0.00

Table 3 shows the comparison between the simulation results of the molar composition of the top and bottom products and the literature. It can be seen that the molar composition of the top and bottom product obtained in this work presented a great agreement with the literature, except for the smaller part, and the simulation results were accurate and reliable.

Table 3. Comparison of product composition of Example 1.

		Simulated	Zhang et al. [6]	δ1%
	IPOH	0.0022	0.0023	4.35
	HAc	0.0048	0.0047	2.13
Mole Fraction (top tower)	IPAc	0.9905	0.9907	0.02
	H2O	0.0023	0.0022	4.55
	DMSO	0.0001	0.0001	0.00
	IPOH	0.0016	0.0017	5.88
	HAc	0.1886	0.1898	0.63
Mole Fraction (bottom tower)	IPAc	0.0060	0.0057	5.26
	H2O	0.2628	0.2609	0.73
	DMSO	0.5410	0.5419	0.17

4.2. Example 2: Propadiene Hydrogenation

Taking the depropanization column in the propylene recovery system of the ethylene production process as an example. The Inside–Out method was used to simulate the depropanizer column, and the results were compared with the measured values. The top of the depropanizer column was equipped with a catalyst. In the reaction section, propadiene was hydrogenated into propylene and propane. The conversion rate of the first reaction was 0.485, the conversion rate of the second reaction was 0.015. The chemical reaction equation were as follows:

$$C_3H_4 + H_2 \rightarrow C_3H_6$$
$$C_3H_4 + 2H_2 \rightarrow C_3H_8$$

The improved Inside–Out method was used to simulate the depropanization column. The property method was Soave–Redlich–Kwong (SRK). The depropanizer column had 3 feeds, and the feed stream parameters are shown in Table 4. The operating conditions of the depropanization column are shown in Table 5.

Table 4. The parameters of feed stream for the depropanization column.

Variables	Feed 1	Feed 2	Feed 3
Temperature/°C	21.0	45.0	56.5
Pressure/kPa	3043.32	2200.32	2333.32
Feed stage	31	33	37
Flow/kmol·h^{-1}	42.01	1329.96	706.03
Mole fraction			
H2	0.9574	0.0000	0.0000
N2	0.0010	0.0000	0.0000
CH4	0.0416	0.0000	0.0000
C2H6	0.0000	0.0005	0.0001
C3H6	0.0000	0.7031	0.7025
C3H8	0.0000	0.0266	0.0551
C3H4-1	0.0000	0.0257	0.0694
C4H8-1	0.0000	0.0874	0.0763
C4H6-4	0.0000	0.1194	0.0950
C4H10-1	0.0000	0.0025	0.0014
C5H10-2	0.0000	0.0256	0.0002
C6H12-3	0.0000	0.0012	0.0000
C6H6	0.0000	0.0069	0.0000
C7H14-7	0.0000	0.0007	0.0000
C7H8	0.0000	0.0003	0.0000

Table 5. Operational conditions for the depropanization column.

Variables	Specifications
Numbers of stages	42
Reaction section	13–stages
Condenser pressure	20.2 atm
Stage pressure drop	1.325 kPa
Bottoms rate	854.8 kmol·h^{-1}
Reflux ratio	0.98
Condenser type	Partial-Vapor-Liquid
Reboiler type	Kettle

Table 6 shows the comparison between the simulation results of the depropanization column and the measured value. It can be seen that the temperatures in the top and bottom products, as calculated by the Inside–Out method were close to the measured value, and the relative error was about 1%. The liquid product flow rate at the top and bottom of the tower coincided with the measured value, only the relative deviation of the vapor product flow at the top of the tower was 2.74%. Table 7 shows the comparison between the mass composition of the top and bottom liquid products and the measured value. The simulation results of the product mass compositions were basically consistent with the measured value, except for the smaller composition values.

Table 6. Comparison of the simulation results of Example 2.

Variables	Simulated	Measured	δ%
Temperature of top tower/°C	49.0	48.5	1.03
Temperature of bottom tower/°C	79.4	79.0	0.51
Condenser heat duty/kW	−12,167.0		
Reboiler heat duty/kW	11,486.7		
Distillate rare of liquid/kg·h^{-1}	49,515.5	49,625.0	0.22
Distillate rare of vapor/kg·h^{-1}	248.7	255.7	2.74
Bottoms product flow/kg·h^{-1}	42,523.3	42,507.0	0.04

Table 7. Comparison of the mass compositions of the product flow of Example 2.

Component	Mass Fraction (Distillate Liquid)			Mass Fraction (Bottom)		
	Simulated	Measured	$\delta 1\%$	Simulated	Measured	$\delta 2\%$
H2	0.0002	0.0002	1.53	0.0000	0.0000	0.00
N2	0.0000	0.0000	0.00	0.0000	0.0000	0.00
CH4	0.0005	0.0005	3.42	0.0000	0.0000	0.00
C2H6	0.0004	0.0004	1.57	0.0000	0.0000	0.00
C3H6	0.9575	0.9596	0.22	0.3225	0.3204	0.66
C3H8	0.0413	0.0392	5.36	0.0313	0.0322	2.80
C3H4-1	0.0001	0.0000	0.00	0.0521	0.0502	3.78
C4H8-1	0.0000	0.0000	0.00	0.2263	0.2243	0.89
C4H6-4	0.0000	0.0000	0.00	0.2854	0.2873	0.66
C4H10-1	0.0000	0.0000	0.00	0.0054	0.0060	10.00
C5H10-2	0.0000	0.0000	0.00	0.0546	0.0564	3.19
C6H12-3	0.0000	0.0000	0.00	0.0030	0.0032	6.25
C6H6	0.0000	0.0000	0.00	0.0162	0.0170	4.71
C7H14-7	0.0000	0.0000	0.00	0.0022	0.0022	0.91
C7H8	0.0000	0.0000	0.00	0.0009	0.0009	7.12

5. Conclusions

Based on the process principle of reactive distillation, a mathematical model of reactive distillation process was established. The improved Inside–Out method was provided in this work for the solution of a reactive distillation process. In view of the nonlinear enhancement of the reactive distillation mathematical model and the difficulty of convergence, the calculation of the K value of the approximate thermodynamic model was improved to simplify the calculation process. The initial value estimation method suitable for the calculation of reactive distillation was proposed, which increased the possibility of convergence.

The algorithm was verified by using an isopropyl acetate reactive distillation column and a depropanization column as examples. In Example 1, the reaction extent calculated by the initial value estimation method was close to the simulation results, it facilitated the convergence process of the solution algorithm. The simulation results obtained in this work were compared with Zhang et al. [6], presenting great agreement with the reference, only the relative deviation of the reboiler heat duty reached 2.57%. In Example 2, the simulation results of the depropanization column presented a good agreement with the measured value, except for the smaller composition value. The results showed that the improved Inside–Out method calculation results were accurate and reliable.

Author Contributions: Conceptualization, X.S.; Data curation, J.W.; Formal analysis, S.X.; Project administration, L.X.; Writing—original draft, L.W.;Writing—review & editing, L.W. All authors have read and agreed to the published version of the manuscript.

Funding: This research was funded by Major Science and Technology Innovation Projects in Shandong Province, grant number [2018CXGC1102]; and The APC was funded by the Science and Technology Department of Shandong Province.

Conflicts of Interest: The authors declare no conflict of interest.

Abbreviations

A:B	Parameters of the Approximate K-Value Model
C	Molar volume concentration
c,d,e,f	Parameters of the approximate enthalpy model
DMSO	Dimethyl sulfoxide
ETBE	Ethyl tert-butyl ether
E	Activation energy, $J \cdot mol^{-1}$
F	Feed flow, $mol \cdot s^{-1}$
H	vapor phase enthalpy, $J \cdot mol^{-1}$

h	Liquid phase enthalpy, J·mol^{-1}
ΔH	Vapor phase enthalpy difference, J·mol^{-1}
Δh	Liquid phase enthalpy difference, J·mol^{-1}
HAc	Acetic acid
IPAc	Isopropyl acetate
IPOH	Isopropanol
k^+	Forward reaction rate constants
k^-	Reverse reaction rate constants
K	Phase equilibrium constant
K_b	Phase equilibrium constant of reference component b
L	Liquid flow, mol·s^{-1}
MESH	Material balance, Phase equilibrium, Molar fraction summation, enthalpy balance
MESHR	Material balance, Phase equilibrium, Molar fraction summation, enthalpy balance, Chemical reaction rate
MTBE	Methyl tert-butyl ether
N	The number of distillation stages
NRTL	Non-random two liquid
R	Reaction extents, mol·s^{-1}
R_L	Liquid stripping factor
R_V	Vapor stripping factor
r	Reaction rate
S_b	Stripping factor for reference component b
T	Temperature, K
T^*	Reference temperature, K
TAME	Methyl tert-pentyl ether
U	Liquid side product flow, mol·s^{-1}
V	Vapor flow, mol·s^{-1}
W_j	Vapor side product flow, mol·s^{-1}
x	Liquid mole fraction
y	Vapor mole fraction
z	Feed mole fraction
υ	Stoichiometric coefficient
Ω	Liquid holdup, m^3
α	Relative volatility
ε	Tolerance
ω	Weighting factor
i	Component i
j	Stage j

References

1. Gao, X.; Zhao, Y.; Li, H.; Li, X.G. Review of basic and application investigation of reactive distillation technology for process intensification. *CIESC J.* **2018**, *69*, 218–238.
2. Li, C.L.; Duan, C.; Fang, J.; Li, H. Process intensification and energy saving of reactive distillation for production of ester compounds. *Chin. J. Chem. Eng.* **2019**, *27*, 1307–1323. [CrossRef]
3. Ling, X.M.; Zheng, W.Y.; Wang, X.D.; Qiu, T. Advances in technology of reactive dividing wall column. *Chem. Ind. Eng. Prog.* **2017**, *36*, 2776–2786.
4. Ma, J.H.; Liu, J.Q.; Li, J.T.; Peng, F. The progress in reactive distillation technology. *Chem. React. Eng. Technol.* **2003**, *19*, 1–8.
5. Steffen, V.; Silva, E.A. Numerical methods and initial estimates for the simulation of steady-state reactive distillation columns with an algorithm based on tearing equations methodology. *Therm. Sci. Eng. Prog.* **2018**, *6*, 1–13. [CrossRef]
6. Zhang, Q.; Yan, S.; Li, H.Y.; Xu, P. Optimization and control of a reactive and extractive distillation process for the synthesis of isopropyl acetate. *Chem. Eng. Commun.* **2019**, *206*, 559–571. [CrossRef]

7. Tao, X.H.; Yang, B.L.; Hua, B. Analysis for fields synergy in the reactive distillation process. *J. Chem. Eng. Chin. Univ.* **2003**, *17*, 389–394.

8. Lee, J.H.; Dudukovic, M.P. A Comparison of the Equilibrium and Nonequilibrium Models for a Multi-Component Reactive Distillation Column. *Comput. Chem. Eng.* **1998**, *23*, 159–172. [CrossRef]

9. Zhang, R.S.; Han, Y.U. Hoffmann. Simulation procedure for distillation with heterogeneous catalytic reaction. *CIESC J.* **1989**, *40*, 693–703.

10. Zhou, C.G.; Zheng, S.Q. Simulating reactive-distillation processes with partially newton method. *Chem. Eng. (China)* **1994**, *3*, 30–36.

11. Qi, Z.W.; Sun, H.J.; Shi, J.M.; Zhang, R.; Yu, Z. Simulation of distillation process with chemical reactions. *CIESC J.* **1999**, *50*, 563–567.

12. Juha, T.; Veikko, J.P. A robust method for predicting state profiles in a reactive distillation. *Comput. Chem. Eng.* **2000**, *24*, 81–88.

13. Wang, C.X. Study on hydrolysis of methyl formate into formic acid in a catalytic distillation column. *J. Chem. Eng. Chin. Univ.* **2006**, *20*, 898–903.

14. Steffen, V.; Silva, E.A. Steady-state modeling of reactive distillation columns. *Acta Sci. Technol.* **2012**, *34*, 61–69. [CrossRef]

15. Boston, J.F.; Sullivan, S.L. A new class of solution methods for multicomponent multistage separation processes. *Can. J. Chem. Eng.* **1974**, *52*, 52–63. [CrossRef]

16. Boston, J.F. Inside-out algorithms for multicomponent separation process calculations. *ACS Symp. Ser.* **1980**, *9*, 135–151.

17. Russell, R.A. A flexible and reliable method solves single-tower and crude-distillation-column problems. *Chem. Eng. J.* **1983**, *90*, 53–59.

18. Saeger, R.B.; Bishnoi, P.R. A modified "Inside-Out" algorithm for simulation of multistage multicomponent separation process using the UNIFAC group-contribution method. *Can. J. Chem. Eng.* **1986**, *64*, 759–767. [CrossRef]

19. Jelinek, J. The calculation of multistage equilibrium separation problems with various specifications. *Comput. Chem. Eng.* **1988**, *12*, 195–198. [CrossRef]

20. Lei, Z.; Lu, S.; Yang, B.; Wang, H.; Wu, J. Multiple steady states analysis of reactive distillation process by excess-entropy production (I) Establishment of stability model on steady state. *J. Chem. Eng. Chin. Univ.* **2008**, *22*, 563–568.

21. Fang, Y.J.; Liu, D.J. A reactive distillation process for an azeotropic reaction system: Transesterification of ethylene carbonate with methanol. *Chem. Eng. Commun.* **2007**, *194*, 1608–1622. [CrossRef]

22. Schubert, L.K. Modification of a quasi-Newton method for nonlinear equations with a sparse Jacobian. *Math. Comput.* **1970**, *24*, 27–30. [CrossRef]

23. Liu, S.P. Compact Representation of Schubert Update. *Commun. Appl. Math. Comput.* **2001**, *1*, 51–58.

24. Kong, S. *The New Technology of Isopropyl Acetate Using Catalytic Distillation*; Yantai University: Yantai, China, 2014; pp. 39–42.

Article

A Hybrid Inverse Problem Approach to Model-Based Fault Diagnosis of a Distillation Column

Suli Sun [1], Zhe Cui [2], Xiang Zhang [3] and Wende Tian [2,*]

[1] College of Marine Science and Biological Engineering, Qingdao University of Science & Technology, Qingdao 266042, China; qdsunsuli@qust.edu.cn
[2] College of Chemical Engineering, Qingdao University of Science & Technology, Qingdao 266042, China; cuizhequst@126.com
[3] Wanhua Chemical Rongwei Polyurethane CO., LTD, Yantai 264000, China; zhangxiangd@whchem.com
* Correspondence: tianwd@qust.edu.cn

Received: 29 November 2019; Accepted: 30 December 2019; Published: 2 January 2020

Abstract: Early-stage fault detection and diagnosis of distillation has been considered an essential technique in the chemical industry. In this paper, fault diagnosis of a distillation column is formulated as an inverse problem. The nonlinear least squares algorithm is used to evaluate fault parameters embedded in a nonlinear dynamic model of distillation once abnormal symptoms are detected. A partial least squares regression model is built based on fault parameter history to explicitly predict the development of fault parameters. With the stripper of Tennessee Eastman process as example, this novel approach is tested for step- and random-type faults and several factors affecting its efficiency are discussed. The application result shows that the hybrid inverse problem approach gives the correct change of fault parameter at a speed far faster than the base approach with only a nonlinear model.

Keywords: fault diagnosis; distillation; inverse problem; parameter estimation

---

1. Introduction

Distillation is a widely used energy consuming unit operation in modern petroleum and chemical industries. Its separation on raw materials, intermediate products, and crude products exerts a strong influence on the energy consumption and product quality of the industrial process involved. The fault diagnosis technique benefits from catching the deterioration symptom of critical parameters in time and predicting their deterioration trend effectively when a distillation column enters an abnormal state. To improve the reliability of control systems, the problem of fault detection and diagnosis has been paid more attention over the past two decades. A robust fault diagnostic method for multiple open-circuit faults and current sensor faults in three-phase permanent magnet synchronous motor drives has been proposed by Jlassi [1]. The proposed observer-based algorithm relies on an adaptive threshold for fault diagnosis. A composite fault tolerant control (CFTC) with disturbance observer scheme has been considered for a class of stochastic systems with faults and multiple disturbances [2]. The problem of fault diagnosis for a class of nonlinear systems has been investigated via the hybrid method of an observer-based approach and homogeneous polynomials technique [3]. As one of the quantitative model-based fault diagnosis methods for the distillation process, the parameter estimation method identifies the process parameters of physical meaning, such as the heat transfer coefficient, thermal resistance, etc., and hereby explains abnormal reasons based on the relationship between input and output signals. From the point of view of control theory, the parameter estimation method provides a closed loop structure with a good computational stability and convergence since parameters are input into the reasoning model again after estimation. Based on a nonlinear dynamic model, parameter estimation, however, becomes a nonlinear optimization problem, whose heavy computational load is a serious bottleneck limiting its application. Some improvements have therefore emerged in the last decade to simplify both the model and algorithm. For example, a hybrid fault detection and diagnosis scheme was implemented in a two-step pattern, that is, neural networks were activated to deduce the root reason for a fault state after the fault-related section of a plant was located by a Petri net [4]. A fault diagnosis technique was proposed based on multiple linear models, in which several linear perturbation models suitable for various operation regimes were identified by a Bayesian approach and then combined with a generalized likelihood ratio method to perform fault identification tasks [5]. A fault detection and diagnosis scheme, which uses one tier of a nonlinear rigorous model and another tier of a linear simplified model to monitor the distillation process and identify abnormal parameters, respectively, was developed to consider the accuracy and speed of nonlinear and linear models simultaneously [6]. A three-layer nonlinear Gaussian belief network was constructed and trained to extract useful features from noisy process data, where the absolute gradient was monitored for fault detection and a multivariate contribution plot was generated for fault diagnosis [7].

The determination of fault parameters on the basis of known input and output signals of distillation can be approached as the solution of an inverse problem [8–10]. The most widely used method to solve such a problem is its least squares (LSQ) formulation as the minimization of an error function between the real measurements and their calculated values, similar to the above improved parameter estimation methods. Meanwhile, meta-heuristics for LSQ optimization are popular due to their inherent advantages, like their global optimum and the few requirements for problem formulation [11,12]. However, the running speed of the LSQ-based method is slow owing to its time-consuming iterative optimization of fault parameters. During iterations, the process of solving the forward problem and then adjusting its parameters is repeated until the difference between the measured values and the calculated ones reaches a minimum. For this reason, a direct derivation of fault parameters from input and output signals instead of trial and error in forward problems is a very current topic, such as the decomposition of solution space followed by polynomial approximation [13], usage of artificial neural networks (ANN) for the automated reconstruction of an inhomogeneous object as pattern recognition [14], and inverse regression between the disturbance and characteristic distances based on single variable perturbation [15].

Based on the LSQ-based fault diagnosis work [16], a hybrid inverse problem approach that uses partial least squares (PLS) to fit and forecast trajectories of the fault parameters generated by LSQ is proposed in this paper to accelerate the model-based fault diagnosis process. PLS is a popular tool for key performance monitoring, quality control, and fault diagnosis in large-scale chemical industry [17,18]. It has been improved from relative contributions of process variables or blocks on faults [19,20], and the orthogonal decomposition of measurement space before deducing new specific statistics with non-overlapped domains [21].

This work aims to test the feasibility of an LSQ and PLS combined hybrid inverse problem approach for model-based fault diagnosis. In the following sections, the proposed hybrid inverse problem solving approach is explained and its positive action in terms of speeding up the diagnosis process is proved via a case study of a stripper simulator in the Tennessee-Eastman process (TEP) compared to the base approach with LSQ only. The effect of initial values, iteration times, calling period, etc. on the performance of the proposed method is also analyzed.

2. Hybrid Fault Diagnosis Structure

Figure 1 shows the structure of fault diagnosis formulated as a parameter-estimation inverse problem solved by the least squares optimization algorithm. The first step is fault detection, in which the system outputs estimated by dynamic simulation with a nonlinear model are compared with those measured online from a plant to check whether the present state adheres to its theoretical estimation. The difference is defined by statistic Q. When Q is greater than its threshold Q_α, the process is considered as deviating from its predefined state and the second step, fault diagnosis, is conducted. Otherwise, the fault detection step continues. Before fault diagnosis, a dynamic simulation based on the process model should be firstly performed to check its coincidence with real measurements from a plant under normal conditions. In this procedure, the model is manually calibrated to guarantee the later detected anomaly coming from a fault occurring in the plant other than an error of the model [16]. During fault diagnosis, fault parameters are obtained as a solution of an inverse problem with LSQ and PLS. PLS regression of fault parameters generated by LSQ is utilized to predict fault parameters. PLS also runs when the statistic Q lies within its threshold Q_α to give continuous fault parameter estimation. Therefore, the aforementioned hybrid inverse problem approach to parameter estimation is composed of one complex optimization part with a nonlinear model and another simple regression part with a linear model.

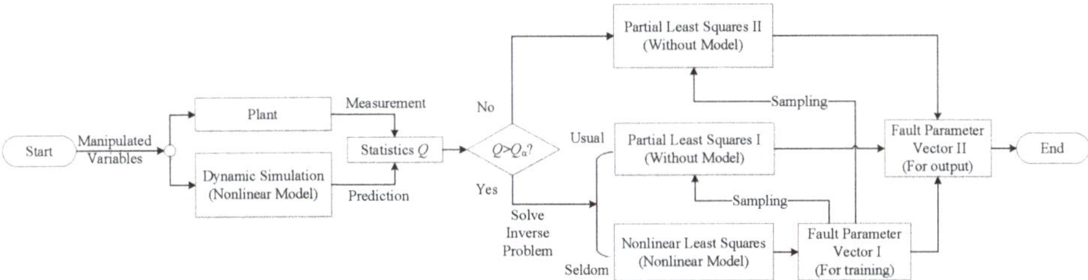

Figure 1. Hybrid inverse problem-solving process for fault diagnosis.

In Figure 1, the nonlinear model is solved once for dynamic simulation at one sampling interval to detect any fault, but is solved many times for fault diagnosis. Therefore, fault diagnosis consumes more computation time than dynamic simulation based on the same nonlinear model. The hybrid inverse problem-solving strategy replaces LSQ with PLS as much as possible since PLS calculates fault parameters directly after fitting the relationship of system outputs and fault parameters. Therefore, such a hybrid inverse problem approach can be expected to reduce the calculation load of fault diagnosis greatly.

2.1. Obtaining Fault Parameters by the LSQ Algorithm

The nonlinear fault parameters can be obtained rigorously by minimizing the deviation of measurable variables from their unsteady-state simulation values. The deviation **r** is defined in Equation (1) with a normalized version, Equation (2), where $\mathbf{y}_{meas}$ and $\mathbf{y}_{sim}$ represent data measured and simulated, respectively. It indicates an anomaly when its aggregated index **Q** exceeds the corresponding threshold Q_α, as defined in Equations (3) and (4). In this phase, fault parameters θ are solved based on an optimization formulation (LSQ) of fault parameters about the mechanism model of distillation composed of measurable variables **y**, manipulated variables **u**, disturbance ω, and state variables **x**, as shown in Equation (5).

$$\mathbf{r} = \mathbf{y}_{meas} - \mathbf{y}_{sim} \tag{1}$$

$$\mathbf{r}^* = \frac{\mathbf{r} - \mathbf{r}_{mean}}{\sigma_r} \tag{2}$$

$$\mathbf{Q} = \mathbf{r}^{*T}\mathbf{r}^* \tag{3}$$

$$Q_\alpha = \chi_\alpha^2(m) \tag{4}$$

$$\min_{\theta_t} \mathbf{Q}$$
$$\text{s.t.} \, \mathbf{y} = f(\mathbf{u}, \omega, \mathbf{x}, \theta) \tag{5}$$

As LSQ is essentially nonlinear, it is time-consuming and should not be performed frequently in practice. In a small enough range of one time point, fault parameter θ can be considered as a linear function of measurable variable **y** (see Equations (6) and (7)). In the present study, **y** and θ are scalar variables. Therefore, a revised multiple linear regression method (PLS) is used in this paper to obtain the explicit correlation between fault parameters and measurable variables.

$$y = f(\theta) \approx f(\theta_0) + f'(\theta_0)(\theta - \theta_0) \tag{6}$$

$$\theta = \frac{1}{f'(\theta_0)}[y - f(\theta_0)] + \theta_0 \tag{7}$$

2.2. Obtaining Fault Parameters by the PLS Algorithm

Most PLS methods are applied to regression modeling, replacing the general multivariate regression and principal component regression to a large extent. Comparatively, PLS can not only exclude the correlation of original variables, but also filter the noise of both independent variables and dependent variables. Its prediction ability is stronger and more stable because it uses fewer characteristic variables to describe the regression model.

Firstly, the data $X = [y \ u]^T \in R^{l \times n}$ and $Y = \theta \in R^{l \times c}$ are normalized and decomposed, respectively, where T, P, and E denote the score, load, and residual matrix of X, respectively; U, Q, and F denote the score, load, and residual matrix of Y, respectively; and a and n denote the number of PLS components and variables, respectively. The external relations are obtained as Equations (8) and (9).

$$X = T_a P_a^T + E, a < n \tag{8}$$

$$Y = U_a Q_a^T + F, a < n \tag{9}$$

Then, their internal relationship is determined as Equation (10).

$$U_a = T_a B, \tag{10}$$

where B is the internal regression matrix.

The PLS model is finally obtained as Equation (11).

$$Y = T_a B Q_a^T + F \tag{11}$$

When the independent variable X is known, PLS can be used to predict the dependent variable Y. The calculation procedure is given in Equations (12) and (13).

$$\hat{t}_h = E_{h-1} w_h \tag{12}$$

$$E_h = E_{h-1} - \hat{t}_h p_h^T, \tag{13}$$

where t and p represent the element vector in the score and load matrix, respectively; w represents the weight vector; h denotes the component index; and $E_0 = X$.

Therefore, dependent variables can be predicted using Equation (14).

$$Y = \sum_{h=1}^{a} v_h \hat{t}_h q_h^T \tag{14}$$

2.3. Correcting PLS by LSQ

Because PLS extracts linear features of fault parameters, it should be corrected continuously by LSQ. The correction framework is shown in Figure 2, where the red color loop represents the correction process. The rigorous iterative LSQ is performed once to supply one accurate value of fault parameters for the PLS training set when an anomaly is detected, not enough sampling points are collected, and correction is needed. The main contribution of this work lies in the frequent usage of fast PLS prediction of fault parameters instead of slow LSQ in the fault diagnosis algorithm. However, compared to LSQ, PLS' accuracy is limited because of its linear regression essence, so it should be corrected periodically by rigorous LSQ results. Sufficient LSQ results are needed to form the training set of the PLS model before PLS correction. Therefore, if not enough sampling points of fault parameters are given by LSQ for PLS training, LSQ should be activated correspondingly to supply one sampling point of fault parameters into the PLS training set to meet the periodical correction requirements of PLS. In this case, the boundary value θ_0 in Equation (7) is kept stable and accurate, and the prediction accuracy of PLS is guaranteed as a consequence.

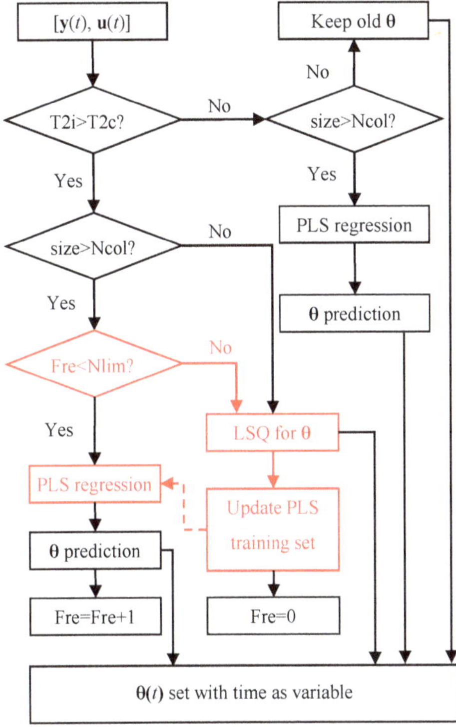

Figure 2. Correction framework of partial least squares (PLS) by least squares (LSQ).

3. Case Study

The stripper of the TEP simulator [22] is used as an example to test the feasibility of the proposed hybrid inverse problem approach to fault diagnosis. The fault set related to this stripper includes two types, that is, a step and a random type. This work chooses fault 7 and 8 as typical examples for these two types. The advantages and disadvantages of this approach are illustrated in comparison with the base approach with LSQ only.

3.1. Solving the LSQ Inverse Problem with Different Initial Values

LSQ is greatly affected by the initial values of iterated variables, so the first step is to discuss different setting methods for initial values of the LSQ algorithm. The following test is firstly based on the fault diagnosis process under fault 7. Fault 7 occurs at 8 h when the header pressure loss of the stripping stream entering the stripper bottom decreases abruptly. The fault diagnosis result using LSQ only is given in Figures 3 and 4, which has been published in our early work [16], henceforth referred to as the base case for comparison. In this base case, the overall running time and quantity of function evaluations (QFE) are 539 seconds and 1826, respectively. In the fault diagnosis algorithm, the boundary state at $t = 0$ is considered a normal state, this is, without anomaly. Therefore, the pressure loss coefficient is set as 1 uniformly in Figure 3. When $t > 0$, the fault diagnosis algorithm is activated to obtain the pressure loss coefficient in a timely manner based on the measurements of a plant.

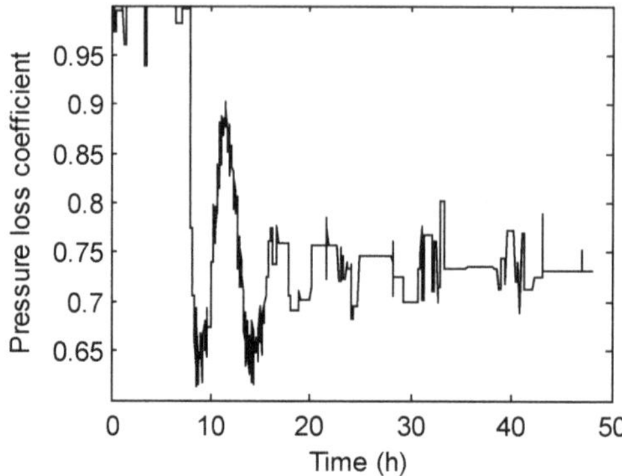

Figure 3. Fault parameter in the base case of fault 7.

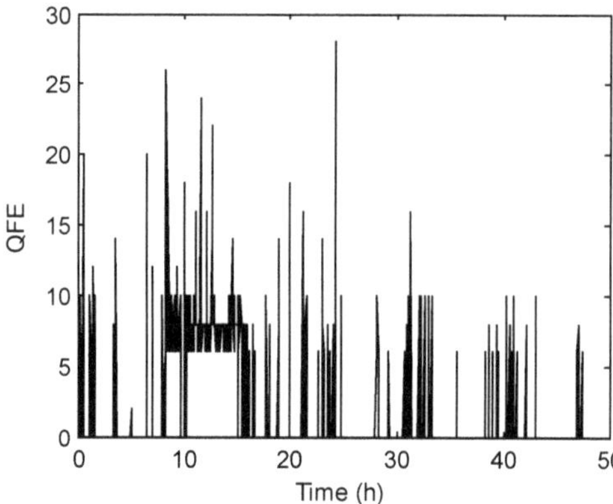

Figure 4. Function evaluation in the base case of fault 7.

(1) Set the initial value with the previous time point

The simplest value-setting method for the initial value is to directly use one from a previous time point. If there are only small changes in the fault parameter between two consecutive time points, the initial value given can thus be accepted. This is also the initial value-setting method adopted by the base case.

(2) Set the initial value with the linear fitting method

Figure 5 shows the diagnosis result with initial values given by the linear fitting method. It indicates that QFE decreases greatly and becomes stochastically stable before the fitted point number equals 30. After that, a rising function evaluation number curve is seen because of the growing time lag of the fitted line. For this reason, 5 or 8 is chosen as the candidate for the optimal fitting point number. Because more fitting points will definitely lead to a heavier computational load for the fitting operation, 5 is finally chosen as the optimal fitting point number. Despite this, QFE is only cut down by 2%, contributing little to the computational efficiency of the diagnosis algorithm.

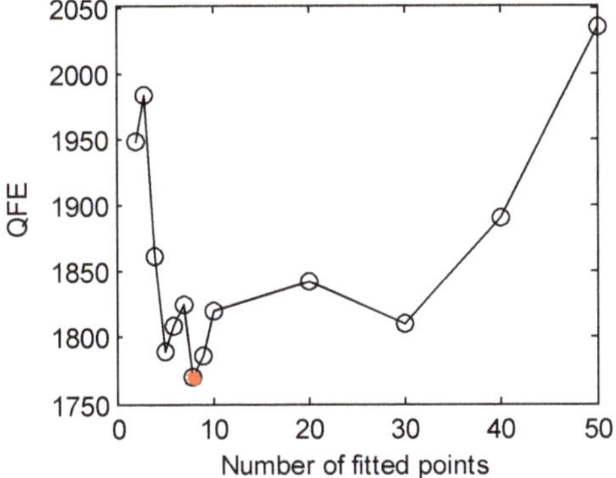

Figure 5. Fault diagnosis of fault 7 with linear fitted initial values.

(3) Set the initial value with the parabolic fitting method

Following the linear fitting method presented above, the parabolic fitting method, the simplest nonlinear fitting method, is utilized here to predict the initial values. Figure 6 shows the diagnosis result with this method, revealing a larger QFE than the base case. This reflects the essentially linear change of the fault parameter between neighboring time points. Therefore, the nonlinear fitting method does not achieve the satisfactory goal of reducing QFE.

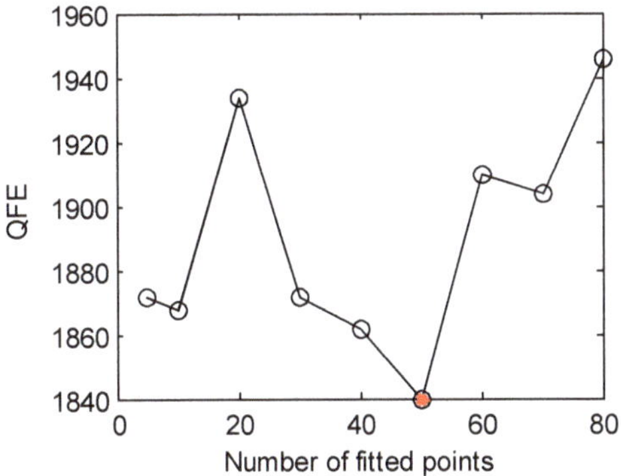

Figure 6. Fault diagnosis of fault 7 with parabola fitted initial values.

(4) Set the initial value with the grey model

Different from the above regression methods, grey system theory uses partial description information of a system to generate a grey sequence revealing the potential rule of data with the goal of system whitening [23]. It has advantages such as a small amount of data, fast operation, easy iteration, and high accuracy [24]. Figure 7 gives the diagnosis result with the grey model-predicted initial values. It shows that QFE decreases by 4.7% at most when adopting 15 as the number of sampling points. So far, the highest reduction ratio of function evaluations to the base case has been obtained by the grey model, which is, therefore, the best method for estimating initial values. At the same time, the limited improvement given by the grey model shows the need for further improvement of the diagnosis algorithm based on other factors.

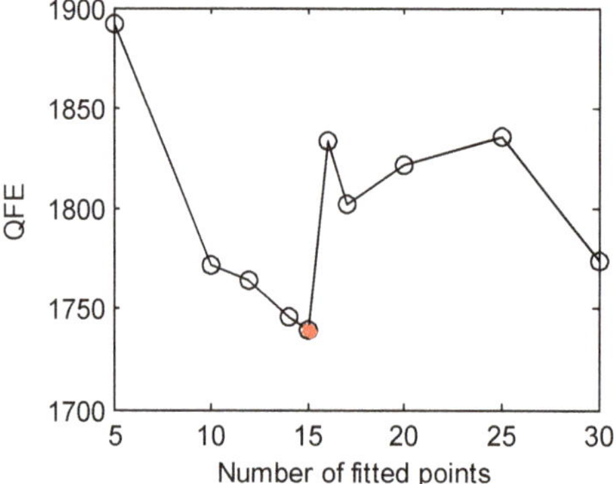

Figure 7. Fault diagnosis of fault 7 with grey model-predicted initial values.

3.2. Solving the LSQ Inverse Problem with Different Numbers of Iterations

As one popular optimization algorithm, LSQ requires a large number of iterations to obtain accurate fault parameters at each sampling time, so its high computational cost is its main disadvantage. In fact, the aim of fault diagnosis is to find the abnormal trend of fault parameters in a timely manner during a given time interval. In this process, completely converged calculation at each time point is not necessary. Based on the idea of tracking approximation, the proposal of the present paper distributes the inner iterative computation into an outer integration progress to decrease the maximum number of iterations at each sampling point. Figure 8 shows the diagnosis result with different maximum numbers of iterations. It presents a great decrease of QFE when reducing the maximum number of iterations. In particular, when the iteration number equals 1, QFE decreases by 55% compared to the base case, being far greater than the value obtained by the grey model.

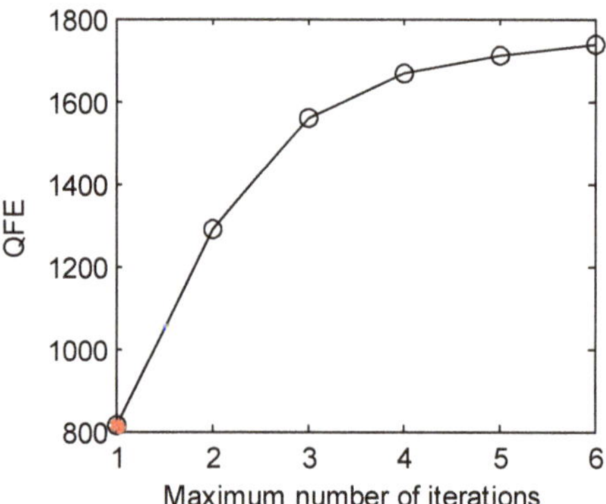

Figure 8. Fault diagnosis of fault 7 with different numbers of iterations.

The fault diagnosis result obtained by this strategy is shown in Figure 9, in contrast with that obtained by the base case. The exact coherence of fault parameters between these two cases evidences no loss of accuracy with this fast algorithm. Meanwhile, minor parameter fluctuation is observed due to the insufficient iterative computation of this algorithm.

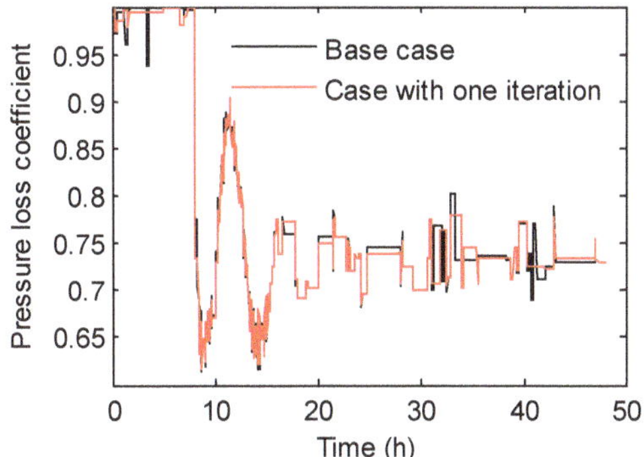

Figure 9. Fault diagnosis of fault 7 with one iteration only.

3.3. Hybrid Inverse Problem-Solving Strategy

In the above sections, two kinds of improvements—increasing the prediction accuracy of initial values and decreasing the number of iterations—were conducted for the least squares algorithm. The computational results show that the latter has a significant effect on the fault diagnosis speed. Generally, these algorithms use the passive trial and error method to solve the inverse problem of fault diagnosis. Fault parameters are defined as input variables for the system model used by LSQ, different from their output variable role defined in the inverse problem. In the following, an alternative inverse problem model using a direct mapping of fault parameters from measurements will be considered to avoid the time-consuming model solving process.

In an information view of fault diagnosis, the inverse problem defined herein is a typical multiple input-multiple output (MIMO) system in which measurable/controllable variables and fault parameters constitute input and output parts, respectively. The linear MIMO model is given by the PLS method in this work owing to the small data change for both input and output variables in a short sampling interval. Furthermore, periodic correction for this linear model by LSQ is necessary to preserve its accuracy.

Figure 10 shows the comparison of fault diagnosis for the base case and the case using a hybrid strategy. It proves the feasibility and accuracy of this strategy, but indicates larger fluctuations of the fault parameter with the hybrid algorithm. Therefore, PLS is suitable for replacing LSQ, but its application should be controlled properly. Factors affecting the efficiency of this strategy will be discussed hereafter.

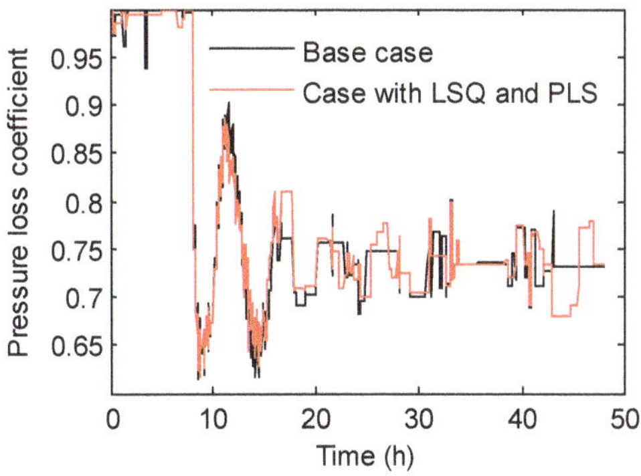

Figure 10. Fault diagnosis of fault 7 with the hybrid algorithm.

(1) Number of PLS components

PLS is something of a cross between multiple linear regression and principal component analysis. It constructs components as linear combinations of the original variables, while allowing for correlation between independent and dependent variables. The number of components is therefore of primary importance to the accuracy of the PLS model. Figure 11 depicts the percent of variance explained in the dependent variable as a function of the component number. A maximum of 16 components is assumed in Figure 11 because the independent variables consist of a total of 16 variables for the stripper in TEP. It can be seen that more than 95 percent variance was explained by the first three components, which were, accordingly, chosen as the principle components in the following PLS modeling process.

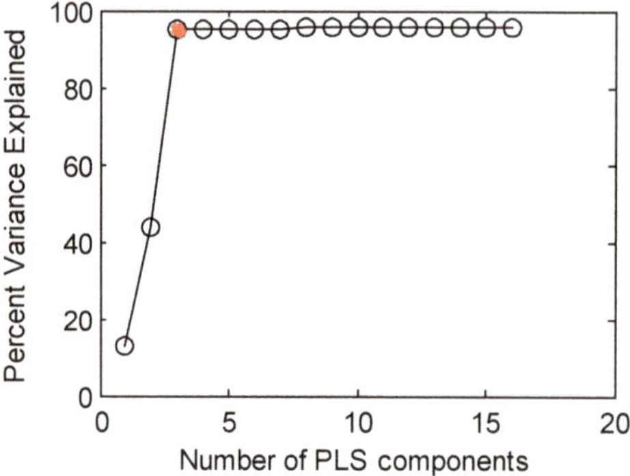

Figure 11. Number analysis of PLS components under fault 7.

(2) Sampling data set for PLS modeling

The training data sets for PLS modeling were composed of 12 measured variables, 4 manipulated variables, and 1 fault parameter. As shown in Figure 1, the fault parameter may be obtained from LSQ or PLS, so the PLS model can be built on LSQ-generated fault parameter sets (Vector I) or mixed sets (Vector II).

Figure 12a,b show the fault diagnosis results with Vector II as training data sets for PLS, whereas Figure 12c,d give results with Vector I as training data sets. The root mean square error (RMSE) of fault parameters between the base case and the proposed approach was also calculated and adhered to Figure 12a–d. As illustrated in Figure 1, a second PLS (PLS II) was performed with the aim of keeping the continuity of fault parameters when no abnormal signals were detected. Figure 12 shows the fault parameters obtained with (b) and without (a) in this second PLS based on Vector II. The worse result obtained with PLS than without PLS to predict the fault parameter in normal states evidences an adverse propagation effect of PLS prediction error on the PLS model itself. The fact that diagnosis results with Vector I as training data sets for PLS coincide exactly with the base case no matter whether they are with (d) or without (c) the second PLS further proves this conclusion. Consequently, PLS modeling should be conducted based on the fault parameters generated by the LSQ algorithm to preserve its accuracy.

Generally, the time-saving prediction of fault parameters with the PLS method may be equal to several sampling periods before each time-consuming LSQ in Figure 1. Although this scheme can cut down the running time of fault diagnosis greatly, the using frequency of PLS should be limited to an allowable range since the PLS accuracy strongly depends on new fault parameters generated by LSQ. In other words, it is crucial to correct the PLS model with LSQ-generated data after consecutive calls of PLS.

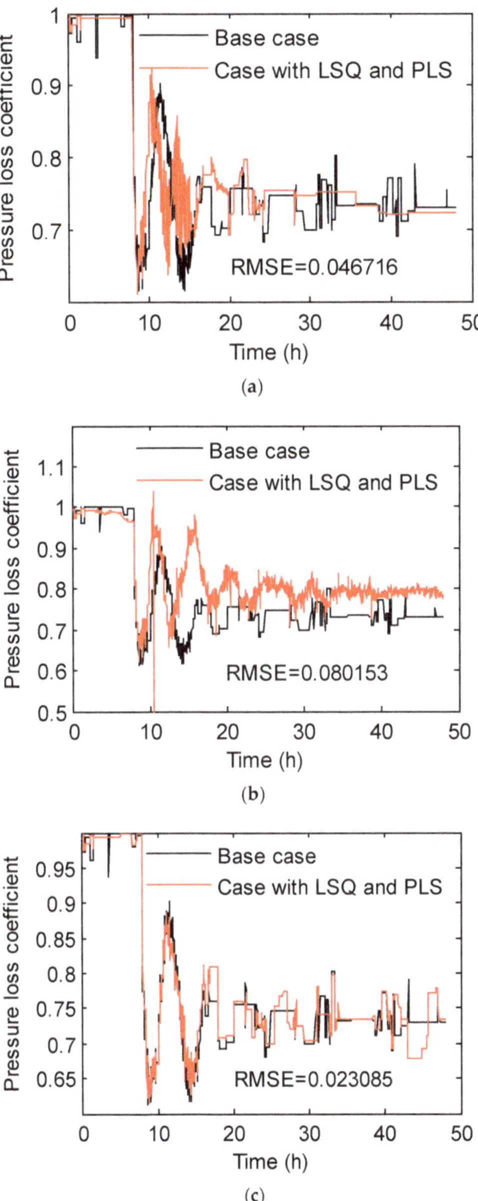

Figure 12. *Cont.*

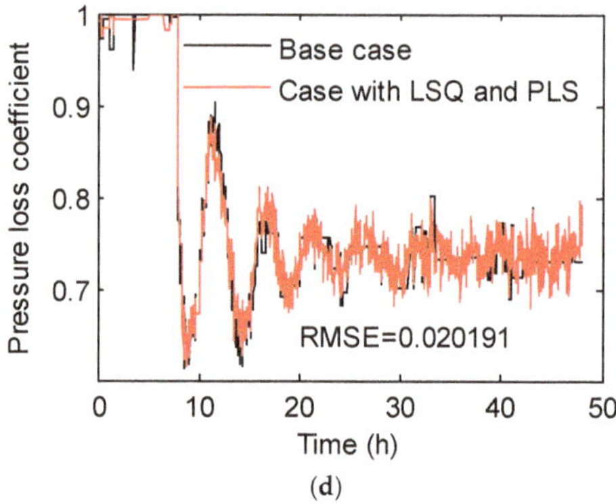

Figure 12. Fault diagnosis under fault 7 (**a**) based on Vector II data, but without second PLS; (**b**) based on Vector II data and with second PLS; (**c**) based on Vector I data, but without second PLS; (**d**) based on Vector I data and with second PLS.

Figure 13 shows the effect of the correction interval on QFE, the running time, and RMSE, respectively. We can see from Figure 13a,b that QFE decreases, as does the running time of the fault diagnosis process, when increasing the correction interval. In particular, their decreasing magnitude becomes small when the correction interval exceeds 5. However, there appears to be a rapid growth of calculation error, as can be seen from Figure 13c. Although the effect of correction interval 4 or 5 on RMSE is not significant, both QFE and the running time will increase for the correction interval of 4 compared with 5. Therefore, 5 is the appropriate calling number of PLS in a correction interval. With this calling number, QFE decreases by 81.60% and the running speed increases about 1.7 times compared to the base case with this calling number.

Table 1 summarizes the approaches that can effectively reduce QFE in fault diagnosis. The best results obtained are indicated in boldface. It leads to the conclusion that the approach proposed in this paper evaluates the fault parameter markedly faster than the pure LSQ-based algorithm used in [16].

Table 1. Algorithm improvement concerning function evaluations.

No.	Improvement Approach	Quantity of Function Evaluations	Cut Ratio (%)
0	base case	1826	-
1	linearly fitted initial values	1790	1.97
2	initial values provided with grey model	1740	4.71
3	one iteration only	816	55.31
4	partial least squares (PLS) mixed algorithm	336	81.60

The above feasibility test was implemented based on a step-type fault, with fault 7 as an example. Next, another type of fault, a random type with fault 8 as an example, will be tested with the aforementioned hybrid structure. In the case of fault 8, the composition of the feed stream (containing component A, B, and C only) changes randomly from the 8 h point. The essentially random sampling values form this fault are fed to the hybrid fault diagnosis algorithm as a preset input from an outside battery. It is satisfactory to discriminate the fault type from fault diagnosis results, with no need for a statistics analysis of random sampling values. Its fault diagnosis result with LSQ only is shown in Figures 14 and 15, also published in our early work [16]. The overall running time and QFE for this diagnosis process are 5103 seconds and 19,256, respectively. This case is referred to as the base case for fault 8 in the following discussions.

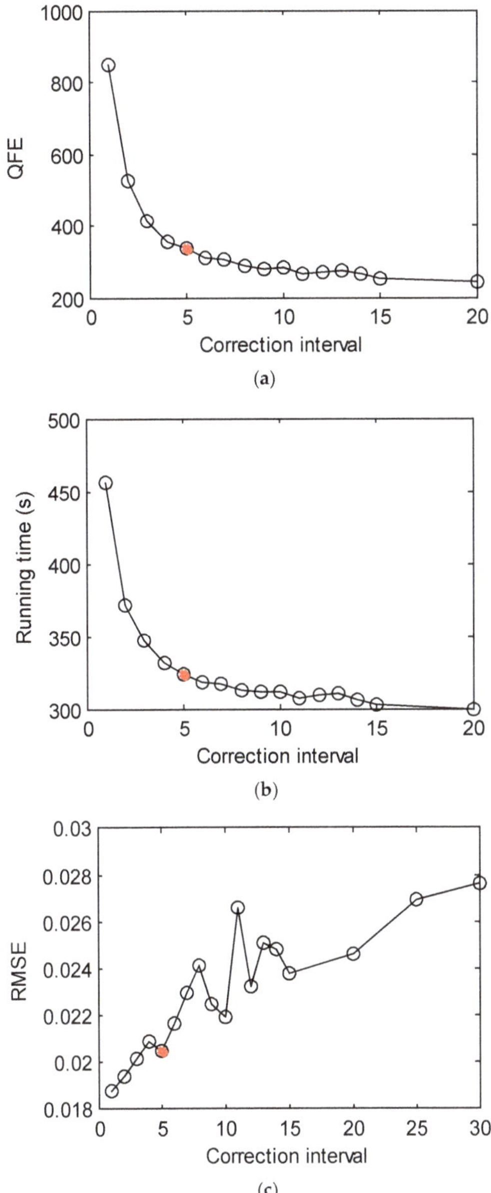

Figure 13. Fault diagnosis with different correction intervals of PLS under fault 7 with respect to (**a**) quantity of function evaluations (QFE); (**b**) the running time; (**c**) the root mean square error (RMSE).

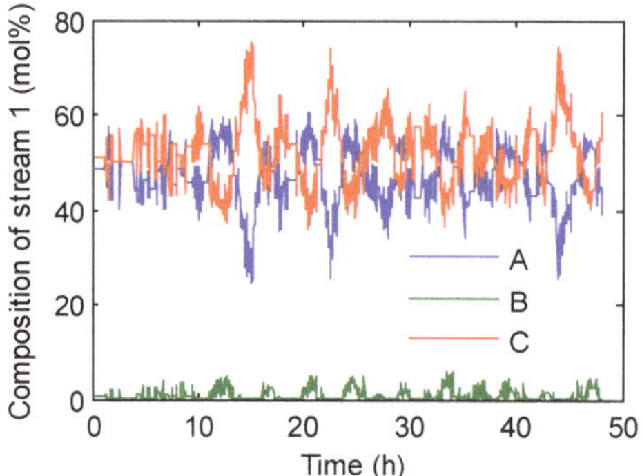

Figure 14. Fault parameter in the base case of fault 8.

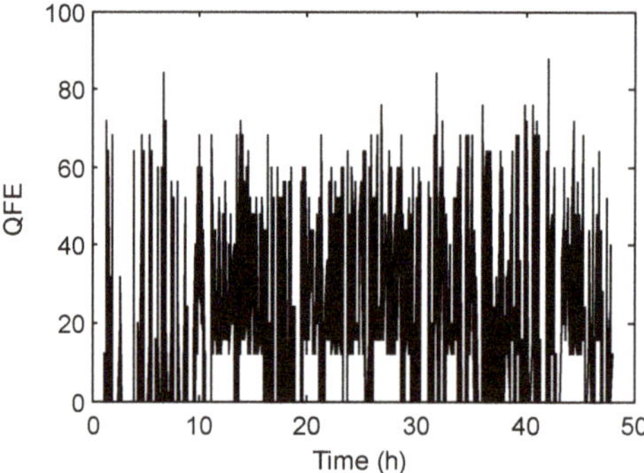

Figure 15. Function evaluation in the base case of fault 8.

Similar to Figure 11, Figure 16 shows the percent of variance explained by independent components. The first two components make more than 80% contributions and were thus selected as the components in PLS modeling under fault 8.

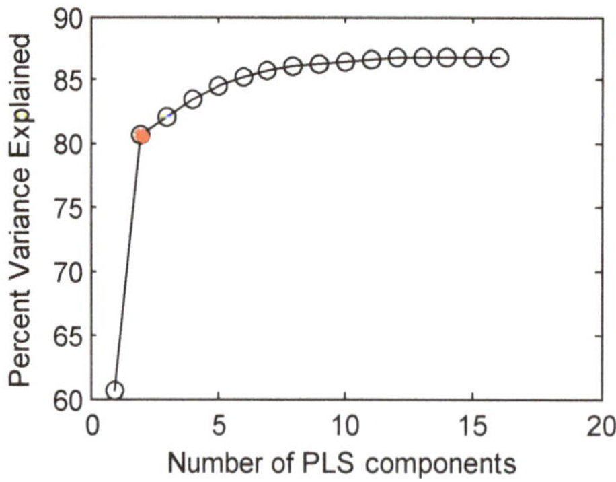

Figure 16. Number analysis of PLS components under fault 8.

Figure 17 shows the effect of the correction interval on QFE (a), the running time (b), and RMSE (c) under fault 8. We can see that QFE and the running time decrease, but RMSE increases, when increasing the correction interval. 5 is chosen as the optimal correction interval since the former two indices do not decrease significantly, while RMSE remains small under this choice. Besides, larger values of the former two indices than fault 7 are observed in Figure 17, indicating that a random-type fault consumes more time than a step-type fault due to its stochastic computing load.

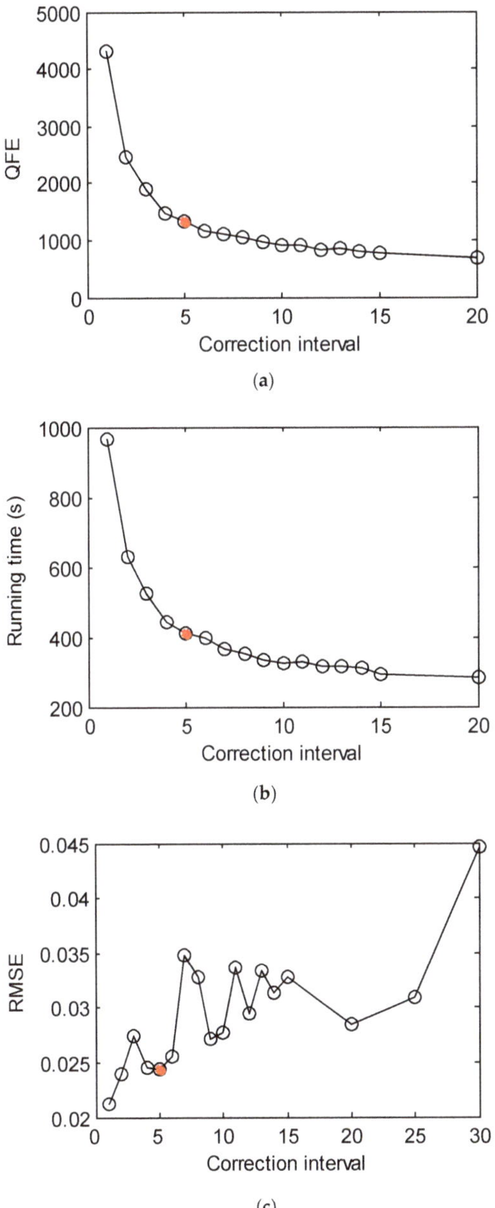

Figure 17. Fault diagnosis with different correction intervals of PLS under fault 8 with respect to (**a**) QFE; (**b**) the running time; (**c**) RMSE.

With 5 as the correction interval, the diagnosis result of three compositions in the bottom feed of the stripper is shown in Figure 18. It indicates nearly the same composition trajectories for the hybrid approach and base case, and proves the feasibility and accuracy of our proposed approach. Finally,

QFE decreases by 92.31% and the running speed increases about 13 times compared to the base case in this situation.

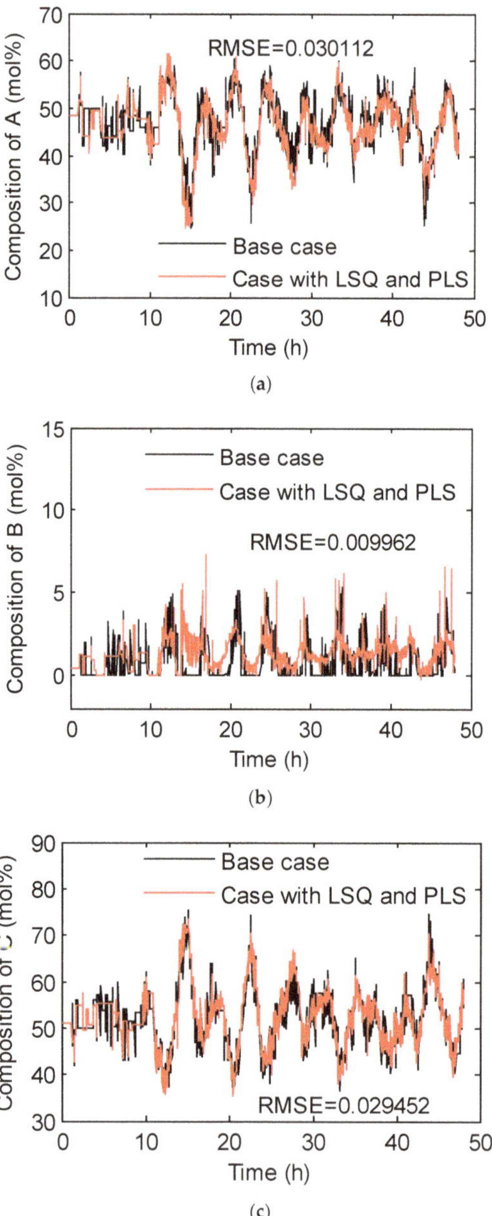

Figure 18. Fault diagnosis with a hybrid scheme under fault 8 for the (**a**) composition of A; (**b**) composition of B; (**c**) composition of C.

4. Conclusions

In this paper, an LSQ and PLS combined hybrid inverse problem approach has been proposed to realize model-based diagnosis for the distillation process. LSQ is used to identify parameters that best-represent an abnormal state of distillation on the basis of a nonlinear dynamic model. PLS regression is then used to fit these parameters with input/output signals and forecast their developing trajectories. The correction interval of PLS significantly affects the speed and accuracy of the fault diagnosis process. The approach has been carried out to successfully identify stripper-related faults in the TEP benchmark process. For fault 7, QFE decreases by 81.60% and the running speed increases about 1.7 times compared to the base case. For fault 8, QFE decreases by 92.31% and the running speed increases about 13 times compared to the base case. Therefore, it has been proven to be a computationally efficient scheme for model-based diagnosis. In conclusion, compared with a single nonlinear LSQ-based approach, the presented hybrid inverse problem approach enables a trade-off between accurate LSQ and fast PLS and is more suitable for real-time fault diagnosis.

In the future, it would be helpful to combine this approach with some process history-based approaches, like a bond graph [25], to enhance its vital ability to locate fault-specific sections prior to fault diagnosis.

Author Contributions: All authors participated to the elaboration of the manuscript. Investigation, methodology and writing-original draft, S.S.; investigation and software, Z.C.; data curation, X.Z.; methodology and supervision, W.T. All the authors discussed the results. All authors have read and agreed to the published version of the manuscript.

Funding: Financial support for carrying out this work was provided by the National Natural Science Foundation of China (Grant No. 21576143).

Conflicts of Interest: The authors declare no conflicts of interest.

Nomenclature

a	Number of PLS components
B	Internal regression matrix of PLS
c	Output variable size of PLS
E, F	Residual matrix after input and output decomposition of PLS, respectively
h	Component index of PLS
l	Sample size
m	Number of measurable variables
n	Variable size of PLS
P, Q	Load vector of PLS input and output matrix, respectively
q	Element in load matrix of PLS output
Q	Fault detection statistic
r	Fault detection deviation
t, p	Element in score and load matrix of PLS input, respectively
T, U	Score matrix of PLS input and output matrix, respectively
u	Manipulated variable vector
v	Regression coefficient of component
w	Weight vector
x	State variable vector
X	Input matrix for PLS
y	Vector of measurable variables
Y	Output matrix for PLS
Greek Symbols	
θ	Fault parameter vector
α	Confidence
χ	Chi square distribution
ω	Disturbance vector

Superscripts

*	Normalized
T	Transposition
ˆ	Prediction value

Subscripts

meas	Measured value
sim	Simulated value

Abbreviations

ANN	Artificial neural networks
LSQ	Least squares
MIMO	Multiple input-multiple output
PLS	Partial least squares
QFE	Quantity of function evaluations
RMSE	Root mean square error

References

1. Jlassi, I.; Estima, J.O.; El Khil, S.K.; Bellaaj, N.M.; Cardoso, A.J.M. A Robust Observer-Based Method for IGBTs and Current Sensors Fault Diagnosis in Voltage-Source Inverters of PMSM Drives. *IEEE Trans. Ind. Appl.* **2017**, *53*, 2894–2905. [CrossRef]
2. Sun, S.; Wei, X.; Zhang, H.; Karimi, H.R.; Han, J. Composite fault-tolerant control with disturbance observer for stochastic systems with multiple disturbances. *J. Franklin Inst.* **2018**, *355*, 4897–4915. [CrossRef]
3. Ge, H.; Yue, D.; Xie, X. Observer-Based Fault Diagnosis of Nonlinear Systems via an Improved Homogeneous Polynomial Technique. *Int. J. Fuzzy Syst.* **2017**, *20*, 403–415. [CrossRef]
4. Power, Y.; Bahri, P.A. A two-step supervisory fault diagnosis framework. *Comput. Chem. Eng.* **2004**, *28*, 2131–2140. [CrossRef]
5. Mahmoud, H.; Abdallh, A.A.; Bianchi, N.; El-Hakim, S.M.; Shaltout, A.; Dupre, L. An Inverse Approach for Interturn Fault Detection in Asynchronous Machines Using Magnetic Pendulous Oscillation Technique. *IEEE Trans. Ind. Appl.* **2016**, *52*, 226–233. [CrossRef]
6. Tian, W.; Sun, S.; Guo, Q. Fault detection and diagnosis for distillation column using two-tier model. *Can. J. Chem. Eng.* **2013**, *91*, 1671–1685. [CrossRef]
7. Yu, H.; Khan, F.; Garaniya, V. Nonlinear Gaussian Belief Network based fault diagnosis for industrial processes. *J. Process Control* **2015**, *35*, 178–200. [CrossRef]
8. Díaz, C.A.; Echevarría, L.C.; Prieto-Moreno, A.; Neto, A.J.S.; Llanes-Santiago, O. A model-based fault diagnosis in a nonlinear bioreactor using an inverse problem approach and evolutionary algorithms. *Chem. Eng. Res. Des.* **2016**, *114*, 18–29. [CrossRef]
9. Lyu, T.; Xu, C.; Chen, G. Health state inversion of Jack-up structure based on feature learning of damage information. *Eng. Struct.* **2019**, *186*, 131–145. [CrossRef]
10. Ramos, A.R.; de Lazaro, J.M.B.; Prieto-Moreno, A. An approach to robust fault diagnosis in mechanical systems using computational intelligence. *J. Intell. Manuf.* **2019**, *30*, 1601–1615. [CrossRef]
11. Echevarría, L.C.; Velho, H.F.d.; Becceneri, J.C.; Neto, A.J.d.; Santiago, O.L. The fault diagnosis inverse problem with Ant Colony Optimization and Ant Colony Optimization with dispersion. *Appl. Math. Comput.* **2014**, *227*, 687–700. [CrossRef]
12. Parolin, R.d.S.; Neto, A.J.d.S.; Rodrigues, P.P.G.W.; Santiago, O.L. Estimation of a contaminant source in an estuary with an inverse problem approach. *Appl. Math. Comput.* **2015**, *260*, 331–341. [CrossRef]
13. Tarokh, M. Solving inverse problems by decomposition, classification and simple modeling. *Inf. Sci.* **2013**, *218*, 51–60. [CrossRef]
14. Sever, A. A neural network algorithm to pattern recognition in inverse problems. *Appl. Math. Comput.* **2013**, *221*, 484–490. [CrossRef]
15. Deshpande, A.P.; Patwardhan, S.C. Online Fault Diagnosis in Nonlinear Systems Using the Multiple Operating Regime Approach. *Ind. Eng. Chem. Res.* **2008**, *47*, 6711–6726. [CrossRef]
16. Tian, W.; Guo, Q.; Sun, S. Dynamic simulation based fault detection and diagnosis for distillation column. *Korean J. Chem. Eng.* **2012**, *29*, 9–17. [CrossRef]

17. Yin, S.; Zhu, X.; Kaynak, O. Improved PLS focused on key-performance-indicator-related fault diagnosis. *IEEE Trans. Ind Electron.* **2015**, *62*, 1651–1658. [CrossRef]
18. Vitale, R.; Palaci-Lopez, D.; Kerkenaar, H.H.M. Kernel-Partial Least Squares regression coupled to pseudo-sample trajectories for the analysis of mixture designs of experiments. *Chemom. Intell. Lab. Syst.* **2018**, *175*, 37–46. [CrossRef]
19. Choi, S.W.; Lee, I.-B. Multiblock PLS-based localized process diagnosis. *J. Process Control* **2005**, *15*, 295–306. [CrossRef]
20. Zhang, Y.W.; Zhou, H.; Qin, S.J.; Chai, T.Y. Decentralized fault diagnosis of large-scale processes using multiblock kernel partial least squares. *IEEE Trans. Ind. Inf.* **2010**, *60*, 3–10. [CrossRef]
21. Godoy, J.L.; Vega, J.R.; Marchetti, J.L. A fault detection and diagnosis technique for multivariate processes using a PLS-Decomposition of the measurement space. *Chemom. Intell. Lab. Syst.* **2013**, *128*, 25–36. [CrossRef]
22. Downs, J.J.; Vogel, E.F. A plant-wide industrial process control problem. *Comput. Chem. Eng.* **1993**, *17*, 245–255. [CrossRef]
23. Kayacan, E.; Ulutas, B.; Kaynak, O. Grey system theory-based models in time series prediction. *Expert Syst. Appl.* **2010**, *37*, 1784–1789. [CrossRef]
24. Tserng, H.P.; Tserng, T.L.; Chen, P.C.; Tran, L.Q. A Grey System Theory-Based Default Prediction Model for Construction Firms. *Comput.-Aided Civ. Infrastruct. Eng.* **2015**, *30*, 120–134. [CrossRef]
25. Yu, M.; Xiao, C.Y.; Jiang, W.H.; Yang, S.L.; Wang, H. Fault diagnosis for electromechanical system via extended analytical redundancy relations. *IEEE Trans. Ind. Inf.* **2018**, *14*, 5233–5244. [CrossRef]

Article

Application of Transformation Matrices to the Solution of Population Balance Equations

Vasyl Skorych [1,*], Nilima Das [2], Maksym Dosta [1], Jitendra Kumar [2] and Stefan Heinrich [1]

[1]   Institute of Solids Process Engineering and Particle Technology, Hamburg University of Technology, Denickestrasse 15, 21073 Hamburg, Germany

[2]   Department of Mathematics, Indian Institute of Technology Kharagpur, Kharagpur 721302, India

*   Correspondence: vasyl.skorych@tuhh.de; Tel.: +49-(0)-40-42-878-2811

Received: 25 July 2019; Accepted: 11 August 2019; Published: 14 August 2019

Abstract: The development of algorithms and methods for modelling flowsheets in the field of granular materials has a number of challenges. The difficulties are mainly related to the inhomogeneity of solid materials, requiring a description of granular materials using distributed parameters. To overcome some of these problems, an approach with transformation matrices can be used. This allows one to quantitatively describe the material transitions between different classes in a multidimensional distributed set of parameters, making it possible to properly handle dependent distributions. This contribution proposes a new method for formulating transformation matrices using population balance equations (PBE) for agglomeration and milling processes. The finite volume method for spatial discretization and the second-order Runge–Kutta method were used to obtain the complete discretized form of the PBE and to calculate the transformation matrices. The proposed method was implemented in the flowsheet modelling framework Dyssol to demonstrate and prove its applicability. Hence, it was revealed that this new approach allows the modelling of complex processes involving materials described by several interconnected distributed parameters, correctly taking into consideration their interdependency.

Keywords: population balance equation; dynamic flowsheet simulation; transformation matrix; process modelling; agglomeration; milling; solids; multidimensional distributed parameters

---

1. Introduction

Bulk materials typically consist of individual non-uniform particles having different parameters, which vary in a certain range. These parameters are referred to as distributed and include, for example, the diameters of particles, their densities, or porosities. Although these parameters can physically be fairly accurately described by continuous distribution functions, such a representation is not always convenient for numerical analysis and modelling. Therefore, in practice, the continuously distributed material parameter space is discretized into shorter intervals, usually called classes. Each class is assigned a quantity of material, whose parameters fall into this particular interval. In this case, the distribution of the parameter of the entire material in this class is narrowed down from a continuous distribution function to a single value, for example, the average value of this class. Thus, it is assumed that all the particles inside have the same value of the parameter. Therefore, a simulation system can operate using a finite number of parameter values, which (if the number of classes is sufficient) accurately describe the original distribution.

Despite the fact that most models in solids processing technology consider particle size distribution as one of the main material parameters, many of them additionally incorporate various other attributes. For example, the shape and orientation of primary particles are important in crystallization processes [1]; the size and yield strength of the colliding granules can be considered as parameters for wet granulation models [2]; the distributions of granules by porosity and saturation are important for breakage rate in some apparatuses [3]; and the moisture content of particles plays a significant role in dryers [4].

Such a diversity of models and their parameters makes the simulation process much more challenging when trying to combine their different types in a single flowsheet. The main problem here is that each unit is usually being developed to consider only a limited number of distributed parameters which are important for its model, neglecting others. Moreover, the usual approach to develop models involves a direct calculation of the distributed parameters at the outlet of a unit. As a result, if the material is additionally described by some secondary distributed parameters, information about them and their dependencies may be lost during the simulation.

For example, consider a flowsheet where a screen and a dryer are connected in series, and the solid phase is distributed according to particle size and moisture content. In this case, the screen unit, usually performing separation of particles only according to size, will fail to determine their moisture properly, which is important for the dryer.

To handle this issue, Pogodda [5] introduced an approach using transformation matrices and implemented it into the SolidSim simulation environment. Later, Dosta [6] and Skorych et al. [7] extended this technique for the dynamic flowsheet simulation using the Euler method. Transformation matrices in this approach describe the laws of mass transfer between all classes of the distribution. Their application, instead of explicit calculation of output flows, enables the usage of more material parameters in a proper way and implicitly preserves information about all secondary distributions and their dependencies at any time point in the unit.

1.1. Flowsheet Simulation of Solid Phase Processes

Nowadays, flowsheet simulation software is commonly used to aid with design, planning, optimization, and testing in chemical engineering [8]. Usually it can help with the minimization of operation costs, troubleshooting, increasing throughput and product quality, and generally provides a better understanding of the process. However, despite the fact that flowsheet simulation of fluid processes has already been state of the art for many decades [9], generally applicable systems for modelling granular materials began to appear only recently and are now being actively developed. This arises from the complexity of description and handling of the solid phase. In contrast to fluids, it requires treatment of distributed parameters, such as particle size, internal porosity or moisture content. Moreover, these distributions can be interrelated, which requires even more advanced methods to treat them properly and preserve their interconnection during the simulation.

Currently, flowsheet modelling systems that can work with the solid phase are being actively developed. Among them may for example be mentioned:

- Aspen Plus [10], which is actively used in chemical industry as well as for simulation of polymers, minerals, metals etc.
- gPROMS FormulatedProducts [11], as a part of gPROMS platform, especially designed to investigate solid phase processes.
- JKSimMet [12], which is a flowsheeting software for simulation of comminution and classification circuits in the mineral processing industry.
- The HSC Sim [13] module of HSC Chemistry software intended for modelling various processes in chemistry, metallurgy, and mineralogy.
- CHEMCAD [14], which was developed to simulate chemical processes with limited consideration of the solid phase.
- Dyssol [7], a flowsheet simulation system designed to simulate complex dynamic processes in solids processing technology, developed within a research collaboration founded by the German Research Foundation.

Active development of new dynamic models and usage of flowsheet simulations to study various solid phase processes, such as continuous granulation [15,16], chemical looping combustion [17], crystallization [1], milling processes [18], separation [19], tablet manufacturing [20,21], and agglomeration [22] are of increasing interest in this area.

1.2. Use of Population Balance Equations for Particulate Processes

Population balance equations (PBE) appear in various branches of engineering where particles continuously change their properties, like in pharmaceutical industries and in the processing of minerals, food, paints, and ink. For example, they are used to describe time-dependent processes in crystallizers, fluidized beds, liquid–liquid, gas–liquid contactors, or polymer reactors. In many particulate processes, one-dimensional PBEs with particle size[1] as the property coordinate are used. In this article, the size is referred to as the volume of particles, unless explicitly stated otherwise. These equations describe a time-dependent change of particle size distribution (PSD) in a specific volume, which is caused by various processes, such as agglomeration, breakage, nucleation, growth, or shrinkage. This contribution considers only agglomeration and breakage processes. The time evolution of the particle size distribution $u(x, t) \geq 0$ in these processes, at time $t \geq 0$, for particles of size $x \geq 0$ can be represented by a special type of nonlinear integro-differential equation, namely a population balance equation [23,24]

$$\frac{\partial u(x,t)}{\partial t} = \frac{1}{2} \int_0^x \beta(x', x-x')u(x',t)u(x-x',t)dx' - \int_0^\infty \beta(x,x')u(x,t)u(x',t)dx'$$
$$+ \int_x^\infty b(x,y)S(y)u(y,t)dy - S(x)u(x,t) + \dot{Q}_{in}(t) - \dot{Q}_{out}(t). \tag{1}$$

Here, the first and the third terms of the right hand side of Equation (1) represent the birth events (appearance of new particles) due to agglomeration and breakage, respectively. The death events (disappearance of existing particles) owing to agglomeration and breakage process are described by the second and the fourth terms, respectively. Terms $\dot{Q}_{in}(t)$ and $\dot{Q}_{out}(t)$ denote the input and the output in a continuous process, accordingly. The function $\beta(x', x)$ is the agglomeration kernel, which represents the rate at which a particle of size x agglomerates with a particle of size x'. The breakage function $b(x, y)$ in the third term represents the fraction of fragments of size x, which appear when a particle of size y breaks. The selection function $S(x)$ describes the rate at which a particle of size x is selected to break.

In general, the breakage function is described by the following dependencies:

$$\begin{cases} b(x,y) = 0, & \forall x \geq y \\ \int_0^y b(x,y)dx = \vartheta(y), & \forall y > 0 \end{cases} \tag{2}$$

and

$$\int_0^y xb(x,y)dx = y, \quad \forall y > 0. \tag{3}$$

Equation (2) designates that the breakage function $b(x, y)$ will be equal to zero when the size of the resulting particle x is greater than the size of the initial particle y. When the initial particle of size y disintegrates to a particle of size x during the breakage process, the total number of particles can be represented by the function $\vartheta(y)$ and, obviously, $\vartheta(y) \geq 1$. Moreover, when a particle of size y splits into daughter particles, it generally satisfies the local mass conservation law, which is given by Equation (3).

Various numerical methods have been applied to solve Equation (1). Examples for the most commonly used methods are: the method of moments [25–28], the fixed pivot technique [29,30], the Monte Carlo simulation method [31–34], the finite element method [35–37], approaches based on a separable approximation of the kernel function with the fast Fourier transformation [38–40], the cell average technique [41,42], and the finite volume method [43,44]. In this work, the finite volume method [43,44] is applied for the spatial discretization, and the second-order Runge–Kutta method is used for the time variable to get the full discretized form of the PBE (1), suitable for formulating transformation matrices.

2. Mathematical Formulations

2.1. Transformation Matrices

With regard to the flowsheet simulation, mathematical models that describe time-dependent changes occurring at individual stages in the process can be represented by the following general equation [45]:

$$Y(t) = F(t, X(t), c(t), p(t), r). \tag{4}$$

Here F is a model function that depends on:

- $X(t)$, $Y(t)$—input and output stream variables;
- $c(t)$—control variables that describe the operating conditions of the process and are usually controlled in certain ranges;
- $p(t)$—model parameters, which are customizable settings of the model itself; they may or may not change over time;
- r—design variables that represent structural features of apparatuses, such as physical dimensions and shapes, and usually do not change during the simulation.

Considering this, the information flow between a flowsheet and a dynamic model can be simplistically represented by the scheme shown in Figure 1. $\{X\}$, $\{Y\}$, and $\{H\}$ are sets of time-dependent state variables, which describe inlet, outlet, and holdup of the model, respectively. They are composed of distributed and concentrated properties of bulk material, such as size, humidity, or composition of particles. The input variables $X(t)$, the control variables $c(t)$, and the model parameters $p(t)$ from Equation (4) together describe the influence of the process environment on the model during the simulation. $c(t)$ and $p(t)$ are not shown on the scheme for simplicity, whereas all $X(t)$ at each moment of time are denoted as $\{X\}$. Emphasizing the dynamic nature of the model, dependent variables $Y(t)$ are divided into two assemblies: $\{H\}$ is a set of variables describing the internal state, or holdup, of the model at each time point, while $\{Y\}$ is a set of output parameters which describe material leaving the model. The model function F from Equation (4) is also divided into the two sets $\{F_H\}$ and $\{F_Y\}$ to describe changes in material parameters occurring in the holdup and in the outlet streams of the model, respectively. In contrast to all input time-dependent parameters, design variables r do not change during the whole simulation. Therefore, they can be considered as model constants for each individual process and, thereby, as internal parameters of $\{F_H\}$ and $\{F_Y\}$ for a particular simulation.

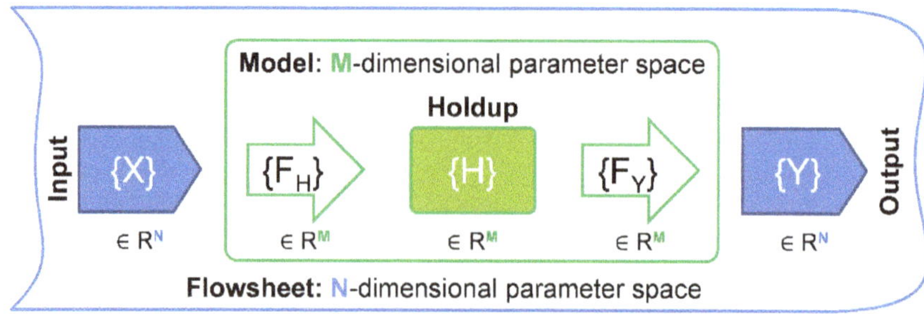

Figure 1. Simplified scheme of embedding a model into a flowsheet.

According to Dosta [6], there are two strategies to calculate holdup and outlet streams for steady-state and dynamic models:

(1) Explicitly, by direct calculation of all state variables in holdups and outlet streams.

(2) Implicitly, through the application of movement (transformation) matrices, which describe the transfer of material between discrete classes in multidimensional parameter space.

The explicit approach is most commonly used in flowsheet software. In the case of an explicit calculation of a dynamic model using the Euler integration scheme, Equation (4) can be reformulated for models, discretized with respect to time, as:

$$\begin{cases} H(t + \Delta t) = F_H(t + \Delta t, \{X(t + \Delta t)\}, \{H(t)\}, \{Y(t)\}) \\ Y(t + \Delta t) = F_Y(t + \Delta t, \{H(t + \Delta t)\}) \end{cases}, \tag{5}$$

where F_H and F_Y are functions to describe changes occurring in holdup and outlet streams. Since steady-state models do not include holdups, their models have a simplified form:

$$Y(t) = F_Y(t, \{X(t)\}). \tag{6}$$

To describe a continuous distribution of material through a specific property, a discretized representation is used. Each property d is described by a certain number of discrete classes L_d. Then, if the solid material is distributed over N property coordinates, the total number of state variables is defined as:

$$L = \prod_{d=1}^{N} L_d. \tag{7}$$

$\{F_H\}$ and $\{F_Y\}$ are usually defined to be fully flexible related to the number of discrete classes (L_d) for each property coordinate. Moreover, to make the simulation system generally applicable, the number of the property coordinates N and their types should also be flexible. This means that the final set of parameters may not be known during model development. However, if the model is designed applying the explicit approach, the number of considered distributed properties M is always fixed. That means that for the general case the model functions $\{F_H\}$, $\{F_Y\}$, and the holdups $\{H\}$ are defined in the M-dimensional parameter space. Since the output distribution $\{Y\}$ is calculated based on the internal states $\{H\}$ and the model functions $\{F_Y\}$, it will also be defined in the M–parameter space:

$$\{X\} \in R^N$$
$$\{F_H\}, \{F_Y\} \in R^M \rightarrow \{H\}, \{Y\} \in R^M \tag{8}$$

for dynamic units or

$$\{X\} \in R^N$$
$$\{F_Y\} \in R^M \rightarrow \{Y\} \in R^M \tag{9}$$

for steady-state units.

Then three cases are possible:

(1) $R^M = R^N$. The model considers in its equations all the distributed parameters given in the inlet streams. In this case, the explicit solution works well, since all distributed parameters can be calculated directly.

(2) $R^M \notin R^N$. The model will not work, since it requires more information about distributed parameters than is available in the flowsheet.

(3) $R^M \subset R^N$. The explicit scheme will provide proper results in the M-dimensional parameter space, but will fail to deliver correct values for R^N/R^M, since $\{F_H\}$ and $\{F_Y\}$ cannot be properly applied for other dimensions, beyond those for which they were defined.

In the third case, all the disadvantages and limitations of the explicit approach become apparent:

- The whole set of possible distributed parameters must be known during the development of the model and they all should be considered in its equations.
- If the number or composition of the distributed parameters alters in an individual simulation, the model itself should track these changes and react to them properly.
- The model must ensure setting all distributions defined at the input to its holdup and output, even those that are not explicitly considered in the model.
- Considering only those distributed parameters that are necessary for the model leads to the loss of information about the remaining ones, despite the fact that there may be enough information to calculate them.

Thus, the use of the method based on transformation matrices is an option for simulation systems with an unlimited number of distributions, since it allows treating all variables correctly, regardless of the number and type of defined distributed parameters. In this case, the model functions are used not directly, but to derive the laws θ of material transition between all classes:

$$\begin{cases} T_H(t) = \theta_H[F_H(t)] \\ T_Y(t) = \theta_Y[F_Y(t)] \end{cases},$$

(10)

where T is the transformation matrix.

Define $\{I^N\}$ as a set of indices to address any value in the N-dimensional space, so that

$$\{I^N\} = i_1 i_2 \ldots i_N, \quad i_d \in [1 : L_d], \quad d \in [1 : N].$$

(11)

The dimension of the transformation matrix T is twice the number of input dimensions taken into account. Each element $T_{\{I^N\},\{J^N\}}$ of T denotes a fraction of material moving from the $\{I^N\}$–th cell of the input distribution to the $\{J^N\}$–th cell of the output distribution. The output values are obtained by applying the transformation matrix to the corresponding input. For example, for a dynamic unit, at each time step, the holdup is calculated as the transformation of the previous state of the unit while considering the inlet and the outlet streams. The output, in turn, is calculated by applying the transformation laws to the holdup. Accordingly, Equation (5) takes the form

$$\begin{cases} H(t + \Delta t) = T_H(t + \Delta t) \otimes H(t) \\ Y(t + \Delta t) = T_Y(t + \Delta t) \otimes H(t + \Delta t) \\ \{T_H\}, \{T_Y\} \in R^M \rightarrow \{H\}, \{Y\} \in R^N, \end{cases}$$

(12)

where denotes the operation of applying the transformation laws. Due to the method of calculating and applying the transformation matrix, this method provides the correct conversion between the N– and M–parameter spaces.

Application of transformation laws for steady-state units is a direct transformation of input variables into output variables, so Equation (6) becomes

$$Y(t) = T_Y(t) \otimes X(t)$$
$$\{T_Y\} \in R^M \rightarrow \{Y\} \in R^N.$$

(13)

Consider the steady-state case (13) with any $X \in \{X(t)\}$ and $Y \in \{Y(t)\}$. For a simple 1D case with $N = 1$ and $R^M = R^N$, the operation of applying the transformation matrix can be written as

$$\forall j_1 \in [1 : L_1] : Y_{j_1} = \sum_{i_1=1}^{L_1} X_{i_1} \cdot T_{i_1, j_1}.$$

(14)

Then, similarly for the general case with $R^M = R^N$:

$$\forall j_1 \in [1 : L_1], \forall j_2 \in [1 : L_2], \ldots, \forall j_N \in [1 : L_N] :$$

$$Y_{j_1 j_2 \ldots j_N} = \sum_{i_1=1}^{L_1} \sum_{i_2=1}^{L_2} \cdots \sum_{i_N=1}^{L_N} X_{i_1 i_2 \ldots i_N} \cdot T_{i_1 i_2 \ldots i_N, j_1 j_2 \ldots j_N} \tag{15}$$

or using Equation (11):

$$\forall \{J^N\} : Y_{\{J^N\}} = \sum_{\{I^N\}} X_{\{I^N\}} \Delta T_{\{I^N\},\{J^N\}}. \tag{16}$$

Then for the case $R^M \subset R^N$, Equation (16) becomes the following form:

$$\forall \{J^N\} : Y_{\{J^N\}} = \sum_{\{I^N\}} X_{\{I^N\}} \cdot T_{\{I^M\},\{J^M\}}. \tag{17}$$

For example, for a three-dimensional case, if N is defined for $\{L_1, L_2, L_3\}$ and M is defined for $\{L_2\}$, Equation (17) can be written as

$$\forall j_1 \in [1 : L_1], \forall j_2 \in [1 : L_2], \forall j_3 \in [1 : L_3] : Y_{j_1 j_2 j_3} = \sum_{i_1=1}^{L_1} \sum_{i_2=1}^{L_2} \sum_{i_3=1}^{L_3} X_{i_1 i_2 i_3} \cdot T_{i_2, j_2}. \tag{18}$$

In this way, the transformation matrix calculated for the M-dimensional unit can be applied to calculate the N-dimensional output from the N-dimensional input, even if $R^M \subset R^N$. Equations (14)–(18) can be similarly derived for dynamic units (Equation (12)).

Usually, models in solids processing technology are described using equations that directly compute output distributions. Therefore, obtaining the transformation laws θ (Equation (10)) is an additional and nontrivial task that may require significant reformulations of existing models. This paper shows how this can be accomplished for agglomeration and breakage processes.

2.2. Agglomeration

The finite volume method developed by Kumar et al. [44] is used to get the discretized form of the pure agglomeration population balance equation. Taking $S(x) = 0$ and $b(x, y) = 0$ in Equation (1), the general agglomeration PBE for batch process can be written as

$$\frac{\partial u(x,t)}{\partial t} = \frac{1}{2} \int_0^x \beta(x', x - x\prime) u(x', t) u(x - x', t) dx\prime - \int_0^\infty \beta(x, x') u(x, t) u(x', t) dx' \tag{19}$$

with initial data $u(x, 0) = u_0(x) \geq 0$.

Take the whole domain as $[0, R]$. Divide the domain into L_1 cells with boundaries $x_{i-\frac{1}{2}}$ and $x_{i+\frac{1}{2}}$ for $i = 1, 2, \ldots, L_1$ where $x_{\frac{1}{2}} = 0$ and $x_{L_1+\frac{1}{2}} = R$. Take the representative of the i–th cell $\left[x_{i-\frac{1}{2}}, x_{i+\frac{1}{2}}\right]$ as $x_i = \left(\frac{x_{i+\frac{1}{2}} + x_{i-\frac{1}{2}}}{2}\right)$ and cell length $\Delta x_i = x_{i+\frac{1}{2}} - x_{i-\frac{1}{2}}$.

Define the initial approximation $u_0(x)$ as

$$u(0, x) = \frac{1}{\Delta x_i} \int_{x_{i-\frac{1}{2}}}^{x_{i+\frac{1}{2}}} u_0(x) dx. \tag{20}$$

To get the discretized form of the agglomeration Equation (19), calculate the total birth and death in each cell. For that, collect all those particles from the lower cell which is going to a particular higher cell. This is done by defining some set of indices. Here, the index set is calculated as follows

$$\mathbb{I}^i = \left\{ (j,k) \in L_1 \times L_1 : x_{i-\frac{1}{2}} < \left(x_j + x_k \right) \leq x_{i+\frac{1}{2}} \right\}. \tag{21}$$

Integrating Equation (19) over each cell $\left[x_{i-\frac{1}{2}}, x_{i+\frac{1}{2}} \right]$, and introducing weights as given by Kumar et al. [44], the discretized form can be obtained

$$\frac{du_i}{dt} = \frac{1}{2} \sum_{(j,k) \in \mathbb{I}^i} \beta_{jk} u_j u_k \frac{\Delta x_j \Delta x_k}{\Delta x_i} w_{jk}^b - \sum_{j=1}^{L_1} \beta_{ij} u_i u_j \Delta x_j w_{ij}^d. \tag{22}$$

where w_{jk}^b and w_{ij}^d are the weights responsible for preservation of number and conservation of mass, respectively. These weights are defined as

$$w_{jk}^b = \begin{cases} \frac{x_j + x_k}{2x_{l_{jk}} - (x_j + x_k)}, & x_j + x_k \leq R \\ 0, & x_j + x_k > R \end{cases} \tag{23}$$

and

$$w_{ij}^d = \begin{cases} \frac{x_{l_{ij}}}{2x_{l_{ij}} - (x_i + x_j)}, & x_i + x_j \leq R \\ 0, & x_i + x_j > R \end{cases}. \tag{24}$$

Here l_{ij} is the symmetric index of the cell, where the agglomerate particle $\left(x_i + x_j \right)$ falls. Rewrite Equation (22) as

$$\frac{d\Delta x_i u_i}{dt} = \frac{1}{2} \sum_{(j,k) \in \mathbb{I}^i} \beta_{jk} u_j u_k \Delta x_j \Delta x_k w_{jk}^b - \sum_{j=1}^{L_1} \beta_{ij} u_i u_j \Delta x_i \Delta x_j w_{ij}^d. \tag{25}$$

Take $g_i = \Delta x_i u_i$, then Equation (25) gives

$$\frac{dg_i}{dt} = \frac{1}{2} \sum_{(j,k) \in \mathbb{I}^i} \beta_{jk} g_j g_k w_{jk}^b - \sum_{j=1}^{L_1} \beta_{ij} g_i g_j w_{ij}^d. \tag{26}$$

To get the particle number at some certain time point τ, the time span $[0, \tau]$ is divided into K subintervals $[t, t + \Delta t]$, where $t_1 = 0$, $t_K = \tau$ and Δt is the time step.

Now the particle number $g(t + \Delta t)$ at time point $t + \Delta t$ using transformation matrix $T(t, \Delta t)$ is given by

$$g(t + \Delta t) = g(t) \cdot T(t, \Delta t). \tag{27}$$

Here $g(t)$ is a $1 \times L_1$ row matrix with i-th element $g_i(t)$, $i = 1, 2, 3, \ldots, L_1$ at time t, and $T(t, \Delta t)$ is a $L_1 \times L_1$ transformation matrix, calculated at time t to obtain the distribution after time step Δt. The elements of the transformation matrix are given by

$$T_{ij}(t, \Delta t) = \begin{cases} 1 + \left(\frac{1}{2} \sum\limits_{(i,k) \in \mathbb{I}^i} \beta_{ik} g_k(t) w_{ik}^b - \sum\limits_{j=1}^{L_1} \beta_{ij} g_j(t) w_{ij}^d \right) \Delta t, & i = j \\ \frac{1}{2} \sum\limits_{(i,k) \in \mathbb{I}^i} \beta_{ik} g_k(t) w_{ik}^b \Delta t, & i < j \\ 0, & i > j \end{cases}. \tag{28}$$

To improve the accuracy of the scheme (27), the Runge–Kutta method is proposed:

$$g(t + \Delta t) = g(t) \cdot \widetilde{T}(t, \Delta t) \tag{29}$$

where

$$\widetilde{T}(t, \Delta t) = \left[T(t, \Delta t) \cdot T\left(t + \Delta t, \frac{\Delta t}{2}\right) + \left(I - T\left(t, \frac{\Delta t}{2}\right)\right) \right]. \tag{30}$$

Here I is the identity matrix. The method given by Equations (29) and (30) gives the second order accuracy.

Since the scheme is explicit, there would be some restriction on the time step to get a non-negative solution of the scheme (Equation (29)). Therefore, the Courant-Friedrichs-Lewy (CFL) condition [46] on the time step is given by

$$\Delta t \leq \min_i \left| \frac{2g_i(t)}{Q(t)} \right|, \tag{31}$$

where

$$Q(t) = W_1(g(t)) - W_2(g(t)) + W_1(g^*(t)) - W_2(g^*(t)),$$

$$W_1(g(t)) = \frac{1}{2} \sum_{(j,k) \in \mathbb{I}^i} \beta_{jk} g_j(t) g_k(t) w_{jk}^b,$$

$$W_2(g(t)) = \sum_{j=1}^{L_1} \beta_{ij} g_i(t) g_j(t) w_{ij}^d, \tag{32}$$

$$g_i^*(t) = g_i(t) + \Delta t \cdot W_1(g(t)) - W_2(g(t))$$

2.3. Breakage

To get the equation for the pure breakage, put $\beta(x, x') = 0$ in Equation (1). Hence, the breakage process can be written as the following linear differential equation

$$\frac{\partial u(x, t)}{\partial t} = \int_x^\infty b(x, y) S(y) u(y, t) dy - S(x) u(x, t). \tag{33}$$

So, integrating Equation (33) over the cell $\left[x_{i-\frac{1}{2}}, x_{i+\frac{1}{2}} \right]$, can be obtained

$$\frac{du_i}{dt} = B_i - D_i, \tag{34}$$

where

$$B_i = \frac{1}{\Delta x_i} \int_{x_{i-\frac{1}{2}}}^{x_{i+\frac{1}{2}}} \int_x^{x_{L_1+\frac{1}{2}}} b(x, \epsilon) S(\epsilon) u(\epsilon, t) d\epsilon dx \tag{35}$$

and

$$D_i = \frac{1}{\Delta x_i} \int_{x_{i-\frac{1}{2}}}^{x_{i+\frac{1}{2}}} S(x) u(x, t) dx, \tag{36}$$

where L_1 is the number of discrete size classes.

Take the approximation of the initial data as

$$u(x, 0) = \frac{1}{\Delta x_i} \int_{x_{i-\frac{1}{2}}}^{x_{i+\frac{1}{2}}} u_0(x) dx. \tag{37}$$

Proceeding in the same way as Kumar et al. [43], the discretized form with two weight functions can be obtained as

$$\frac{du_i}{dt} = \frac{1}{\Delta x_i} \sum_{k=i}^{L_1} \varphi_k^b S_k u_k \Delta x_k B_{ik} - \psi_i^d S_i u_i, \tag{38}$$

where the weighted parameters for birth and death φ_i^b and ψ_i^d are given by

$$\varphi_i^b = \frac{x_i(\vartheta(x_i) - 1)}{\sum_{k=1}^{i-1}(x_i - x_k)B_{ki}} \tag{39}$$

and

$$\psi_i^d = \frac{\varphi_i^b}{x_i} \sum_{k=1}^{i} x_k B_{ki}, \quad i = 2, 3, 4, \dots, L_1 \ . \tag{40}$$

Take $\varphi_1^b = 0$, $\psi_1^d = 0$ and

$$B_{ik} = \int_{x_{i-\frac{1}{2}}}^{p_k^i} b(x, x_k)dx \tag{41}$$

with

$$p_k^i = \begin{cases} x_i, & k = i \\ x_{i+\frac{1}{2}}, & k \neq i \end{cases} \tag{42}$$

and

$$\vartheta(x_i) = \int_0^{x_i} b(x, x_i)dx. \tag{43}$$

Multiplying Δx_i on both sides of Equation (38), one obtains

$$\frac{d\Delta x_i u_i}{dt} = \sum_{k=i}^{L_1} \varphi_k^b S_k u_k \Delta x_k B_{ik} - \psi_i^d S_i u_i \Delta x_i. \tag{44}$$

Take $g_i = \Delta x_i u_i$. Then, Equation (44) can be rewritten as

$$\frac{dg_i}{dt} = \sum_{k=i}^{L_1} \varphi_k^b S_k g_k B_{ik} - \psi_i^d S_i g_i. \tag{45}$$

To get the particle number at some certain time point τ, the time span $[0, \tau]$ is divided into K subintervals $[t, t + \Delta t]$, where $t_1 = 0$, $t_k = \tau$ and Δt is the time step.

The particle number $g(t + \Delta t)$ at time point $t + \Delta t$ using transformation matrix $T(t, \Delta t)$ is given by

$$g(t + \Delta t) = g(t) \cdot T(t, \Delta t), \tag{46}$$

where $g(t)$ is a $1 \times L_1$ row matrix (vectors) with elements denoted by $g_i(t)$, and $T(t, \Delta t)$ is a $L_1 \times L_1$ matrix whose elements are denoted by $T_{ij}(t, \Delta t)$. The elements of the transformation matrix can be written with the help of the discretized form of Equation (33) as follows

$$T_{ij}(t, \Delta t) = \begin{cases} 1 + \left(\varphi_i^b B_{ii} - \psi_i^d\right)S_i \Delta t, & i = j \\ S_i B_{ji}\varphi_i^b \Delta t, & i > j \\ 0, & i < j \end{cases} \tag{47}$$

The scheme (Equation (46)) gives the first-order accuracy. To get a higher order accuracy,

$$g(t + \Delta t) = g(t) \cdot \widetilde{T}(t, \Delta t), \tag{48}$$

is proposed using the Runge–Kutta method. Here

$$\widetilde{T}(t, \Delta t) = \left[T(t, \Delta t) \cdot T\left(t + \Delta t, \frac{\Delta t}{2} \right) + \left(I - T\left(t, \frac{\Delta t}{2} \right) \right) \right]. \tag{49}$$

I is the identity matrix. The method given by Equations (48) and (49) gives the second order accuracy.

Consider a positive initial data $g_i(0)$ for all i. However, the solution $g_i(t)$ may become negative. In order to ensure positivity, a condition for the time step must be defined:

$$\Delta t \leq \min_i \left| \frac{2g_i(t)}{Q(t)} \right|, \tag{50}$$

where

$$Q(t) = W_1(g(t)) - W_2(g(t)) + W_1(g^*(t)) - W_2(g^*(t)),$$

$$W_1(g(t)) = \sum_{k=i}^{L_1} \varphi_k^b S_k g_k(t) B_{ik} - \psi_i^d S_i g_i(t), \tag{51}$$

$$W_2(g(t)) = \psi_i^d S_i g_i(t),$$

$$g_i^*(t) = g_i(t) + \Delta t \cdot W_1(g(t)) - W_2(g(t)).$$

3. Implementation Details

3.1. Dynamic Flowsheet Simulation in Dyssol

In this work, the Dyssol modelling framework [7] was used to perform simulations using the newly proposed methods. The system is being developed within the priority program of the German Research Foundation (DFG) SPP 1679 "Dynamic simulation of interconnected solids processes DYNSIM-FP" [47].

Dyssol is written entirely in the C++ programming language [48] using a modular architecture and an object-oriented approach [49]. This made it possible to split the entire program into separate, relatively independent modules, to facilitate their development. They are integrated into a single system using specially designed software interfaces. In particular, each model (e.g. agglomerator, mill, or screen) is an independent software library that can be created separately, and then connected to the simulation program. This greatly increases the flexibility and the extensibility of the entire system and simplifies its development and maintenance.

Due to the mathematical complexity and diversity of models in the area of solids processing technology, the calculations in Dyssol are carried out according to the sequential-modular approach [50]. Each model is calculated independently of the others, using its own most appropriate equations solvers and computational methods. The task of the simulation system itself in this case is to manage the information and material flows between the models in accordance with the flowsheet structure and to ensure convergence. The sequential-modular approach simplifies the development of new models, since they are less limited to the implementation and functionality of the modelling framework. Moreover, this method can be effectively applied even at the scale of individual process units [51].

Application of the modular approach imposes some restrictions on the flowsheet structures. In particular, schemes containing recycle flows cannot be calculated directly and require additional treatment. Before starting the simulation, all such flows must be identified and initialized with some predefined values in order to break the cycles [52]. After that, all units inside the recycle loop are calculated iteratively until convergence. To speed up this process, the initial parameters of recycles are calculated anew at each iteration using convergence methods [7,53].

Reaching convergence over a relatively large time interval can be very difficult and can become a resource-intensive task. To overcome this problem, dynamic modelling of flowsheets with recycle flows is performed using an approach based on the waveform relaxation method [54]. Here, iterative calculations are performed only for a certain short time interval [6,15], where convergence can be reached much faster. When the calculation of the current time interval is completed, the system uses data interpolation methods [7] to initialize the next time window and proceeds to its calculation.

3.2. Application of Transformation Matrices

From a mathematical point of view, the application of a full transformation matrix (Equation (16)) is a relatively simple operation and can be described as the regular multiplication of matrices. However, with a large number of matrix dimensions and a large number of classes, such a naive calculation will be very demanding in terms of computational capabilities and memory consumption. Moreover, the need to work with time-dependent parameters imposes additional restrictions on data structures. Therefore, all distributed parameters in Dyssol are stored in sparse format, represented by specially developed tree data structures [7], as is schematically shown in Figure 2 for a single time point.

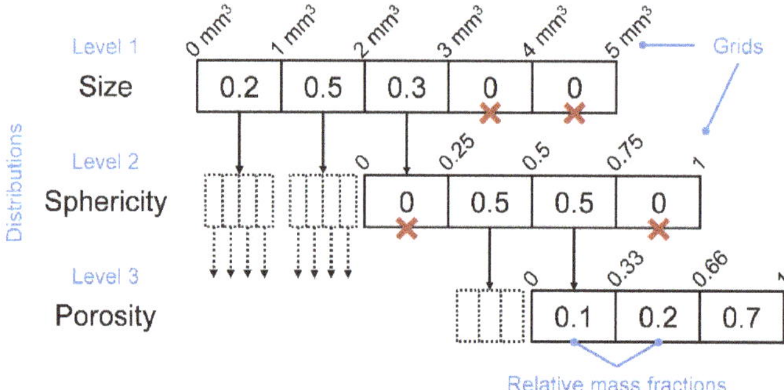

Figure 2. Tree data structure for storing interdependent multidimensional distributed parameters of solids in Dyssol.

Here, each hierarchy level describes the distribution of material over a specific parameter, and each entry is the mass fraction of material that has a specific combination of parameters. Since distributions often contain a lot of zero values, such a structure can drastically reduce the amount of data, which needs to be stored and processed. In addition, this data structure allows the use of relative mass fractions for each level of the hierarchy, instead of operating with absolute mass fractions. Thus, from Figure 2 it follows that there are 30% of particles with a size of $(2 \text{ mm}^3 + 3 \text{ mm}^3)/2 = 2.5 \text{ mm}^3$; of these, 50% have a sphericity of $(0.5 + 0.75)/2 = 0.625$; and from them, 20% have a porosity of $(0.33 + 0.66)/2 = 0.5$. The total percentage of particles with this combination of parameters is $0.3 \times 0.5 \times 0.2 = 3\%$.

The use of absolute mass fractions and tree structures allows for working with individual distributions without changing information in the others. So, transformation matrices can be effectively applied for each level separately and this will not affect all distributions located at higher hierarchical levels.

Take the N–dimensional distribution represented by N hierarchy levels, where L_d is the number of classes in d–th dimension and X_q^p and Y_q^p are relative mass fractions in the q–th class of the p–th level in the initial and the transformed distribution, respectively (Figure 3).

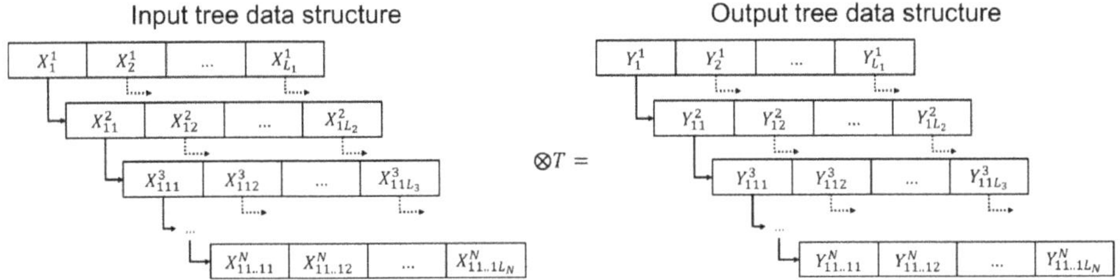

Figure 3. The scheme of application of the transformation matrix for the N–dimensional distribution.

If the transformation matrix T is calculated for M–dimensional parameter space, its application will consist of three stages:

1. Apply T to the lowest M–th hierarchy level of input distribution, using Equation (17) in the form of

$$Y^M_{j_1 j_2 \cdots j_{M-1} j_M} = \sum_{i_1=1}^{L_1} \sum_{i_2=1}^{L_2} \cdots \sum_{i_{M-1}=1}^{L_{M-1}} \sum_{i_M=1}^{L_M} X^M_{i_1 i_2 \cdots i_M} \cdot X^{M-1}_{i_1 i_2 \cdots i_{M-1}} \cdot \ldots \cdot X^2_{i_1 i_2} \cdot X^1_{i_1} \cdot T_{i_1 i_2 \cdots i_M, j_1 j_2 \cdots j_M}. \tag{52}$$

For example, for $M = 2$, the second level of the hierarchy should be calculated as

$$Y^2_{j_1 j_2} = \sum_{i_1=1}^{L_1} \sum_{i_2=1}^{L_1} X^2_{i_1 i_2} \cdot X^1_{i_1} \cdot T_{i_1 i_2, j_1 j_2} \tag{53}$$

2. Extract the transformation laws for the previous level $(M-1)$ from T and apply them using Equation (52). Repeat it up to the hierarchy level 1. For example, for $M = 2$, the first level will be calculated as

$$Y^1_{j_1} = \sum_{i_1=1}^{L_1} X^1_{i_1} \cdot T_{i_1, j_1}, \tag{54}$$

which corresponds to Equation (14).

3. Apply T to calculate the remaining levels K below M as

$$Y^K_{j_1 j_2 \cdots j_{K-1} j_K} = \frac{\sum_{i_1=1}^{L_1} X^K_{i_1 j_2 \cdots j_K} \cdot X^{K-1}_{i_1 j_2 \cdots j_{K-1}} \cdot \ldots \cdot X^2_{i_1 j_2} \cdot X^1_{i_1} \cdot T_{i_1, j_1}}{Y^{K-1}_{j_1 j_2 \cdots j_{K-1}} \cdot Y^{K-2}_{j_1 j_2 \cdots j_{K-2}} \cdot \ldots \cdot Y^2_{j_1 j_2} \cdot Y^1_{j_1}}, \tag{55}$$

$$M < K \leq N$$

For example, if $M = 1$ and $K = 2$:

$$Y^2_{j_1 j_2} = \frac{\sum_{i_1=1}^{L_1} X^2_{i_1 j_2} \cdot X^1_{i_1} \cdot T_{i_1, j_1}}{Y^1_{j_1}}. \tag{56}$$

Thus, using Equations (52) and (55), as well as sorting the levels in the hierarchical data structure, one can apply the transformation matrix of any complexity to any input distribution.

Depending on the unit type, the conventional algorithm for applying transformation matrices varies:

- For steady state units: copy the distribution from the input to the output, then apply the transformation matrix to the output.

- For dynamic units: use the input distribution to calculate the holdup, apply the transformation matrix to the holdup, and then calculate the output.

In the latter case, because of time discretization, the mixing of the input material with the holdup and the application of the transformation matrix are performed sequentially.

3.3. Implementation of Units

To verify and investigate the proposed new methods, the Dyssol simulation system has been extended with several models that have been implemented using the provided program interfaces.

3.3.1. Agglomerator

The new method of solving the agglomeration PBE was implemented as a new model of the Dyssol simulation system according to Equation (29). To simplify the model, it was assumed that the mass within the apparatus remains constant and there is no classification by size at the output. Therefore, the mass flow of material leaving the apparatus $\dot{m}_{out}(t)$ equals to the mass flow entering it $\dot{m}_{in}(t)$, and the distribution of the material at the output $Y(t)$ equals to the distribution in the holdup $H(t)$.

Since the applicability of the proposed method does not depend on the chosen agglomeration kernel, to describe the agglomeration rate in all case studies, a frequently used physically relevant kernel based on a Brownian motion [55,56] was applied in the following from:

$$\beta(x, x') = \beta_0 \beta^* \left(x^{\frac{1}{3}} + x'^{\frac{1}{3}} \right) \left(x^{-\frac{1}{3}} + x'^{-\frac{1}{3}} \right),$$

$$\beta^* = \begin{cases} 1, & x + x' \le x_{a,min} \\ 1 - \frac{(x+x')-x_{a,min}}{x_{a,max}-x_{a,min}}, & x_{a,min} < x + x' < x_{a,max} \\ 0, & x + x' \ge x_{a,max} \end{cases} \tag{57}$$

Here $\beta(x, x')$ denotes the agglomeration rate between particles x and x'; β_0 is the size-independent agglomeration rate constant, dependent on operating conditions; β^* is the size-dependent agglomeration rate constant; and $x_{a,min}$ and $x_{a,max}$ are minimum and maximum sizes of resulting agglomerates.

The agglomerator unit was implemented according to Equations (28) and (29) to calculate and apply the transformation matrix for the particles size. To consider secondary distributions, it was extended to perform aggregation with averaging of the properties distributed over the second dimension, which is calculated as follows.

Let particles of two sizes x_{src1} and x_{src2} aggregate into a particle of size x_{dst}. Then $T_{src1,dst}$ and $T_{src2,dst}$ are the corresponding entries of the transformation matrix, calculated by Equation (28). The summarized mass fraction of particles from size-class i is distributed over L_2 classes of the secondary dimension, so that

$$c_i = \sum_{j=1}^{L_2} c_{i,j}, \tag{58}$$

where $c_{i,j}$ is a mass fraction of granular material with size x_i and value of the secondary dimension falling into class j. If particles from two classes $(src1, j1)$ and $(src2, j2)$ aggregate, the result lands into some class (dst, j). The size-class dst is directly determined from the agglomeration mechanism and obtained from the transformation matrix. The mass fraction of each j–th class of the secondary distribution is calculated as

$$c_{dst,j} = \sum_{j1=1}^{L_2} \sum_{j2=1}^{L_2} \frac{\left(c_{src1,j1} T_{src1,dst} \right) \cdot \left(c_{src2,j2} \cdot T_{src2,dst} \right)}{\sum_{k=1}^{S_2} c_{src2,k} \cdot T_{src2,dst}} \tag{59}$$

whereas the class index j for equidistant grid can be obtained as

$$j = \frac{j1 \cdot c_{src1,j1} \cdot T_{src1,dst} + j2 \cdot c_{src2,j2} \cdot T_{src2,dst}}{c_{src1,j1} \cdot T_{src1,dst} + c_{src2,j2} \cdot T_{src2,dst}}. \tag{60}$$

3.3.2. Mill

The mill model, developed in Dyssol, implements Equations (47) and (48) to calculate transformation laws and Equations (52) and (55) to apply them. To calculate Equations (41) and (43), the adaptive Simpson's method of numerical integration [57] was applied. The criterion used to determine when to terminate interval subdivision was chosen according to Lyness [58], and its desired tolerance was set to 10^{-15}.

For simplification, the holdup mass within the unit stays constant and no classification by size occurs at the outlet.

As can be seen from Section 2.3, the proposed approach does not depend on the chosen selection or breakage functions. Therefore, for testing purposes, the following physically relevant models were taken from literature. To describe the selection rate in all case studies, the selection function based on the model proposed by King [59] was applied:

$$S(x) = s_0 \cdot \begin{cases} 0, & x \leq x_{b,min} \\ 1 - \left(\frac{x_{b,max} - x}{x_{b,max} - x_{b,min}}\right)^n, & x_{b,min} < x < x_{b,max} \\ 1, & x \geq x_{b,max} \end{cases} \tag{61}$$

where $S(x)$ is the mass fraction of particles of size x that are crushed; s_0 introduces a size-independent selection rate factor; $x_{b,min}$ and $x_{b,max}$ are critical sizes of particles; and n is a power law exponent of the King's selection function.

The breakage function was implemented according to Vogel et al. [60] in a number-based distributed form:

$$B(x,y) = \begin{cases} 0.5 \frac{q}{y} \left(\frac{x}{y}\right)^{q-2} \left(1 + tanh\left(\frac{y - y'}{y'}\right)\right), & y \geq x \\ 0, & y < x \end{cases}. \tag{62}$$

Here $B(x,y)$ denotes the mass fraction of particles with the size of y, which after breakage get the size less or equal to x; y' is the minimum fragment size that can be achieved after milling; and q is a power law exponent of the Vogel breakage function.

3.3.3. Screen

The screen apparatus is a steady-state unit, which is described in terms of the grade efficiency $G(x)$, calculated according to the model of Plitt [61] as

$$G(x) = 1 - exp\left(-0.693\left(\frac{x}{x_{cut}}\right)^\alpha\right). \tag{63}$$

$G(x)$ determines the mass fraction of material of size x, which leaves the screen through an outlet for coarse particles; x_{cut} is the cut size of the classification model; α is the separation sharpness.

For proper consideration of the distributed parameters, each deck of the screen unit formulates two diagonal transformation matrices for the particle size distribution (PSD): for coarse (T^c) and for fines (T^f) output, so that

$$\begin{aligned} T^c_{i,i} &= G(x_i) \\ T^f_{i,i} &= 1 - G(x_i). \end{aligned} \tag{64}$$

The total mass flows of coarse $\dot{m}^{c}_{out}$ and fines $\dot{m}^{f}_{out}$ outlets can be calculated as

$$\dot{m}^{c}_{out} = \dot{m}_{in} \sum_{i=1}^{L_1} G(x_i) \cdot c_i$$

$$\dot{m}^{f}_{out} = \dot{m}_{in} \left(1 - \sum_{i=1}^{L_1} G(x_i) \cdot c_i \right), \tag{65}$$

where $\dot{m}_{in}$ is the mass flow at the inlet, c_i is the mass fraction of particles of size x_i, and L_1 is the number of size classes.

Application of the transformation matrices was performed according to Equations (52) and (55).

4. Simulation Examples

All simulations were performed in the dynamic flowsheet simulation system Dyssol using the developed models, described in Section 3.3.

4.1. Agglomeration

A simple pharmaceutical process of agglomerating blends with two different concentrations of the active pharmaceutical ingredient (API) was simulated. The process structure is illustrated in Figure 4. The solid material is described using two interdependent distributed parameters:

- Particle size representing the volume of particles ranging from 0 to 4×10^{-3} mm^3 and distributed over 100 equidistant classes; and
- the API concentration, which describes the mass content of an active ingredient from 0 to 10% divided into 500 equidistant classes.

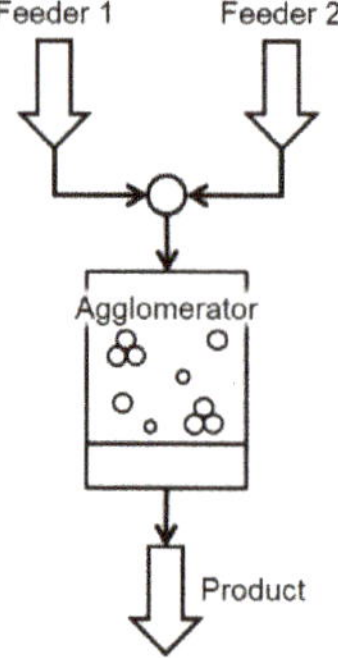

Figure 4. Flowsheet structure of the agglomeration process.

The initial blend entering the agglomerator is a mixture of materials from two feeders (Figure 5):

- Feeder 1 supplies smaller particles with a lower API concentration;
- Feeder 2 supplies larger particles with a higher API concentration.

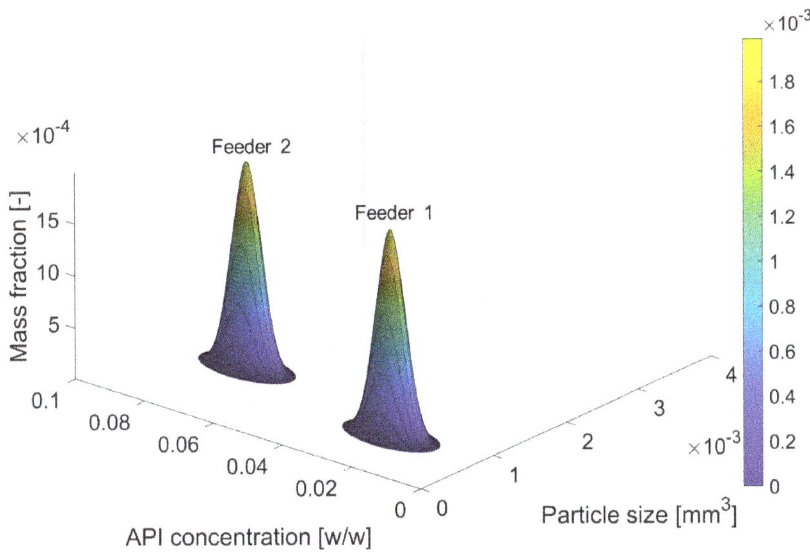

Figure 5. Input distribution for the agglomeration process.[2] To improve readability, in all the distributions shown here and further, values of mass fractions less than 10^{-5} were cut off.

Both particle sizes and API concentrations are given in terms of the Gaussian normal distribution with mean values μ and standard deviations σ shown in Table 1. Feeders continuously supply the material at a rate of 0.25 kg/s each. The final distribution after their mixing is shown in Figure 5. The agglomerator is initially filled with 200 kg of the same blend. All model parameters used to simulate the agglomeration process are listed in Table 1.

Table 1. Model parameters used for the agglomeration process.

Feeder 1		
Mass Flow	$\dot{m}$	0.25 kg/s
Mean value of the particle size distribution	μ_{size}	0.78×10^{-3} mm^3
Standard deviation of the particle size distribution	σ_{size}	0.8×10^{-4} mm^3
Mean value of the API concentration distribution	μ_{conc}	0.025
Standard deviation of the API concentration distribution	σ_{conc}	0.004
Feeder 2		
Mass Flow	$\dot{m}$	0.25 kg/s
Mean value of the particle size distribution	μ_{size}	1.18×10^{-3} mm^3
Standard deviation of the particle size distribution	σ_{size}	0.8×10^{-4} mm^3
Mean value of the API concentration distribution	μ_{conc}	0.075
Standard deviation of the API concentration distribution	σ_{conc}	0.004
Agglomerator		
Holdup mass	m	200 kg
Size-independent rate constant	β_0	2×10^{-16}
Minimum agglomeration size	$x_{a,min}$	0.4×10^{-3} mm^3
Maximum agglomeration size	$x_{a,max}$	3.75×10^{-3} mm^3

Figures 6 and 7 show the distribution of the product stream in steady-state, which was reached after 25 min of process time. Clearly visible individual peaks, which were formed as a result of the agglomeration of particles with different sizes and different API concentrations. Since the initial blend is supplied continuously at a constant rate during the entire agglomeration process, the initial distributions *A* and *B* (Figure 7) are still distinguishable in the final mixture. Agglomeration of material

B with itself gives F, which has a doubled mean size and the same API concentration as B, as expected. In turn, agglomeration of B with F (having the same concentrations) gives a peak at M, which does not alter API content and has a particle size equal to $(B + F)$. The same is true for the material from feeder 1, agglomerating with itself according to the schema: $(A + A) = C$, $(A + C) = E$, and $(A + E) + (C + C) = H$.

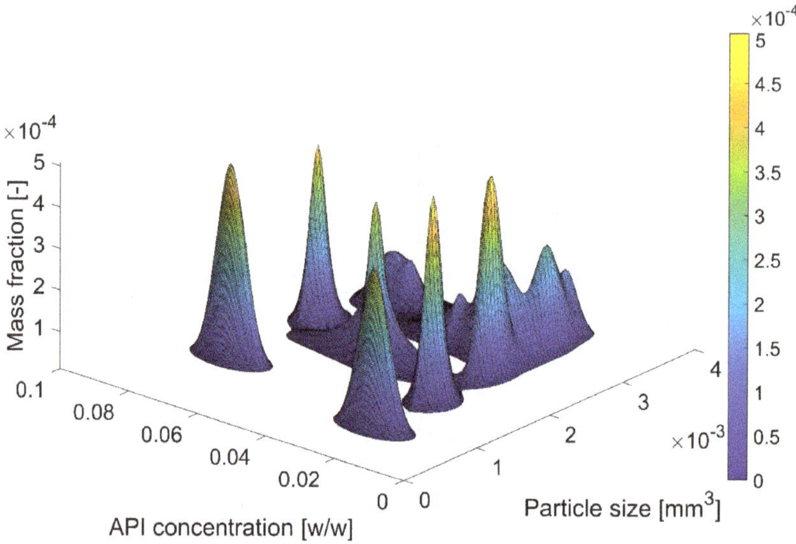

Figure 6. Simulation results of the agglomeration process.

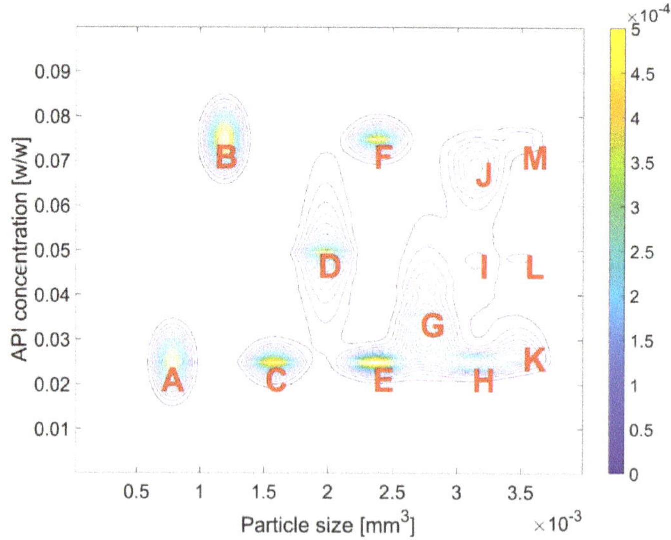

Figure 7. Agglomeration diagram.

More complex behaviour is observed if particles with different API concentrations agglomerate. In this case, the particle sizes are added up and the concentration is averaged according to Equation (59). For example, agglomeration of particles from A and B gives D. Further, G is formed as a mixture of $(A + D)$ and $(B + C)$. The result of the agglomeration of B and C is shifted below the concentration value of 0.05 due to the mutual influence of two factors. First, the introduction of the size-dependent growth rate β^* (Equation (57)) leads to faster growth of smaller particles. Second, averaging by the second distribution takes into account the actual amount of material being agglomerated in selected classes (see Equations (59) and (60)).

The complete scheme of material transitions during the agglomeration, shown in Figure 7, is represented in Table 2.

Table 2. Material transitions during agglomeration.

+	A	B	C	D	E	F	G
A	C	D	E	G	H	I	K
B		F	G	J	L	M	-
C			H	K	-	-	-

Thus, information on the change in particle size during the agglomeration process, obtained from the generated transformation matrix, made it possible to take into account and correctly process the dependent distributed parameters of the agglomerated material.

4.2. Breakage

To verify the proposed new method for solving the breakage PBE, a simplified process of milling a two-component blend was simulated. Figure 8 shows its flowsheet structure. Two feeders supply material with different particle sizes and API concentrations into the system. Their mixture, shown in Figure 9, enters the mill unit to be mixed with the holdup material and crushed afterwards.

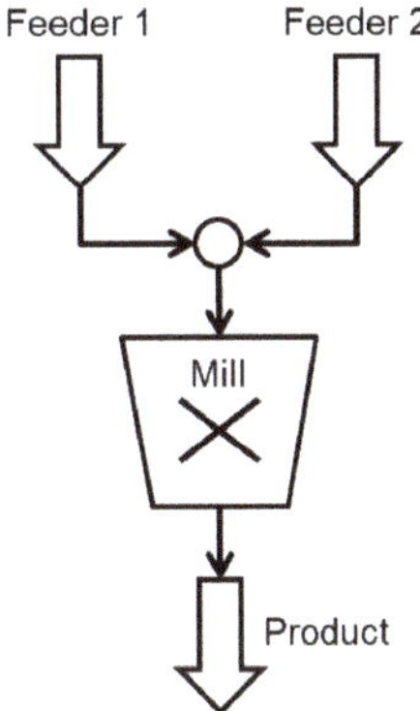

Figure 8. Flowsheet structure of the breakage process.

The initial blend entering mill is a mixture of smaller particles with a higher API concentration and larger particles with a lower API concentration. Both distributed parameters are given as Gaussian normal distributions with parameters from Table 3. The mill is initially filled with 50 kg of the same blend (Figure 9).

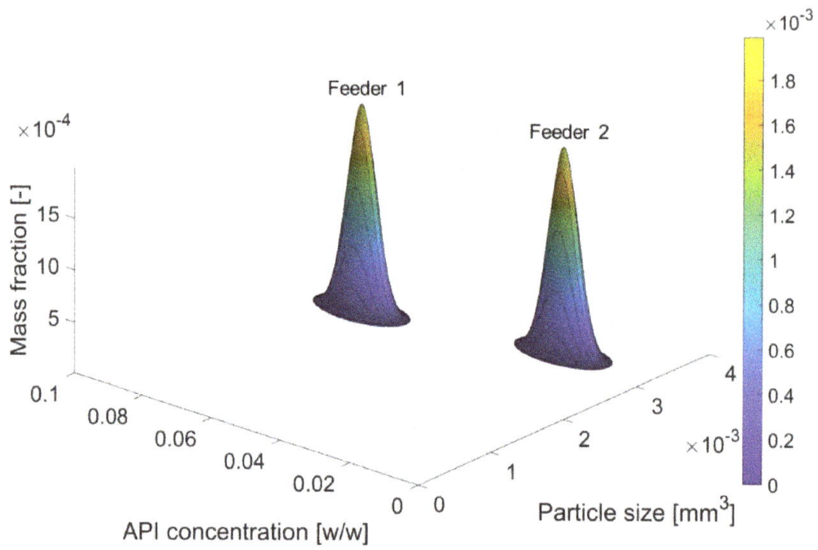

Figure 9. Initial distribution for the breakage process.

Table 3. Model parameters used for breakage process.

Feeder 1		
Mass Flow	$\dot{m}$	0.25 kg/s
Mean value of the particle size distribution	μ_{size}	3.18×10^{-3} mm^3
Standard deviation of the particle size distribution	σ_{size}	0.8×10^{-4} mm^3
Mean value of the API concentration distribution	μ_{conc}	0.025
Standard deviation of the API concentration distribution	σ_{conc}	0.004
Feeder 2		
Mass Flow	$\dot{m}$	0.25 kg/s
Mean value of the particle size distribution	μ_{size}	2.78×10^{-3} mm^3
Standard deviation of the particle size distribution	σ_{size}	0.8×10^{-4} mm^3
Mean value of the API concentration distribution	μ_{conc}	0.075
Standard deviation of the API concentration distribution	σ_{conc}	0.004
Mill		
Holdup mass	m	50 kg
Selection rate factor	s_0	3×10^{-2}
Minimum breakage size	$x_{b,min}$	1.5×10^{-3} mm^3
Maximum breakage size	$x_{b,max}$	4×10^{-3} mm^3
Selection power law exponent	n	3.1
Minimum fragment size	$y\prime$	0.5×10^{-3} mm^3
Breakage power law exponent	q	5.5

The particle size is described between 0 to 4×10^{-3} mm^3 with 100 equidistant classes, whereas the mass content of an active ingredient is distributed over 500 equidistant classes ranging from 0 to 10%.

All model parameters used to simulate breakage process are listed in Table 3.

The steady-state is reached in 25 min of the process time. Figures 10 and 11 show the distribution of the product after this time point.

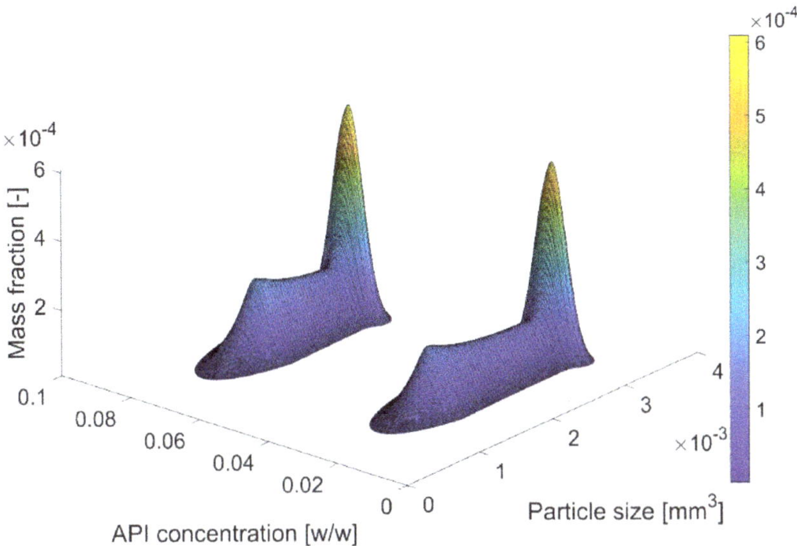

Figure 10. Simulation results of the breakage process.

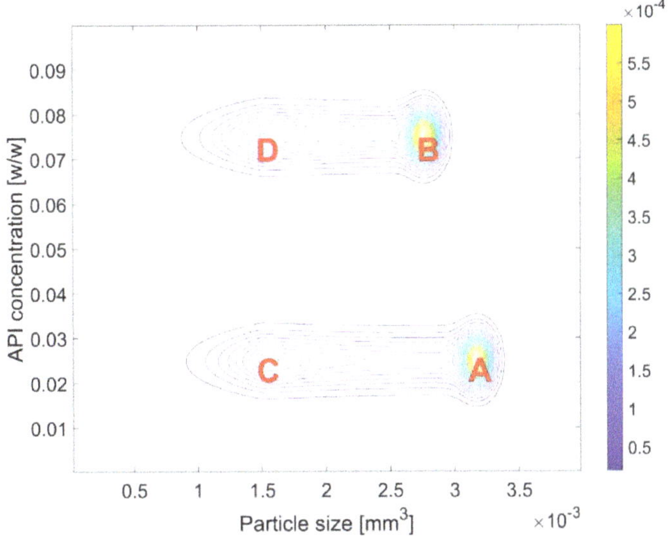

Figure 11. Breakage diagram.

Due to the constant supply of material during the whole process time, initial distributions *A* and *B* (Figure 11) are still clearly visible in the product. At the same time, it is evident from Figure 10 that the reduction in particle size caused by the breakage process does not lead to changes in the content of the active component: the distribution of the API concentration in the outlet stream remains the same as in the initial distribution. Both *A* and *B* do not mix, but only change particle size gradually grinding down from *A* to *C* and from *B* to *D*.

Thus, with the help of transformation matrices, it was possible to avoid mixing the secondary parameters, despite the fact that the model of the mill unit was developed considering only the particle size distribution.

4.3. Coupled Agglomeration and Breakage

Having both models of agglomerator and mill, it is possible to simulate part of a complex pharmaceutical process with a closed circuit and an external classification of the material, as it is shown in Figure 12. New material enters the process through two feeders, supplying particles with different sizes and concentrations of API, which mixture is shown in Figure 13. This blend is mixed with the holdup inside the agglomerator, where the growth of particles occurs. The two-compartment screen unit separates material leaving the agglomerator into three fractions. Oversized particles are crushed using the mill unit, and afterwards combined with the undersized particles and sent back to the agglomerator by mixing with external material from feeders. The middle-sized agglomerates are considered as product. Initially, both agglomerator and mill are filled with the same material as it comes from feeder 1.

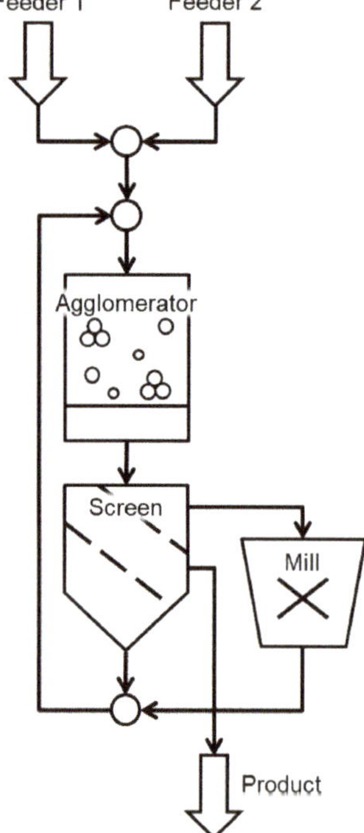

Figure 12. Flowsheet structure of a coupled agglomeration and breakage process.

All model parameters used to simulate the process from Figure 12 can be found in Table 4. Initial distributions of material are given by Gaussian function.

Table 4. Model parameters used for coupled agglomeration and breakage process.

Feeder 1		
Mass Flow	$\dot{m}$	0.15 kg/s
Mean value of the particle size distribution	μ_{size}	0.78×10^{-3} mm^3
Standard deviation of the particle size distribution	σ_{size}	2.0×10^{-4} mm^3
Mean value of the API concentration distribution	μ_{conc}	0.03
Standard deviation of the API concentration distribution	σ_{conc}	0.007
Feeder 2		
Mass flow	$\dot{m}$	0.35 kg/s
Mean value of the particle size distribution	μ_{size}	1.58×10^{-3} mm^3
Standard deviation of the particle size distribution	σ_{size}	3.2×10^{-4} mm^3
Mean value of the API concentration distribution	μ_{conc}	0.08
Standard deviation of the API concentration distribution	σ_{conc}	0.005
Agglomerator		
Holdup mass	m	200 kg
Size-independent rate constant	β_0	2×10^{-16}
Minimum agglomeration size	$x_{a,min}$	0.4×10^{-3} mm^3
Maximum agglomeration size	$x_{a,max}$	3.75×10^{-3} mm^3
Mill		
Holdup mass	m	50 kg
Selection rate factor	s_0	3×10^{-2}
Minimum breakage size	$x_{b,min}$	1.5×10^{-3} mm^3
Maximum breakage size	$x_{b,max}$	3.2×10^{-3} mm^3
Selection power law exponent	n	3.1
Minimum fragment size	$y\prime$	10^{-3} mm^3
Breakage power law exponent	q	5.5
Screen deck 1		
Cut size	x_{cut}	3.2×10^{-3} mm^3
Separation sharpness	α	13
Screen deck 2		
Cut size	x_{cut}	2.3×10^{-3} mm^3
Separation sharpness	α	13

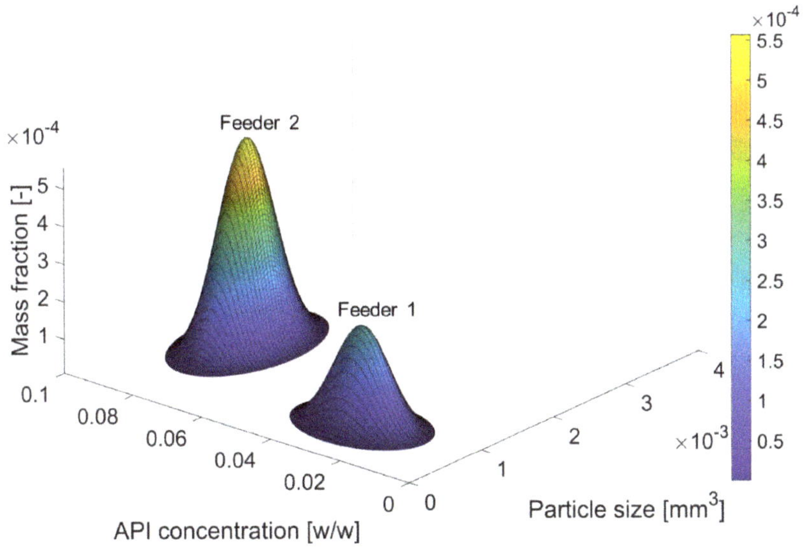

Figure 13. Initial distribution for the coupled agglomeration and breakage process.

The flowsheet was simulated until a steady state was reached, which occurred after 40 min of the process time. The distribution of the product is shown in Figure 14. There are two peaks for particles with a higher API concentration. One is mainly due to the presence of the source blend from feeder 2 (its coarse part obtained after the sieving), mixed with new particles formed in agglomerator and mill. The second peak is the result of agglomeration of particles with the same high API concentration with each other. For a low API content, only one peak is observed, since all smaller fractions (from agglomeration of the primary material from feeder 1 and milling results) were cut off by the screen unit. At the same time, a mixture of two initial blends, formed as a result of the agglomeration of materials with different API content, is present in a significant amount, but does not dominate the material with initial concentrations.

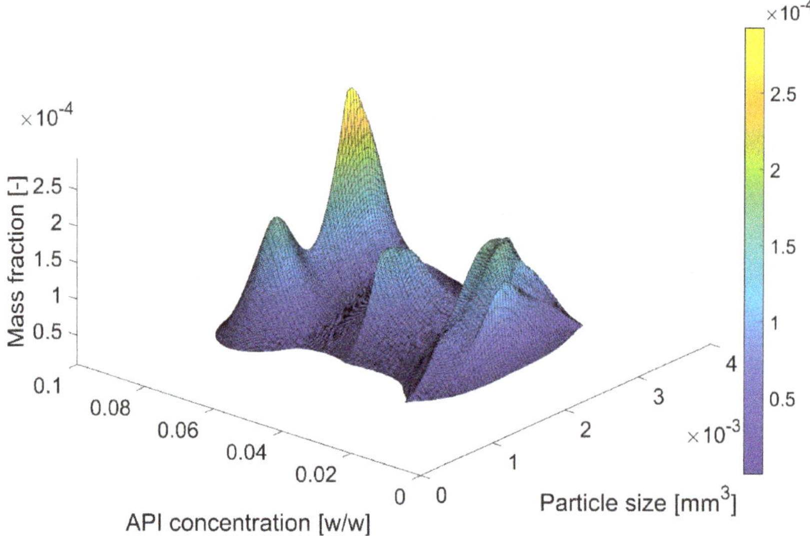

Figure 14. Distribution of the material in the product at steady-state.

In general, using the description of granular materials with multidimensional distributions and units developed on the basis of transformation matrices, it is possible to perform simulations of complex processes involving materials distributed over a large number of dimensions. Thus, in this example, it is possible to assess the homogeneity of the material that appears at the output of the simulated process, since this parameter is crucial in pharmaceutical production.

5. Conclusions

The use of transformation matrices is a prerequisite for working with multidimensional distributed parameters of solids. This approach allows for proper handling of all parameters of granular materials, even those which are not directly handled by the model. Consequently, the number of distributed parameters in the model and in its environment may not match. At the same time, the development of each individual model becomes more complicated due to the fact that in contrast to the traditional approach, when the output distribution is explicitly calculated and set to the outlet, it is required to derive the laws of material transformation between inputs and outputs for each time step.

In this article, the applicability of transformation matrices to the description of processes of agglomeration and crushing, which are usually formulated by the population balance equations, was shown. The proposed new methods allow for extracting information about the transformation laws from PBEs and calculating the transformation matrices on that basis. The approach uses the finite volume method for spatial discretization and the second-order Runge–Kutta method for obtaining the complete discretized form of the population balance equations.

To perform the test studies, the derived methods have been implemented in the dynamic flowsheet simulation environment Dyssol. Some exemplary production processes and their parts were simulated using the derived models to prove the applicability and advantages of presented approaches for modelling of granular materials described by multidimensional distributed parameters.

As case studies have shown, the use of transformation matrices enables the correct calculation of the dependent secondary parameters, avoiding their mixing during the simulation. The proposed method can be extended with additional laws for dependent parameters, for example, for mixing of materials. This will significantly expand its application scope and will allow simulating more sophisticated processes. In general, the proposed approach allows usage of a more complex description of materials with the help of several distributed parameters, without the need to adjust each specific model, and transferring the task of their correct calculation to the modelling system. In this way, the new method of reformulating population balance equations greatly increases the possibilities for creating generally applicable systems for the simulation of solid phase processes.

Author Contributions: Conceptualization: M.D. and J.K.; methodology: M.D., N.D., and V.S.; software: V.S.; validation: V.S. and N.D.; formal analysis: V.S.; investigation: V.S. and N.D; data curation: V.S.; writing—original draft preparation: V.S. and N.D.; writing—review and editing: V.S., N.D., M.D., S.H., and J.K.; visualization: V.S.; supervision: S.H., J.K., and M.D.; project administration: M.D.; funding acquisition: S.H., M.D., and J.K.

Funding: Authors gratefully acknowledge the financial support of the Deutsche Forschungsgemeinschaft (DFG, German Research Foundation) within the priority program SPP 1679 "Dynamic simulation of interconnected solids processes DYNSIM-FP"; of the German Academic Exchange Service (DAAD) via Research Grants—Bi-nationally Supervised Doctoral Degrees, 2017/18 (57299293); of the Alexander von Humboldt Foundation (Research Group Linkage Programme); and of the DFG—Projektnummer 392323616 and the Hamburg University of Technology (TUHH) via the funding programme "Open Access Publishing".

Notation

$b(x : y)$	fraction of particles of size x formed due to the breakage of a particle of size y
b_{ij}	fraction of particles of size-class i formed due to the breakage of a particle of size-class j
c	control variables of a unit
$c_{i,j}$	mass fraction of granular material falling into class (i, j) of two-dimensional distribution
d	index of a distributed property
F_H, F_Y	model functions of a unit for calculating holdup and output, respectively
G	grade efficiency of the screen unit
H	unit's holdup variables
I	identity matrix
L	total number of discrete classes for all distributed properties
L_d	number of discrete classes for distributed property d
$\dot{m}_{in}, \dot{m}_{out}$	mass flow at the inlet and at the outlet streams of a unit, respectively
M	number of distributed properties in a model
n	power law exponent of the King's selection function
N	number of distributed properties in a flowsheet
p	model parameters of a unit
q	power law exponent of the Vogel breakage function
$\dot{Q}_{in}, \dot{Q}_{out}$	distribution of particles by size at the input and at the output of a unit, respectively
r	design or structural parameters of a unit
R	maximum size of particles in the considered interval
R^N	parameter space formed by N distributed parameters
s_0	size-independent selection rate factor

$S(y)$	breakage rate of a particle of size y
S_i	breakage rate of a particle from size-class i
t	time
$T(t, \Delta t)$	transformation matrix, calculated at time t to obtain the distribution after time step Δt
T_{ij}	particular entry of a transformation matrix with specific indices
T_H, T_Y	transformation matrices to calculate holdups and output of a unit, respectively
$u(x)$	mass fraction of particles of size x
$u_0(x)$	mass fraction of particles of size x at the initial moment of time
u_i	mass fraction of particles from size-class i
w^b, w^d	weighting parameters for agglomeration, for birth and death of particles, respectively
x, x'	particle sizes
$x_{a,min}, x_{a,max}$	critical sizes of particles for agglomeration
$x_{b,min}, x_{b,max}$	critical sizes of particles for breakage
x_{cut}	cut size of the screen unit
x_i	size of particles in a discrete class i
X	unit's input variables
X_q^p	relative mass fraction, stored in the q–th class of the p–th hierarchical level in the initial distribution
y	particle size
y'	minimum fragment size of the Vogel breakage function
Y	unit's output variables
Y_q^p	relative mass fraction, stored in the q–th class of the p–th hierarchical level in the transformed distribution
α	separation sharpness of the screen unit
$\beta(x, x')$	agglomeration rate for particles of sizes x and x'
β_{ij}	agglomeration rate for particles from size-classes i and j
β_0	size-independent agglomeration rate constant
β^*	size-dependent agglomeration rate constant
Δx_i	length of size-class i
$\vartheta(y)$	total number of particles of size y appearing after breakage
θ_H, θ_Y	model functions of a unit in terms of the laws of material transition
μ_{conc}	mean value of the normal distribution function describing API concentration
μ_{size}	mean value of the normal distribution function describing particle sizes
σ_{conc}	standard deviation of the normal distribution function describing API concentration
σ_{size}	standard deviation of the normal distribution function describing particle sizes
φ^b, ψ^d	weighting parameters for breakage, for birth and death of particles, respectively
	operation of applying the transformation matrix
$\{I^N\}$	set of indices to address any value in the N–dimensional parameter space
$\mathbb{I}^i$	set of indices to address particles going from a lower to a higher cell

References

1. Kovačević, T.; Wiedmeyer, V.; Schock, J.; Voigt, A.; Pfeiffer, F.; Sundmacher, K.; Briesen, H. Disorientation angle distribution of primary particles in potash alum aggregates. *J. Cryst. Growth* **2017**, *467*, 93–106. [CrossRef]
2. Liu, L.X.; Litster, J.D.; Iveson, S.M.; Ennis, B.J. Coalescence of deformable granules in wet granulation processes. *AIChE J.* **2000**, *46*, 529–539. [CrossRef]
3. Iveson, S.M.; Beathe, J.A.; Page, N.W. The dynamic strength of partially saturated powder compacts: The effect of liquid properties. *Powder Technol.* **2002**, *127*, 149–161. [CrossRef]
4. Alaathar, I.; Hartge, E.-U.; Heinrich, S.; Werther, J. Modeling and flowsheet simulation of continuous fluidized bed dryers. *Powder Technol.* **2013**, *238*, 132–141. [CrossRef]
5. Pogodda, M. *Development of an Advanced System for the Modeling and Simulation of Solids Processes*; Shaker Verlag: Aachen, Germany, 2007.

6. Dosta, M. *Dynamic Flowsheet Simulation of Solids Processes and Its Application to Fluidized Bed Spray Granulation*; Cuvillier Verlag: Göttingen, Germany, 2013.

7. Skorych, V.; Dosta, M.; Hartge, E.-U.; Heinrich, S. Novel system for dynamic flowsheet simulation of solids processes. *Powder Technol.* **2017**, *314*, 665–679. [CrossRef]

8. Hartge, E.U.; Pogodda, M.; Reimers, C.; Schwier, D.; Gruhn, G.; Werther, J. Flowsheet simulation of solids processes. *KONA Powder Part. J.* **2006**, *24*, 146–158. [CrossRef]

9. Werther, J.; Heinrich, S.; Dosta, M.; Hartge, E.-U. The ultimate goal of modelling—Simulation of system and plant performance. *Particuology* **2011**, *9*, 320–329. [CrossRef]

10. Aspentech—Aspen Plus. Available online: https://www.aspentech.com/en/products/engineering/aspen-plus (accessed on 24 July 2019).

11. PSE Products—gFORMULATE—gPROMS FormulatedProducts—Solids Processing. Available online: https://www.psenterprise.com/products/gproms/formulatedproducts (accessed on 24 July 2019).

12. JKSimMet—JKTech Simulation of Comminution and Classification Circuits. Available online: https://jktech.com.au/jksimmet (accessed on 24 July 2019).

13. HSC Sim—Process Simulation Module. Available online: https://www.outotec.com/products/digital-solutions/hsc-chemistry/hsc-sim-process-simulation-module (accessed on 24 July 2019).

14. CHEMCAD—Chemical Engineering Simulation Software by Chemstations. Available online: https://www.chemstations.com/CHEMCAD (accessed on 24 July 2019).

15. Dosta, M.; Heinrich, S.; Werther, J. Fluidized bed spray granulation: Analysis of the system behaviour by means of dynamic flowsheet simulation. *Powder Technol.* **2010**, *204*, 71–82. [CrossRef]

16. Neugebauer, C.; Palis, S.; Bück, A.; Diez, E.; Heinrich, S.; Tsotsas, E.; Kienle, A. Influence of mill characteristics on stability of continuous layering granulation with external product classification. *Comput. Aided Chem. Eng.* **2016**, *38*, 1275–1280.

17. Haus, J.; Hartge, E.-U.; Heinrich, S.; Werther, J. Dynamic flowsheet simulation for chemical looping combustion of methane. *Int. J. Greenh. Gas Control* **2018**, *72*, 26–37. [CrossRef]

18. Koeninger, B.; Hensler, T.; Romeis, S.; Peukert, W.; Wirth, K.-E. Dynamics of fine grinding in a fluidized bed opposed jet mill. *Powder Technol.* **2018**, *327*, 346–357. [CrossRef]

19. Sander, S.; Gawor, S.; Fritsching, U. Separating polydisperse particles using electrostatic precipitators with wire and spiked-wire discharge electrode design. *Particuology* **2018**, *38*, 10–17. [CrossRef]

20. Boukouvala, F.; Niotis, V.; Ramachandran, R.; Muzzio, F.J.; Ierapetritou, M.G. An integrated approach for dynamic flowsheet modeling and sensitivity analysis of a continuous tablet manufacturing process. *Comput. Chem. Eng.* **2012**, *42*, 30–47. [CrossRef]

21. Rogers, A.J.; Inamdar, C.; Ierapetritou, M.G. An integrated approach to simulation of pharmaceutical processes for solid drug manufacture. *Ind. Eng. Chem. Res.* **2014**, *53*, 5128–5147. [CrossRef]

22. Skorych, V.; Dosta, M.; Hartge, E.-U.; Heinrich, S.; Ahrens, R.; Le Borne, S. Investigation of an FFT-based solver applied to dynamic flowsheet simulation of agglomeration processes. *Adv. Powder Technol.* **2019**, *30*, 555–564. [CrossRef]

23. Sastry, K.V.S. Similarity size distribution of agglomerates during their growth by coalescence in granulation or green pelletization. *Int. J. Miner. Process.* **1975**, *2*, 187–203. [CrossRef]

24. Hill, P.J.; Ng, K.M. Statistics of multiple particle breakage. *AIChE J.* **1996**, *42*, 1600–1611. [CrossRef]

25. Hulburt, H.M.; Katz, S. Some problems in particle technology: A statistical mechanical formulation. *Chem. Eng. Sci.* **1964**, *19*, 555–574. [CrossRef]

26. Barrett, J.; Jheeta, J. Improving the accuracy of the moments method for solving the aerosol general dynamic equation. *J. Aerosol Sci.* **1996**, *27*, 1135–1142. [CrossRef]

27. Madras, G.; McCoy, B.J. Reversible crystal growth-dissolution and aggregation-breakage: Numerical and moment solutions for population balance equations. *Powder Technol.* **2004**, *143*, 297–307. [CrossRef]

28. Marchisio, D.L.; Fox, R.O. Solution of population balance equations using the direct quadrature method of moments. *J. Aerosol Sci.* **2005**, *36*, 43–73. [CrossRef]

29. Kumar, S.; Ramkrishna, D. On the solution of population balance equations by discretization—I. A fixed pivot technique. *Chem. Eng. Sci.* **1996**, *51*, 1311–1332. [CrossRef]

30. Vale, H.M.; McKenna, T.F. Solution of the population balance equation for two-component aggregation by an extended fixed pivot technique. *Ind. Eng. Chem. Res.* **2005**, *44*, 7885–7891. [CrossRef]

31. Kruis, F.E.; Maisels, A.; Fissan, H. Direct simulation Monte Carlo method for particle coagulation and aggregation. *AIChE J.* **2000**, *46*, 1735–1742. [CrossRef]

32. Lee, K.; Matsoukas, T. Simultaneous coagulation and break-up using constant-N Monte Carlo. *Powder Technol.* **2000**, *110*, 82–89. [CrossRef]

33. Lin, Y.; Lee, K.; Matsoukas, T. Solution of the population balance equation using constant-number Monte Carlo. *Chem. Eng. Sci.* **2002**, *57*, 2241–2252. [CrossRef]

34. Smith, M.; Matsoukas, T. Constant-number Monte Carlo simulation of population balances. *Chem. Eng. Sci.* **1998**, *53*, 1777–1786. [CrossRef]

35. Mahoney, A.W.; Ramkrishna, D. Efficient solution of population balance equations with discontinuities by finite elements. *Chem. Eng. Sci.* **2002**, *57*, 1107–1119. [CrossRef]

36. Nicmanis, M.; Hounslow, M. A finite element analysis of the steady state population balance equation for particulate systems: Aggregation and growth. *Comput. Chem. Eng.* **1996**, *20*, S261–S266. [CrossRef]

37. Rigopoulos, S.; Jones, A.G. Finite-element scheme for solution of the dynamic population balance equation. *AIChE J.* **2003**, *49*, 1127–1139. [CrossRef]

38. Hackbusch, W. On the efficient evaluation of coalescence integrals in population balance models. *Computing* **2006**, *78*, 145–159. [CrossRef]

39. Le Borne, S.; Shahmuradyan, L.; Sundmacher, K. Fast evaluation of univariate aggregation integrals on equidistant grids. *Comput. Chem. Eng.* **2015**, *74*, 115–127. [CrossRef]

40. Matveev, S.A.; Smirnov, A.P.; Tyrtyshnikov, E.E. A fast numerical method for the Cauchy problem for the Smoluchowski equation. *J. Comput. Phys.* **2015**, *282*, 23–32. [CrossRef]

41. Kumar, J.; Peglow, M.; Warnecke, G.; Heinrich, S. An efficient numerical technique for solving population balance equation involving aggregation, breakage, growth and nucleation. *Powder Technol.* **2008**, *182*, 81–104. [CrossRef]

42. Mostafaei, P.; Rajabi-Hamane, M. Numerical solution of the population balance equation using an efficiently modified cell average technique. *Comput. Chem. Eng.* **2017**, *96*, 33–41. [CrossRef]

43. Kumar, J.; Saha, J.; Tsotsas, E. Development and convergence analysis of a finite volume scheme for solving breakage equation. *SIAM J. Numer. Anal.* **2015**, *53*, 1672–1689. [CrossRef]

44. Kumar, J.; Kaur, G.; Tsotsas, E. An accurate and efficient discrete formulation of aggregation population balance equation. *Kinet. Relat. Models* **2016**, *9*, 373–391.

45. Schwier, D.; Hartge, E.-U.; Werther, J.; Gruhn, G. Global sensitivity analysis in the flowsheet simulation of solids processes. *Chem. Eng. Process.* **2010**, *49*, 9–21. [CrossRef]

46. Courant, R.; Friedrichs, K.; Lewy, H. Über die partiellen Differenzengleichungen der mathematischen physik. *Math. Ann.* **1928**, *100*, 32–74. [CrossRef]

47. DFG-Priority Programme SPP 1679. Available online: https://www.dynsim-fp.de/en/home (accessed on 24 July 2019).

48. Stroustrup, B. *The C++ Programming Language*, 4th ed.; Addison-Wesley: Boston, MA, USA, 2013.

49. Stroustrup, B. *Tour of C++*, 4th ed.; Addison-Wesley: Boston, MA, USA, 2014; pp. 23–57.

50. Hillestad, M.; Hertzberg, T. Dynamic simulation of chemical engineering systems by the sequential modular approach. *Comput. Chem. Eng.* **1986**, *10*, 377–388. [CrossRef]

51. Dosta, M. Modular-based simulation of single process units. *Chem. Eng. Technol.* **2019**, *42*, 699–707. [CrossRef]

52. Towler, G.; Sinnott, R. *Chemical Engineering Design: Principles, Practice and Economics of Plant and Process Design*, 2nd ed.; Butterworth-Heinemann: Kidlington, Oxford, UK, 2013; pp. 223–236.

53. Dimian, A.C.; Bildea, C.S.; Kiss, A.A. Dynamic Simulation. *Comput. Aided Chem. Eng.* **2014**, *35*, 127–156.

54. Lelarasmee, E.; Ruehli, A.E.; Sangiovanni-Vincintelli, A.L. The Waveform Relaxation Method for Time-Domain Analysis of Large Scale Integrated Circuits: Theory and Applications. Ph.D. Thesis, University of California, Berkeley, Berkeley, CA, USA, 1982.

55. Smoluchowski, M. Versuch einer mathematischen Theorie der Koagulationskinetik kolloider Losungen. *Z. Phys. Chem.* **1916**, *17*, 129–168. [CrossRef]

56. Aldous, D.J. Deterministic and stochastic models for coalescence (aggregation and coagulation): A review of the mean-field theory for probabilists. *Bernoulli* **1999**, *5*, 3–48. [CrossRef]

57. Kuncir, G.F. Algorithm 103: Simpson's rule integrator. *Commun. ACM* **1962**, *5*, 347. [CrossRef]

58. Lyness, J.N. Notes on the adaptive Simpson quadrature routine. *J. ACM* **1969**, *16*, 483–495. [CrossRef]

59. King, R.P. *Modeling and Simulation of Mineral Processing Systems*; Butterworth & Heinemann: Oxford, UK, 2001; p. 158.
60. Vogel, L.; Peukert, W. Modelling of grinding in an air classifier mill based on a fundamental material function. *KONA Powder Part. J.* **2003**, *21*, 109–120. [CrossRef]
61. Plitt, L.R. The analysis of solid-solid separations in classifiers. *CIM Bull.* **1971**, *64*, 42–47.

 *processes* 

Article

A Numerical Approach to Solve Volume-Based Batch Crystallization Model with Fines Dissolution Unit

Safyan Mukhtar [1,*], Muhammad Sohaib [2] and Ishfaq Ahmad [2]

[1] Department of Basic Sciences, Deanship of Preparatory Year, King Faisal University, 31982 Hofuf, Al Ahsa, Saudi Arabia
[2] Department of Mathematics & Statistics, Bacha Khan University, 24461 Palosa, Charsadda Khyber Pakhtunkhwa, Pakistan
* Correspondence: smahmad@kfu.edu.sa; Tel.: +966-552825413

Received: 11 June 2019; Accepted: 12 July 2019; Published: 15 July 2019

Abstract: In this article, a numerical study of a one-dimensional, volume-based batch crystallization model (PBM) is presented that is used in numerous industries and chemical engineering sciences. A numerical approximation of the underlying model is discussed by using an alternative Quadrature Method of Moments (QMOM). Fines dissolution term is also incorporated in the governing equation for improvement of product quality and removal of undesirable particles. The moment-generating function is introduced in order to apply the QMOM. To find the quadrature abscissas, an orthogonal polynomial of degree three is derived. To verify the efficiency and accuracy of the proposed technique, two test problems are discussed. The numerical results obtained by the proposed scheme are plotted versus the analytical solutions. Thus, these findings line up well with the analytical findings.

Keywords: volume-based population balance model with fines dissolution; quadrature method of moments; orthogonal polynomials

1. Introduction

Population balance models (PBMs) show a significant role in different areas of science and engineering. These models have numerous applications in high-energy physics, geophysics, biophysics, meteorology, pharmacy, food science, chromatography, chemical engineering, civil engineering, and environmental engineering. These models are used in the process of cell dynamics, polymerization, cloud formation, and crystallization. In biophysics, these models are concerned with population of various kinds of cells. Population balance models are also used in the formation of ceramics mixtures and nanoparticles which have a lot of applications.

In 1964, Hulburt & Katz [1] were first to discuss the PBMs in chemical engineering. A detailed description of these models is given in [2]. The main components in the models are the process of nucleation, growth, aggregation, breakage, inlet, outlet, growth, and dissolution. The mathematical model of population balance equations are partial integro-differential equations (PIDEs). Analytical solutions of these PIDEs are very rare except for a few simple cases. Therefore, researchers are interested in developing numerical solutions for these equations. Numerous numerical methods are accessible in the literature to solve certain kinds of PBMs; see, for example, [3–11].

The Quadrature Method of Moments (QMOM) for solving the governed models was first introduced by McGraw [12]. The direct QMOM was proposed by Fan et al. [4]. In addition, a new QMOM has been proposed by Gimbun [5]. Qamar et al. [13] introduced an alternative QMOM for solving length-based batch crystallization models telling crystals nucleation, size-dependent growth, aggregation, breakage, and dissolution of small nuclei below certain critical size. Safyan et al. [14] followed the technique of [13] for solving volume-based batch crystallization model nucleation, size-dependent growth, aggregation, breakage. In this article, the QMOM is used to solve volume-based batch crystallization models with fines dissolution. The fines dissolution unit improves the product quality and removes unwanted particles.

In the proposed method, orthogonal polynomials, taken from the lower-order moments, are used to find the quadrature points and weights. For confirming accuracy of the scheme, a third-order orthogonal polynomial, employing the first six moments, was chosen to estimate the quadrature points (abscissas) and equivalent quadrature weights. Hence, it was essential to solve at least a six-moment system (i.e., $0, \dots, 5$). This type of polynomial provides a three-point Gaussian quadrature rule which usually produces precise output for polynomials of degree five or less. The expression for the preferred orthogonal polynomial can be derived analytically and the mandatory number of moments rises with order of polynomial. It is important to mention that the calculation of six moments to generate each time the third-order polynomial is compulsory. However, the technique itself is not limited to the calculation of every listed number of moments. The calculation of additional moments is conducted by adding ordinary differential equation.

This article is divided into two parts: In the first part, the mathematical model of volume-based Population Balance Equation (PBE) with fines dissolution is presented. In the second part, the proposed QMOM is derived for PBEs. Furthermore, the mathematical outcomes of QMOM are compared with the analytical outcomes that are accessible in literature.

2. Materials and Methods

Suppose u_d represents the number density function, then a general PBE is given by [1,2]:

$$\frac{\partial u_d(T,V_p)}{\partial T} = -\underbrace{\frac{\partial[G(T,V_p)u_d(T,V_p)]}{\partial V_p}}_{\text{Growth-term}} + \underbrace{Q_{nuc}(T,V_p)}_{\text{Nucleation-term}} + \underbrace{Q_{agg}^{\pm}(T,V_p)}_{\text{Aggregation-term}} + \underbrace{Q_{break}^{\pm}(T,V_p)}_{\text{Breakage-term}} + \underbrace{Q_{diss}(T,V_p)}_{\text{Dissolution-term}} \quad (1)$$

$$u_d(0,V_p) = u_{d0}(V_p) \quad (T,V_p) \in R_+^2$$

where $R_+ = (0, \infty)$. The variable T represents the time and V_p may be size, length, or composition. In this article, it represents the particle volume. In the above equation, each term has its specific definition. These terms are given by

$$Q_{agg}^{\pm}(T,V_p) = B_{agg}(T,V_p) - D_{agg}(T,V_p) \quad (2)$$

where

$$B_{agg}(T,V_p) = \frac{1}{2} \int_0^{V_p} \beta(T, V_p - V_p', V_p')u_d(T, V_p - V_p')u_d(T, V_p')dV_p'$$

and

$$D_{agg}(T,V_p) = \int_0^{\infty} \beta(T, V_p, V_p')u_d(T, V_p)u_d(T, V_p')dV_p'$$

Here, $B_{agg}(T,V_p)$ represents the birth of particles of volume V_p resulting for amalgamation of two particles with respective volume V_p' and $V_p - V_p'$, where $V_p' \in (0, V_p)$ and $D_{agg}(T,V_p)$ is the death of particles, describing the decrease in particle volume V_p by aggregation with other particles of any volume. The aggregation kernel $\beta(T, V_p, V_p')$ is the rate at which aggregation of two particles V_p and V_p' produces a particle volume $V_p + V_p'$.

$$Q_{break}^{\pm}(T,V_p) = B_{break}(T,V_p) - D_{break}(T,V_p) \quad (3)$$

where

$$B_{\text{break}}\left(T, V_p\right) = \int_{V_p}^{\infty} b\left(T, V_p, V_p'\right) S\left(V_p'\right) u_d\left(T, V_p'\right) dV_p'$$

and

$$D_{\text{break}}(T, V_p) = S(V_p) u_d(T, V_p)$$

Here, $B_{\text{break}}\left(T, V_p\right)$ represents the birth of new particles during the breakage process and the breakage function $b\left(T, V_p, V_p'\right)$ is the probability density function for the formation of particle volume V_p from particle volume V_p', whereas, $D_{\text{break}}\left(T, V_p\right)$ is the death of particles, and $S\left(V_p'\right)$ is the selection function describing the rate at which the particles are selected to break.

$$Q_{\text{diss}}\left(T, V_p\right) = \frac{\dot{V}}{V_c} h\left(V_p\right) u_d\left(T, V_p\right) \tag{4}$$

$Q_{\text{diss}}\left(T, V_p\right)$ represents the dissolution term, V_c is the volume of crystallizer, $\dot{V}$ is the volumetric flow rate from the crystallizer to dissolution unit, $h\left(V_p\right)$ is the dissolution function describing dissolution of small particles below some critical size. The population balance equation can be simplified through introducing the moment function. The kth moment of the population density function is mathematically written in the form:

$$m_k(T) = \int_0^{\infty} V_p^k u_d(T, V_p) dV_p \tag{5}$$

The first and second moments $m_0(T)$, $m_1(T)$ denote particle population and volume, respectively, at any instant T. Multiply V_p^k to left- and right-hand side of Equation (1) and at that moment integrate it over the volume V_p, so we obtain the following equation:

$$\frac{dm_k(T)}{dT} = \overline{G_{\text{growth}}}(T, V_p) + \overline{Q_{\text{nucleation}}}(T, V_p) + \overline{Q_{\text{aggregation}}}(T, V_p) + \overline{Q_{\text{breakage}}}(T, V_p) + \overline{Q_{\text{dissolution}}}(T, V_p) \tag{6}$$

where

$$\overline{G_{\text{growth}}}(T, V_p) = \int_0^{\infty} V_p^k \frac{\partial[G(T, V_p) u_d(T, V_p)]}{\partial V_p} dV_p$$

$$\overline{Q_{\text{nucleation}}}(T, V_p) = \int_0^{\infty} V_p^k Q_{nuc}(T, V_p) dV_p$$

$$\overline{Q_{\text{aggregation}}}(T, V_p) = \overline{B_{\text{agg}}}(T, V_p) - \overline{D_{\text{agg}}}(T, V_p)$$

The aggregation terms $\overline{B_{\text{agg}}}(T, V_p)$ and $\overline{D_{\text{agg}}}(T, V_p)$ are mathematically defined as

$$\overline{B_{\text{agg}}}\left(T, V_p\right) = \frac{1}{2} \int_0^{\infty} V_p^k \int_0^{V_p} \beta\left(T, V_p - V_p', V_p'\right) u_d\left(T, V_p - V_p'\right) u_d\left(T, V_p'\right) dV_p' dV_p$$

$$\overline{D_{\text{agg}}}\left(T, V_p\right) = \int_0^{\infty} V_p^k u_d\left(T, V_p\right) \int_0^{\infty} \beta\left(T, V_p, V_p'\right) u_d\left(T, V_p'\right) dV_p' dV_p$$

$$\overline{Q_{\text{breakage}}}(T, V_p) = \overline{B_{\text{break}}}(T, V_p) - \overline{D_{\text{break}}}(T, V_p)$$

whereas $\overline{B_{break}}(T, V_p)$ and $\overline{D_{break}}(T, V_p)$ are birth and death functions because of breakage term and are given by

$$\overline{B_{break}}(T, V_p) = \int\limits_0^\infty V_p^k \int\limits_{V_p}^\infty b(T, V_p, V_p')S(V_p')u_d(T, V_p')dV_p'dV_p$$

$$\overline{B_{break}}(T, V_p) = \int\limits_0^\infty V_p^k S(V_p)u_d(T, V_p)dV_p$$

$$\overline{Q_{dissolution}}(T, V_p) = \frac{\dot{V}}{V_c} \int\limits_0^\infty V_p^k h(V_p)u_d(T, V_p)dV_p$$

Due to aggregation, the upper limit in the birth function is not infinity; therefore, we cannot apply quadrature method of moment for this integral. To apply the QMOM, we have to convert the upper limit to infinity. Here, we introduce a Heaviside step function $H(V_p - V_p')$ to solve the integral such that $H(V_p - V_p') = 0$ when $V_p - V_p' < 0$ and $H(V_p - V_p') = 1$ otherwise. As a result, we will find the limits of integration over V_p' from $(0, V_p)$ to $(0, \infty)$. Thus, by applying it to the birth function, we obtain the following equation:

$$\begin{aligned}
\tfrac{1}{2} \int\limits_0^\infty V_p^k dV_p \int\limits_0^\infty H(V_p - V_p')\beta(T, V_p - V_p', V_p')u_d(T, V_p - V_p')u_d(T, V_p')dV_p' \\
= \tfrac{1}{2} \int\limits_0^\infty u_d(T, V_p')dV_p' \int\limits_0^\infty (u + V_p')^k \beta(T, u, V_p')u_d(T, u)dV_p
\end{aligned} \tag{7}$$

On the right side of Equation (7), we have switched the order of integration and made the replacement $u = V_p - V_p'$. Lastly, we have replaced V_p for u with no damage of generality and caught the necessary outcome for birth due to aggregation which is given by

$$\overline{B_{agg}}(T, V_p) = \frac{1}{2} \int\limits_0^\infty u_d(T, V'_p) \int\limits_0^\infty (V_p + V'_p)^k \, \beta(T, V_p, V'_p)u_d(T, V_p)dV_p dV'_p$$

After substituting all of the above terms in Equation (6), we get the system of differential equations:

$$\begin{aligned}
\frac{dm_k(T)}{dT} &= \int\limits_0^\infty kV_p^{k-1}G(T, V_p)u_d(T, V_p)dV_p + \int\limits_0^\infty V_p^k Q_{nuc}(T, V_p)dV_p \\
&+ \tfrac{1}{2} \int\limits_0^\infty u_d(T, V_p') \int\limits_0^\infty (V_p + V_p')^k \beta(T, V_p, V_p')u_d(T, V_p)dV_p dV_p' \\
&- \int\limits_0^\infty V_p^k u_d(T, V_p) \int\limits_0^\infty \beta(T, V_p, V_p')u_d(T, V_p')dV_p'dV_p \\
&+ \int\limits_0^\infty V_p^k \int\limits_{V_p}^\infty b(T, V_p, V_p')S(V_p')u_d(T, V_p')dV_p'dV_p \\
&- \int\limits_0^\infty V_p^k S(V_p)u_d(T, V_p)dV_p + \tfrac{\dot{V}}{V_c} \int\limits_0^\infty V_p^k h(V_p)u_d(T, V_p)dV_p
\end{aligned} \tag{8}$$

In addition, with solid phase, the liquid phase yields ordinary differential equation for the solute mass in the form:

$$\frac{dm(T)}{dT} = \dot{m}_{in}(T) - \dot{m}_{out}(T) - 3\sigma_c k_v \int\limits_0^\infty V_p^2 G(T, V_p)u_d(T, V_p)dV_p \tag{9}$$

with $m(0) = m_0$ where σ_c is the density of crystals and k_v is a volume shape factor. $\dot{m}_{\text{in}}(T) - \dot{m}_{\text{in}}(T)-$ is the incoming mass flux from dissolution unit to crystallizer and outgoing mass flux from crystallizer to dissolution unit, respectively, which is mathematically defined by

$$\dot{m}_{\text{out}}(T) = \frac{m(T)}{m(T) + m_{\text{solv}}(T)} \sigma_{sol}(T)\dot{V} \tag{10}$$

where $m_{\text{solv}}(T)$ is the mass of solvent and $\sigma_{sol}(T)$ is the density of the solution. $T_p = \frac{V_p}{V} \geq 0$ is the residence time in the dissolution unit, where V_p represents volume of the pipe.

Here, we will consider the Gaussian quadrature method to solve the complicated integral terms appearing in Equations (8) and (9). Consequently, a closed-form system of moments is found. Further, we have accurately and efficiently solved the closed-form system with an ODE solver. The approximation of definite integral with the help of quadrature rule is an important numerical aspect. For this purpose, first we calculate the function at definite points and then use the formula of weight function which gives approximation of the definite integral. To approximate a definite integral given in Equations (8) and (9), first we find the values of the function at a set of equidistant points. After evaluating, the weight function is multiplied to approximate the integrals. In this rule, there is no limitation of choosing abscissas and weights. This rule also works for the points which are not likewise spread out. Let us assume the integral of the formula $\int_a^b \psi(V_p)u_d(V_p)dV_p$. We can find the weights w_j and abscissas V_{kp_i} by approximating the definite integral as

$$\int_a^b \psi(V_p)u_d(V_p)dV_p = \sum_{j=1}^N w_j u_d(V_{p_j}) \tag{11}$$

where provided it is exact and $u_d(V_p)$ is smooth function. A set of orthogonal polynomials is necessary for making certain n^{th} order orthogonal polynomial and it should contain only one polynomial of order j for $j = 1, 2, 3, \ldots$ We can define the scalar product of the two functions $k(V_p)$ and $h(V_p)$ over a weight function $\psi(V_p)$ by

$$< h\backslash k >= \int_a^b \psi(V_p)h(V_p)k(V_p)dV_p \tag{12}$$

If we find the zero scalar product of any two functions h and k, then these functions are termed as orthogonal functions. It is also noted that supplementary information will be required for obtaining abscissas and weights if the classical weight function $\psi(V_p)$ is not provided. For this purpose, we have used the moment:

$$m_i(T) = \int_0^\infty V_p^i u_d(T, V_p)dV_p \approx \sum_{j=1}^N V_{p_j}^j w_j \tag{13}$$

In the above equation, $u_d(T, V_p)$ is used as a weight function $\psi(V_p)$. V_p^i given in Equation (13) denotes the polynomial $u_d(V_p)$ of ith order. To find n abscissas weights, we need $2n$ moments from m_0 to m_{2n-1}. For $i \leq 2n - 1$ this approximation will be exact. For simplicity and accurate approximations, we set $n = 3$. As a result, six-moment system for $m_0, \ldots\ldots, m_5$ is calculated. To discuss the procedure, we consider ODEs (8) and (9). Using Equation (13) in Equations (8) and (9), we obtain the following equations:

$$
\begin{aligned}
\frac{dm_k(T)}{dT} &= k \sum_{i=1}^{N} \left(V_{p_i}\right)^{k-1} w_i G\!\left(T,\, V_{p_i}\right) + a_{nuc}^{(k)} \\
&\quad + \tfrac{1}{2} \sum_{i=1}^{N} w_i \sum_{j=1}^{N} \left(V_{p_i} + V_{p_j}\right)^{k} \beta\!\left(T, V_{p_i}, V_{p_j}\right) w_j \\
&\quad - \sum_{i=1}^{N} \left(V_{p_i}\right)^{k} w_i \sum_{j=1}^{N} \beta\!\left(T, V_{p_i}, V_{p_j}\right) w_j \\
&\quad + \sum_{i=1}^{N} \left(V_{p_i}\right)^{k} w_i \sum_{j=1}^{N} \beta\!\left(T, V_{p_i}, V_{p_j}\right) S\!\left(V_{p_j}\right) w_j \\
&\quad + \tfrac{\dot{V}}{V_c} \sum_{i=1}^{N} w_i \left(V_{p_i}\right)^{k} h\!\left(V_{p_i}\right) \qquad k = 0, 1, 2, \ldots\ldots
\end{aligned}
\tag{14}
$$

where $a_{nuc}^{(k)} = \int_{0}^{\infty} V_p^{\,k} Q_{nuc} dV_p$ and

$$
\frac{dm(T)}{dT} = \dot{m}_{in}(T) - \dot{m}_{out}(T) - 3\sigma_c k_v \sum_{i}^{N} w_i V_p^{2} G\!\left(T,\, V_{p_i}\right)
\tag{15}
$$

Our next step is computing the quadrature points V_{pi}, V'_{pj} and the quadrature weights w_i and w_j. The quadrature points V_{pi}, V'_{pj} are obtained from the roots of orthogonal polynomials. The construction of orthogonal polynomials is described as follows:

$$
p_{-1} = 0,\, p_0 = 1,\, p_j = (V_p - \alpha_j) p_{j-1} - \beta_j p_{j-1}
\tag{16}
$$

with

$$
\alpha_j = \frac{< V_p p_{j-1}/p_{j-1} >}{< p_{j-1}/p_{j-1} >},\, j = 1, 2, \ldots .
\tag{17}
$$

$$
\beta_j = \frac{< p_{j-1}/p_{j-1} >}{< p_{j-2}/p_{j-2} >},\, j = 2, 3, \ldots .
\tag{18}
$$

Since the function $u_d(T, V_p)$ is used as a weight function $\psi(V_p)$, so from Equation (12) we have

$$
< p_j/p_j >= \int_{a}^{b} u_d(T, V_p) p_j^2 dV_p
$$

Using all of the above definitions, the orthogonal polynomials of any order can be obtained. Our first step is to find the roots of the nth order polynomial. The quadrature points V_p are then obtained from these roots. To explain the procedure of deriving these orthogonal polynomials, we have calculated $p_1(V_p)$, $p_2(V_p)$, $p_3(V_p)$. $p_1(V_p)$ as

$$
p_1(V_p) = (V_p - \alpha_1) p_0 = (V_p - \alpha_1)
$$

First, α_1 will be calculated, which is given by

$$
\alpha_1 = \frac{< V_p p_0/p_0 >}{< p_0/p_0 >} = \frac{\int_{0}^{\infty} V_p u_d(T, V_p) p_0^2 dV_p}{\int_{0}^{\infty} u_d(T, V_p) p_0^2 dV_p} = \frac{m_1(T)}{m_0(T)}
$$

so

$$
p_1(V_p) = V_p - \frac{m_1}{m_0}
\tag{19}
$$

Now

$$p_2(V_p) = (V_p - \alpha_2)p_1 - \beta_2 p_1 \tag{20}$$

Again from Equations (17) and (18) we have

$$\alpha_2 = \frac{<V_p p_1/p_1>}{<p_1/p_1>} = \frac{\int_0^\infty V_p u_d(T,V_p)p_1^2 dV_p}{\int_0^\infty u_d(T,V_p)p_1^2 dV_p} = \frac{\int_0^\infty V_p u_d(T,V_p)\left(V_p - \frac{m_1}{m_0}\right)^2 dV_p}{\int_0^\infty u_d(T,V_p)\left(V_p - \frac{m_1}{m_0}\right)^2 dV_p} = \frac{m_3 m_0^2 - 2m_0 m_1 m_2 + m_1^3}{m_2 m_0^2 - m_0 m_1^2} \tag{21}$$

and

$$\beta_2 = \frac{<p_1/p_1>}{<p_0/p_0>} = \frac{\int_0^\infty V_p u_d(T, V_p)\left(V_p - \frac{m_1}{m_0}\right)^2 dV_p}{\int_0^\infty u_d(T, V_p)dV_p} = \frac{m_2 m_0 - m_1^2}{m_0^2} \tag{22}$$

so from Equation (20) we have

$$p_2(V_p) = \frac{V_p^2(m_0 m_2 - m_1^2) + V_p(m_1 m_2 - m_0 m_3) + m_1 m_3 - m_2^2}{m_0 m_2 - m_1^2} \tag{23}$$

Proceeding in a similar way, we can calculate the polynomial of higher order. The third-order polynomial $p_3\left(V_p\right)$ is given by the following equation:

$$
\begin{aligned}
p_3(V_p) \;=\; & V_p^{\,3} + \frac{(m_2 m_4 m_1 - m_0 m_4 m_3 + m_2 m_0 m_5 + m_3^2 m_1 - m_5 m_1^2 - m_2^2 m_3)V_p^2}{m_2^3 - m_2 m_4 m_0 - 2m_2 m_3 m_1 + m_3^2 m_0 + m_4 m_1^2} \\
& + \frac{(m_2 m_5 m_1 + m_0 m_4^2 - m_0 m_5 m_3 - m_4 m_3 m_1 - m_2^2 m_4 + m_3^2 m_2)V_p}{m_2^3 - m_2 m_4 m_0 - 2m_2 m_3 m_1 + m_3^2 m_0 + m_4 m_1^2} + \frac{2m_2 m_4 m_3 - m_2^2 m_5 - m_3^3 - m_4^2 m_1 + m_5 m_3 m_1)}{m_2^3 - m_2 m_4 m_0 - 2m_2 m_3 m_1 + m_3^2 m_0 + m_4 m_1^2}
\end{aligned}
\tag{24}
$$

The roots of the selected polynomial will give us the abscissas V_p. Next, the weights w_i will be calculated. According to Press et al. [15], the expression for the weight function is given by

$$w_i = \frac{\langle p_{N-1}/p_{N-1}\rangle}{p_{N-1}\left(x_j\right)p_N'\left(x_j\right)} \qquad i = 1, 2, \ldots \ldots \ldots N \tag{25}$$

where N is the order of the selected polynomial. At last, the resulting system of ODEs is then solved by any standard ODE solver in MatLab.

3. Results

Test Problem 1: Aggregation with Fines Dissolution

Here, the proposed scheme is analyzed for aggregation with fines dissolution (see Figure 1) problem encountered in several particulate methods (i.e., fluidized beds, formation of rain droplets, and manufacture of dry powders). The effects of further procedures, such as breakage, growth, and nucleation, are negligible. During the aggregation process, the total mass of particles m_1 is conserved and the amount of particles m_0 reduces during the processing time. The aggregation kernel is held to be constant and is defined as $\beta(V_p, V'_p) = \beta_0$, where $\beta_0 = 1$. The exponential initial particle size distribution is given by

$$u_d\left(0, V_p\right) = \frac{N_0}{V_{P_0}}exp\left(-V_p/V_{P_0}\right) \tag{26}$$

where $N_0 = 1$ and $V_{p0} = 1$. The dissolution term $h(V_p)$ explains the dissolution of particles below certain critical size, that is, $2 \times 10^{-4} m^3$. The analytical solution in terms of the number density $u_d(T, V_p)$ is given by Scott [16]:

$$u_d(T, V_p) = \frac{4N_0}{V_{P_0}(\tau + 2)^2} exp(-2V'_p/\tau + 2) \tag{27}$$

where $\tau = N_0 \beta_0 t$ and $V'_p = V/V_{p0}$. The plot in Figure 2 shows normalized moments. The outcomes of our QMOM are in decent agreement with analytical outcomes. It is also observed from Figure 2 that during the aggregation process, the number of particles $m_0(T)$ decreases while the volume of the particles $m_1(T)$ remains constant.

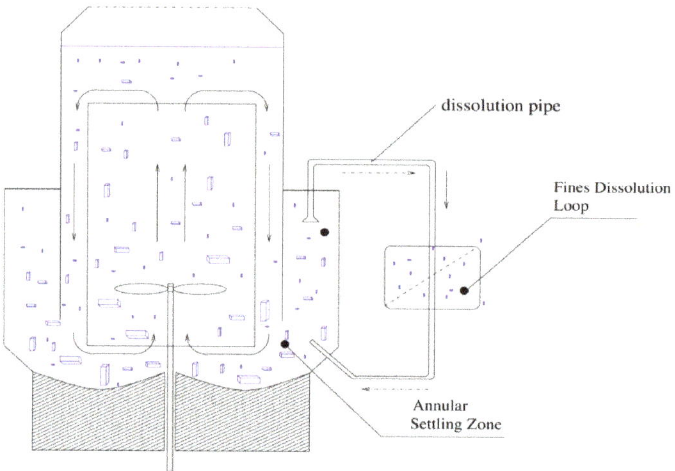

Figure 1. Single batch setup process with fines dissolution.

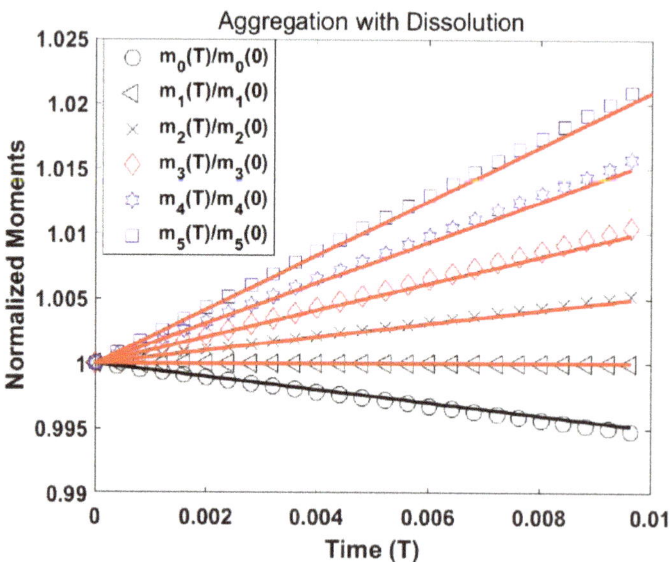

Figure 2. Aggregation with fines dissolution.

Test Problem 2: Aggregation and Breakage with Fines Dissolution

In this problem, we take a batch crystallizer in which aggregation and breakage are the main occurrences and which is connected with a fines dissolver. The growth and nucleation terms are neglected in this process. The initial distribution is given by

$$u_d\left(0, V_p\right) = M_0 \cdot \frac{M_0}{M_1} exp\left(-\frac{M_0}{M_1} V_p\right) \tag{28}$$

where, $M_0 = \frac{2N_0}{2+\beta_0 N_0 t}$ and $M_1 = N_0 V_{p0}\left[1 - \frac{2G_0}{\beta_0 N_0 V_{p0}} \ln\left(\frac{2}{2+\beta_0 N_0 t}\right)\right]$ represent the zero and first moments, respectively. A constant aggregation term, $\beta(V_p, V'_p) = \beta_0 = 1$, a breakage kernel $b(V_p, V'_p) = 2/V'_p$, and uniform daughter distribution $S(V_p) = V_p$ are taken. The analytical solution is given by Patel [17]:

$$u_d\left(T, V_p\right) = \frac{M_0^2}{M_1} exp\left(-\frac{M_0}{M_1} V_p\right) \tag{29}$$

The numerical results are displayed in Figure 3. The moments of the numerical system are in good agreement with those taken from the analytical solution. It is also observed from Figure 3 that during the aggregation and breakage process, the number of particles $m_0(T)$ decreases while the volume of the particles $m_1(T)$ remains constant.

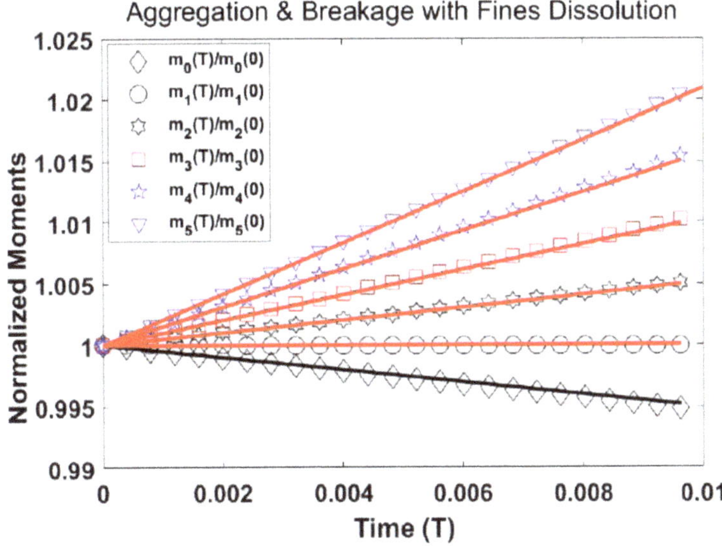

Figure 3. Aggregation and breakage with fines dissolution.

4. Conclusions

The moment-generating function was used to convert the governing partial differential equation into a system of ordinary differential equations. The mathematical term for fines dissolution was incorporated in the model for improving the quality of the product. The Gaussian quadrature method was implemented to solve the complicated integrals in this research. An orthogonal polynomial of degree three, which utilizes the first six moments, was used for better accuracy of the proposed scheme. To check the significance of the scheme, two case studies were discussed. The results of the proposed scheme are in perfect agreement with the available analytical results. No dissipation was observed in the results. This work is extendable for batch preferential crystallization models with fines dissolution.

Furthermore, the developed scheme could be applicable to solve multidimensional batch crystallization models with fines dissolution.

Author Contributions: Conceptualization: S.M. and I.A.; Methodology: S.M.; Software: I.A. and M.S.; Validation: S.M. and I.A.; Formal Analysis: S.M.; Investigation: S.M.; Writing-Original Draft Preparation: S.M.; Writing-Review & Editing: S.M., M.S., and I.A.

Funding: Safyan Mukhtar is thankful to the Deanship of Scientific Research, King Faisal University, for research grant through the Nasher track (186122).

Conflicts of Interest: The authors declare no conflict of interest.

References

1. Hulburt, H.M.; Katz, S. Some problems in particle technology. *Chem. Eng. Sci.* **1964**, *19*, 555–574. [CrossRef]
2. Randolph, A.D.; Larson, M.A. *Theory of Particulate Processes*, 2nd ed.; Academic Press: Cambridge, MA, USA, 1988.
3. Barrett, J.C.; Jheeta, J.S. Improving the accuracy of the moments method for solving the aerosol general dynamic equation. *J. Aerosol Sci.* **1996**, *27*, 1135–1142. [CrossRef]
4. Fan, R.; Marchisio, D.L.; Fox, R.O. Application of the direct quadrature method of moments to polydisperse gas-solid fluidized beds. *J. Powder Technol.* **2004**, *139*, 7–20. [CrossRef]
5. Gimbbun, J.; Nagy, Z.K.; Rielly, C.D. Simultaneous quadrature method of moments for the solution of population balance equations using a differential algebraic framework. *Ind. Chem. Eng.* **2009**, *48*, 7798–7812. [CrossRef]
6. Gunawan, R.; Fusman, I.; Braatz, R.D. High resolution algorithms for multidimensional population balance equations. *AICHE J.* **2004**, *50*, 2738–2749. [CrossRef]
7. Kumar, S.; Ramkrishna, D. On the solution of population balance equations by discretization-I. A fixed pivot technique. *Chem. Eng. Sci.* **1996**, *51*, 1311–1332. [CrossRef]
8. Lim, Y.I.; Lann, J.-M.L.; Meyer, L.M.; Joulia, L.; Lee, G.; Yoon, E.S. On the solution of population balance equation (PBE) with accurate front tracking method in practical crystallization processes. *Chem. Eng. Sci.* **2002**, *57*, 3715–3732. [CrossRef]
9. Madras, G.; McCoy, B.J. Reversible crystal growth-dissolution and aggregation breakage: Numerical and moment solutions for population balance equations. *J. Powder Technol.* **2004**, *143*, 297–307. [CrossRef]
10. Marchisio, D.L.; Vigil, R.D.; Fox, R.O. Quadrature method of moments for aggregation-breakage processes. *J. Colloid Interface Sci.* **2003**, *258*, 322–334. [CrossRef]
11. Gahn, C.; Mersmann, A. Brittle fracture in crystallization processes Part A. Attrition and abrasion of brittle solids. *Chem. Eng. Sci.* **1999**, *54*, 1273–1282. [CrossRef]
12. McGraw, R. Description of aerosol dynamics by the quadrature method of moments. *Aerosol Sci. Tech.* **1997**, *27*, 255–265. [CrossRef]
13. Qamar, S.; Mukhtar, S.; Seidel-Morgenstern, A.; Elsner, M.P. An efficient numerical technique for solving one-dimensional batch crystallization models with size-dependent growth rates. *Chem. Eng. Sci.* **2009**, *64*, 3659–3667. [CrossRef]
14. Mukhtar, S.; Hussain, I.; Ali, A. Quadrature method of moments for solving volume-based population balance models. *World Appl. Sci. J.* **2012**, *20*, 1574–1583.
15. Press, W.H.; Teukolsky, S.A.; Vetterling, W.T.; Flannery, B.P. *Numerical Recipes: The Art of Scientific Computing*, 3rd ed.; Cambridge University Press: Cambridge, UK, 2007.
16. Scott, W.T. Analytic studies of cloud droplet coalescence I. *J. Atmos. Sci.* **1968**, *25*, 54–65. [CrossRef]
17. Patel, D.P.; Andrews, J.R.G. An analytical solution to continuous population balance model describing coalescence and breakage. *J. Chem. Eng. Sci.* **1998**, *53*, 599–601. [CrossRef]

 *processes*

Article

Numerical Simulation of a Flow Field in a Turbo Air Classifier and Optimization of the Process Parameters

Yun Zeng [1], Si Zhang [1], Yang Zhou [2,3,]* and Meiqiu Li [1]

[1] Institute for Strength and Vibration of Mechanical Structures, Yangtze University, Jingzhou 434023, China; mechanicszy@163.com (Y.Z.); medeka@163.com (S.Z.); limeiqiu@sina.com (M.L.)
[2] School of Mechatronic Engineering and Automation, Shanghai University, Shanghai 200444, China
[3] Shanghai Institute of Intelligent Science and Technology, Tongji University, Shanghai 200444, China
* Correspondence: saber_mio@shu.edu.cn

Received: 31 January 2020; Accepted: 14 February 2020; Published: 19 February 2020

Abstract: Due to the rapid development of powder technology around the world, powder materials are being widely used in various fields, including metallurgy, the chemical industry, and petroleum. The turbo air classifier, as a powder production equipment, is one of the most important mechanical facilities in the industry today. In order to investigate the production efficiency of ultrafine powder and improve the classification performance in a turbo air classifier, two process parameters were optimized by analyzing the influence of the rotor cage speed and air velocity on the flow field. Numerical simulations using the ANSYS-Fluent Software, as well as material classification experiments, were implemented to verify the optimal process parameters. The simulation results provide many optimal process parameters. Several sets of the optimal process parameters were selected, and the product particle size distribution was used as the inspection index to conduct a material grading experiment. The experimental results demonstrate that the process parameters of the turbo air classifier with better classification efficiency for the products of barite and iron-ore powder were an 1800 rpm rotor cage speed and 8 m/s air inlet velocity. This research study provides theoretical guidance and engineering application value for air classifiers.

Keywords: turbo air classifier; process parameters; numerical simulation; particle trajectory; relative classification sharpness index

---

1. Introduction

Currently, ultrafine powders are widely used in various fields, and the powder separation technique has gradually occupied an important position in industry. The main production equipment for ultrafine powders is the turbo air classifier. The classification performance of the classifier directly affects the efficiency of powder production. Therefore, many researchers [1–7] have conducted extensive studies on the theoretical analysis, flow field simulation, structural optimization, and other aspects of pneumatic grading equipment, and have made progress by obtaining many valuable results and providing the basis for the optimization of classifiers, performance enhancements, and fine separations. The main factors affecting the classification sharpness index and performance during the classification process are the rotor cage speed and the air inlet velocity inside the classifier [6–8]. According to the principle of classification, a material is subjected to inertial centrifugal force and air drag force at the same time during the classification process. Some researchers analyzed the effect of the rotor cage rotary speed on the classification sharpness index, using the Fluent software, and obtained a reasonable parameter combination for classification. Gao, Yu, and Liu [9,10] found that increasing the rotor cage rotary speed resulted in a finer product, but the higher speed caused the flow field to become uneven, and increased the classification sharpness index. Through the study of the classifier airflow velocity, Diao et al. [11] found that increasing the air inlet velocity could improve the classification efficiency. However, the inertial centrifugal forces of small- and large-sized materials are different. The distribution of small-diameter materials in the flow field is relatively uniform, but large-particle-sized materials are easily moved to the outside of the flow field. This leads to a high concentration outside the flow field and reduces the classification performance.

Based on above researches, in order to further study the classification performance, many scholars have found that it is meaningful to conduct in-depth research on the evaluation index of classification performance. Some scholars [12–14] pointed out that the Whiten's efficiency curve equation needs to be revised. Hence, the parameters in the Whiten's equation were correlated with the operating conditions of the air classifier as well as the material characteristics. The fish-hook phenomenon was demonstrated in a circulating-air classifier. Based on the experimental data, a process model was developed to predict the bypass fraction within the classifier [15]. Xing et al. [16] measured and analyzed the vortex swirling between rotor blades, using the particle image velocimetry (PIV) technique. They found changes in the regulation of the classification efficiency and cut-size, and optimized the operating parameters to achieve the minimum cut-size. However, in the actual classification process, the agglomeration and inclusion of fine particles in the coarse particles would cause a decrease of the classification sharpness index. Some researchers [17] demonstrated secondary airflow and found that when the ratio of the secondary airflow to the main airflow was maintained at 0.168, the classification was optimum. Based on the narrow particle size distribution experimental system, the best rotor cage speed difference between two turbo air classifiers was found, and the results showed that with a decreasing rotation speed difference, the productivity of the narrow-level product decreased and the uniformity increased. Nevertheless, many evaluation indexes of classification performance can accurately judge the grading performance, but it is very troublesome in actual production. Therefore, it is especially important to propose an efficient and simple evaluation index [18–22].

Due to limitations, such as the processing cost and other factors, it is difficult for enterprises and research institutes to produce a variety of grading wheels for structural parameters. Therefore, in the production process, one or more optimal processes are obtained through manual adjustment and matching of various process parameters. In the above studies, scholars mainly studied the effect of a single factor (rotation speed, airflow velocity, etc.) on the classification performance; however, research on the influence of various factors on the classification performance is rare. Consequently, this study applied the combination of process parameters as variables in numerical simulations and material experiments to obtain the optimum process parameters of the KFF ('KFF' is a code for a vertical turbo air classifier type) series turbine air classifier. Thus, in the production process, using the different manually controlled process parameters, one or more optimal process parameters can be obtained. In addition, a new evaluation index, the relative classification sharpness index, is proposed. The test results showed that it is the same as other classification performance evaluation indicators. It can be used to determine whether the classification status is good, simple, and easy, and has a certain guiding effect on industrial production.

2. Details of the Calculation Methodology

2.1. Description of the Equipment

The equipment for the experiment comprised of a KFF series turbo air classifier, high-pressure induced-draft fan, cyclone collector, pulse bag-filter, and electrical control system. The sketch of the KFF series turbo air classifier is shown in Figure 1a. The schematic diagrams of the vertical turbo air classifier with corresponding geometric parameters are shown in Figure 1b,c. The air supply system consisted of induced-draft fans. The induced-draft fan is "pumping" at the end of the turbo air classifier, providing the transport power for the particles. Firstly, the material is sent to the main classifier by the feeding system, and effective classification of the material is achieved by adjusting the rotor cage speed and matching it with a reasonable secondary air inlet velocity. Under the action of centrifugal force, the coarse powder is collected along the wall of the cylinder and the fine powder continues to be classified by the airflow into the next grading machine so that a reasonably super-fine material is collected when it enters the cyclone separator or dust removal system.

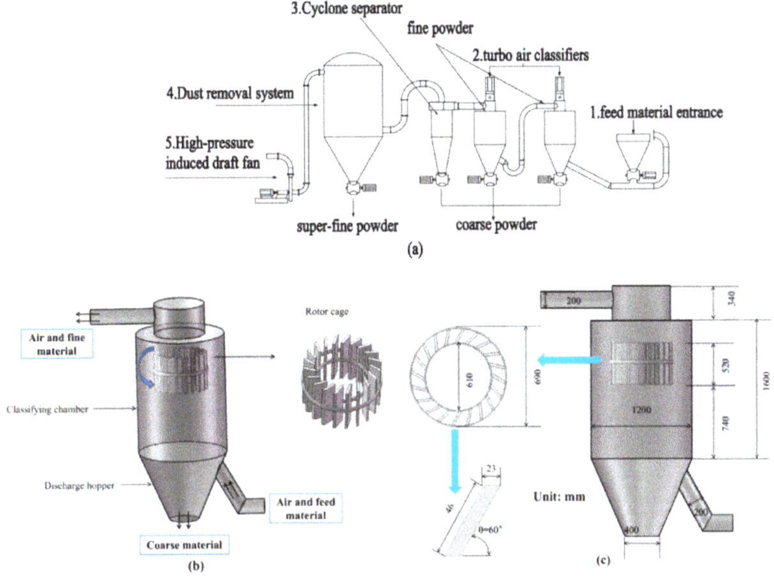

Figure 1. Diagram of the experiment equipment (**a**) and the 3D view of the geometry (**b**) and dimensions (**c**) of the turbo air classifier.

2.2. Particle Cutting Size Calculations

The force diagram at the edge of the classifier rotor cage after the material particles enter the classifier is shown in Figure 2. According to the theoretical method in the literature [15,16], we know that the particle is mainly influenced by the centrifugal and drag forces when it is delivered at the inlet edge of the rotor cage. In this experiment, the radial velocity of the airflow moving around the rotor cage is denoted as V_r, and the rotational speed of the rotor cage generating the centrifugal force is n. It was assumed that the tangential and radial velocities in the circumferential and vertical directions are uniform and that the particles are spherical. The mathematical definitions of the forces are given in Equations (1)–(3):

$$m\frac{D\mathbf{V}_{pr}}{Dt} = F_c - F_d \tag{1}$$

$$F_d = \frac{1}{2}C_D(\mathrm{Re}) * \rho_{air} * \pi * \left(\frac{d_p}{2}\right)^2 * (\mathbf{V}_{pr} - \mathbf{V}_r) * \left|\mathbf{V}_{pr} - \mathbf{V}_r\right| \tag{2}$$

$$F_c = m\frac{\mathbf{V}_T^2}{r} = \frac{4}{3} * \pi * \left(\frac{d_p}{2}\right)^3 * \rho_p * \frac{\mathbf{V}_T^2}{r} \tag{3}$$

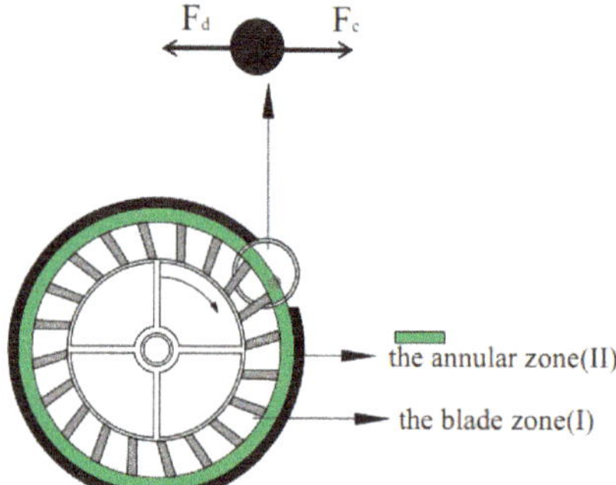

Figure 2. Particle-influencing forces at the inlet of the rotor cage.

It was assumed that there is no slip between the particle and the air tangential velocity. When the centrifugal and fluid drag forces reach equilibrium on the particle at the outer periphery of the rotor cage, and if the radial velocity of the particle is zero, the size of the particle is called the cut size (d_{50}). The d_{50} can be expressed as follows from Equations (1)–(3):

$$d_{50} = \frac{3C_D\rho_{air}\mathbf{V}_r^2 r}{4\mathbf{V}_T^2\rho_p} \tag{4}$$

By testing the classifier inlet air volume, the corresponding airflow radial velocity can be calculated:

$$\mathbf{V}_r = \frac{Q}{120\pi rh} \tag{5}$$

Using the known rotor cage speed, the tangential velocity of the particles at the outer edge of the grading wheel can be calculated:

$$\mathbf{V}_T = \frac{2\pi rn}{60} \tag{6}$$

The following equation can be derived from Equations (4)–(6):

$$d_{50} = \frac{3}{64}\frac{C_D\rho_{air}Q^2}{\pi^4 r^3 n^2 h^2 \rho_p} \tag{7}$$

where:

d_p: Particle diameter (μm);
$\mathbf{V}_r$: Radial velocity of airflow at the outer cylindrical periphery (m/s);
$\mathbf{V}_T$: Tangential velocity of airflow at the outer cylindrical periphery of the rotor cage (m/s);
$\mathbf{V}_{pr}$: Particle radial velocity (m/s);
ρ_{air}: Density of airflow (kg/m^3);
ρ_p: Particle density (kg/m^3);
r: Radius of rotor cage (mm);
h: Blade height (mm);
m: Mass of particle (kg/m^3);
n: Rotor cage speed (rpm);
Q: Total volumetric flow rate of air (m^3/s);
C_D: Drag coefficient;
Re: Reynolds number; and
d_{50}: Cut size of classification (μm).

2.3. Mathematical Model

2.3.1. Continuous Phase Governing Equations

The three-dimensional steady simulation was performed using ANSYS-FLUENT 15.0. For the case of incompressible flow, the mass and momentum equations are as follows:

$$\frac{\partial u_i}{\partial x_i} = 0 \tag{8}$$

$$\frac{\partial}{\partial t}(\rho u_i) + \frac{\partial}{\partial x_j}(\rho u_i u_j) = -\frac{\partial p}{\partial x_i} + \frac{\partial}{\partial x_j}(\mu \frac{\partial u_i}{\partial x_j} - \overline{\rho u'_i u'_j}) + S_i \tag{9}$$

Where u_i, x_i, ρ, P, and μ represent the fluid velocity, position, time, constant fluid density, static pressure, and gas viscosity, respectively. $-\overline{\rho u'_i u'_j}$ is the Reynolds stress term. Choosing a suitable turbulence model in the case is of paramount importance. Furthermore, the RNG k-e model has been proven to be an appropriate model to describe the turbulence of turbo air classifier flow [21]. The turbulent kinetic energy and turbulent dissipation rate are expressed as follows:

$$\rho\frac{dk}{dt} = \frac{\partial}{\partial x_i}\left[(\alpha_k \mu_{eff})\frac{\partial k}{\partial x_i}\right] + G_k + G_b - \rho\varepsilon - Y_M \tag{10}$$

$$\rho\frac{d\varepsilon}{dt} = \frac{\partial}{\partial x_i}\left(\alpha_\varepsilon \mu_{eff}\frac{\partial \varepsilon}{\partial x_i}\right) + C_{1\varepsilon}\frac{\varepsilon}{k}(G_k + C_{3\varepsilon}G_b) - C_{2\varepsilon}\rho\frac{\varepsilon^2}{k} - R \tag{11}$$

where G_k and G_b represent the components of the turbulent kinetic energy caused by the average velocity gradient and buoyancy. Y_M is the effect of compressible turbulent pulsation expansion on the total dissipation rate. The values of the constant are $\alpha_\varepsilon = 0.7692$, $\alpha_k = 1$, $C_{1\varepsilon} = 1.44$, $C_{2\varepsilon} = 1.92$, $C_{3\varepsilon} = 0.09$.

The turbulent viscosity coefficient can be calculated as:

$$\mu_t = \rho C_\mu \frac{k^2}{\varepsilon} \tag{12}$$

where $C_\mu = 0.0845$.

2.3.2. Discrete Phase Governing Equations

The choice of a multiphase flow model is mainly determined by the particle volume loading rate, κ, and mass loading rate, υ, which are demonstrated in the following two parts:

(1) Volume loading rate is the ratio of the particle volume to gas volume per unit time in a space. It can be expressed as follows:

$$\kappa = \frac{\alpha_p}{\alpha_f} = \upsilon\frac{\rho_f}{\rho_p} \tag{13}$$

where:

α_p, α_f—Particle volume and gas volume passing through the effective section at per unit time; and

ρ_p, ρ_f—Particle density and air density.

Using the particle volume loading rate, the dimensionless distance between particles and particles in the particle phase can be calculated:

$$D = \frac{L}{d_p} = \frac{\pi}{6}(\frac{1+\kappa}{\kappa})^{\frac{1}{3}} \tag{14}$$

where:

L—The distance between particle and particle; and

d_p—Particle diameter.

(2) Mass loading rate is the ratio of particle mass to gas mass across an effective section in a certain space per unit time. It is expressed as follows:

$$v = \frac{m_p}{m_f} = \frac{\alpha_p \rho_p}{\alpha_f \rho_f} \tag{15}$$

In this study, the airflow and material are fed into the classifier by a circular section inlet. Thus, the gas volume passing through the effective section per unit time can be calculated:

$$\alpha_f = 2\pi (r_x)^2 V_0 \tag{16}$$

where V_0 is the air inlet velocity, which was set in the range from 6 to 12 m/s in this study. r_x is the radius of the section at the inlet. The feeding speed is 240 kg/h. Combined with Equations (13)–(15), the range of the volume loading rate can be calculated from 2.624×10^{-6} to 4.34×10^{-6}. Meanwhile, the range of the mass loading rate can be calculated from 0.0257 to 0.0342. These values are very small. However, the dimensionless distance between particles and particles ranged from 32.09 to 37.9. According to the calculated results, the value of the particle volume loading rate and mass loading rate is very small, and the value of the dimensionless distance between particles and particles is very large. Therefore, the particle phase is considered to be highly sparse, which satisfies the DPM calculation conditions. Furthermore, it can also be considered that the coupling between the particles and the gas phase is unidirectional. Namely, only the influence of gas on the particles is taken into account, rather than the influence of particles on the gas.

Through the DPM of FLUENT, the trajectory of a discrete phase particle can be calculated in a Lagrangian reference frame by integrating the force balance on the particle. This force balance equation can be written in Cartesian coordinates:

$$\frac{du_P}{dt} = F_D(u - u_P) + \frac{g_x(\rho_P - \rho)}{\rho_P} + F_x \tag{17}$$

$$F_D = \frac{18\mu}{\rho_P d_P^2} \frac{C_D Re}{24} \tag{18}$$

$$Re_P = \frac{\rho d_P |u_P - u|}{\mu} \tag{19}$$

where $F_D(u - u_P)$ is the drag force per unit particle mass, μ is the fluid phase velocity, u_P is the particle velocity, μ is the kinematic viscosity of fluids, ρ is the fluid density, ρ_P is the particle density, d_P is the particle diameter, Re is the relative Reynolds number (particle Reynolds number) (the define of Particle Reynolds number can be calculated by Equation (19)) C_D is the drag coefficient, and F_x is an additional acceleration (force/unit particle mass) term.

2.4. Boundary Conditions and Parameter Setting

The model was designed to be imported into the ANSYS-Fluent software for numerical calculations. There is one entrance and two exits in the model. A "velocity inlet" boundary condition was used at the air-inlet, the air velocity was assumed to be uniformly distributed at the air inlet section, and its direction is normal to the air-inlet boundary. The boundary condition at the turbo air classifier was prescribed as a fully developed pipe flow and treated as "outflow". A no-slip boundary condition was used on the wall boundary and the near wall treatment was a standard wall function. The SIMPLEC algorithm was adopted for the pressure–velocity coupling, and the QUICK difference scheme were used for the convection and diffusion. The convection terms of the discrete equations were all in the default format. An insufficient relaxation factor empirical selection was used. In total, 2000 steps were iterated to set the solution accuracy at 1e–03.

3. Simulation Results and Analysis

According to Equation (7), the two main factors affecting the particle size grading are the tangential velocity, V_T, and radial velocity, V_r, of the gas flow, when the other parameters, such as the radius and height of the runner, are quantitative. The tangential velocity of the airflow is directly related to the rotor cage speed, and the radial velocity of the airflow is related to the total volumetric flow rate of air. Increasing the rotor cage speed can result in particles with finer particle sizes, but this can cause an uneven distribution of the flow field, affect the classification sharpness index, and reduce the classification performance. When the air volume of the system is changed, if the airflow rate is too low, the feeding force of the raw materials becomes too low, resulting in grading failure; if the main airflow is too high, the compulsive action of the grading wheel may be invalidated, and the coarse particles are rejected. The airflow is transmitted into the fine powder, and when the classifier cuts the particle size, the classification effectiveness worsens. Therefore, changing the rotor cage speed or the air inlet velocity of the system alone cannot help obtain the optimum process parameters. Only by simultaneously controlling the air inlet velocity of the system and the corresponding rotor cage speed to perform particle grading can one or more sets of optimal process parameter combinations be obtained

Numerical simulation experiments were performed using multiple groups of variables by setting different simulation parameters to achieve quantitative changes in V_T and V_r, and the particle velocity map of the impeller surface was used as a reference to judge the better process parameter. Owing to the actual process parameters, the limited range of the rotation speed of the rotor wheel was 500–3000 rpm and the range of air volume was 2700–5500 m³/h. Calculated according to Equation (7) and the actual working condition limit, four groups of the rotor cage speed were set: 1800, 2000, 2200, and 2400 rpm. The air inlet velocity was set in four groups: 6, 8, 10, and 12 m/s. The above four groups of rotor cage speed and air inlet velocity were combined to perform an orthogonal numerical simulation test.

3.1. The Tangential Velocity Distribution in the Classifier.

There are two important grading functional zones in the classifier: One of them is the flow region between the blades (I: blade zone, it called the separation functional zone), another is an annular region surrounding the inlet boundary of the rotor cage (II: annular zone, it called the decentralized separation functional zone) [18]. The rotation of the rotor cage causes turbulence in the annular zone, and the turbulence in the annular zone has some influence on the classification effect, which causes a decrease in the classification efficiency and classification sharpness index. In addition, if the tangential velocity in the blade zone is much larger than the tangential velocity in the annular zone, the fine particles can also move to the annular zone under strong centrifugal force, and finally settle along the side wall to become coarse powder, which causes a decrease in the classification efficiency and classification sharpness index. Therefore, it is necessary to study the tangential velocity distribution of zone I and zone II.

Figure 3 shows the tangential velocity details of the rotor cage section. The grading wheel surface (select a section at half the height of the rotor cage, Y = 260 mm) tangential velocity contour diagrams for 16 groups and tangential velocity contrast under different radial distances from the axis are shown in Figure 3.

According to the above analysis, the more uniform tangential velocity distribution of the airflow in the annular zone (II) and blade zone (I), the more stable the flow field. Based on Figure 3a, under the condition that the air inlet velocity remains unchanged at 6 m/s, a comparison of the tangential speed profiles of the blade zone and annular zone under different process parameters indicates that the tendency of the tangential distribution at rotor cage speeds of 1600 and 2200 rpm is more stable than that at rotor cage speeds of 1800 and 2000 rpm. Based on Figure 3b, under the condition that the air inlet velocity remains unchanged at 8 m/s, a comparison of the tangential speed profiles of the blade zone and annular zone under different process parameters indicates that the tendency of the tangential distribution at he rotor cage speeds of 1800 rpm is more stable than that at rotor cage speeds of 1600, 2000, and 2200 rpm. Nevertheless, the tendency of the tangential change gradient amplitude at rotor cage speeds of 2200 rpm is slower than that at rotor cage speeds of 1600 and 2000 rpm. Based on Figure 3c, under the condition that the air inlet velocity remains unchanged at 10 m/s, it can be clearly seen that the tendency of the tangential distribution at rotor cage speeds of 1800 and 2000 rpm is more stable than that at rotor cage speeds of 1600 and 2200 rpm. Based on Figure 3d, under the condition that the air inlet velocity remains unchanged at 12 m/s, it can be clearly seen that the tendency of the tangential distribution at rotor cage speeds of 2000 and 2200 rpm is more stable than that at rotor cage speeds of 1600 and 1800 rpm.

From the above discussion, the conclusion can be reached that eight groups of process parameters (V_{in}-n,6-1600,6-2200,8-1800,8-2200,10-1800,10-2000,12-2000,12-2200) may be better than others.

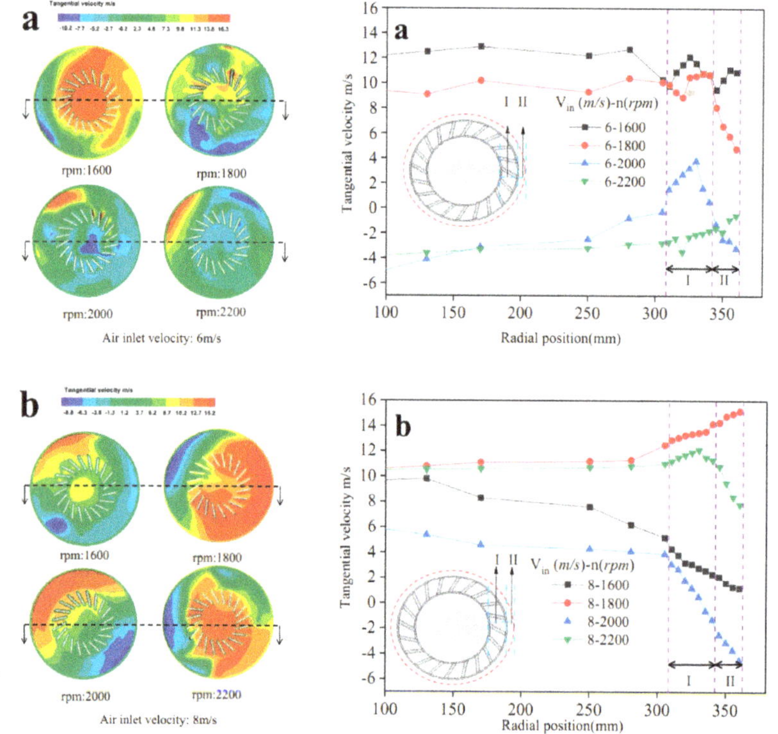

Figure 3. *Cont.*

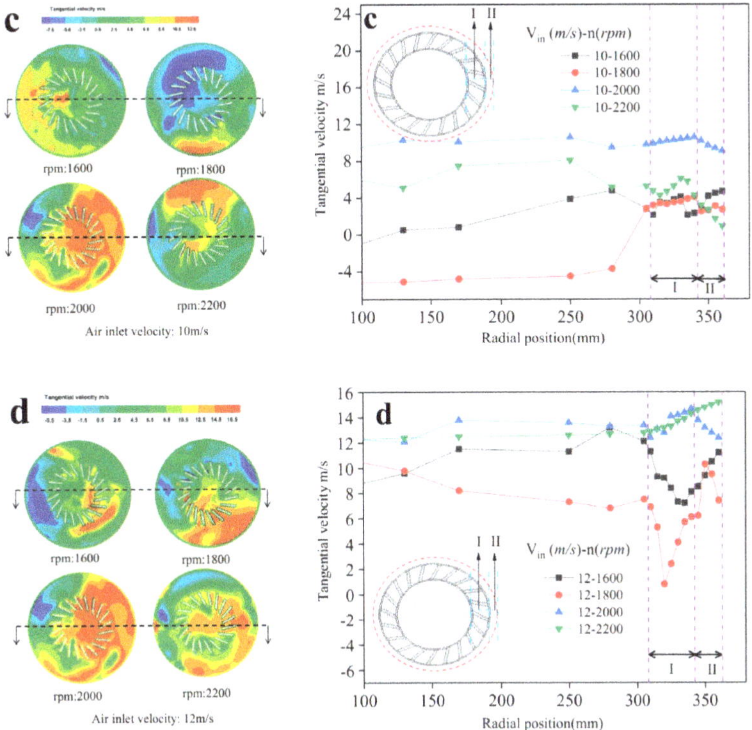

Figure 3. Contours of the tangential velocity distribution and air tangential velocity contrast under different radial distances from the axis at different process parameters (305–345 mm is the area between the rotor blades, 345–365 mm is the annular zone). (**a**) Air inlet velocity 6 m/s and 4 different rotor cgae speed. (**b**) Air inlet velocity 8 m/s and 4 different rotor cgae speed. (**c**) Air inlet velocity 10 m/s and 4 different rotor cgae speed. (**d**) Air inlet velocity 12 m/s and 4 different rotor cgae speed.

3.2. The Radial Velocity Distribution in the Classifier

In the process of grading, the blade zone (I) is an important grading functional zone, and its main function is to transport fine powder and separate the coarse and fine powder. Furthermore, after the airflow enters the rotor cage, on the one hand, it is driven by the rotation of the rotating blades, and on the other hand, it moves inward along the blade under the action of the central negative pressure. Finally, inertial anti-vortex will be formed between the blades. The material is transported by radial airflow in zone I. As the radial airflow in the passage between adjacent blades is affected by the inertial anti-vortex, it leads to uniformity of the radial velocity distribution. The fine powder that has entered the cage will leave the cage under the influence of the anti-vortex. Finally, the cutting particle size is dispersed, which decreases the classification sharpness index. In addition, if the radial velocity distribution is not uniform in the blade zone, the coarse particles are also collected by the airflow into fine powder, which affects the classification sharpness index. Therefore, the radial velocity distribution of the airflow in blade zone (I) must be studied.

Figure 4 shows the radial velocity details of the rotor cage section. The grading wheel surface (select a section at half the height of the rotor cage, Y = 260 mm) radial velocity contour diagrams for 16 groups and radial velocity contrast under different radial distances from the axis are shown in Figure 4.

In the classification process, the vane flow velocity near the "surface of the advance blade" (the surface that is in the blade facing in the direction of rotation) is larger than the "surface of the back blade" (the surface that in the blade is back to the direction of rotation). Therefore, it can be observed whether the positive vortex and the anti-vortex exist between the blades to judge whether the flow field between the rotor cage blades is uniform. As is shown in Figure 4a–d, the size of the vortex and the velocity gradient between the blades can be easily and intuitively found, as eight groups of process parameters (V_{in}-n,6-1600,6-2200,8-1800,8-2200,10-1800,10-2000,12-2000,12-2200) may be better than others. The conclusion is the same as the results of the tangential velocity.

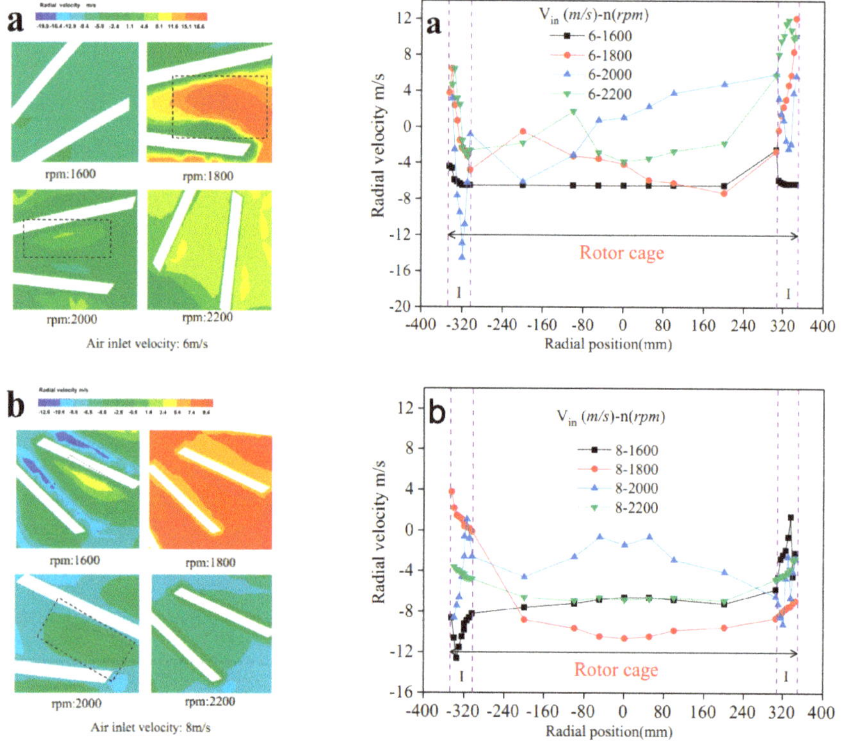

Figure 4. *Cont.*

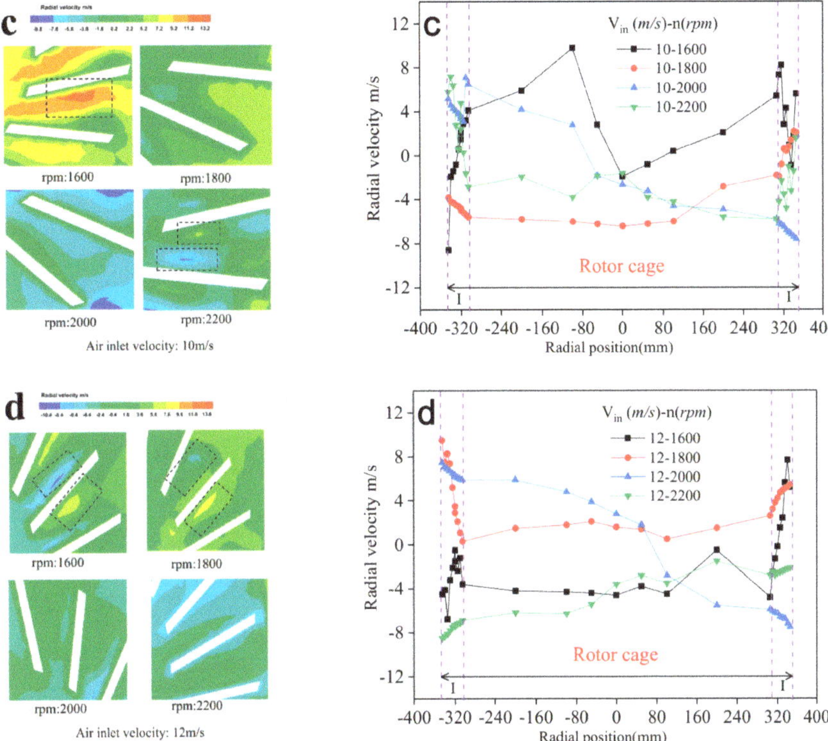

Figure 4. Contours of the air radial velocity distribution between rotor blades and radial velocity contrast under different radial distances from the axis at different process parameters (−345–305mm, 305–345mm is the area between the rotor blades, −305–305 mm is the central region).(**a**) Air inlet velocity 6 m/s and 4 different rotor cgae speed. (**b**) Air inlet velocity 8 m/s and 4 different rotor cgae speed. (**c**) Air inlet velocity 10 m/s and 4 different rotor cgae speed. (**d**) Air inlet velocity 12 m/s and 4 different rotor cgae speed.

3.3. Discrete-Phase Simulated Results and Analysis

In the particle classification process, the movement process of particles is intuitively described and is revealed by the particle trajectory, which can also explain the particle separation mechanism. Consequently, the discrete phase model was established to simulate the particle trajectory, and the particle motions for eight groups of process parameters were contrasted.

3.3.1. Simulated Results and Analysis of Single Particle

Iron ore powder particle (12 μm) was chosen as the material, and steady flow simulation was set up. According to the continuous phase simulation, eight groups of process parameters were better than others. Therefore, these eight groups of process parameters were selected for discrete phase simulation. The particle tracks cloud diagrams for the eight groups are as follows:

As is shown in Figure 5, comparing the particle tracks under the process parameters for the eight groups in the numerical simulations, as seen from the Figure 5a–h, it can be found that the number of particle tracks(a), (c), (f), (h) are more than (b–e), (g). Therefore, it can be simply inferred that four process parameters can be estimated as better than the others for producing the particle size of 12 μm, rotor speed of 1600 rpm with the air inlet velocity at 6 m/s, rotor speed of 1800 rpm with the air inlet velocity at 8 m/s, rotor speed of 2000 rpm with the air inlet velocity at 10 m/s, and rotor speed of 2200 rpm with the air inlet velocity at 12 m/s.

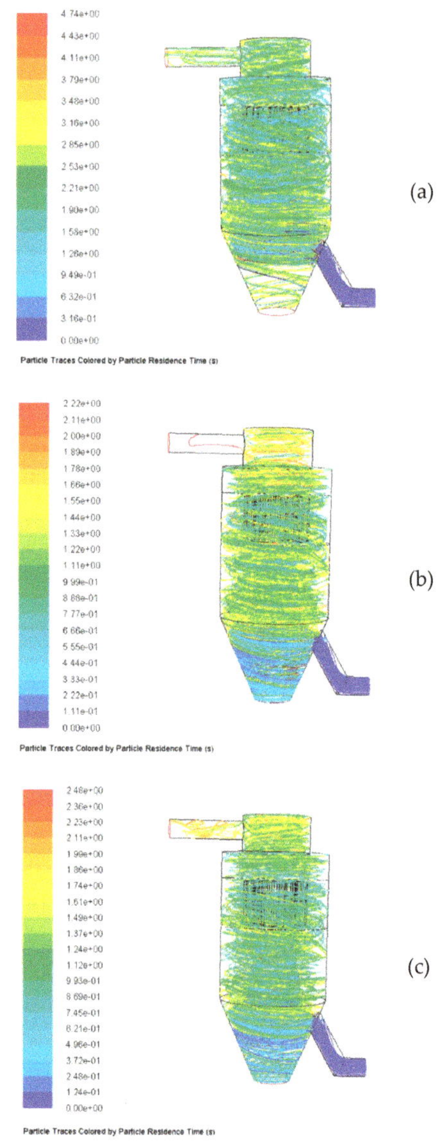

Figure 5. *Cont.*

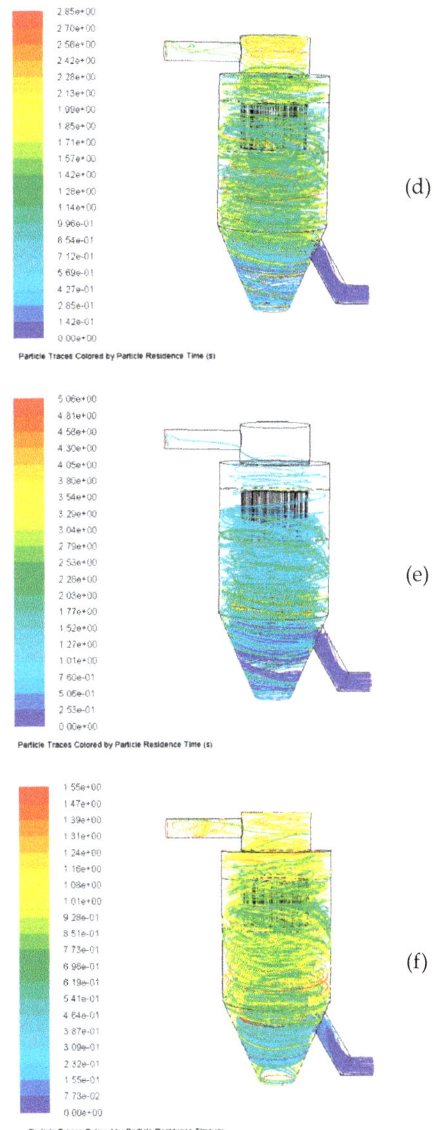

Figure 5. *Cont.*

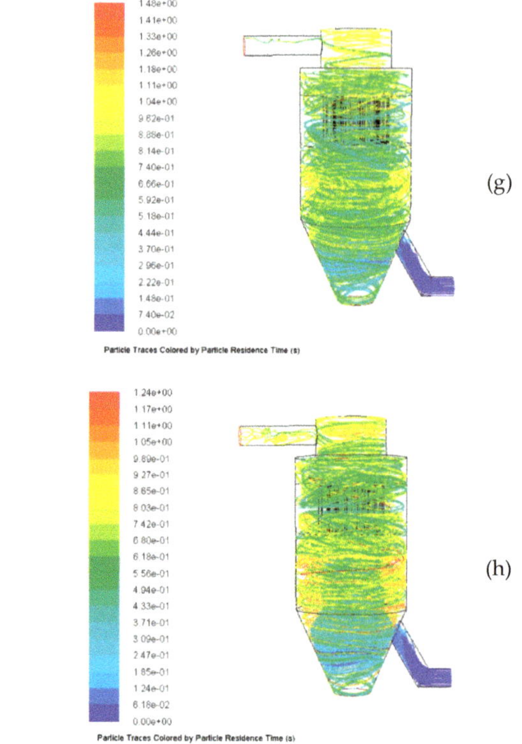

Figure 5. Particle tracks for 8 groups of different rotor speeds and air inlet velocity. (**a**) Rotor speed 1600 rpm, air inlet velocity 6 m/s. (**b**) Rotor speed 2200 rpm, air inlet velocity 6 m/s. (**c**) Rotor speed 1800 rpm, air inlet velocity 8 m/s. (**d**)Rotor speed 2200 rpm, air inlet velocity 8 m/s. (**e**) Rotor speed 1800 rpm, air inlet velocity 10 m/s. (**f**) Rotor speed 2000 rpm, air inlet velocity 10 m/s. (**g**) Rotor speed 2000 rpm, air inlet velocity 12 m/s. (**h**) Rotor speed 2200 rpm, air inlet velocity 12 m/s.

3.3.2. Simulated Results and Analysis of Multi-Particle

The authors used the ANASYS-fluent discrete phase model for numerical simulation, and investigation of gas flow behaviors in the turbo air classifier. In order to obtain the simulated Tromp curve at different process parameters, the authors set up 13 different particle sizes for each set of process parameters. In order for better comparison with the electronic test report, the specific particle size parameters were set as follows:

Iron ore fines: 1, 2, 4, 5.13, 6.21, 7.51, 8, 10, 11, 12.66, 16.62, 19.5, and 23.6 μm

Barite powder: 4, 5.13, 6.21, 7.51, 8, 11, 12.66, 13.31, 16.62, 19.5, 23.6, 28.56, 32, and 41.8 μm

The number of particles escaped and trapped was calculated by the numerical simulation, and the upper limit of the particle calculation step was 20,000. Finally, the tromp curve was drawn the light of the percentage of escape particles in the total number of particles. The pictures are shown in Figure 6a,b. The purpose of this study was to research the combination of two process parameter variables. Therefore, according to the result of the numerical simulation, the cut size under the combination of the rotor speed and air flow rate can be roughly estimated. The results are shown in Figure 6c,d. The effect of the rotor speed and air inlet velocity on the cut size can be found.

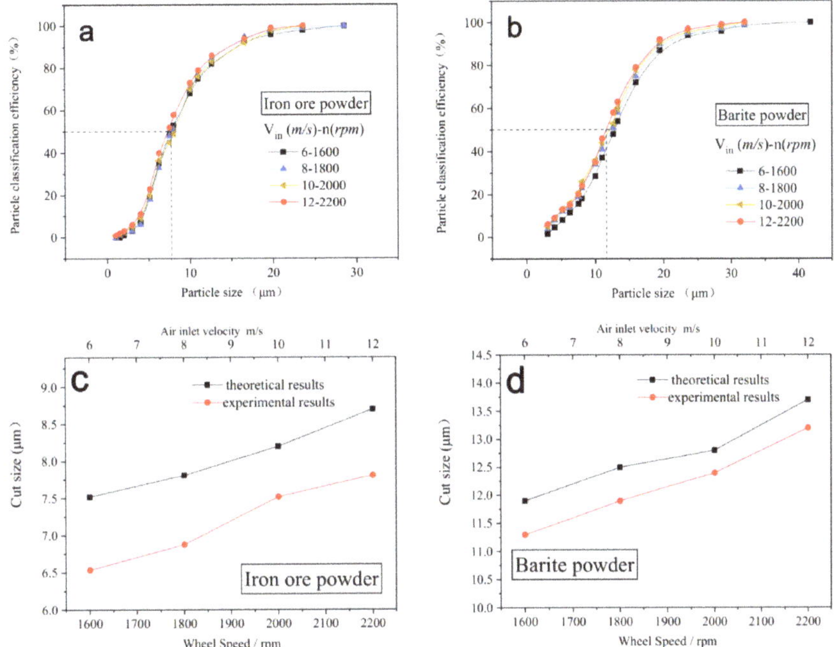

Figure 6. The numerical partial classification efficiency curves and cut size of four different process parameters. (**a**) Iron ore powder numerical Tromp curves (**b**) Barite powder numerical Tromp curves (**c**) Iron ore powder numerical cut size (**d**) Barite powder numerical cut size

Figure 6a,b shows the schematic diagram of the numerical simulated Tromp curve by different process parameters. It is not easy to obtain the classification sharpness K ($K = d_{75}/d_{25}$) under different process parameters through observation of the tromp curve. However, it can be roughly judged by the cut size by different process parameters. According to Equation (7), it can be easily inferred that with the air inlet velocity increase, the cut size will increase, or with the rotor speed increase, the cut size will decrease. However, when the air inlet velocity and the rotor speed increase simultaneously, we cannot conclude whether the cut size is increased or decreased. From Figure 6, it can be easily found that the effect of the rotor speed on the cut size is more than the air inlet velocity. In addition, in order to select which combination-type process parameter is better than others, the optimum process parameter was verified using a material grading experiment.

4. Classification Experiment and Discussion

The raw materials for this experiment were iron ore fines and barite, with material densities of 7.83 and 4.3 g/cm^3, respectively (the densities of materials were detected by an HX-TD-type true density tester). According to the numerical simulation results, four process parameters (1#rotation speed of 1600 rpm and an air inlet speed of 6 m/s; 2# rotation speed of 1800 rpm and an air inlet speed of 8 m/s; 3# rotation speed of 2000 rpm and an air inlet speed of 10 m/s; 4# rotation speed of 2200 rpm and an air inlet speed of 12 m/s) were better than others.

4.1. Particle Size Distribution and the Classification Efficiency

Therefore, the iron ore fines and barite raw materials were divided into two groups and subjected to grading tests under four kinds of process parameters. The grading of the particle size distribution maps are as follows:

According to the powder particle size distribution diagrams in Figure 7a,b, it can be found that the particle size distributions of the product by four process parameters are evidently different.

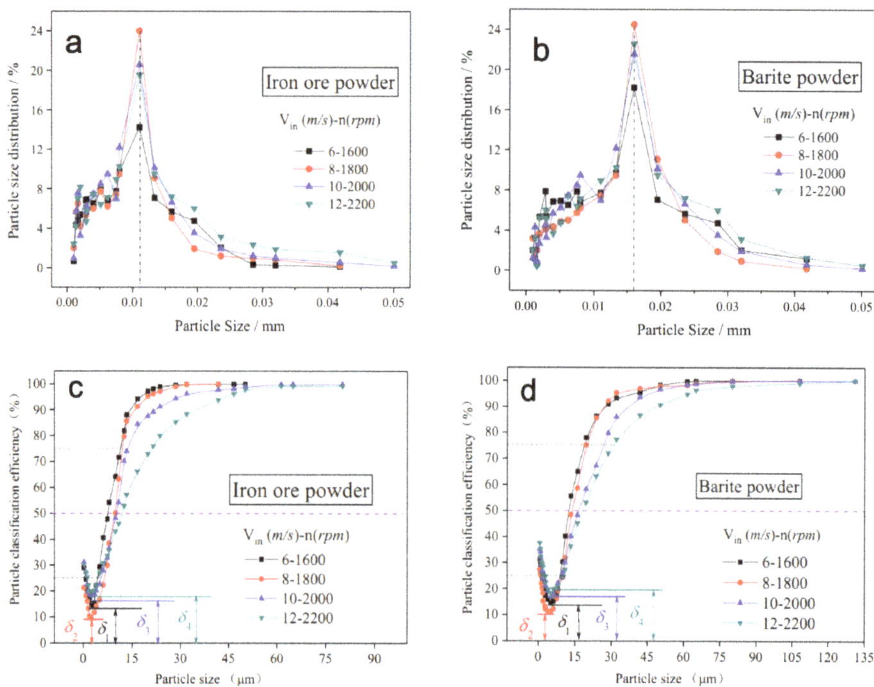

Figure 7. The particle size distribution and the classification efficiency of two different materials. (**a**) Iron ore powder experiments distribution curves (**b**) Barite powder experiments distribution curves (**c**) Iron ore powder experiments Tromp curves (**d**) Barite powder experiments Tromp curves.

Firstly, iron ore powder with a particle size distribution ranges from 9 to 13 µm is mainly produced. It can be found that the second group of process parameters has the highest existence of the particle size distribution ranges from 8 to 12 µm. In addition, the highest existence of the barite particle size distribution ranges from 11 to 16 µm, and the best process parameters are in the second group. The case studies of the particle classification efficiency are shown in Figure 7c,d. These curves depict the same changing tendency of the distorted S-shape. As the particle size decreases, the classification efficiency of the particles first decreases and then increases. This phenomenon is commonly referred to as the hook effect. With the increase of the rotor cage speed and air inlet velocity, the fish-hook effect is found to be enhanced. Generally, the hook effect will decrease corresponding to the increase of the air inlet velocity. However, with the rotor cage speed increases, the particle cut size decreases, the d_{50} also decreases correspondingly. The consequence is that the Tromp curve will move to the left, causing the agglomeration to advance. At last, the fish-hook effect will increase. Therefore, considering these two factors of the air inlet velocity and rotor cage speed, it can be found that the hook effect is increasing. On the other hand, the bypass value (δ) is an important index to evaluate the performance of classification. As is shown in Figure 7c,d, the bypass value ($\delta_2 < \delta_1 < \delta_3 < \delta_4$) of the second process parameters is the smallest. It can be considered that the performance of the second process parameters' classification is better than the others.

4.2. Coarse Powder Yield and Newton Efficiency

For most airflow grading equipment, the powder is classified according to a certain cutting particle size. The large particle portion after classification is referred to as coarse products, and the small particle portion is referred to as fine products. Commonly, Newton's classification efficiency is a main classification performance index. It comprehensively examines the degree of separation of coarse and fine powder particles. The expression is:

$$\eta_N = Y_c - (1 - Y_f) = Y_c + Y_f - 1 \tag{20}$$

$$Y_c = \frac{x_a A}{x_c F} \tag{21}$$

$$Y_f = \frac{(1 - x_b)B}{(1 - x_c)F} \tag{22}$$

where Y is the coarse powder yield, Y_f is the fine powder yield, x_c is the mass percentage of coarse particle in raw material, A and B are the mass of the collected coarse fraction and collected fine fraction, F is the mass of the raw material, x_a and x_b are the mass percentage of coarse particles ($d_p > d_{50}$) in the collected coarse fraction.

Figure 8a,b, shows the effects of four process parameters on the iron ore coarse powder yield (Y_c) and Newton efficiency (η_N). It can be found that with the rotor cage speed and air inlet velocity increase, the collection efficiency of the coarse particle gradually decreases. The reason is that the increase in the amount of air entering the main air causes the air drag to rise and more and more coarse particles are taken away; therefore, the value of Y_c decreases. In addition, as the rotation speed of the runner and the air intake speed increase, Newton's classification efficiency increases initially, but as the rotation speed and air intake speed of the runner continue to increase, Newton's classification efficiency decreases. It means that an excessive rotor cage speed and air inlet velocity is not good for classification. Finally, according to the results of Newton's classification efficiency, it can be considered that the performance of the second process parameters' classification is better than the others.

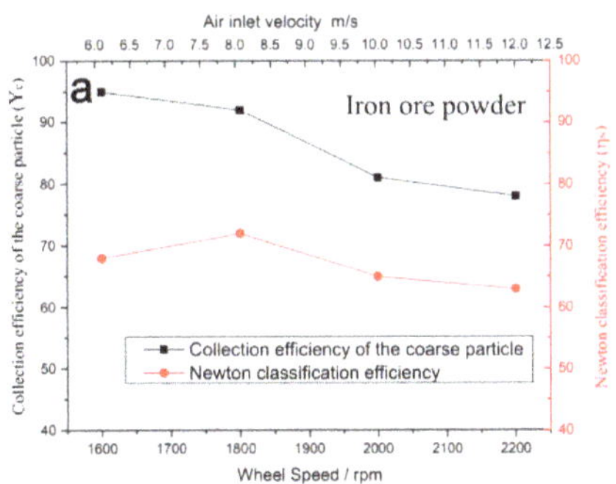

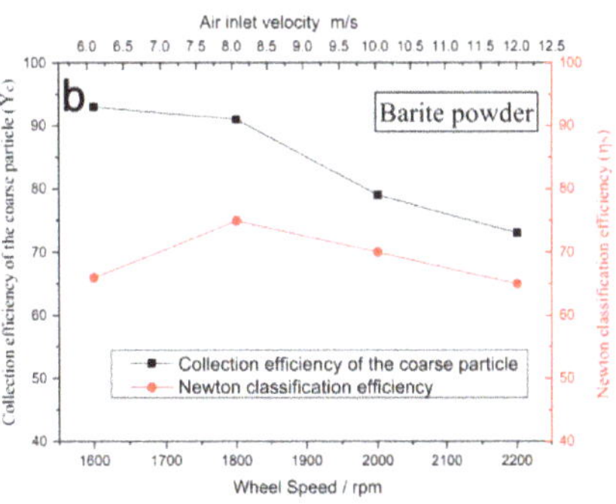

Figure 8. Effects of four process parameters on two different materials of the coarse powder yield (Y_c) and Newton efficiency (η_N). (**a**) Iron ore powder (**b**) Barite powder.

4.3. Classification Sharpness Index (K)

There are roughly three kinds of indicators for evaluating the grading effectiveness [19–22]. The representative index is the classification sharpness index. The index proposed by Germany's Leschonski is $K = d_{75}/d_{25}$. Currently, in an effective grading apparatus for a commercial field, if the sharpness index is close to 1, the performance of the grading apparatus is considered perfect. In an actual industry experiment, if the value of the K is in the range from 1.4 to 2.0, the performance of the grading apparatus is considered good; if the value of the K is in the range from 1.0 to 1.4, the performance of the grading apparatus is considered excellent. According to the classification experiment results, the effects of four process parameters on the classification sharpness index are shown in Figure 9. It can be easily found that the classification sharpness index (K) value of the second process parameters is closest to 1. Therefore, the performance of the second process parameters' classification is better than others.

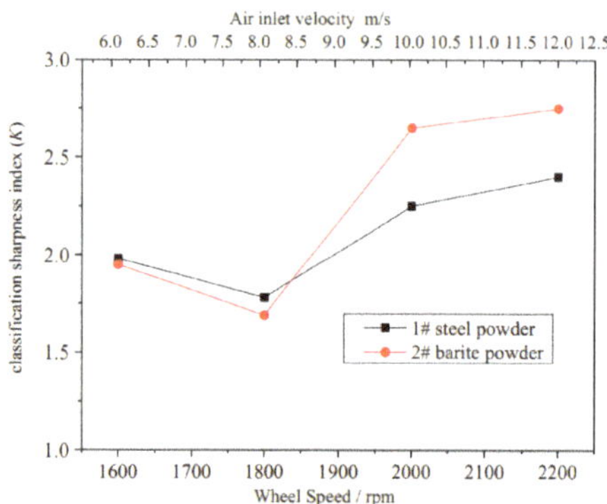

Figure 9. Effects of four process parameters on the classification sharpness index (K).

4.4. Relative Classification Sharpness Index

The author proposes a relative classification sharpness index as the examination index for this experiment. As shown in Figure 10a,b, the curves f, a, and b represent the raw material particle size distribution, the classified fine powder particle size distribution, and the classified coarse powder particle size distribution, respectively. $d_{10coarse}$ indicates a particle size of 10% for the cumulative content in the coarse powder after classification, d_{90fine} indicates a particle size of 90% for the cumulative content in the fine powder after classification, and d' indicates the distribution frequency of the fine powder in the coarse powder and the coarse powder in the fine powder is equivalent. The expression for the relative grading sharpness index is:

$$\delta = \frac{d_{10coarse} - d_{90fine}}{d'} \tag{23}$$

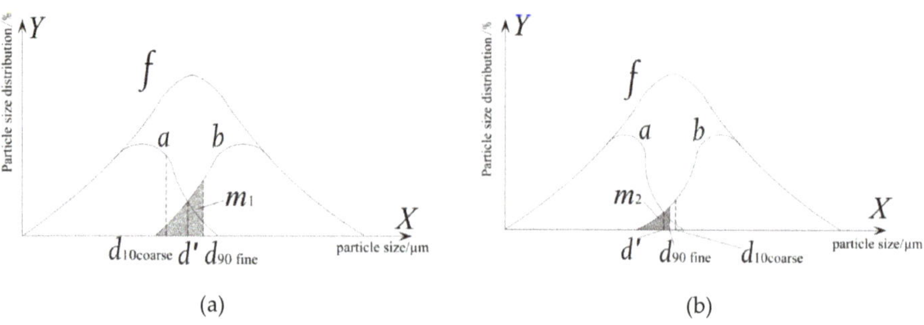

(a)   (b)

Figure 10. Coarse- and fine-grade product size distribution curve. (**a**) $d_{10coarse} < d_{90fine}$. (**b**) $d_{10coarse} > d_{90fine}$.

The value of the relative classification sharpness index, δ, is the index for testing this experiment. A larger δ indicates a better grading effectiveness and a smaller δ indicates a poorer grading effectiveness.

Judging from the numerical simulation results, the authors chose four sets of process parameters for the actual experiment, with the powder particle size distribution as the inspection index, to investigate the effectiveness of the four sets of process parameters on the relative classification sharpness index of the powder.

The test instrument selected was the *LS-C (IIA)* laser particle size analyzer (error $\leq 3\%$).

Based on the data in Table 1, a graph of δ changing with the process parameters was plotted, as shown in Figure 11. As the rotation speed of the runner and the air intake speed increase, the relative classification sharpness index increases initially, but as the rotation speed and air intake speed of the runner continue to increase, the relative classification sharpness index decreases. According to the results of the relative classification sharpness index, it can be found that the performance of the second process parameters' classification is better than the others.

Table 1. Relative classification accuracies under different process parameters.

Material Name	Number	Wheel Speed Rpm	Air inlet Velocity m/s	Particle Size/μm $d_{10coarse}$	d_{90fine}	d'	δ
Steel Powder	1#	1600	6	8.66	7.08	7.58	0.208
	2#	1800	8	8.78	6.36	7.25	0.334
	3#	2000	10	9.61	10.23	9.98	−0.621
	4#	2200	12	9.43	10.49	9.92	−0.106
Barite Powder	1#	1600	6	14.62	13.36	13.87	0.091
	2#	1800	8	14.95	11.01	12.36	0.318
	3#	2000	10	15.73	16.66	15.95	−0.583
	4#	2200	12	16.28	17.69	16.89	−0.083

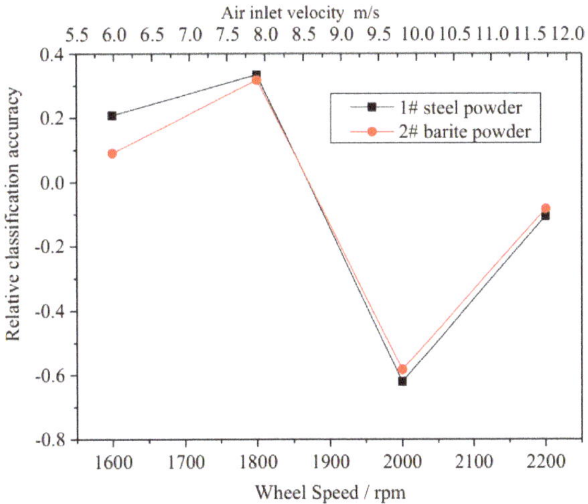

Figure 11. Comparison of relative classification accuracies for different process parameters.

5. Conclusions

This study was based on the kinetics of single particles. The trajectory of the particles was quantitatively analyzed under different rotation speeds and intake air volumes. The following conclusions were obtained:

(1)   The grading experiment results for iron ore fines and barite powder materials indicate that the better process parameter combination for the production of 12-μm particles using the KFF series turbo air classifier is a 1800 rpm rotor speed and 8 m/s air inlet velocity.

(2) For the same process parameters, when the same grain size of 12-μm barite and iron ore fines are produced, the relative classification sharpness index is different, indicating that the grading effectiveness for the smaller density particles is higher.

(3) Numerical simulation experiments showed that the air inlet velocity has an effect on the grading effectiveness rather than the rotor cage speed.

(4) The proposed new evaluation index, the relative classification sharpness index, could accurately evaluate the classification performance.

Author Contributions: The author S.Z. provided help in the preliminary investigation of this article. M.L. provided resources such as experimental equipment. Y.Z. (Yang Zhou) conducted in-depth research on the evaluation index of classification performance, then proposed the concept of the relative classification sharpness index, which could be applied in the classification performance tests of actual production. Y.Z. (Yun Zeng) was in charge of the entire research experiment, statistics all the data and sorted it out. Finally wrote and revised the manuscript. All authors have read and agreed to the published version of the manuscript.

Funding: This research received no external funding.

Acknowledgments: This project was supported financially by the National Key R&D project (2016YFC0303703), the National Natural Science Foundation of China (No. 51674040 and No. 51904181), and the National Natural Science Foundation of Hubei province (No. 2016CFC740). The authors would like to thank all the members of the project team for their support.

Conflicts of Interest: The authors declare no conflict of interest.

References

1. Ren, W.; Liu, J.; Yu, Y. Design of a rotor cage with non-radial arc blades for turbo air classifiers. *Powder Technol.* **2016**, *292*, 46–53. [CrossRef]

2. Xiong, D.; Li, S.; Huang, P. Effect of feeding type on classification performance of superfine classifier. *CIESC J.* **2012**, *63*, 3818–3825.

3. Huang, Q.; Liu, J.; Yu, Y. Turbo air classifier guide vane improvement and inner flow field numerical simulation. *Powder Technol.* **2012**, *226*, 10–15. [CrossRef]

4. Shapiro, M.; Galperin, V. Air classification of solid particles: Are view. *Chem. Eng. Process.* **2005**, *44*, 279–285. [CrossRef]

5. Morimoto, H.; Shakouchi, T. Classification of ultrafine powder by a new pneumatic type classifier. *Powder Technol.* **2003**, *131*, 71–79. [CrossRef]

6. Liu, R.; Liu, J.; Yu, Y. Effects of axial inclined guide vanes on a turbo air classifier. *Powder Technol.* **2015**, *280*, 1–9. [CrossRef]

7. Sun, Z.; Sun, G.; Yang, X.; Yuan, Y.; Wang, Q.; Liu, J. Effects of fine particle outlet on performance and flow field of a centrifugal air classifier. *Chem. Eng. Res. Des.* **2017**, *117*, 139–148. [CrossRef]

8. Sun, Z.; Sun, G.; Liu, J.; Yang, X. CFD simulation and optimization of the flow field in horizontal turbo air classifiers. *Adv. Powder Technol.* **2017**, *28*, 1474–1485. [CrossRef]

9. Gao, L.; Yu, Y.; Liu, J. Effect of rotor cage rotary speed on classification accuracy in turbo air classifier. *CIESC J.* **2012**, *63*, 1056–1062.

10. Gao, L.; Yu, Y.; Liu, J. Study on the cut size of a turbo air classifier. *Powder Technol.* **2013**, *237*, 520–528. [CrossRef]

11. Xiong, D.; Li, S.; Huang, P. Numerical Simulation of Distribution Characteristics of Particle Concentration in Inlet Tube of Superfine Classifier. *Chin. J. Process. Eng.* **2011**, *11*, 729–735.

12. Okay, A.; Nurettin, A.T.; Hakan, B.; Ozgun, D. Multi component modelling of an air classifier. *Miner. Eng.* **2016**, *93*, 50–56.

13. Okay, A.; Hakan, B. Selection and mathematical modelling of high efficiency air classifiers. *Powder Technol.* **2014**, *264*, 1–8.

14. Benzer, H.; Ergun, L.; Lynch, A.J.; Oner, M.; Gunlu, M.; Celik, I.B.; Aydogan, N.A. Modelling cement grinding circuits. *Miner. Eng.* **2001**, *14*, 1469–1482. [CrossRef]

15. Eswaraiah, C.; Angadi, S.I.; Mishra, B.K. Mechanism of particle separation and analysis of fish-hook phenomenon in a circulating air classifier. *Powder Technol.* **2012**, *218*, 57–63. [CrossRef]

16. Xing, W.; Wang, Y.; Zhang, Y.; Yoshiyuki, Y.; Saga, M.; Lu, J.; Zhang, H.; Jin, Y. Experimental study on velocity field between two adjacent blades and gas–solid separation of a turbo air classifier. *Powder Technol.* **2015**, *286*, 240–245. [CrossRef]

17. Zeng, C.; Liu, C.H.; Chen, H.Y.; Zhang, M.; Fu, Y.; Wang, X. Effects of secondary air on the classification performances of LNJ-36A air classifier. *Chem. Ind. Eng. Prog.* **2015**, *34*, 3859–3863.

18. Liu, J.; Xia, J.; He, T. Air flow field characteristics analyzing and classification process of the turbo classifier. *J. Chin. Ceram. Soc.* **2003**, *31*, 485–489.

19. Tao, Z.; Zheng, S. *Powder Engineering and Equipment*; Chemical Industry Press: Beijing, China, 2010.

20. Zhang, S.; Chen, Y.; Li, S. Effects of process parameters on particle size distribution and productivity of narrow level product in turbo air classifier. *Chem. Ind. Eng. Prog.* **2015**, *33*, 1113–1117+1155.

21. Napier-Munn, T.J. *Mineral Comminution Circuits*; Julius Kruttschnitt Mineral Research Centre: Queensland city, Australia, 1996.

22. Tirado, J.M.; Kumar, S.; Vandewinckel, J. Ladder Shelf System. U.S. Patent 10,167,669, 1 January 2002.

Article

Investigation of Plume Offset Characteristics in Bubble Columns by Euler–Euler Simulation

Yixuan Cheng [1], Qiong Zhang [1,2], Pan Jiang [1], Kaidi Zhang [1] and Wei Wei [1,*]

[1]   School of Energy and Power Engineering, Wuhan University of Technology, Wuhan 430063, China; cyx1995@whut.edu.cn (Y.C.); 09094501@wit.edu.cn (Q.Z.); river@whut.edu.cn (P.J.); 279809@whut.edu.cn (K.Z.)

[2]   Communist Youth League, Wuhan Institute of Technology, Wuhan 430205, China

*   Correspondence: wei_wei@whut.edu.cn; Tel.: +86-27-86581992

Received: 16 June 2020; Accepted: 6 July 2020; Published: 7 July 2020

Abstract: Based on low-cost and easy to enlarge, the bubble column device has been widely concerned in chemical industry. This paper focuses on bubble plumes in laboratory-scale three-dimensional rectangular air-water columns. Static behavior has been investigated in many experiments and simulations, and our present investigations consider the dynamic behavior of bubble plume offset in three dimensions. The investigations are conducted with a set of closure models by the Euler–Euler approach, and subsequently, literature data for rectangular bubble columns are analyzed for comparison purposes. Moreover, the transient evolution characteristics of the bubble plume in the bubble column and the gas phase distribution in sections are introduced, and the offset characteristics and the oscillation period of the plume are analyzed. In addition, the distributions of the vector diagram of velocity and vortex intensity in the domain are given. The effects of different fluxes and column aspect ratios on bubble plumes are studied, and the offset and plume oscillation period (POP) characteristics of bubbles are examined. The investigations reveal quantitative correlations of operating conditions (gas volume flux) and aspect ratios that have not been reported so far, and the simulated and experimental POP results agree well. An interesting phenomenon is that POP does not occur under conditions of a high flux and aspect ratio, and the corresponding prediction values for the conditions with and without POP are given as well. The results reported in this paper may open up a new way for further study of the mass transfer of bubble plumes and development of chemical equipment.

Keywords: CFD; bubble plume; oscillation and offset characteristic; bubble; gas–liquid flow

1. Introduction

The bubble column is widely used in the chemical industry [1,2]. It is a typical gas–liquid two-phase flow system [3]. Among them, the water phase exists as a continuous phase, and the air phase as a discrete phase rises from the reactor in the form of bubbles [4]. Bubbles are easy to observe and much attention has been paid to instrument measurement. Bubbles, namely, discrete elements, have always been a hot and challenging topic in both experiments and simulations [5]. Difficulties in bubble columns stem from the fact that the disperse phase is characterized by a complex behavior [6]. Consequently, the design and optimization of bubbly flow equipment present a great challenge to improve the mass transfer efficiency [7]. Although the industrial application of bubble columns has been extensively used, there are still some unsolved problems in their design and amplification, mainly due to the unknown problems in the hydrodynamics of gas–liquid flow [8]. Therefore, we focus our attention on the bubble plume. The bubble plume is the movement of bubbles in a two-phase flow, which is caused by momentum exchange. The bubble plume can also disturb the ambient flow field, and the main objectives are to promote the mixing of the liquids and to enhance the mass transfer.

However, due to the complexity of two-phase flow, especially the appearance of bubble motion, the basic properties of gas–liquid hydrodynamics and the characteristics of a bubble plume are still limited [9].

There are many works in the literature on the study of bubble plumes in a bubble column by numerical simulation via bubbly flow observation [10–12]. Different numerical methods have been proposed to describe a bubble plume, mainly including the bubble shape, state and plume oscillation [13,14]. The dependence of the period of bubble plume oscillations, global gas hold-up and local phase velocity on the superficial gas velocity and aspect ratio have previously been studied [15]. Diaz, Montes [16] focused on an oscillating plume in terms of the aspect ratio and gas flow rate. It was found that the oscillation period did not depend on the aspect ratio if the latter exceeded a value of 2. They investigated the dependency of the superficial gas velocity and aspect ratio on the time-averaged gas hold-up as well. With the exception of surface tension, Cachaza, Elena Díaz [17] showed that although there are differences in the mechanism of gas–liquid interface properties, the state of bubble flow, including the plume oscillation period, can be unified modeling and analysis. These are the critical problems in the design of bubble column equipment. Bannari, Kerdouss [18] effectively obtain bubble rise and oscillation characteristics, which were in good agreement with experimental results with the aspect ratio of 2.25 and a superficial gas velocity of 0.73 cm/s, and the oscillation characteristics of bubble plume in aspect ratio of 4.5 are analyzed and predicted. Gupta and Roy [19] captured bubble plume oscillation by radioactive particle tracking, and added a series of various interfacial forces (drag, lift, and virtual mass) to the Euler–Euler model and conducted a comparative analysis of different combinations. They emphasized the importance of low superficial gas velocity in the prediction of plume oscillations in transient simulations. Masood and Delgado [20] compared different drag models to bubble plume oscillation in 3D columns. Moreover, the drag model will affect the predicted oscillation period in the aspect ratio of 3. Previous studies have also closely considered interaction and turbulence simulation models to effectively simulate bubble plume, each with a different closure relation model. Liu and Luo [21] found the effect of surface tension on the oscillation behavior of bubble plume is obvious. In addition, the gas velocity is also very important for the prediction of plume oscillation. The velocities and turbulence characteristics of bubble columns with different aspect ratios were studied, and the snapshots of oscillation and turbulent kinetic energy at different flow rates and aspect ratios were provided, and they offered available data for theoretical studies [22]. Fleck and Rzehak [23] compiled an extensive summary of the plume oscillation characteristics and successful prediction of the gas–liquid two-phase characteristics in the bubble column with an aspect ratio of 2.5, and a qualitative study of the plume dynamics. The effects of superficial gas velocity, aspect ratio, reactor and liquid viscosity on low frequency oscillations were studied [24]. When the aspect ratio is greater than 2.0, the plume oscillation frequency remains almost unchanged, and as the aspect ratio decreases from 2.0 to 1.375, the plume oscillation frequency decreases significantly. Different CFD-PBM frameworks have been proposed to describe effectively simulate the plume oscillation period [25,26]. The polydispersed bubble size is used to effectively predict the plume oscillation period with the aspect ratio of 2.5 [27]. Shang, Ng [28] made use of two sets of momentum closures, and it was found that the momentum closure in the numerical simulation has a dramatic effect on the bubble plume oscillation characteristics. Moreover, there have also been reports of experiments to study the oscillation characteristics of bubble plumes [29]. Several available measuring techniques and instruments (i.e., laser Doppler anemometry (LDA), particle image velocimetry (PIV), laser-induced fluorescence (LIF) and computer-aided radioactive particle tracking (CARPT)) to investigate bubble rise and processes of bubble plume oscillations have been applied [11,30,31].

By sorting out a lot of previous research work, we found that more attention has been paid to the oscillation period of a bubble plume, and few studies focus on the offset characteristics of the bubble behavior. Furthermore, we also noticed that there are many comparisons between simulations and experiments, and the emphasis is on constructing numerical simulation methods for bubble plumes. For instance, different models for the description of turbulence have previously been studied [32], which also showed a notable dependence of the κ-ε turbulence model. Yang, Zhang [33] suggested that the resistance correction has a strong ability to simulate the plume oscillation, which should be based on the size distribution of bubbles and the local gas holdup. The application of bubble offset is of great significance to the further study of gas–liquid mixing and mass transfer behavior, while the plume oscillation and offset characteristics correspond to each other. Based on these, the offset characteristics of bubble plume fully developed in bubble column are analyzed in detail.

These issues are illustrated in this work. The main purpose of the present work is to analyze the mixing of gas–liquid and oscillation characteristics of dynamic bubbly flows (periodic bubble plumes). In particular, the dynamic rise process of bubbles and the offset behavior of bubbly flow in a rectangular column are examined. The feasibility of the bubbly flow simulation method is demonstrated with a practical case in rectangular bubble columns. Based on available experimental data and our own simulations, correlations are proposed for plume offset characteristics in dimensionless form. In addition, a mechanism for generating no-plume oscillation period (POP) is given.

In this paper, previous experimental data are first described in Section 2. In Section 3, numerical setup implementation details are provided, all the hydrodynamic equations are derived and computational models are reviewed. Then, the simulation results are presented and compared to experimental data in Section 4. Changes in the plume oscillation position over time and space are provided as well. Finally, conclusions are drawn in Section 5.

2. Experimental Data in the Literature

A series of experiments investigated bubbly flow in rectangular bubble columns, where tap water for the liquid phase and air for the gas phase were used. These experiments are listed in Table 1, and experimental data for different column sizes and aspect ratios, reactor structures (including a simplified structure in simulations) and operating conditions were gathered from the literature. Columns [10] were maintained at 25 °C and atmospheric pressure, while most of the experiments were performed on gas spargers with several holes, including pitches. The holes were located at the center position of the sparger. Many of the spargers produced a broad bubble size distribution (1–10 mm) with a mean size of 5 mm. The gas volume flux was measured by volumetric flow meters, varying from 3 to 745.5 liters per hour (LPH). All the experimental data were considered to examine the bubbly flow oscillation and offset characteristics.

Table 1. Details of the experimental data in the literature.

Refs.	W-D-H, cm	Aspect Ratio	Gas Distributor	Simplified Structure, mm	Gas Volume Flux, L/h	D_B, mm
[10]	20-5-45	2.25	sparger (8 holes)	rectangle (24 × 12)	20–90	1–10
[4]	20-4-45	2.25	single-orifice hole	diameter of 1	48	1–10
[34]	20-5-120	6	sparger (8 holes)	rectangle (18 × 6)	56–296	1–10
[16]	20-4-180	1.25–2.25	sparger (8 holes)	none	69–613	1–10
[13]	20-4-45	2.25	sparger (8 holes)	rectangle (18 × 6)	69–613	1–10
[22]	20-5-120	1.05–4	sparger (8 holes)	none	48–600	1–10
[35]	26.7-1.5-50	1.87	needle	diameter of 0.4	3–12	1.5–2.5

By summarizing the above experiments, we have a more specific understanding of the bubble plume in the bubble column. The size of bubble column varies, and the volume flux is also different. However, the range of bubble size in the domain is roughly the same. Of course, it is related to the reactor and flux, which is significant for the selection of bubble size below. In order to study the offset characteristics of plume oscillation more accurately, we control the flux range of this work within the experimental data, which has more practical significance. This is also one of the purposes of gathering experimental data. However, in the experiment, the snapshot of the gas phase is only taken instantaneously, which is not useful in this paper. We extract the period of plume oscillation in the literature and compare it with the results. In summary, we can get the proper flux, bubble size distribution, size of device and distributor, and part data of plume oscillation periods from these experiments.

3. Models and Numerical Details

Different TFMs are repeatedly cited in many previous works. Therefore, a concise description is provided here. References to the original works are presented as well. The conservation equations of two-phase flow are listed in Table 2. The surface tension and material properties of tap water and air at atmospheric pressure and temperature are summarized in Table 3. Numerical details suitable for this work are introduced later.

Table 2. Summary of the governing equations for two-phase flow.

Models	Equations	Refs./Remarks
Conservation equations	$\frac{\partial}{\partial t}(\rho_i \alpha_i) + \nabla \cdot (\rho_i \alpha_i \mathbf{u}_i) = 0$ $\frac{\partial}{\partial t}(\rho_i \alpha_i \mathbf{u}_i) + \nabla \cdot (\rho_i \alpha_i \mathbf{u}_i \mathbf{u}_i) = \nabla \cdot (\alpha_i \mathbf{T}_i) - \alpha_i \nabla p_i + \alpha_i \rho_i \mathbf{g} + \mathbf{F}_i^{inter}$ $\mathbf{T}_i = \mu_i^e \left(\nabla \mathbf{u}_i + (\nabla \mathbf{u}_i)^T \right)$	$i = g, l$
Turbulence Equations for water phase	$\frac{\partial}{\partial t}(k_l \rho_l) + \nabla \cdot (k_l \rho_l \alpha_l \mathbf{u}_l) = \frac{\partial}{\partial x_i}\left[\alpha_l \left(\mu + \frac{\mu_t}{\sigma_k} \right) \frac{\partial k}{\partial x_i} \right] + \alpha_l (G_k + G_b - \rho\varepsilon - Y_M)$ $\frac{\partial}{\partial t}(\varepsilon_l \rho_l) + \nabla \cdot (\varepsilon_l \rho_l \alpha_l \mathbf{u}_l) = \frac{\partial}{\partial x_i}\left[\alpha_l \left(\mu + \frac{\mu_t}{\sigma_\varepsilon} \right) \frac{\partial \varepsilon_l}{\partial x_i} \right] + \alpha_l G_{1\varepsilon} \frac{\varepsilon_l}{k_l}(G_k + C_{3\varepsilon}G_b)$ $-\alpha_l C_{2\varepsilon}\rho \frac{\varepsilon_l^2}{k_l}$	$C_{1\varepsilon} = 1.44, C_{2\varepsilon} = 1.92,$ $C_{3\varepsilon} = 0.09, \sigma_k = 1.0,$ $\sigma_\varepsilon = 1.3$
Turbulence viscosity of air phase	$\mu_l^e = \mu_{t,l} + \mu_{lam,l} = \rho C_u \frac{k_l^2}{\varepsilon_l} + \mu_{lam,l}$ $\mu_g^e = \mu_{t,g} + \mu_{lam,g} = \mu_{t,l} \times \rho_g / \rho_l + \mu_{lam,g}$	-
Interphase forces	$\mathbf{F}_l^{inter} = -\mathbf{F}_g^{inter}, \mathbf{F}_i^{inter} = \mathbf{F}_{Di}^{inter} + \mathbf{F}_{Li}^{inter} + \mathbf{F}_{Wi}^{inter} + \mathbf{F}_{Ti}^{inter} + \mathbf{F}_{VMi}^{inter}$	$i = g, l$
Drag forces	$\mathbf{F}_D^{inter} = \frac{3}{4}\alpha_g \rho_l C_D \frac{1}{d_g} \left| \mathbf{u}_g - \mathbf{u}_l \right| (\mathbf{u}_g - \mathbf{u}_l)$ $C_D = \mathrm{MAX}\left[\frac{24}{Re}(1 + 0.15Re^{0.687}), \frac{8}{3}\left(\frac{Eo}{Eo+4} \right) \right]$	[36]
Transverse lift forces	$\mathbf{F}_L^{inter} = -C_l \rho_l (\mathbf{u}_g - \mathbf{u}_l) rot \mathbf{u}_l$ $C_l = \begin{cases} \min[0.288\tanh(0.121Re), f(Eo_d)] & Eo_d < 4 \\ f(Eo_d) & 4 < Eo_d < 10.7 \\ -0.27 & Eo_d > 10.7 \end{cases}$ $f(Eo_d) = 0.00105 Eo_d^3 - 0.0159 Eo_d^2 - 0.0204 Eo_d + 0.474$ $Eo_d = \frac{g(\rho_l - \rho_g)d_H^2}{\sigma}$	[37]
Wall lubrication forces	$\mathbf{F}_W^{inter} = -\left(C_{W_1} + C_{W_2}\frac{D_s}{y_w} \right)\frac{\left[(\alpha_g \rho_l \mathbf{u}_g - \mathbf{u}_l) - ((\mathbf{u}_g - \mathbf{u}_l)\cdot \mathbf{n}_\omega)\mathbf{n}_\omega \right]^2}{D_s},$ $C_{W_1} = 0.0064., C_{W_2} = 0.016$	[6]
Turbulent dispersion forces	$\mathbf{F}_T^{inter} = C_{TD}C_D \frac{\mu_{tg}}{\sigma_{tg}}\left(\frac{\tilde{N}\alpha_l}{\alpha_l} - \frac{\tilde{N}\alpha_g}{\alpha_g} \right), C_{TD} = 1, \sigma_{tg} = 0.9$	[38]
Virtual mass forces	$\mathbf{F}_{VM}^{inter} = C_{VM}\rho_l \alpha_g \left(\frac{D\mathbf{u}_g}{Dt} - \frac{D\mathbf{u}_l}{Dt} \right)$	$C_{VM} = 0.5$
Inlet	$\mathbf{u} = x\mathbf{i} + y\mathbf{j} + z\mathbf{k} = C\mathbf{j}$	C is Constant value
Outlet	$\frac{\partial u}{\partial y} = 0, \frac{\partial p}{\partial y} = 0$	-
Wall	$\mathbf{u} = 0$	-
Convergence criterion	$\left| \frac{\varphi(n+1) - \varphi(n)}{\varphi(n+1)} \right| \leq 0.001$	n, n + 1 are the steps of iterations

Table 3. Details of the physical properties and environmental conditions.

Materials	Density, kg/m^3	Viscosity, Pa·s	Temperature, °C	Surface Tension, N/m
air	1.225	1.79×10^{-5}	25	0.0725
tap water	998.2	1.01×10^{-3}		

With regard to the conservation equations, each phase is assumed to be incompressible, and the volume fractions of the two phases always add up to 1 under all circumstances. Interphase forces are exchanged for momentum transfer between the two phases. For the liquid phase, the turbulent dynamic energy and energy rate are calculated from the standard κ-ε turbulence (SKE) model. To obtain a reliable solution of the turbulence equations, the dispersed method is adopted, and standard wall functions are applied to describe the wall–turbulence interactions.

The simulations refer to a bubble column [10] 0.2 m wide, 1.8 m high and 0.04 m deep, as previously tested [13]. The height of the apparatus can be arbitrarily adjusted. The column is filled with water up to a certain height (such as 0.45 m) from the bottom, and the water volume is regarded as the computational domain. Air is injected through a sparger in the center position of the bottom of the domain containing eight holes (diameter: 1 mm) and a 6 mm pitch. Moreover, the sparger is simplified as a rectangular region of 18×6 mm size in the simulations. An inlet boundary condition is prescribed on the simplified area, while the holes are treated as a whole. An outlet boundary condition is applied at the top of the domain, and the pressure at the outlet is uncertain. On the remainder of the domain boundary area, a free-slip condition is employed for the air phase and a no-slip condition for the water phase. The water phase is fully turbulent (at a turbulence intensity equal to 5%), and the hydrodynamic diameter is set as the width of the column. For the dispersed phase (i.e., the air phase), the gas velocity is converted by the gas volume flux through the inlet region, and no backflow occurs in the domain. The bubble size is set to 5.0 mm for the sparger on the basis of previous experiments.

The convergence criterion is set to 0.001 for the simulation cases to ensure a high calculation accuracy. Spatial discretization schemes are critical for the solution process. Therefore, the second order upwind scheme is adopted for κ and ε. The time step is set to 0.01 s, which guarantees that the Courant–Friedrichs–Lewy (CFL) number always remains below 0.5. All the simulations are performed with commercial software ANSYS FLUENT 16.0 and conducted on a 2.10-GHz platform (56-core CPUs) with 192 GB of RAM.

The simulated bubble column is shown in Figure 1. The position of the maximum lateral displacement is indicated by the yellow bubble. The partial working conditions chosen for the simulations are listed in Table 4. Six values are chosen for the aspect ratio (ξ_i) and gas volume flux (R_i), which yields 36 groups of test cases by cross-combination. The dynamic flow properties during the plume oscillation period are closely considered in all simulations.

Table 4. Details of the partial working conditions adopted in the simulation cases.

Test Case (Partial)	Aspect Ratio	Gas Volume Flux, l/h	D_B, mm	d_B, mm	U_g, mm/s
$\xi_1 R_1$	2	136	1–10	5	3.77
$\xi_2 R_2$	2.5	226	1–10	5	6.26
$\xi_3 R_3$	3	317	1–10	5	8.78
$\xi_4 R_4$	3.5	407	1–10	5	11.28
$\xi_5 R_5$	4	497	1–10	5	13.78
$\xi_6 R_6$	4.5	588	1–10	5	16.29

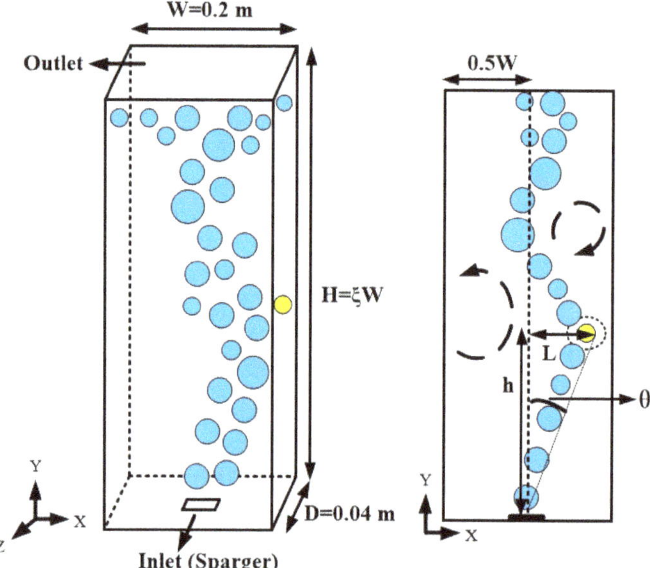

Figure 1. Schematic of the column geometry: (**left**) 3D representation and (**right**) 2D view.

The position of the knee point of the first oscillation during bubbly flow is used to illustrate the offset characteristics of the plume. The maximum offset distance and angle of the bubble plume in dimensionless form are calculated by using the following equations:

$$\eta = \frac{2L}{W} \tag{1}$$

$$\theta = \arctan\frac{L}{h} \tag{2}$$

where L is the maximum offset distance of the bubble, h is the vertical distance of the maximum offset, and η and θ are the maximum offset distance and angle, respectively. As a result, the calculated η value is not more than 1.

4. Simulation Results and Discussion

All simulations are presented in this section, and they have been run in the transient mode across the full 3D domain of the column. The results are calculated over a simulated physical time of 300 s, which is long enough to ensure full domain development, and the resulting domain no longer exhibits notable fluctuations. Since the different studies described in Section 2 provide different sizes and parameters, the gas inlet type must adopt the volume flux to ensure that the simulations are meaningful while enabling a comparison to experiments which is as accurate as possible.

4.1. Model Validation

As mentioned above, a grid independence study is required to obtain the optimal balance between the computational time and numerical accuracy. Non-uniform hexahedral grids are considered in this work. The results of the grid independence study are not provided, and further details can be found in [13]. Calculations were conducted on fine grids to ensure an adequate resolution. The simulation case against which the approach was validated [10] compared velocity profiles to experimental profiles. The POP data are also assessed via a comparison to the simulation results.

Figure 2 shows the results obtained by using the test case compared to the experimental data with R = 48 L h^{-1}. It is worth noting here that the depth of the experiment is 0.05 m, and the depth of this model is approximately the same. Regardless of whether the model is an experimental or simulation model, it is not a real 3D model (since the depth of the column is very small). Therefore, a comparison is feasible. The agreement between the experiments and simulations is quite good.

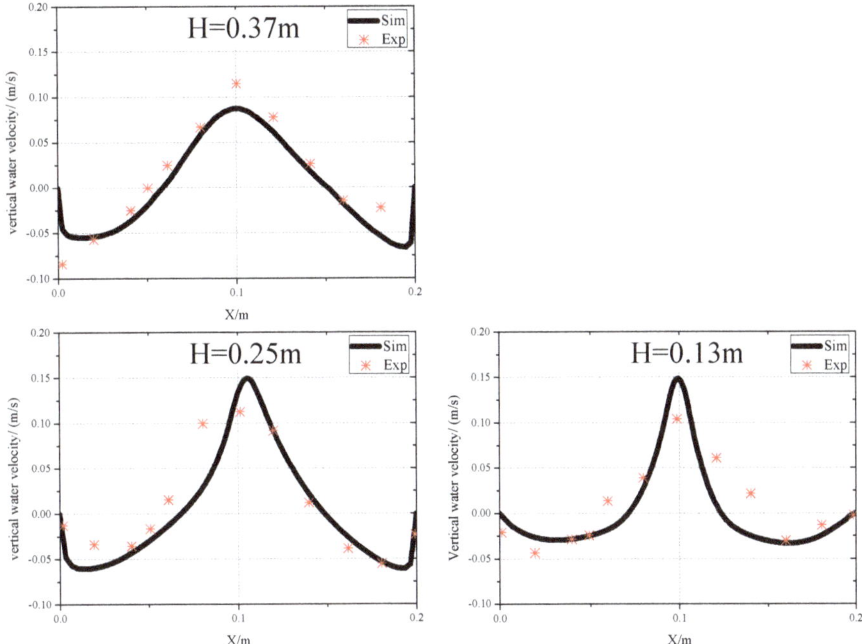

Figure 2. Comparison of the experimental and simulated long time-averaged vertical liquid velocity profiles.

The predicted values of the middle position are slightly different, but the overall trend is consistent, which is due to the intense momentum exchange between the gas and liquid phases in the intermediate region. Furthermore, the POP obtained with this gas volume flux is 14.6 s, and the prediction model captures this feature, while the calculated POP almost agrees with the predicted values in previous studies.

4.2. Transient Evolution of Bubble Plume

In handling the dynamics of bubbly flow, in contrast to the time-averaged flow problem, the rise process of bubbles is required for two-phase flow. Few studies have addressed the variations in bubble plume oscillation with simulations and measurements. Furthermore, most are two-dimensional dynamic studies. To better understand the continuous behavior at different times, the performance of the bubble plume has been determined. Predicted values are obtained through simulations with the use of three-dimensional rectangular columns.

The instantaneous gas fraction ranging from 0 to 0.1 is shown in the test case of ε3R3, and the liquid velocity fields are also shown in Figure 3. The POP in this case is 4.9 s, and the POP start time is 12.5 s. A clear change is observed between the results of the four different physical moments, which are obtained at 5, 10, 15 and 20 s. Due to an insufficient calculation time, the bubbly flow rises vertically, as shown in Figure 3a. An asymmetric gas fraction is observed in the bubble plume. Transient oscillation occurs and is implicitly contained in the gas fraction contour, as shown in Figure 3b.

Thereafter, POP occurs, and a periodic state is attained. Two vortices emerge on the two sides of the bubbly flow. The momentum inequality between the two vortices results in the development of an asymmetric flow. Compared to the results of adjacent periods, more bubble plume oscillation details are similar, as shown in Figure 3c,d. Note that the gas fraction changes over time and space. Through the observation of the gas fraction, it is found that the bubbly flow spirals to the outlet of the top domain. Moreover, the gas fraction distribution is not fully developed before the onset of POP. The bubbly flow shape varies since the bubbles flowing in and out of the domain are not in dynamic equilibrium. Some bubbles become separated from the main bubbly flow zone. A steady bubble plume oscillation phenomenon can be observed in the bubble column after 12.5 s, and the gas fraction distribution continuously changes during this period.

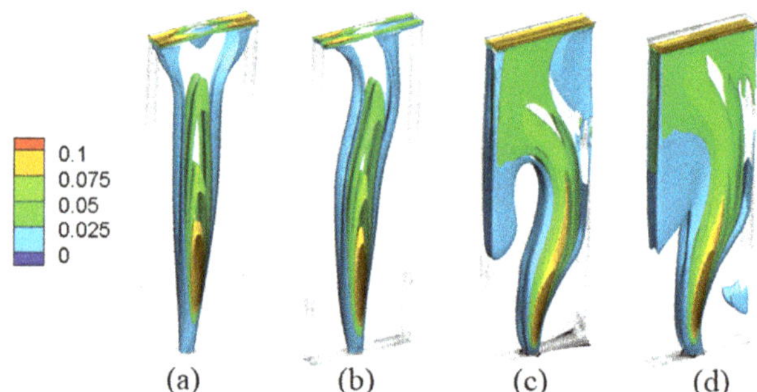

Figure 3. Instantaneous gas fraction from 0 to 0.1 and liquid velocity fields in the test case of Ɛ3R3 at (**a**) 5 s, (**b**) 10 s, (**c**) 15 s, and (**d**) 20 s.

In a word, from the instantaneous gas fraction of bubbles, we can clearly see the oscillation and offset characteristics of bubbles, especially in the period of oscillation.

In the first 10 s, the offset of bubble plume is very weak, and the gas–liquid mixing is uniform, but as the plume oscillation period begins, the offset of bubble increases, resulting in the flow field disorder and the gas–liquid mixture is inhomogeneous, but the scope and intensity of momentum exchange increase constantly.

In the test case of Ɛ3R3, contours of the gas fraction are generated at five sections, as shown in Figure 4, which are located at different Z-coordinates. On both sides of the sections (Z = −0.02, 0.02), the simulated contours are not notable. This effect becomes more pronounced at higher positions, where notable deviations are observed from the central plane. At the other sections (Z = −0.01, 0, 0.01), the contours in this case agree very well, except for a small difference at the sparger. With increasing distance from the central plane, the gas fraction notably decreases. However, the profile simultaneously changes from a full bubbly flow to a partial bubbly flow. The symmetrical profiles always remain the same. The higher the bubbly flow position is, the smaller the gas volume fraction is. This illustrates that the central surface (Z = 0) suitably represents the column to analyze bubble plume oscillation.

In the test case of Ɛ3R3, the POP was acquired by averaging the time interval of the X-component of the water velocity at a certain point in the center plane at a height of y = 0.25 m. The trajectory of the maximum offset position is shown in Figure 5 at heights of Y = 0.32 m and Z = 0 since the maximum offset can be determined in terms of the height from the two-dimensional perspective of the central surface during POP. In addition, we also discover that the Z-component of the locations of the maximum gas fraction always occurs at the center position (Z = 0) at different times, except for the moments at 2 and 11 s (the Z-components are −0.0016 and −0.0012, respectively). This may occur due to bubble plume instability, but most of the moments are observed in the center. A spline curve is applied to connect all the moments sequentially, which only reveals the trend.

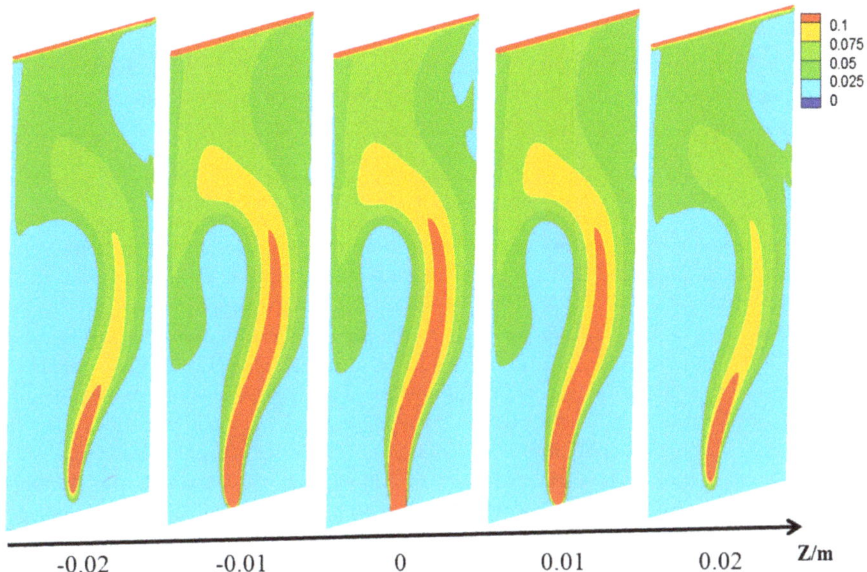

Figure 4. Contours of the gas hold-up over time (half-period, 15 s) for the five sections (Z = −0.02, −0.01, 0, 0.01 0.02).

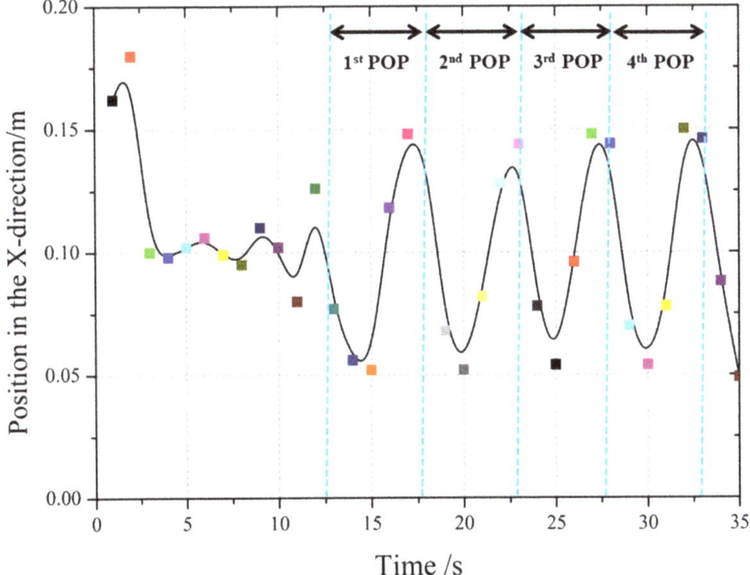

Figure 5. Temporal evolution of the maximum gas fraction locations at Y = 0.32 m and Z = 0; the colors indicate the time.

Before the onset of POP, an unstable fluctuation of the maximum gas fraction is observed since the fluid fields are still developing. Thereafter, the locations exhibit regular changes, and the agreement is good for each POP, especially in regard to the peaks. Moreover, it is suggested that this happens due to the balance between bubble coalescence and breakup. This explanation matches the appearance of the bubble plume.

We can find that the period of bubble oscillation is synchronous with the offset characteristic. This is beneficial for understanding the offset time, but the time at which the maximum offset position is located is not necessarily the start or end of the oscillation period. This provides a basis for determining the location and time of mass transfer and gas–liquid mixing.

4.3. Analysis of the Vorticity Distribution and Velocity

Similarly, we choose the test case of ξ3R3 as an example for analysis. The vorticity indicates the velocity and orientation of local rotation, which is adopted to represent the vortex characteristics. To determine the influence of bubbly flow in the column, it is of significance to investigate the vortex intensity and vorticity distribution. Figure 6 shows the distribution of various vortices of different magnitudes in the column at 15 s under the condition of ξ = 3. It is clear that the number and intensity of the vortices in the Y-component are very low, and we thus ignore them here. As shown in Figure 6a, consistent with the distribution of the liquid velocity field, a low-vorticity magnitude area (≤ 5 s^{-1}) is located in the bubbly flow in the column. High-vorticity magnitude areas (>5 s^{-1}) are located near the corner, top surface and maximum oscillation positions. The results also indicate that the vorticity distribution of the X-component is highly dispersed along the column, which corresponds to long-distance two-phase flow in the X-direction. Figure 6b shows the vorticity magnitude distribution of the Z-component. Larger-scale vortex structures are captured in the column. The results indicate that the position in the Z-direction and momentum exchange are limited, but consistent with the gas fraction distribution, more intense vortices are clearly encountered. High-vorticity magnitude areas occur near the walls and at the inlet.

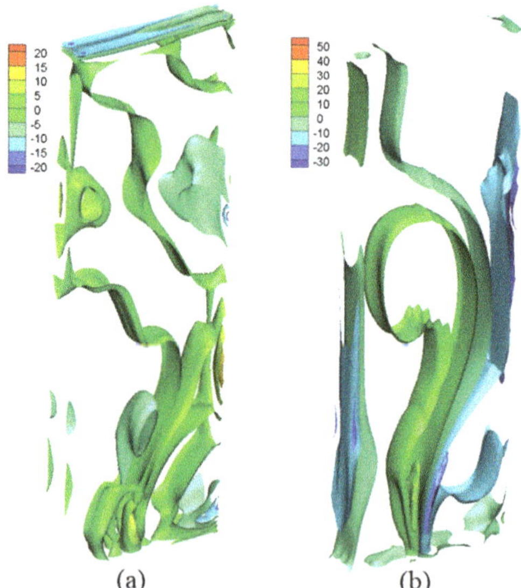

Figure 6. Distribution of various vortices of different magnitudes in the column: (**a**) X-component, (**b**) Z-component.

From the distribution of vortex intensity in directions, the greater the vortex intensity, the more obvious the mass transfer. The best mixing is around the maximum offset position, the second is above the reactor, and the worse is above the maximum offset position. Therefore, in order to better design the bubble column device, the device height can be reduced effectively.

Figure 7 shows the velocity vector of a particular three-dimensional simulation. The dynamic movement of the bubble plume is similar to the experiments. Vortices in the opposite direction are also shown surrounding the bubbly flow. The result of the section clearly captures bubble oscillation, and the velocity magnitudes are mainly concentrated in the middle of the domain. This is explained by the turbulent viscosity of the continuous phase, which decreases the bubble movement, and the size of one vortex is slightly larger than that of the other.

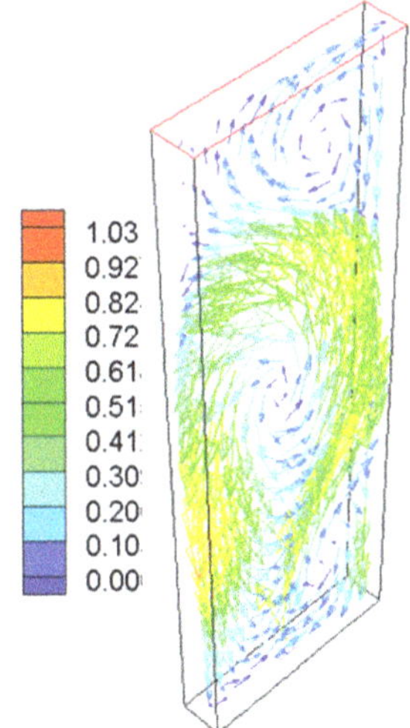

Figure 7. Contour of the velocity vector of the central surface in the column.

Vectors of velocity and vortex distribution are interrelated as well. In Figure 7, we find that the gas–liquid mixing at the vortex on both sides of the bubble flow is very good, and the liquid phase is surrounded by the gas phase in the middle area, we get the gas–liquid momentum exchange and mass transfer concentrated in this region.

4.4. Effect of the Gas Volume Flux on the Offset Characteristics of the Plume

The offset characteristics of bubbly flow are the main feature of the bubble column, and are captured by acquiring model predictions as a function of η and θ. Moreover, the central surface (Z = 0) is applied to analyze the offset characteristics since it effectively represents the whole domain and the column depth is very small. The characteristics are approximately the same in each POP, and the 20th period is consequently chosen for unified analysis. It is worth noting that there is no oscillation period in the case of a high gas flux and aspect ratio, but the first oscillation of bubbly flow reaches the left or right wall of the column within a short time, and the height of the knee point of the first oscillation tends to remain unchanged. Hence, it has no impact on our data. An interesting phenomenon occurs at a dimensionless maximum oscillation distance of 1, which is the condition in this situation, since the column width is relatively small, and the gas flux is sufficiently high that the bubbly flow cannot be fully expanded, resulting in the bubbly flow not exhibiting a POP.

To investigate the effect of the gas volume flux on the bubbly flow offset behavior in columns of different aspect ratios (ξ), test simulations were performed at different aspect ratios and six gas fluxes: 136 LPH (lowest flow rate), 226 LPH (relatively low flow rate), 317 LPH (moderately flow rate), 407 LPH (relatively moderate flow rate), 497 LPH (relatively high flow rate), and 588 LPH (highest flow rate).

Figure 8 shows the dimensionless maximum offset distance at the various gas fluxes, and Figure 9 shows the angles under the corresponding conditions. At the low fluxes, the bubble plume exiting the sparger follows a specific path, and bubbly flow is developed. Depending on the initial condition, this bubbly flow can repeatedly change course within a certain range. However, at the high fluxes, bubble oscillation intensifies, and the domain becomes disordered. It is also observed that the area occupied by the bubble plume rapidly increases until bubbles flow along the column wall. Therefore, a typical no-POP mode of bubbly flow is observed at the high fluxes and aspect ratios. Figure 9 indicates that the liquid surrounding the bubble plume remains relatively stable at the low fluxes, and an increasing trend of the angle is captured. However, at the high fluxes, large angles are observed, and bubbles flow upward close to the containing wall, even along the column.

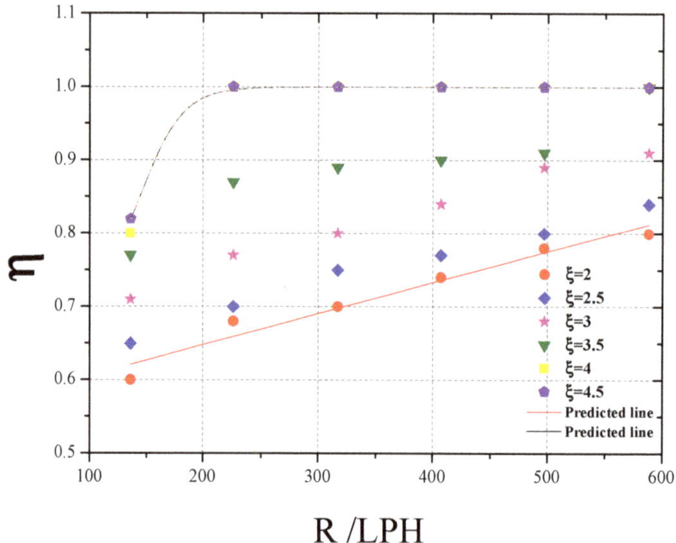

Figure 8. Dimensionless maximum oscillation distance of bubbly flow at the various gas volume fluxes.

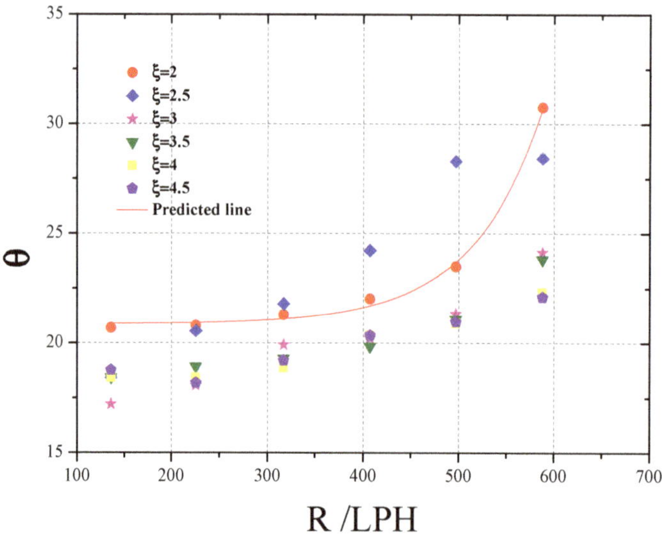

Figure 9. Oscillation angle of bubbly flow at the various gas volume fluxes.

4.5. Effect of the Aspect Ratio on the Offset Characteristics of Plume

Regarding the effect of different aspect ratios, simulations were performed at aspect ratios ranging from 2 to 4.5. Figure 10 shows the dimensionless maximum offset distance of bubbly flow under these conditions. Under each condition, η is plotted at 6 different levels. The results indicate that at a low aspect ratio, η gradually increases, which essentially confirms that bubbly flow occurs along a single path and that its trajectory does not touch the two walls. One can determine the occurrence of POP by the plume distance. At the upper level, the distance continues to increase and even remains unchanged under some conditions. This happens because bubbly flow is limited in the domain by the column width and cannot fully develop, which implies that the momentum exchange between gas and liquid occurs forcefully. Figure 11 shows the plume angle of bubbly flow at the different aspect ratios. Here, the results indicate that at the low levels, the angle decreases. Two main vortices are formed around the bubbly flow in which liquid flows in the opposite direction. It is also found that the bubbly flow progressively moves toward the column walls with increasing aspect ratio. This is observed because while the two vortices are formed, their sizes are not uniform. At the high levels, it can be inferred that some bubble plumes continuously move toward the walls and exhibit POP. Moreover, other bubble plumes move along the walls with no POP, and the liquid flows violently.

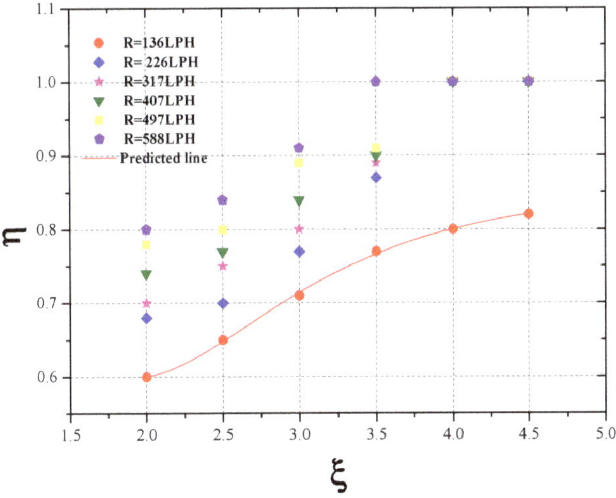

Figure 10. Dimensionless maximum oscillation distance of bubbly flow at the various aspect ratios.

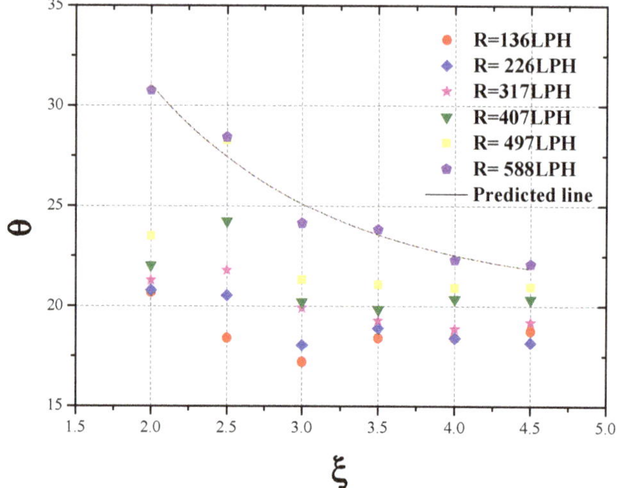

Figure 11. Oscillation angle of bubbly flow at the various aspect ratios.

4.6. Correlations

As already mentioned, correlations for the offset characteristics are derived based on the data in this work and under turbulent flow conditions. In Figures 8–11, the predicted lines are presented as fitting curves. At low aspect ratios (ξ = 2, 2.5, 3), η is linearly fitted with the flux, as shown in Figure 8. The minimum adjusted R-square value is 0.93, and the value for the predicted red line is 0.95. At high aspect ratios (ξ = 4, 4.5), η and the flux are suitably fitted with a lognormal curve, and the minimum adjusted R-square value is 0.96, while the correlation coefficient of the purple line is 0.99. At an aspect ratio of 3, the change in data is irregular in the transition stage, and there is no suitable fitting relationship. However, the fitting correlation between the flux and angle is uniform, both of which exhibit exponential growth. The minimum correlation coefficient is 0.85, and the red line in Figure 9 has a value of 0.99. As shown in Figure 10, the lognormal profile is adapted to the correlation of the aspect ratio and η. The correlation coefficient of the red line is as high as 0.99, and the lowest value is not smaller than 0.95. Finally, at the highest flux (R = 588 LPH), the aspect ratio and angle reveal an exponential decay relationship (adjusted R-square = 0.94). Under the other conditions, the angle change is very slight, and a suitable fitting curve cannot be established.

4.7. Plume Oscillation Period

Comparing the previous data in terms of the POP at the different values of the aspect ratio and gas volume flux, the simulation results obtained agree well with the predicted data of [16]. It was previously noted [22] that no POP occurred at a high flux and high aspect ratio. This conclusion is confirmed by the present simulations, with more detailed results provided. Figure 12 reveals that in some cases, plume oscillation is not observed. In other words, this means that at a high aspect ratio of 4 or 4.5, the bubbly flow does not oscillate at these high fluxes, and according to a particular flow path along the walls, the bubble plume is almost completely governed by the narrow width (and cannot fully develop). POP only happens under certain conditions.

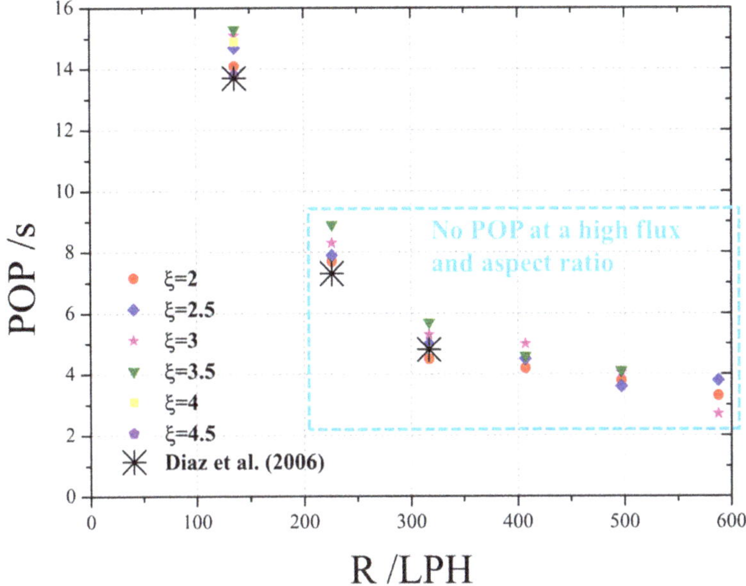

Figure 12. Plume oscillation period (POP) statistics under the different conditions.

5. Conclusions

To sum up, we use numerical simulation method to study the offset characteristics of a bubble plume in a bubble column, in which momentum exchange induced by gas–liquid interaction leads to bubble offset. The detailed flow field shows that the characteristics of bubble offset need time to fully develop. It is similar to the period of bubble oscillation, and also presents periodic range changes. In addition, from the height surface (Y-coordinates unchanged) of the maximum offset position, the offset characteristic exhibits periodic fluctuations in each section. The vortices are mainly concentrated around the bubble flow and on the wall in the domain. Furthermore, we apply the maximum offset position to construct the quantitative analysis of the offset characteristics of the bubble plume, and successfully draw the correlations between the dimensionless offset distance and angle in the oscillation characteristics of different flux and aspect ratio. The fitting degree is good as well. With the increase of volume flux, due to the limited area space, the dimensionless offset distance has an upper limit, so that the value will not change after 1, and the offset angle gradually increases; with the increase of aspect ratio, the dimensionless offset distance shows the same trend, but the angle presents a decreasing trend when the volume flow is large, and other situations fluctuate greatly without obvious regularity. Considering the effect of a limited closed region on the offset of bubble plume, the conditions without oscillation period are found as well.

A series of closure models are applied to the case of bubbly flow, especially to examine bubble plume oscillation based on previous works, with a focus on the dynamic bubble plume process. The simulation results agree quite well with the experimental data, including the velocity profile and POP. Some of the observed deviations could occur because a fixed bubble diameter was used in the simulations, and the change in bubble size was neglected. The models can be further optimized, because the resulting effects are not yet truly reliable. The dynamic details of the bubble plume, including the velocity, volume fraction, vortex intensity and dynamic bubbly flow variations, are observed in detail. It is found that the column is not a real three-dimensional model because its depth is very small. Therefore, the intermediate plane (Z = 0) is chosen to describe the dynamic behavior of bubble oscillation. Furthermore, the influence of the aspect ratio and gas volume flux on the oscillation characteristics is examined based on our simulation results, and qualitative fitting correlations are provided. It is noteworthy that when the dimensionless offset distance (η) is 1, no POP is observed. In other words, POP does not occur in bubble columns at a high gas volume flux and aspect ratio. A more detailed investigation of the specific conditions resulting in the absence of POP has a great potential. Additional research on bubble plume oscillation is required for the development of accurate bubble interaction and turbulence models. The gas–liquid mixing and mass transfer can be enhanced around the maximum offset position. Gas–liquid mixing is mainly concentrated in the middle region. Furthermore, the maximum offset position is related to the aspect ratio and volume flux. These should have a wide application range in the chemical industry.

Author Contributions: Y.C.: Software, Investigation, Writing—original draft preparation. Q.Z.: Resources, Data curation, Visualization. P.J.: Supervision, Project administration. K.Z.: Validation, Formal analysis. W.W.: Conceptualization, Methodology, Funding acquisition, Writing—review and editing. All authors have read and agreed to the published version of the manuscript.

Funding: This research was funded by Natural Science Foundation of China, grant number 51709210 and 51679178.

Conflicts of Interest: The authors declare no conflict of interest.

Nomenclature

Symbol	Denomination (Unit)		
a_p	the center coefficient (-)	ε	turbulent dissipation rate ($m^2\ s^{-3}$)
a_{nb}	influence coefficients for the neighboring cells (-)	κ	turbulent kinetic energy ($m^2\ s^{-2}$)
d_B	bubble diameter (mm)	ρ_i	gas density (kg/m^3)
C_D	coefficient of drag force (-)	μ_i	dynamic viscosity (Pa s)
C_L	coefficient of lift force (-)	σ	surface tension (N/m)
C_{TD}	coefficient of turbulent dispersion force (-)	ϕ	Variable (-)
C_{VM}	coefficient of virtual mass force (-)	ξ	aspect ratio (H/W)
C_W	coefficient of wall lubrication force (-)	Index	Denomination
D_B	bubble diameter distribution (mm)		
D	depth of the column (m)	e	effective value
EO	Eötvös number (-)	g	gas phase
F	interaction force (N)	l	liquid phase
H	height of the column (m)	*inter*	at interface
g	acceleration of gravity ($m\ s^{-2}$)	i	gas and liquid
p	pressure ($N\ m^{-2}$)	*lam*	laminar flow
POP(s)	plume oscillation period (s)	t	turbulent flow
R	gas volume flux (LPH)		
t	time (s)	Abbreviation	Denomination
T	temperature (K)	CARPT	computer-aided radioactive particle tracking
T	stress tensor ($N\ m^2$)	CFD-PBM	Computational fluid dynamics coupled with a population balance model
u	velocity (m/s)	CFL	Courant–Friedrichs–Lewy number
U_g	superficial gas velocity (m/s)	LDA	laser Doppler anemometry
W	width of the column (m)	LIF	laser-induced fluorescence
X	axial coordinate (m)	LPH	liter per hour
Y	vertical coordinate (m)	SKE	standard κ-ε turbulence model
Z	spanwise coordinate (m)	PIV	particle image velocimetry
α_i	gas volume fraction (-)	TFM	two-fluid model

References

1. Tomiyama, A.; Celata, G.; Hosokawa, S.; Yoshida, S. Terminal velocity of single bubbles in surface tension force dominant regime. *Int. J. Multiph. Flow* **2002**, *28*, 1497–1519. [CrossRef]

2. Tomiyama, A.; Tamai, H.; Zun, I.; Hosokawa, S. Transverse migration of single bubbles in simple shear flows. *Chem. Eng. Sci.* **2002**, *57*, 1849–1858. [CrossRef]

3. Antal, S.; Lahey, R.; Flaherty, J. Analysis of phase distribution in fully developed laminar bubbly two-phase flow. *Int. J. Multiph. Flow* **1991**, *17*, 635–652. [CrossRef]

4. Bannari, R.; Kerdouss, F.; Selma, B.; Bannari, A.; Proulx, P. Three-dimensional mathematical modeling of dispersed two-phase flow using class method of population balance in bubble columns. *Comput. Chem. Eng.* **2008**, *32*, 3224–3237. [CrossRef]

5. Becker, S.; De Bie, H.; Sweeney, J. Dynamic flow behaviour in bubble columns. *Chem. Eng. Sci.* **1999**, *54*, 4929–4935. [CrossRef]

6. Besagni, G.; Gallazzini, L.; Inzoli, F. On the scale-up criteria for bubble columns. *Petroleum* **2019**, *5*, 114–122. [CrossRef]

7. Besbes, S.; El Hajem, M.; Ben Aissia, H.; Champagne, J.; Jay, J. PIV measurements and Eulerian–Lagrangian simulations of the unsteady gas–liquid flow in a needle sparger rectangular bubble column. *Chem. Eng. Sci.* **2015**, *126*, 560–572. [CrossRef]

8. Besbes, S.; Gorrab, I.; Elhajem, M.; Ben Aissia, H.; Champagne, J.Y. Effect of bubble plume on liquid phase flow structures using PIV. *Part. Sci. Technol.* **2019**, 1–10. [CrossRef]

9. Buffo, A.; Marchisio, D.L.; Vanni, M.; Renze, P. Simulation of polydisperse multiphase systems using population balances and example application to bubbly flows. *Chem. Eng. Res. Des.* **2013**, *91*, 1859–1875. [CrossRef]

10. Burns, A.D.; Frank, T.; Hamill, I.; Shi, J.M. The Favre Averaged Drag Model for Turbulent Dispersion in Eulerian Multi-Phase Flows. In Proceedings of the 5th International Conference on Multiphase Flow, Yokohama, Japan, 30 May 30–4 June 2004.

11. Buwa, V.V.; Ranade, V.V. Dynamics of gas–liquid flow in a rectangular bubble column: Experiments and single/multi-group CFD simulations. *Chem. Eng. Sci.* **2002**, *57*, 4715–4736. [CrossRef]

12. Cachaza, E.M.; Díaz, M.E.; Montes, F.J.; Galán, M.A. Unified study of flow regimes and gas holdup in the presence of positive and negative surfactants in a non-uniformly aerated bubble column. *Chem. Eng. Sci.* **2011**, *66*, 4047–4058. [CrossRef]

13. Cheung, S.C.P.; Deju, L.; Yeoh, G.H.; Tu, J. Modeling of bubble size distribution in isothermal gas–liquid flows: Numerical assessment of population balance approaches. *Nucl. Eng. Des.* **2013**, *265*, 120–136. [CrossRef]

14. Cheung, S.C.P.; Yeoh, G.H.; Tu, J. Population balance modeling of bubbly flows considering the hydrodynamics and thermomechanical processes. *AIChE J.* **2008**, *54*, 1689–1710. [CrossRef]

15. Díaz, M.E.; Iranzo, A.; Cuadra, D.; Barbero, R.; Montes, F.J.; Galán, M.A. Numerical simulation of the gas–liquid flow in a laboratory scale bubble column. *Chem. Eng. J.* **2008**, *139*, 363–379. [CrossRef]

16. Díaz, M.E.; Montes, F.J.; Galán, M.A. Experimental study of the transition between unsteady flow regimes in a partially aerated two-dimensional bubble column. *Chem. Eng. Process. Process. Intensif.* **2008**, *47*, 1867–1876. [CrossRef]

17. Díaz, M.E.; Montes, F.J.; Galán, M.A. Influence of Aspect Ratio and Superficial Gas Velocity on the Evolution of Unsteady Flow Structures and Flow Transitions in a Rectangular Two-Dimensional Bubble Column. *Ind. Eng. Chem. Res.* **2006**, *45*, 7301–7312. [CrossRef]

18. Fleck, S.; Rzehak, R. Investigation of bubble plume oscillations by Euler-Euler simulation. *Chem. Eng. Sci.* **2019**, *207*, 853–861. [CrossRef]

19. Guo, K.; Wang, T.; Liu, Y.; Wang, J. CFD-PBM simulations of a bubble column with different liquid properties. *Chem. Eng. J.* **2017**, *329*, 116–127. [CrossRef]

20. Gupta, A.; Roy, S. Euler–Euler simulation of bubbly flow in a rectangular bubble column: Experimental validation with Radioactive Particle Tracking. *Chem. Eng. J.* **2013**, *225*, 818–836. [CrossRef]

21. Hallmark, B.; Chen, C.-H.; Davidson, J. Experimental and simulation studies of the shape and motion of an air bubble contained in a highly viscous liquid flowing through an orifice constriction. *Chem. Eng. Sci.* **2019**, *206*, 272–288. [CrossRef]

22. Huang, Z.; McClure, D.D.; Barton, G.; Fletcher, D.; Kavanagh, J. Assessment of the impact of bubble size modelling in CFD simulations of alternative bubble column configurations operating in the heterogeneous regime. *Chem. Eng. Sci.* **2018**, *186*, 88–101. [CrossRef]

23. Kantarci, N.; Borak, F.; Ulgen, K.O. Bubble column reactors. *Process. Biochem.* **2005**, *40*, 2263–2283. [CrossRef]

24. Krepper, E.; Vanga, B.N.R.; Zaruba, A.; Prasser, H.-M.; De Bertodano, M.A.L. Experimental and numerical studies of void fraction distribution in rectangular bubble columns. *Nucl. Eng. Des.* **2007**, *237*, 399–408. [CrossRef]

25. Li, G.; Wang, B.; Wu, H.; DiMarco, S.F. Impact of bubble size on the integral characteristics of bubble plumes in quiescent and unstratified water. *Int. J. Multiph. Flow* **2020**, *125*, 103230. [CrossRef]

26. Liu, L.; Yan, H.; Ziegenhein, T.; Hessenkemper, H.; Li, Q.; Lucas, D. A systematic experimental study and dimensionless analysis of bubble plume oscillations in rectangular bubble columns. *Chem. Eng. J.* **2019**, *372*, 352–362. [CrossRef]

27. Liu, Q.; Luo, Z.-H. Modeling bubble column reactor with the volume of fluid approach: Comparison of surface tension models. *Chin. J. Chem. Eng.* **2019**, *27*, 2659–2665. [CrossRef]

28. Masood, R.; Delgado, A. Numerical investigation of the interphase forces and turbulence closure in 3D square bubble columns. *Chem. Eng. Sci.* **2014**, *108*, 154–168. [CrossRef]

29. McGinnis, D.F.; Lorke, A.; Wüest, A.; Stöckli, A.; Little, J.C. Interaction between a bubble plume and the near field in a stratified lake. *Water Resour. Res.* **2004**, *40*. [CrossRef]

30. Murgan, I.; Bunea, F.; Ciocan, G.D. Experimental PIV and LIF characterization of a bubble column flow. *Flow Meas. Instrum.* **2017**, *54*, 224–235. [CrossRef]

31. Pfleger, D.; Gomes, S.; Gilbert, N.; Wagner, H.-G. Hydrodynamic simulations of laboratory scale bubble columns fundamental studies of the Eulerian–Eulerian modelling approach. *Chem. Eng. Sci.* **1999**, *54*, 5091–5099. [CrossRef]

32. Clift, R.; Grace, J.R.; Weber, M.E. *Bubbles, Drops, and Particles*; A Subsidiary of Harcourr Brace Jovanovic: New York, NY, USA, 1978.

33. Rensen, J.; Roig, V. Experimental study of the unsteady structure of a confined bubble plume. *Int. J. Multiph. Flow* **2001**, *27*, 1431–1449. [CrossRef]

34. Shang, X.; Ng, B.F.; Wan, M.P.; Ding, S. Investigation of CFD-PBM simulations based on fixed pivot method: Influence of the moment closure. *Chem. Eng. J.* **2020**, *382*, 122882. [CrossRef]

35. Silva, M.K.; D'Ávila, M.A.; Mori, M. Study of the interfacial forces and turbulence models in a bubble column. *Comput. Chem. Eng.* **2012**, *44*, 34–44. [CrossRef]

36. Upadhyay, R.K.; Pant, H.J.; Roy, S. Liquid flow patterns in rectangular air-water bubble column investigated with Radioactive Particle Tracking. *Chem. Eng. Sci.* **2013**, *96*, 152–164. [CrossRef]

37. Yang, G.; Zhang, H.; Luo, J.; Wang, T. Drag force of bubble swarms and numerical simulations of a bubble column with a CFD-PBM coupled model. *Chem. Eng. Sci.* **2018**, *192*, 714–724. [CrossRef]

38. Zhang, H.; Sayyar, A.; Wang, Y.; Wang, T. Generality of the CFD-PBM coupled model for bubble column simulation. *Chem. Eng. Sci.* **2020**, *219*, 115514. [CrossRef]

Article

Numerical Simulation Study of Heavy Oil Production by Using In-Situ Combustion

Zhao Yang [1,*], Shuang Han [1] and Hongji Liu [2]

[1] School of Petroleum Engineering, Northeast Petroleum University, Daqing 163000, China; Shuanghan317@163.com

[2] Key Laboratory of Tectonics and Petroleum Resources (China University of Geosciences), Ministry of Education, Wuhan 430074, China; liuhjcug@126.com

* Correspondence: Zhao.yang@nepu.edu.cn; Tel.: +86-1834-666-9952

Received: 17 July 2019; Accepted: 27 August 2019; Published: 14 September 2019

Abstract: An in-situ combustion method is an effective method to enhance oil recovery with high economic recovery rate, low risk, fast promotion and application speed. Currently, in-situ combustion technique is regarded as the last feasible thermal recovery technology to replace steam injection in the exploitation of bitumen sands and heavy oil reservoirs. However, the oil-discharging mechanism during the in-situ combustion process is still not clearly understood. In this paper, the in-situ combustion process has been numerically simulated based on the Du 66 block. The effect of production parameters (huff and puff rounds, air injection speed, and air injection temperature) and geological parameters (bottom water thickness, stratigraphic layering, permeability ratio, and formation thickness) on the heavy oil recovery have been comprehensively analyzed. Results show that the flooding efficiency is positively correlated with the thickness of the bottom water, and negatively correlated with the formation heterogeneity. There exist optimum values for the oil layer thickness, huff and puff rounds, and air injection speed. And the effect of air injection temperature is not significant. The results of this paper can contribute to the understanding of mechanisms during in-situ combustion and the better production design for heavy oil reservoirs.

Keywords: heavy oil reservoir; in-situ combustion; oil recovery; numerical simulation

---

1. Introduction

1.1. Research Status

In-situ combustion technique is also called the inner layer combustion or fire-flooding, which is one kind of enhanced oil recovery (EOR) technology developed in 1930s. There are mainly two ways for in-situ combustion: Dry combustion and wet combustion [1,2]. A lot of studies have been conducted to study the characteristics of these two combustion modes. Wilson and Root proposed relevant calculation formulas through dry forward combustion and wet forward combustion experiments and discussed the main influencing factors [3]. Alexander et al. studied the effect of original oil saturation on the combustion efficiency [4]. Chleh and Gates proposed the methods for estimating the minimum required air flow to maintain oil combustion [5,6]. Thomas proposed a more mature energy conservation equation for the fired oil layers [7]. Parrish et al. conducted a forward wet combustion test, discussed the influencing factors of various parameters, and provided a design method for wet combustion [8]. Penberthy et al. have proposed a relationship between temperature and crude oil saturation distribution, material balance, air demand and oxygen concentration near the combustion front [9]. Garon et al. carried out inverse combustion experiments and discussed the related influencing factors [10]. Suat and Mustafa conducted in-situ combustion technique on Turkish heavy oil reservoirs [11]. In the dry combustion experiment, as the API (A measure of the density of petroleum and petroleum products developed by the American Petroleum Institute.) of the crude oil decreases, the fuel consumption rate becomes faster. In wet combustion, the higher the air-water ratio, the fuel consumption will decrease. Burger published a research report on the fired oil layer and proposed the oxygen demand calculation formula and the ignition time equation [12].

Besides the in-situ combustion technique, there are another two commonly used EOR technologies: Steam assisted gravity drainage (SAGD) and the polymer flooding. SAGD is a cutting-edge technology to improve the recovery of heavy oil, super heavy oil, and high condensate oil. SAGD technology can greatly increase oil recovery [13,14]. Polymer flooding technology is representative of the tertiary oil recovery stage. Through the application of polymer flooding, more supporting techniques can be studied to improve the sweep volume of the injection agent and the recovery factor of the tertiary oil recovery stage [15,16]. In China, in-situ combustion experiments were carried out in Xinjiang, Yumen, Shengli and Fuyu oilfields from 1958 to 1976. Since 1993, Shengli oilfield has listed the burning reservoir as a key pilot experimental project and six field experiments have been carried out [17]. However, due to the limitation of technical conditions, only dry combustion experiments have been carried out, and no wet combustion experiments have been carried out. Since 1999, based on the experience of dry combustion experiments, laboratory research on wet combustion has been carried out, and some achievements have been achieved. In 2001, Cai et al. carried out wet combustion experiments on heavy oil in the Hekou oilfield by using physical simulation technology [18]. The effect of the parameters on the reservoir performance under wet spontaneous combustion was studied, such as fuel consumption, apparent hydrogen-carbon atom ratio, combustion front propulsion speed, and air requirement. The results show that wet combustion can recover heat more effectively than dry combustion. It reduces fuel consumption and air consumption, and also improves oil recovery. In 2005, Guan et al. provided a method to determine the reservoir ignition temperature through laboratory tests [19]. The method was adopted in the Zheng 408 block of the Shengli oilfield, and the ignition temperature was successfully determined to be about 370 °C. Jiang et al. used a combination of physical modeling and numerical simulation to systematically study the mechanism of oil displacement in low-permeability reservoirs [20].

Due to the complexity in the process of in-situ combustion, it is difficult to summarize the general laws of thermal oil displacement in the fired oil layer by relying only on limited combustion experiments. In this case, the numerical simulation method is becoming more important. The numerical simulation of the in-situ combustion reservoir is more complex and difficult than steam injection. Frequent changes of chemical reactions and phase states greatly increase the number of governing equations. Thus, the relatively perfect numerical simulation technique of in-situ combustion appears later than steam injection. The development of the numerical simulation of in-situ combustion reservoir is from one-dimensional to two-dimensional and three-dimensional. The phase number is generally three-phase (gas, oil, water) or four-phase (gas, oil, water, and solid). In addition the nature of reservoir rocks and fluids, the consideration of gravity and capillary force make the numerical simulation of in-situ combustion more complex. Currently, there are some commercial numerical simulators which are suitable for various thermal recovery methods. Through numerical simulation, Bottia et al. found that the delayed ignition indicates high probability to get a spontaneous ignition. Furthermore the distance at which at which ignition occurs can be modified by the air injection rate [21]. Rahnema et al. found that the oil displacement is mainly driven by gravity drainage through the experiment and numerical simulation. Vigorous combustion was observed at the early stages near the heel of the injection well [22]. Pei et al. studied the effect of nitrogen injection on the effectiveness of in-situ conversion process by numerical simulation [23]. Nesterov et al. found that the activation energy of the light fraction in the oil is the most significant factor which affects the possibility of ignition through numerical simulation [24]. However, the effect of production and geological parameters on the heavy oil reservoir production with in-situ combustion has not been systematically studied in previous research.

For the Du 66 block, the fire-flooding is the main development method since 2011. However, with the expanding scale of fire-flooding pilot test in the Du 66 block and the influence of reservoir heterogeneity, there are some problems, such as the difference of the combustion state between the thin interbedded layers, the serious overlap of fire-flooding line in thick interbedded layers, the uneven spread of fire line and the unclear understanding of the combustion state, which affect the efficiency of in-situ combustion. In order to reveal the mechanisms of multi-layer in-situ combustion and understand the characteristics of multi-layer fire wave, and also determine the main factors affecting the oil recovery of the Du 66 block, it is necessary to carry out the numerical simulation study of in-situ combustion in the Du 66 block.

In this paper, the numerical simulation method was used to study the influence of production parameters and geological parameters on the fire-flooding efficiency based on the Du 66 block in the Shuguang oilfield. The rest of this paper is organized as follows: Section 2 presents the geological background of the studied field; Section 3 is the construction process of the geological model; Section 4 shows the results and analysis.

1.2. Geological Background

The Du 66 block of the Shuguang oilfield is structurally located in the northwest of the Shuguang oilfield in the middle part of the western slope of the Western Sag of the Liaohe fault basin. The development target stratum in this area is the Dujiatai reservoir in the upper fourth member of the Shahejie Formation of Paleogene. Up to 2000, the proven oil-bearing area in this area was 9.41 km^2, and the petroleum geological reserve was 5935.2×10^4 t.

The top surface structure of the Dujiatai oil layer in the Du 66 block of the Shuguang Oilfield is generally a monoclinic structure under the slope background, which is inclined from the northwest to the southeast. The dip angle of the stratum is generally $5° \sim 10°$. Reservoir lithology is mainly conglomerate sandstone and unequal-grained sandstone with medium sorting deviation, belonging to medium-high porosity and medium-high permeability reservoirs. The oil layer is mainly composed of a thin to medium–thick layer and the reservoir type is layered edge water reservoir. The density of crude oil at 20 °C is 0.9001–0.9504 g/cm^3, and the viscosity of ground degassed crude oil at 50 °C is 325–2846 mPa·s, which is ordinary heavy oil.

(1) Formation characteristics

The Dujiatai oil layer in the Du 66 block is located in the front of a large fan-delta developed under the condition of slow water inflow. The sand bodies in the upper strata are very well developed. The average thickness of the Du II-1 sandstone is 12.5 m, and that of the Du I-2 sandstone is 6.5 m. The interlayers are relatively well developed with a maximum thickness of 44 m, generally 0.6–20 m. However, the development of different oil layers is different, and the distribution characteristics are also quite different. The maximum thickness of the interlayers from the upper layer to the Du 0 layer is 21.7 m, generally 10 m to 20 m. And the average thickness is 16.3 m. The lithology is brown–gray mudstone and the undeveloped areas of sandstone are often interbedded with oil shale, dolomitic limestone, and brown–grey mudstone. The porosity of the upper layer is generally 15% to 25%, with an average of 20.7%. The permeability is generally 200 to 1200 mD, with an average of 920.6 mD, which is a typical medium–high porosity and medium–high permeability reservoir. The oil layer group division and oil layer thickness statistics of the Du 66 block have been listed in Table 1.

Table 1. Oil layer group division and oil layer thickness statistics of the Du 66 block.

Development Layer	Oil Layer Group	Sandstone Group	Small Layer	Oil Layer Thickness (m)	
				General	Average
Upper system	I	I1	1–2	1.0–11.8	6.5
		I2	3–5	1.4–8.2	4.8
		I3	6–9	1.8–12.5	6.7
Lower system	II	II1	10–13	2.5–14.6	8.4
		II2	14–16	1.7–9.4	3.5
		II3	17–20	2.1–15.8	9.1
	III	III	21–30	1.5–11.6	4.9
total	3	7	30	24–62	44.5

(2) Development characteristics

The Du 66 block was developed in 1985 with a square well pattern. The current well spacing is 100 m after secondary infilling adjustment. The main development mode is in-situ combustion in the upper layer and steam huff and puff in the lower layer. Four development stages can be classified. The first stage is from 1985 to 1989, which can be called the production stage. The annual oil production rises to 45×10^4 t; the oil recovery rate rises to above 1.0%; the stage production rate is 4.4%. The second stage is from 1989 to 1999, which can be called the stable production stage. The annual oil production is more than 45×10^4 t, the oil recovery rate is more than 1.0%, and the stage recovery degree is 13.1%. The third stage is from 1999 to 2005, which can be called the stage of decline in production. The annual oil production drops rapidly, from 45×10^4 t to 16×10^4 t, and the oil recovery rate drops below 0.5%. The fourth stage is from June 2005 to the present, which is called the in-situ combustion development stage. With the continuous expansion of the scale of fire-flooding, the well opening rate has increased from 23.5% to 87.5%. The daily oil production of the single well has increased from 0.7 t/d to 3.2 t/d. The formation pressure has increased from 1.3 MPa to 3 MPa, and the annual oil production has rebounded, reaching more than 20×10^4 t, which means that the oil production by fire-flooding accounts for 80% of the block production.

However, with the expanding scale of the in-situ combustion pilot test in the Du 66 block and the influence of the reservoir heterogeneity, there are some problems, such as the difference of combustion state between the thin interbedded reservoirs, the serious overlap of fire-flooding line in thick interbedded reservoirs, the uneven spread of fire line and the unclear understanding of the combustion state, which affect the efficiency of fire-flooding and the oil recovery. In order to reveal the development mechanisms of the in-situ combustion and better understanding the characteristics of the multi-layer fire wave and determine the main factors affecting the efficiency of fire-flooding, it is necessary to carry out the simulation study of in-situ combustion in the Du 66 block.

2. Materials and Methods

The burning oil layer was a complex chemical process. On the basis of experimental research, this paper reproduced the whole process of the formation and migration of the oil-wall by numerical simulation, and analyzed the changes of temperature, composition content and crude oil characteristics in different stages. Thus, the characteristics of the oil-wall in fired oil layer could be revealed. At the same time, the influencing factors of the in-situ combustion process could be analyzed through sensitivity analysis on the production and geological parameters, which could provide the basis for the preliminary screening of the in-situ combustion reservoir and the production design of the target reservoirs.

(1) Modeling parameters

The preparation of the data was the basis of the reservoir modeling. Based on the study of the formation structure, sedimentary microfacies, and parameter interpretation, the data of wells, layers, and reservoir were collected and sorted. The grid step size was 10 m × 10 m and a 3-dimensional geological model was established for the working area. The total number of grids was $22 \times 21 \times 25 = 11,550$.

(2) Modeling area

The modeling area was 46,200 m^2. Nine vertical wells were modeled: 46-x38p, 46-g37p, 46-g36p, 46-038p, 46-036p, 46-g38p, 45-37p, 45-36p, 46-037AI. Among them, 46-037AI was the injection well, and the remaining 8 wells were production wells.

(3) Construction model

The anti-nine-point well network with the size of 220 × 210 m was adopted in this study and the development method was in-situ combustion. The Du 66 block was inclined from the northwest to the southeast, and the dip angle of the stratum was generally 5°~10°, which can truly reflect the changing trend of the stratum. The injection well was located in the center of the block, and the production wells were distributed at the corners of the block. The 3-dimensional distribution of the original oil saturation and porosity has been shown in Figure 1. From the oil saturation model of the Du 66 block (Figure 1a), the oil saturation was evenly distributed, and the initial oil saturation was 0.69 in average. From the porosity model of the Du 66 block (Figure 1b), there were 25 layers in the stratum. The porosity of the odd layers was 0.21, and the porosity of the even layers was 0.0013. The porosity was evenly distributed in the transverse direction.

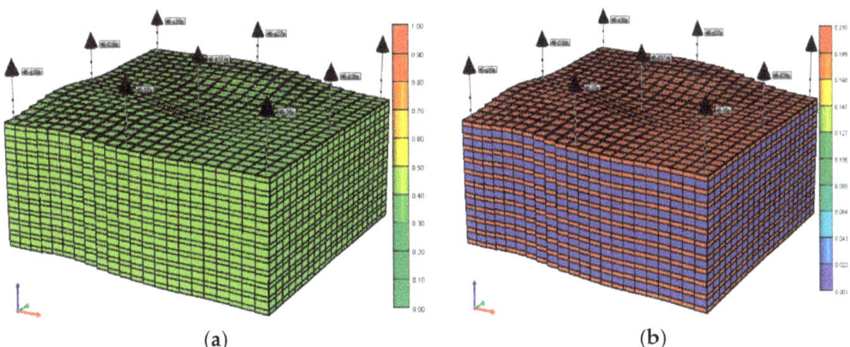

(a)   (b)

Figure 1. Three-dimensional distribution of the initial oil saturation and porosity. (**a**) Initial oil saturation distribution; (**b**) porosity distribution.

3. Results

3.1. Formation Conditions of Oil Wall

3.1.1. Ventilation Intensity

From the simulation results, when the ventilation intensity of the injection well was less than 0.55 m^3/(d.m^2), the formation would not form oil-wall and fire-wall. In addition, the combustion leading edge temperature was lower than 220 °C, showing a low temperature oxidation state. At the end of the production, the maximum temperature of the formation was less than 220 °C. The highest temperature zone existed at the injection well. And there was no tendency to advance to the production well and no fire wall was formed. In the simulation, the pressure difference between the injection and production well was 7.4 MPa, which means the injection well pressure was much higher than the production well. It can be found that if the ventilation intensity and oxygen was not enough, it was difficult to maintain the fuel supply and the formation oxidation reaction. It can be concluded that the air injection speed should be increased to maintain stable combustion and meet the development of the in-situ combustion.

3.1.2. Original Oil Saturation

From the simulation results, when the original oil saturation of the local layer was less than 30%, the formation will not form oil-wall. At the end of production, the maximum temperature of the formation was less than 340 °C, and the highest temperature zone existed at the injection well end. In addition, there was no tendency to advance to the production well. Therefore, no fire wall was generated. It can be found that if the original oil saturation of the formation was too low, which was not conducive to the formation of the oil-walls. The formation of oil-walls was closely related to the geological conditions of the reservoirs.

3.2. Effect of Production Parameters

In this section, we analyzed the effect of the production parameters (huff and puff rounds, air injection speed, and air injection temperature) on the formation of oil-wall and the efficiency of in-situ combustion. Here, we chose different huff and puff rounds (1, 2, 3, 4, 5, 6, 7 rounds), different air injection speeds (1000 m^3/day, 2000 m^3/day, 3000 m^3/day, 5000 m^3/day), different air injection temperatures (20 °C, 60 °C, 80 °C, 100 °C). Based on the simulation results, optimal production parameters were obtained.

3.2.1. Huff and Puff Rounds

Here, we only changed the air huff and puff rounds and kept other the conditions the same to analyze the effect of the huff and puff rounds (one to seven rounds). The bottom pressure of the injection well was kept as 20 MPa; the steam temperature was 270 °C (543.15 K); the steam dryness was 0.70; the injection air temperature was 20 °C (298.15 K); the numerical simulation was carried out for 20 years. Comparison of the production indicators (recovery degree, cumulative oil production, gas-oil ratio, and gas production rate) under different huff and puff rounds are shown in Figure 2.

Dynamic production indicators under different huff and puff rounds have been listed in Table 2. It can be seen that the recovery degree and the cumulative oil production increased with the increase of the huff and puff rounds. However, the relative increase in the yield after five rounds of huff and puff was the most obvious. The more huff and puff rounds, the later the time the inject air burns the formation, and the longer the gas breakthrough the formation. In addition, it could be found that the gas-to-oil ratio was proportional to the huff and puff rounds. For one round huff and puff, the gas-oil-ratio still rose with time and was unstable. However, for other rounds, the ratio of gas to oil increased to the maximum and then gradually decreased and became stable.

The characteristic parameters of the oil-walls under different huff and puff rounds have been listed in Table 3. As can be found from Table 3, when the number of huff and puff rounds increased, the later the oil-wall was generated. The initial formation position was closer to the production well. The time when the thickness of the oil-wall reached a maximum was also delayed. As the huff and puff rounds increased, the formation water saturation increased and the width of the oil-wall gradually narrowed and the average saturation of the oil-wall decreased. In addition, it can be found that the pressure gradient in the oil-wall was generally high. The pressure gradient increased with the increase of the huff and puff rounds. The migration speed of the oil-wall also accelerated. This was because the oil-wall saturation was reduced and the resistance was reduced, which was conducive to the migration of the oil-wall.

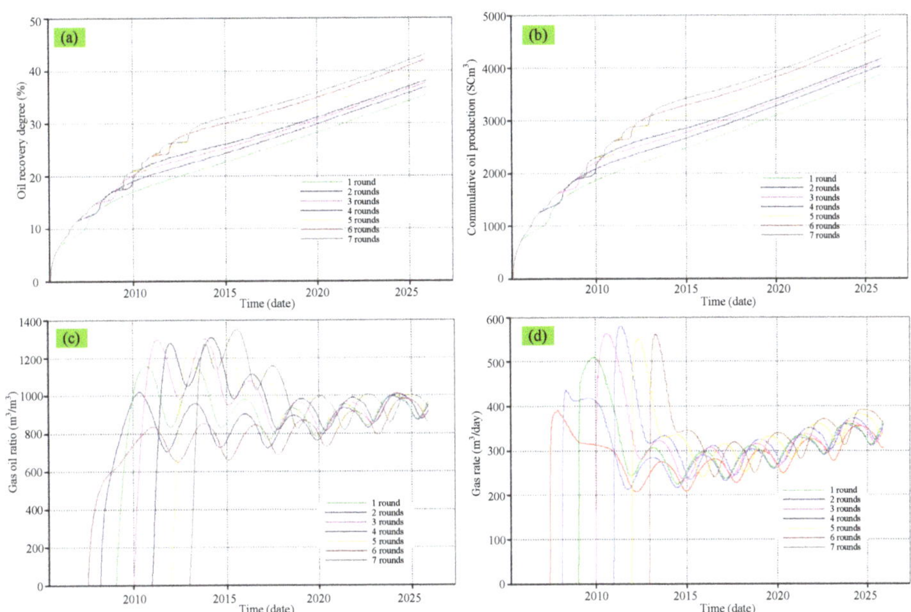

Figure 2. Comparison of the production indicators under different huff and puff rounds. (**a**) Recovery degree; (**b**) cumulative oil production; (**c**) gas-oil ratio; and (**d**) gas production rate.

Table 2. Dynamic production indicators under different huff and puff (HP) rounds.

Sweep Round	Cumulative Steam Injection (m³)	End of Production (%)	The Degree of Recovery at the End of the Fire Drive (%)	Production Oil (m³)	See the Gas Time (days)	Gas to Oil Ratio Peak m³/(m³)
1	14,000 m³	6.69	35.28	3869.43	770	984.84
2	28,000 m³	11.57	36.85	4029.53	1010	1019.57
3	42,000 m³	14.82	37.63	4115.76	1330	1158.95
4	56,000 m³	17.32	38.05	4173.21	1680	1300.2
5	70,000 m³	20.87	40.30	4419.19	2030	1306.69
6	84,000 m³	23.72	42.04	4610.94	2400	1311.4
7	98,000 m³	26.24	43.08	4713.31	2755	1349.99

Table 3. Characteristic parameters of oil-wall under different huff and puff rounds.

Rounds / Condition	1	2	3	4	5	6	7
Oil-wall formation time (days)	810	1030	1340	1620	2010	2450	1980
Oil saturation peak	0.6351	0.6387	0.6397	0.6427	0.6484	0.6356	0.6321
Average oil saturation	0.6043	0.6051	0.6113	0.6131	0.6101	0.6017	0.6168
Oil-wall average width (m)	50	42	38	33	30	30	25
Oil-wall pressure gradient (kPa/m)	66.47	66.39	65.67	59.31	54.32	54.11	52.56
Oil-wall migration speed (m/day)	0.01702	0.01703	0.01707	0.01733	0.01739	0.01835	0.01843
Oil-wall average temperature (°C)	88	82	80	77	75	71	70

The characteristic parameters of the fire-wall under different huff and puff rounds have been listed in Table 4. As can be found from Table 4, after five rounds of huff and puff, the combustion front had the highest temperature, which was up to 710 °C. In addition, the combustion under this condition had the best effect. The position of the fire-wall under different rounds was almost the same. Therefore, the huff and puff rounds will affect the temperature of the fire-wall.

Table 4. Characteristic parameters of the fire-wall under different huff and puff rounds.

Rounds Condition	1	2	3	4	5	6	7
Average temperature of the fire wall (°C)	650	685	704	689	710	702	676
The distance between the fire wall and the oil-wall (m)	100	110	100	100	90	110	115
Fire wall propulsion speed (m/day)	0.034	0.034	0.035	0.036	0.037	0.036	0.035

The temperature change of the firing front and oil saturation change under different huff and puff rounds have been shown in Figure 3a,b, respectively. As can be found from previous figures and tables, after five rounds of huff-and-puff, the increase of the recovery degree and cumulative oil production were the most obvious. In addition, the temperature of combustion front was the highest, and the gas appearing time in the production well was also the latest. Therefore, for this study, five rounds of huff-and-puff was more conducive to in-situ combustion and the formation of oil-walls.

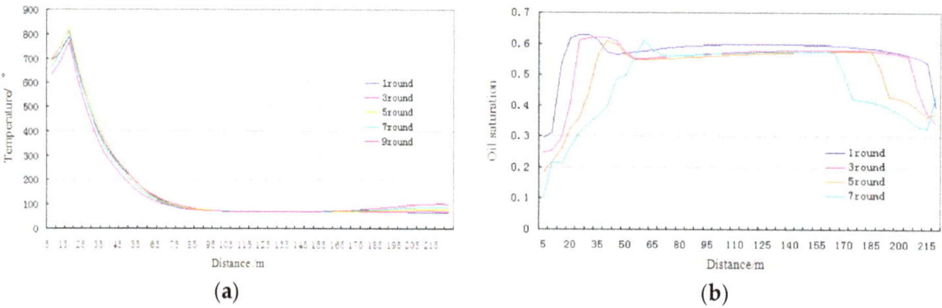

Figure 3. The temperature change of the firing front and oil saturation change under different huff and puff rounds. (**a**) Temperature change of the firing front; (**b**) oil saturation change.

3.2.2. Air Injection Speed

Based on previous analysis, here we set the huff-and-puff round as five in this section of study. We chose different air injection speeds for comparison, which were 1000 m³/d, 2000 m³/d, 3000 m³/d, and 5000 m³/d, respectively. Comparison of production indicators (recovery degree, cumulative oil production, gas-oil ratio, and gas production rate) under different air injection speeds are shown in Figure 4.

The dynamic production index under different air injection speeds have been listed in Table 5. As can be found from Table 5, at the end of the huff and puff, the recovery degree was approximately the same. At the end of the fire-flooding, the recovery degree and cumulative oil production had no significant difference for the four air injection conditions. When the daily air injection volume was 3000 m^3, the recovery degree was the best (40.33%). The cumulative oil production was also the highest (4422.87 m^3). The higher the air injection speed, the larger the air injection amount, the shorter the gas breakthrough time, and the higher the gas-oil ratio was.

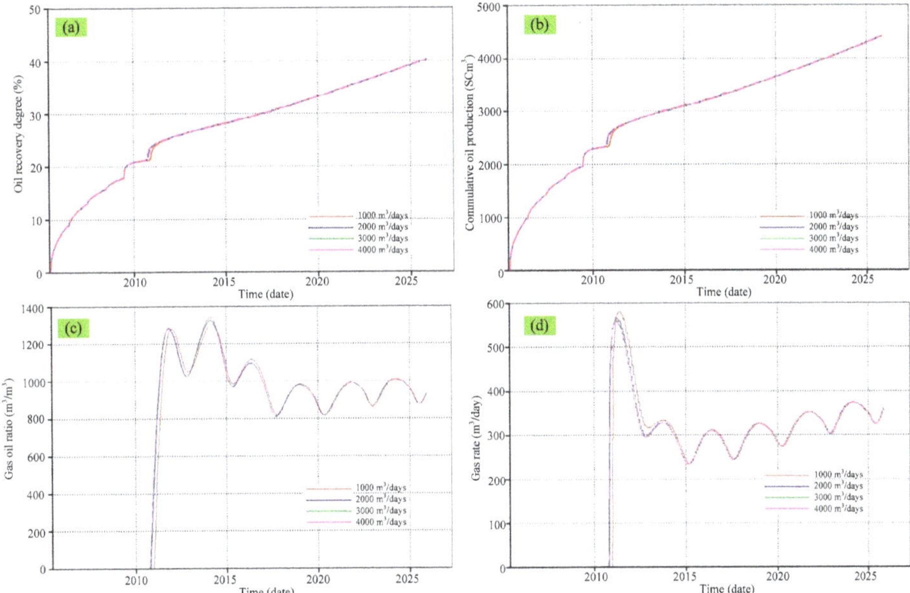

Figure 4. Comparison of production indicators under different air injection speeds. (**a**) recovery degree; (**b**) cumulative oil production; (**c**) gas-oil ratio; and (**d**) gas production rate.

Table 5. Dynamic production index under different air injection speeds.

Air Injection Speed (m^3/day)	End of Production (%)	The Degree of Recovery at the End of the Fire Drive (%)	Production Oil (m^3)	See the Gas Time (days)	Gas to Oil Ratio Peak (m^3/m^3)
1000	19.87	40.30	4419.19	2030	1306.69
2000	19.52	40.26	4416.50	1970	1324.52
3000	19.99	40.33	4422.87	1950	1334.53
5000	19.64	40.28	4417.90	1945	1344.23

The characteristic parameters of oil-wall under different air injection speeds have been listed in Table 6. As can be found from Table 6, when the air injection speed increased, the formation time of oil-wall was shortened and the oil saturation was easy to reach the peak. When the air injection speed was 3000 m^3/day, the average oil saturation value was the highest, and the peak value was higher than other situations. The air injection speed had almost no effect on the migration length of the oil-wall. However, the migration speed of the oil-wall was proportional to the air injection speed.

The characteristic parameters of fire wall under different air injection speeds have been list in Table 7. As can be found from Table 7, the daily air injection has no significant effect on the temperature of the fire wall. The temperature of the combustion front for all situations is about 700 °C.

Table 6. Characteristic parameters of oil-wall with different air injection speed.

Injection Speed (m³/day) / Condition	1000	2000	3000	5000
Oil-wall formation time (days)	2010	1945	1855	1830
Oil saturation peak	0.6245	0.6278	0.6315	0.6310
Average oil saturation	0.6043	0.6051	0.6131	0.6113
Oil-wall average width (m)	45	30	33	33
Oil-wall migration length (m)	95	95	95	95
Oil-wall pressure gradient (kPa/m)	54.32	61.16	62.32	62.15
Oil-wall migration speed (m/day)	0.01639	0.01683	0.01726	0.01778
Oil-wall average temperature (°C)	68	71	69	70

Table 7. Characteristic parameters of different air injection velocity fire wall.

Injection Speed (m³/day) / Condition	1000	2000	3000	5000
Average temperature of the fire wall (°C)	693	704	698	705
The distance between the fire wall and the oil-wall (m)	90	90	90	90
Fire wall propulsion speed (m/day)	0.037	0.037	0.036	0.036

The temperature change of the firing front and oil saturation change under different air injection speeds have been shown in Figure 5a,b respectively. As can be found from previous figures and tables, we can conclude that the air injection speed had little effect on the fire wall. However, it had significant effect on the oil-wall. The daily injection volume of air was 3000 m³, which was conducive to the in-situ combustion and formation of oil-wall.

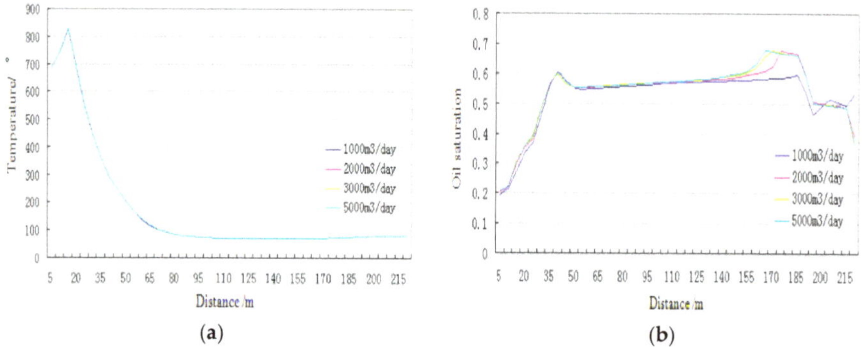

Figure 5. The temperature change of the firing front and oil saturation change under different air injection speeds. (**a**) temperature change of the firing front; (**b**) oil saturation change.

3.2.3. Air Injection Temperature

Here, we set the huff and puff rounds as five and the air injection speed was 3000 m³/day based on a previous study. We changed the air injection temperature to 20 °C (298.15 K), 60 °C (333.15 K), 80 °C (353.15 K), and 100 °C (373.15 K) for comparative analysis. Comparison of production indicators (recovery degree, cumulative oil production, gas-oil ratio, and gas production rate) under different air injection temperatures are shown in Figure 6.

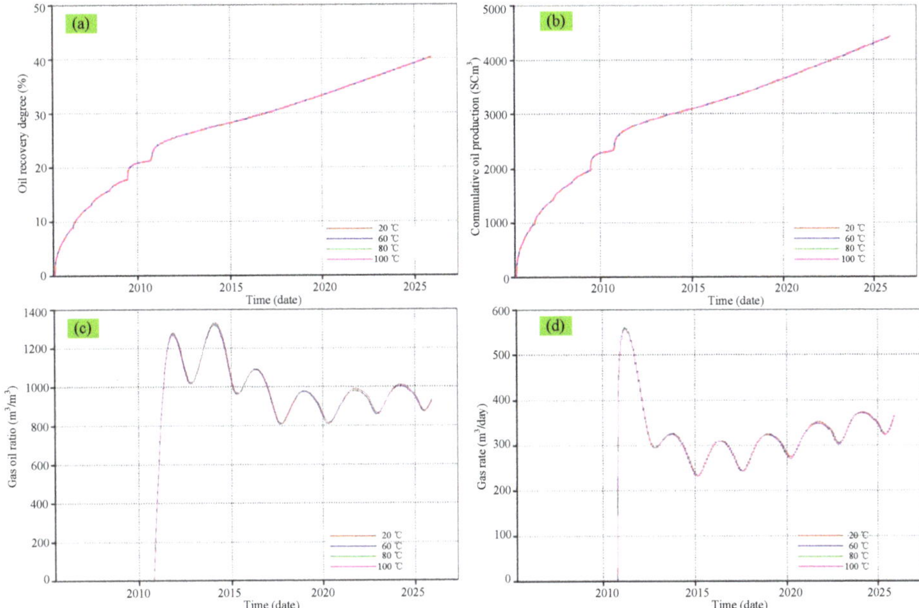

Figure 6. Comparison of production indicators under different air injection temperatures. (**a**) Recovery degree; (**b**) cumulative oil production; (**c**) gas-oil ratio; and (**d**) gas production rate.

The dynamic production index under different air injection temperatures have been listed in Table 8. As can be found from Table 8, the air injection temperature had little effect on the recovery degree and the cumulative oil production. The gas appearance time in the production well and the gas-oil ratio were almost the same at different air injection temperatures.

Table 8. Dynamic production index under different air injection temperatures.

Air Injection Temperature (m^3/day)	The Recovery Degree at the End of HP (%)	The Recovery Degree at the End of the In-Situ Combustion (%)	Cumulative Oil Production (m^3)	Gas Appearance Time (days)	Gas to Oil Ratio Peak (m^3/m^3)
20	20.93	40.33	4422.87	1950	1334.53
60	20.96	40.35	4425.47	1950	1327.20
80	20.98	40.35	4425.61	1950	1323.63
100	21.05	40.37	4427.83	1950	1319.35

The characteristic parameters of the oil-wall under different air injection temperatures have been listed in Table 9. As it can be found from Table 9, the air injection temperature had no effect on the formation time of the oil-wall. The oil saturation at the initial moment of the oil-wall formation was almost the same. When the air injection temperature was 80 °C, the peak of oil saturation was the highest, and the pressure gradient of the corresponding oil-wall was the largest. The effect of the air injection temperature on the migration length and speed of oil-wall was also negligible.

Table 9. Characteristic parameters of oil-wall under different air injection temperatures.

Injection Temperature °C / Condition	20	60	80	100
Oil-wall formation time (days)	1855	1855	1855	1855
Oil saturation peak	0.6389	0.6387	0.6401	0.6337
Average oil saturation	0.6101	0.6101	0.6151	0.6150
Oil-wall average width (m)	30	32	25	33
Oil-wall migration length (m)	95	95	95	95
Oil-wall pressure gradient (kPa/m)	62.32	62.50	67.89	67.82
Oil-wall migration speed (m/day)	0.01739	0.01713	0.01786	0.01778
Oil-wall average temperature (°C)	69	71	72	74

The characteristic parameters of the fire wall under different air injection temperatures have been listed in Table 10. As can be found from Table 10, the temperature at the front of the combustion increased if the air injection temperature increased. When the temperature increase was small, the variation of the fire wall migration velocity was also small.

Table 10. Characteristic parameters of the fire wall under different air injection temperatures.

Injection Temperature °C / Condition	20	60	80	100
Average temperature of the fire wall (°C)	693	708	711	715
The distance between the fire wall and the oil-wall (m)	110	110	110	110
Fire wall propulsion speed (m/day)	0.03064	0.03071	0.03089	0.03085

The temperature change of the firing front and the oil saturation change under different air injection temperatures have been shown in Figure 7a,b, respectively. As can be found from previous figures and tables, when the huff and puff rounds was five and the air injection speed was 3000 m^3/day, the features of the oil-wall were obvious and the fire-flooding efficiency was good. The temperature of the injected air had little effect on the in-situ combustion and oil recovery.

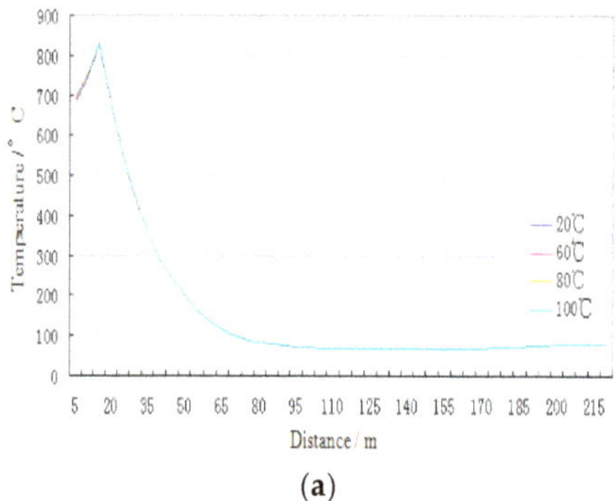

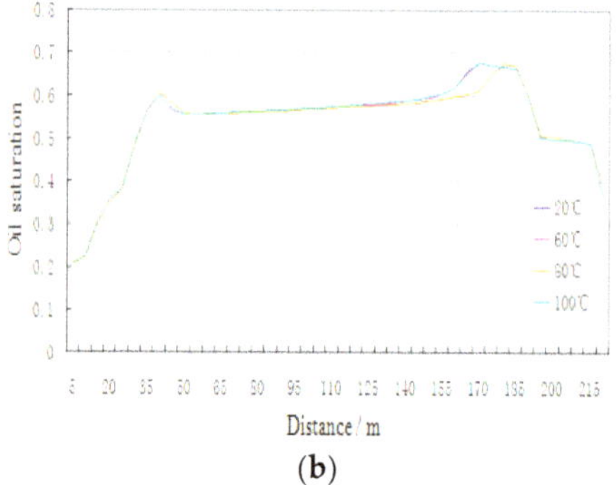

Figure 7. The temperature change of the firing front and oil saturation change under different air injection temperatures. (**a**) Temperature change of the firing front; (**b**) oil saturation change.

3.3. Effect of Geological Parameters on oil-walls

3.3.1. Bottom Water Thickness

When the bottom water was present in the reservoir, different bottom water thicknesses (0 m, 100 m, 200 m, 360 m) were simulated to analyze the influence of the bottom water on the characteristic parameters of the oil-wall. Comparison of the production indicators (recovery degree, cumulative oil production, gas-oil ratio, and gas production rate) under different bottom water thickness are shown in Figure 8.

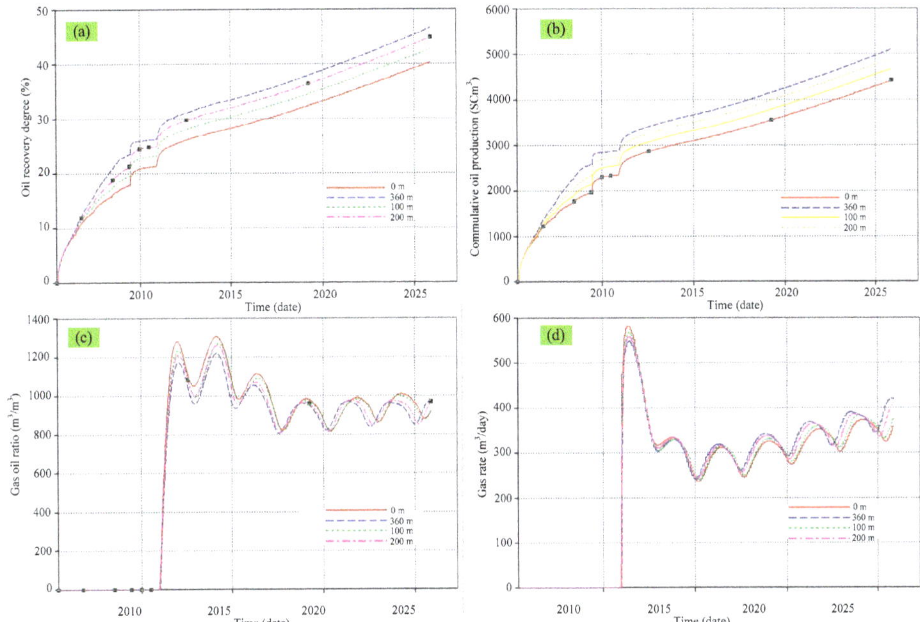

Figure 8. Comparison of the production indicators under different bottom water thickness. (**a**) recovery degree; (**b**) cumulative oil production; (**c**) gas-oil ratio; and (**d**) gas production rate.

The dynamic production indicators under different bottom water thicknesses have been listed in Table 11. As can be found from Table 11, with the increase of the thickness of the bottom water, the oil recovery degree and the accumulated oil production were increasing; the gas appearance time in production well was almost the same; and the peak of gas-oil ratio decreased.

Table 11. Dynamic production indicators under different bottom water thicknesses.

Bottom Water Thickness (m)	The Recovery Degree at the End of HP (%)	The Recovery Degree at the End of the In-Situ Combustion (%)	Cumulative Oil Production (m³)	Gas Appearance Time (days)	Gas to Oil Ratio Peak (m³/m³)
0	21.37	40.24	4242.16	2007	1355.89
100	22.96	43.14	4698.07	2007	1268.41
200	25.28	45.84	4895.55	2006	1222.14
360	27.04	46.37	5191.22	2005	1209.71

The characteristic parameters of oil-wall with different bottom water thickness have been listed in Table 12. As can be found from Table 12, when the thickness of the bottom water was 360 m, the characteristics of the oil-wall were the most obvious: The average oil saturation, oil saturation peak, oil-wall average width, oil-wall migration length, oil-wall pressure gradient, oil-wall migration speed, and oil-wall average temperature were the highest. This was because the bottom water could provide the driving force and assisted the migration of the oil-wall.

Table 12. Characteristic parameters of oil-wall under different bottom water thickness.

Condition / Bottom Water Thickness (m)	0	100	200	360
Oil-wall formation time (days)	1720	1704	1704	1707
Oil saturation peak	0.5096	0.5110	0.5123	0.5126
Average oil saturation	0.5020	0.5076	0.5098	0.5106
Oil-wall average width (m)	8	8	8	10
Oil-wall migration length (m)	150	160	160	170
Oil-wall pressure gradient (kPa/m)	241	256	263	273
Oil-wall migration speed (m/day)	0.026	0.028	0.030	0.034
Oil-wall average temperature (°C)	70.2	70.5	70.6	71.8

The characteristic parameters of the fire wall under different bottom water thickness fire walls have been listed in Table 13. As can be found from Table 13, when the thickness of the bottom water was 200 m, the temperature of the combustion front reached 672 °C, and the average pressure of the fire wall was the highest. Therefore, the fire wall advanced faster and promoted the formation of the oil-wall.

Table 13. Characteristic parameters of the fire wall under different bottom water thickness.

Condition / Bottom Water Thickness (m)	0	100	200	360
Average temperature of the fire wall (°C)	665	662	672	669
The distance between the fire wall and the oil-wall (m)	100	80	90	80
Fire wall propulsion speed (m/day)	0.00263	0.00261	0.00271	0.00261

In summary, the presence of the bottom water in the formation will increase the efficiency of in-situ combustion and improve oil recovery. The greater the thickness of the bottom water, the stronger the force of water flooding is.

3.3.2. Stratigraphic Layering

Stratigraphic layering (SL) was also an important geological parameter which affected reservoir production. The positive SL meant that the permeability of the upper layer was less than the lower layer. The negative SL means that the permeability of upper layer was higher than the lower layer. If the permeability of the upper and lower layer was the same, it could be called "No SL", which also meant that the reservoir was vertically homogenous. In this study, the positive and negative SL models were established, respectively (average permeability is 1000 mD), for the positive SL case, the top layer permeability (600 mD) was lower than the bottom layer permeability (1400 mD). For the negative SL case, the top layer permeability (1400 mD) was higher than the bottom layer permeability (600 mD). Comparison of the production indicators (recovery degree, cumulative oil production, gas-oil ratio, and gas production rate) under different stratigraphic layering are shown in Figure 9.

The dynamic production indicators under different stratigraphic layering (SL) have been listed in Table 14. As can be found from Table 14, the influence of the SL on the recovery degree and cumulative oil production of the reservoir was: no SL > positive SL > negative SL. The positive SL case had the earliest gas appearance in the production well and the no SL case had the latest gas appearance in the production well. It can be seen that the more homogenous of the reservoir, the better the oil recovery for the in-situ combustion scenario.

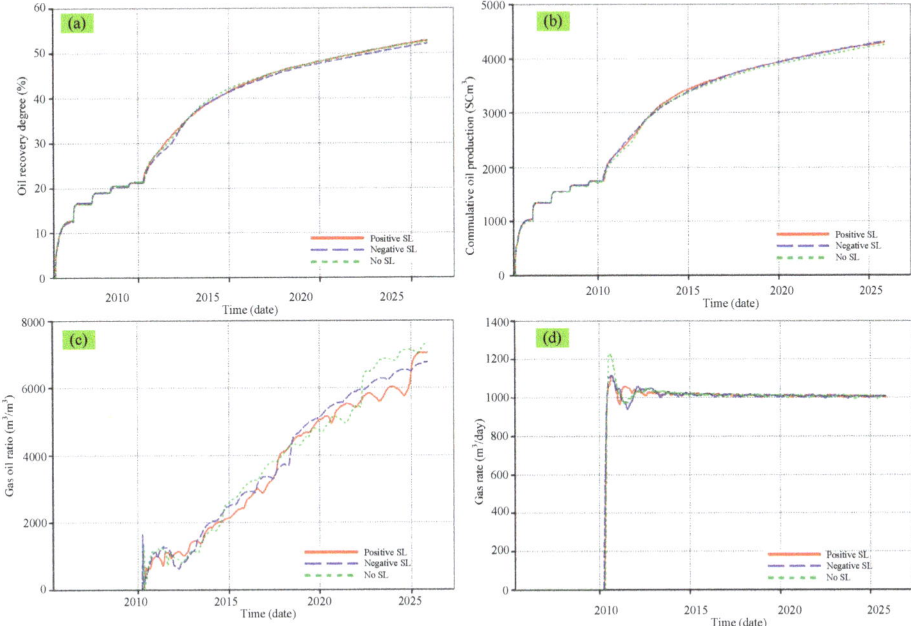

Figure 9. Comparison of production indicators under different stratigraphic layering. (**a**) Recovery degree; (**b**) cumulative oil production; (**c**) gas-oil ratio; and (**d**) gas production rate.

Table 14. Dynamic production indicators under different stratigraphic layering (SL).

Stratigraphic Layering	The Recovery Degree at the End of HP (%)	The Recovery Degree at the End of the In-Situ Combustion (%)	Cumulative Oil Production (m³)	Gas Appearance Time (days)
No SL	21.51	52.84	4298.27	1785
Positive SL	21.43	52.79	4314.16	1758
Negative SL	21.30	52.17	4310.24	1773

3.3.3. Permeability Ratio

Based on a previous study, here we considered the reservoir to be the positive stratigraphic layering case. The effect of the permeability ratio on the formation oil-wall and oil recovery was analyzed, considering the four different permeability ratios: 1, 5, 10, 100. Comparison of the production indicators (recovery degree, cumulative oil production, gas-oil ratio, and gas production rate) under different permeability ratio are shown in Figure 10.

The dynamic production indicators under different permeability ratios have been listed in Table 15. As can be found from Table 15, the greater the permeability ratio, the lower the recovery degree and the cumulative oil production, which means that the permeability difference was not conducive to in-situ combustion. The larger the permeability ratio was, the earlier the gas appeared in the production well. The gas-oil ratio also increased at the end of production.

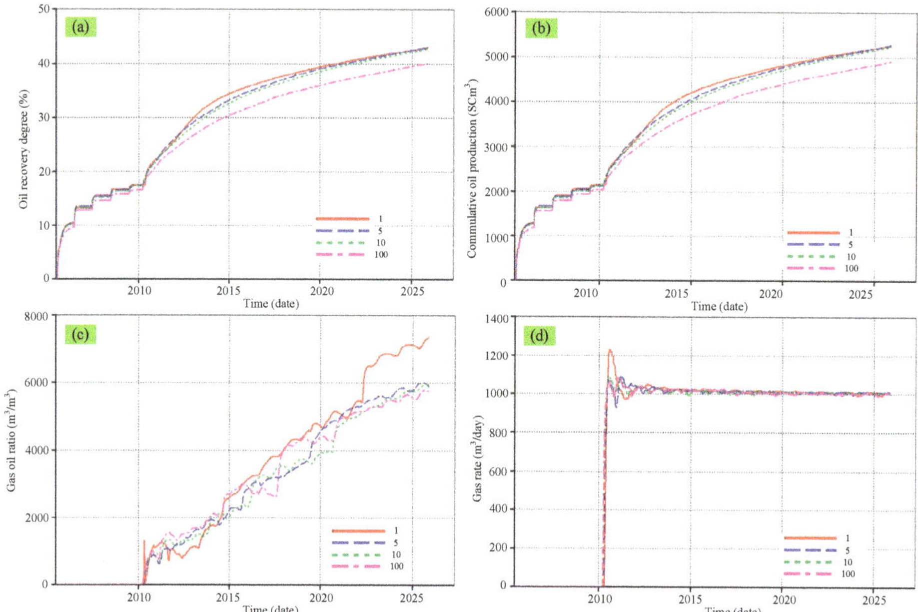

Figure 10. Comparison of production indicators under different permeability ratio. (**a**) recovery degree; (**b**) cumulative oil production; (**c**) gas-oil ratio; and (**d**) gas production rate.

Table 15. Dynamic production indicators under different permeability ratio.

Permeability Ratio	The Recovery Degree at the End of HP (%)	The Recovery Degree at the End of the In-Situ Combustion (%)	Cumulative Oil Production (m^3)	Gas Appearance Time (days)
1	21.92	52.89	4325.81	1785
5	21.50	52.71	4309.40	1773
10	21.38	52.43	4295.58	1760
100	20.38	49.17	4016.40	1754

The characteristic parameters of the oil-wall with different permeability ratios have been listed in Table 16. As can be found from Table 16, the oil saturation peak, average oil saturation, and oil-wall migration length, oil-wall pressure gradient, oil-wall migration speed, and the oil-wall average temperature were the highest for the case with a permeability ratio equal to one. This means that the greater the permeability ratio (the difference in permeability), the more unstable the oil-wall was. When the permeability ratio was larger than 10, there was no oil-wall formed in the relatively low permeability layers. The oil layers were mainly formed in the middle and lower layers.

The characteristic parameters of the fire wall under different permeability ratios have been listed in Table 17. As can be found from Table 17, the greater the difference in permeability, the faster the rate of the fire wall propulsion. However, the breakthrough of the fire wall in one direction was not conducive to the formation of oil-walls and also damaged to the oil recovery.

Table 16. Characteristic parameters of oil-wall with different permeability ratio.

Condition / Permeability Ratio	1	5	10	100
Oil-wall formation time (days)	1670	1670	1670	1670
Oil saturation peak	0.5178	0.5042	0.5115	0.5068
Average oil saturation	0.5092	0.5023	0.5059	0.5041
Oil-wall average width (m)	25	25	20	15
Oil-wall migration length (m)	175	170	170	170
Oil-wall pressure gradient (kPa/m)	111.42	90.51	60.13	75.60
Oil-wall migration speed (m/day)	0.030	0.0292	0.0292	0.0292
Oil-wall average temperature (°C)	72	70	70	69

Table 17. Characteristic parameters of fire wall under different permeability ratio.

Condition / Permeability Ratio	1	5	10	100
Average temperature of the fire wall (°C)	897	954	987	1005
Fire wall propulsion speed (m/day)	0.0082	0.0088	0.0089	0.0092

In summary, the permeability ratio had an important influence on the formation of the oil-wall. When the formation permeability was uniform, the condition for the oil-wall formation was the best.

3.3.4. Formation Thickness

In order to improve the calculation accuracy, the grid was encrypted. The plane grid step size was 1 m × 5 m, and the total number of grids was 220 × 1 × 5 = 1100. Comparison of the production indicators (recovery degree, cumulative oil production, gas-oil ratio, and gas production rate) under different formation thickness are shown in Figure 11.

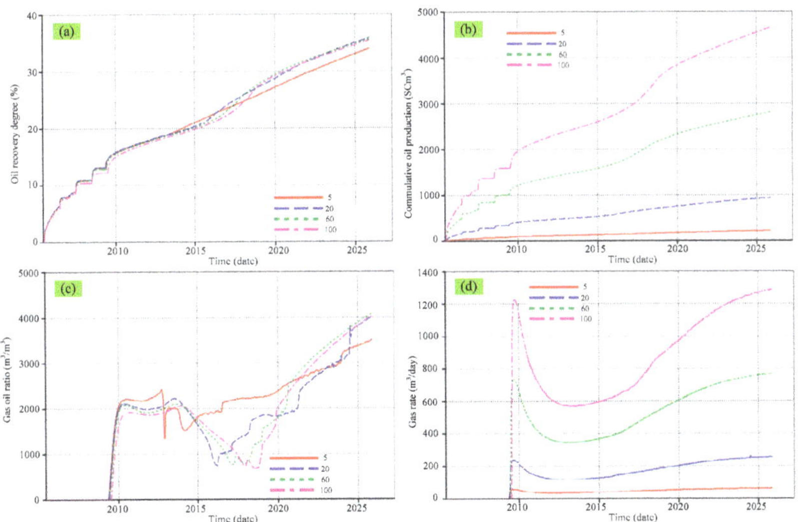

Figure 11. Comparison of production indicators under different oil layer thickness. (**a**) Recovery degree; (**b**) cumulative oil production; (**c**) gas-oil ratio; and (**d**) gas production rate.

The dynamic production indicators under different formation thickness have been listed in Table 18. As can be found from Table 18, the cumulative oil production increased with the formation thickness. This was because when the formation thickness increased, the reserve increased. When the formation thickness was 20 m, the oil recovery degree was the highest. With formation thickness increased, the gas appearance time in production well appeared later. The peaks of the gas-oil ratio showed an opposite trend.

Table 18. Dynamic production indicators under different formation thickness.

Formation Thickness (m)	The Recovery Degree at the End of HP (%)	The Recovery Degree at the End of the In-Situ Combustion (%)	Cumulative Oil Production (m^3)	Gas Appearance Time (days)	Gas to Oil Ratio Peak (m^3/m^3)
5	15.55	33.91	223.26	1443	2431.45
20	15.76	35.77	939.86	1460	2232.01
60	15.66	35.52	2813.19	1475	2107.74
100	14.85	35.36	4642.90	1502	2058.66

The characteristic parameters of the oil-wall with different formation thickness have been listed in Table 19. As can be found from Table 19, the oil-wall formation time was the same for the four cases. When the formation thickness was 100 m, the oil-wall had highest oil saturation. The features of the oil-wall were more obvious. The oil-wall migration speed is the highest.

Table 19. Characteristic parameters of oil-wall with different formation thickness.

Condition / Oil Layer Thickness m	5	20	60	100
Oil-wall formation time (days)	1349	1349	1349	1349
Oil saturation peak	0.5065	0.5083	0.5137	0.5246
Average oil saturation	0.5029	0.5066	0.5070	0.5082
Oil-wall average width (m)	5	5	5	5
Oil-wall migration length (m)	31	61	61	66
Oil-wall pressure gradient (kPa/m)	240.27	107.56	184.59	254.24
Oil-wall migration speed (m/day)	0.078	0.0629	0.045	0.0863
Oil-wall average temperature (°C)	72.5	71.77	71.73	71.77

The characteristic parameters of the fire wall under different formation thickness have been listed in Table 20. As can be found from Table 20, the average temperature of the fire wall, the distance between the fire wall and the oil-wall, and the fire wall propulsion speed were negatively related with the formation thickness. For formations with too small thickness, the heat loss during the fire propulsion process will be relatively large, which may cause the combustion temperature to be very low. Under this situation, the fire will be extinguished, and the in-situ combustion will fail. In summary, the formation thickness had a great influence on the characteristics of the oil-wall and the fire wall. Combining above analysis, it can be concluded that the best option for in-situ combustion in this study was when the formation thickness was 20 m.

Table 20. Characteristic parameters of fire wall under different formation thickness.

Condition / Thickness (m)	5	20	60	100
Average temperature of the fire wall (°C)	816.15	799.3	790.5	782.1
The distance between the fire wall and the oil-wall (m)	31	29	26	24
Fire wall propulsion speed (m/day)	0.00415	0.00382	0.00371	0.00367

4. Conclusions

1. Oil saturation and ventilation strength are the basic conditions for the formation of the oil-walls. The appropriate ventilation intensity is required to form an effective fire line, which can greatly reduce the viscosity of crude oil and enhance the fluidity of crude oil.
2. The effect of production parameters (huff and puff rounds, air injection speed, and air injection temperature) and geological parameters (bottom water layer thickness, stratigraphic layering, permeability ratio, and formation thickness) on the efficiency of in-situ combustion and oil recovery have been analyzed.
3. Through numerical simulation results, the optimum value of huff and puff round and air injection speed has been obtained. The temperature of the injected air has little effect on the in-situ combustion and oil recovery.
4. With the increase of the thickness of the bottom water, the oil recovery degree and the accumulated oil production increasing, formation heterogeneity has a negative effect on oil recovery and formation of the oil-wall. A too thin layer is not suitable for in-situ combustion.

Author Contributions: This is a joint work and the authors were in charge of their expertise and capability: Z.Y. contributed to the conceptualization and methodology and data analysis; S.H. for writing and revision; H.L. for manuscript revision.

Funding: This research received no external funding.

Conflicts of Interest: The authors declare no conflict of interest.

References

1. Kok, M. Progress and recent utilization trends in petroleum recovery: Steam injection and in-situ combustion. *Energy Sources* **2012**, *34*, 2253–2259. [CrossRef]
2. Samimi, A.; Karimi, G. In-situ combustion process, one of IOR methods livening the reservoirs. *Pet. Coal* **2010**, *52*, 342–354.
3. Wilson, L.; Root, P. Cost comparison of reservoir heating using steam or air. *J. Pet. Technol.* **1966**, *18*, 233–239. [CrossRef]
4. Alexander, J.; Martin, W.; John, N. Factors affecting fuel availability and composition during in-situ combustion. *J. Pet. Technol.* **1962**, *14*, 1154–1164. [CrossRef]
5. Cheih, C. State-of-the-art review of fireflood field projects (includes associated papers 10901 and 10918). *J. Pet. Technol.* **1982**, *34*, 19–36. [CrossRef]
6. Gates, C.; Ramey, H. A method for engineering in-situ combustion oil recovery projects. *J. Pet. Technol.* **1980**, *32*, 285–294. [CrossRef]
7. Thomas, G. A study of forward combustion in a radial system bounded by permeable media. *J. Pet. Technol.* **1963**, *15*, 1145–1149. [CrossRef]
8. Parrish, D.; Craig, F. Laboratory study of a combination of forward combustion and waterflooding the cofcaw process. *J. Pet. Technol.* **1969**, *21*, 753–761. [CrossRef]
9. Penberthy, W.; Ramey, H. Design and operation of laboratory combustion tubes. *SPE J.* **1965**, *6*, 183–190. [CrossRef]
10. Garon, A.; Kumar, M.; Lau, K. A laboratory investigation of sweep during oxygen and air fireflooding. *SPE Reserv. Eng.* **1986**, *1*, 565–574. [CrossRef]
11. Suat, B.; Mustafa, V. In-situ combustion laboratory studies of Turkish heavy oil reservoirs. *Fuel Process. Technol.* **2001**, *74*, 65–79.
12. Burger, J. Spontaneous ignition in an oil reservoir. *SPE J.* **1976**, *16*, 73–81. [CrossRef]
13. Zeinab, Z.; Farouq, S. Analytical modelling of steam chamber rise stage of steam-assisted gravity drainage (SAGD) process. *Fuel* **2018**, *233*, 732–742.
14. Amirian, E.; Fedutenko, E.; Yang, C. Artificial neural network modeling and forecasting of oil reservoir performance. In *Applications of Data Management and Analysis*; Springer: Cham, Switzerland, 2018; pp. 43–67.
15. Lee, Y.; Lee, W.; Jang, Y. Oil recovery by low-salinity polymer flooding in carbonate oil reservoirs. *J. Pet. Sci. Eng.* **2019**, *181*, 106211. [CrossRef]

16. Fedutenko, E.; Nghiem, L.; Yang, C. Artificial neural network modeling of compaction dilation data for unconventional oil reservoirs. *Soc. Pet. Eng.* **2019**. [CrossRef]

17. Zhang, X.; Lin, C.; Gu, L. Application and outlook of in-situ combustion for developing heavy oil reservoir. In Proceedings of the 4th International Conference on Sensors, Mechatronics and Automation (ICSMA 2016), Zhuhai, China, 12–13 November 2016.

18. Cai, W.; Xie, Z.; Wang, W. Experimental study of wet combustion. *Spec. Oil Gas Reserv.* **2000**, *8*, 81–85.

19. Guan, W.; Wu, S.; Wang, S. Physical simulation of in-situ combustion of sensitive heavy oil reservoir. In Proceedings of the Asia Pacific Oil Gas Conference Exhibition, Dubai, UAE, 4–6 December 2007.

20. Jiang, Y.; Zhang, Y.; Liu, S. Displacement mechanisms of air injection in low permeability reservoirs. *Pet. Explor. Dev.* **2010**, *37*, 471–476.

21. Bottia, H.; Aguillon, M.; Lizcano, H. Numerical modeling on in-situ combustion process in the chichimene field: Ignition stage. *J. Pet. Sci. Eng.* **2017**, *154*, 462–468. [CrossRef]

22. Rahnema, H.; Barrufet, M.; Mamora, D. Combustion assisted gravity drainage – experimental and simulation results of a promising in-situ combustion technology to recover extra-heavy oil. *J. Pet. Sci. Eng.* **2017**, *154*, 513–520. [CrossRef]

23. Pei, S.; Wang, Y.; Zhang, L. An innovative nitrogen injection assisted in-situ conversion process for oil shale recovery: Mechanism and reservoir simulation study. *J. Pet. Sci. Eng.* **2018**, *171*, 507–515. [CrossRef]

24. Nesterov, I.; Shapiro, A.; Stenby, E. Numerical analysis of a one-dimensional multicomponent model of the in-situ combustion process. *J. Pet. Sci. Eng.* **2013**, *106*, 46–61. [CrossRef]

Article

Performance Comparison of Industrially Produced Formaldehyde Using Two Different Catalysts

Kamran Shakeel [1], Muqaddam Javaid [1], Yusra Muazzam [1], Salman Raza Naqvi [1,*], Syed Ali Ammar Taqvi [2,3], Fahim Uddin [2], Muhammad Taqi Mehran [1], Umair Sikander [1] and M. Bilal Khan Niazi [1]

[1] School of Chemical & Materials Engineering, National University of Sciences & Technology, Islamabad 44000, Pakistan; kshakeel_che08@scme.nust.edu.pk (K.S.); muqaddam96@gmail.com (M.J.); ymuazzam_che08@scme.nust.edu.pk (Y.M.); taqimehran@scme.nust.edu.pk (M.T.M.); umair.sikandar@scme.nust.edu.pk (U.S.); m.b.k.niazi@scme.nust.edu.pk (M.B.K.N.)

[2] Department of Chemical Engineering, NED University of Engineering & Technology, Karachi 75270, Pakistan; aliammar@neduet.edu.pk (S.A.A.T.); fahimuddin@neduet.edu.pk (F.U.)

[3] Chemical Engineering Department, Universiti Teknologi PETRONAS, Perak 32610, Malaysia

* Correspondence: salman.raza@scme.nust.edu.pk

Received: 25 September 2019; Accepted: 21 October 2019; Published: 12 May 2020

Abstract: Formaldehyde is an important industrial chemical that is a strong-smelling and colorless gas. It is used in a number of processes such as making household products and building materials, glues and adhesives, resins, certain insulation materials, etc. Formaldehyde can be produced industrially using air and methanol as raw materials in the presence of metal oxide catalyst or silver-based catalyst. The operating conditions and requirements of the process depend on the type of catalyst used. Therefore, a comparative study of both processes was conducted, and the results were compared. It was observed that the silver-based catalyst process has a compact plant size since the amount of air required is halved as compared to the metal oxide process. Thus, it appears that the silver-based catalyst process is more suitable for small-scale production due to its compact size and reduced utility cost.

Keywords: Formox Perstorp; formalin; fixed catalytic bed reactor; silver catalyst; metal oxide catalyst

1. Introduction

The extensively used formaldehyde is produced by using air and methanol as the raw materials. The reaction occurs in the reactor in the presence of a catalyst. The resulting products of the reaction are formaldehyde and water [1]. Then the mixture of products and unreacted reactants goes to the absorption column where water is showered from the top. The bottom product is formalin i.e., a 37% aqueous solution of formaldehyde [2]. The unreacted reaction mixture is removed from the top [1,3–7]. The extensive range of applications of formaldehyde makes it a valuable chemical. It may be used in different industries such as domestic, medical, cosmetics, and the textile industry [8–10].

The consumption and demand of formaldehyde is increasing. Formaldehyde is the principal component for the production of resins, phenols, urea, and melamine [11]. It is used for weather resistance i.e., in adhesives and wood coatings [12]. In addition, it has a disinfectant property; it is present in soaps as a disinfectant. In medical fields, formaldehyde is used for the sterilization of the surgical instruments. It imparts the resistance to fabric against crumples. In cosmetic products, formaldehyde is used as a preservative since it enhances the effectiveness of products against different microorganisms. It is used in glue production for household use. Formaldehyde is used in the manufacturing of plastics, carpets, and vaccines, etc. In plastic utensils industry, it is the major component [13].

Commercially, formaldehyde is produced mostly from air and methanol as raw materials using three different methods. In the first method, formaldehyde is produced using air and methanol in the presence of molybdenum oxide catalyst present inside the tubes of shell and tube reactor [14]. The reacting mixture enters at tube side to interact with catalyst forming the product [15].

$$CH_3OH + \frac{1}{2}O_2 \rightarrow HCHO + H_2O \quad \Delta H = -156 \text{ kJ} \tag{1}$$

The second method involves the production of formaldehyde in the presence of silver oxide catalyst present in fixed catalytic bed reactor [16].

$$\begin{aligned} CH_3OH + \tfrac{1}{2}O_2 &\rightarrow HCHO + H_2O \quad \Delta H_1 = -156 \text{ kJ} \\ CH_3OH &\rightarrow HCHO + H_2 \quad \Delta H_2 = 85 \text{ kJ} \end{aligned} \tag{2}$$

In third method, formaldehyde is produced using oxidation of methane and other hydrocarbons [17]. The separation processes and reaction mechanism in the above three methods are almost the same. For commercial production of formaldehyde, process optimization is required.

Lefferts et al. studied the production process of formaldehyde through oxidative hydrogenation of methanol in the presence of silver catalyst [18]. They studied the effect of temperature, gas velocity, and concentration of both reactants on the production process. They developed the reaction model based on the experimental data and explained the impact of form and composition of silver catalyst over methanol conversion. Yang et al. used molybdenum oxide catalyst supported over silica for the oxidation of methanol to formaldehyde [19]. Their study was based on the selectivity and activity of N_2O and O_2 used as oxidants. They observed that N_2O is responsible for the oxidation of carbon monoxide. Moreover, the supported molybdenum catalyst has higher activity than the non-supported catalyst. Qian et al. explained the formaldehyde synthesis process using polycrystalline silver catalyst [20]. They compared the water ballast process with the methanol ballast process and observed an increased selectivity of formaldehyde in the absence of water. Moreover, the selectivity of the product is highly temperature dependent. Moreover, Waterhouse et al. used SEM techniques to determine the relationship between morphology of silver catalyst and its performance [21].

In this study, a performance comparison of the industrially produced formaldehyde using two different catalysts is presented. Real-time industrial data are collected from a local industry in Pakistan, and material and energy balances, simulations, and cost analysis are executed. We have compared the two different catalysts based on material and energy balances, the size of the plant, the installation, and utility cost.

2. Materials and Methods

2.1. Material and Energy Balance

The steady-state material balance calculation is done for both processes. It was assumed that there was no accumulation in the system and there was no change with respect to time. The total amount of material in and out is almost the same. The basis for calculations were considered to be one day of operation i.e., 90 tons of formalin is produced in one day.

The energy balance calculations involve calculation of ΔH at both inlet and outlet sides. The reference temperature was 25 °C. Since the entering and leaving streams are mixtures of components, average heat capacity ($C_{p,avg}$) was calculated by the product of mole fraction (x_i) of the component in the mixture and its individual heat capacity (C_p). The change in temperature was calculated by the subtraction of the stream temperature with the reference temperature. The overall enthalpy content was found by taking the difference of the total heat content at the outlet and the inlet. The heat of the reaction was incorporated in the energy balance of the reactor and heat of condensation of formaldehyde was included in the energy balance around the absorption column.

$$\Delta H = mCp_{avg}\Delta T \tag{3}$$

2.2. Cost and Payback Period

The equipment and utility costs were calculated from Aspen Plus®. Then, the different incorporating factors like piping, instrumentation, buildings, and total physical plant cost were calculated. Then, total fixed capital was determined by incorporating some other factors like design and engineering, contractor's fee, and contingency. After that, working cost was calculated, which is 5% of total fixed capital. Then, the total investment was determined by adding working cost and total fixed capital.

$$\text{Total Investement} = \text{Working Capital} + \text{Total Fixed Cost} \tag{4}$$

Revenue was calculated by multiplying the amount of formalin produced by price of formalin.

$$\text{Revenue} = \text{Price of Formalin (kg)} \times \text{Formlain Produced (kg/year)} \tag{5}$$

The cost of raw material and catalyst was determined by multiplying the quantities (units) consumed per year by price per unit.

Then, profit was calculated by subtracting all the expenses from the revenue and payback period was determined.

$$\text{Payback Period} = \frac{\text{Total Investment}}{\text{Profit}} \tag{6}$$

3. ASPEN Flow Sheets

Based on real time industrial data, simulations are performed for the both processes using ASPEN PLUS® V8.8. Methanol, oxygen, nitrogen, silver, and molybdenum were selected as the components for simulation. For both the process, non-random two liquids (NRTL) was selected as the fluid package. The mixture of air and methanol is preheated and then sent to the reactor. The reaction mechanism that produces the formaldehyde depends upon the type of catalyst used. The product stream is sent to the adsorption column where it is treated with water to separate the desired product. The simulation models are explained in Sections 3.1 and 3.2.

3.1. Simulation Model of Molybdenum-Based Formaldehyde Process

Figure 1 presents the simulation of Molybdenum-based process on ASPEN PLUS. Non-random two-liquid model (NRTL) is used as the fluid package. The air and methanol are used as starting materials. Methanol and air both are initially mixed and passed to pre-heater where the temperature of the mixture increases from 27 °C to 107 °C. The mixture is passed to the reactor. The reactor is a shell and tube type reactor. The reaction mixture enters the reactor in catalyst-filled tube sides at 107 °C. The reaction occurs here and water and formaldehyde are produced. The shell side contains DTH (Dowtherm heat transfer media), which is used to extract the excess heat of the reaction i.e., the reaction is exothermic producing 159 KJ/mol of the energy. The boiling point of DTH is around 260 °C and same is the temperature inside the reactor, so the DTH leaves the reactor shell side as vapors. DTH goes to the condenser where the DTH vapors condense back to liquid state and accumulated in DTH tank. Upon requirement, DTH again goes to the reactor and the cycle continues. The mixture (product and unreacted reactants) leaving the tube side is at 280 °C. Since this mixture is at a very high temperature, so the heat of this mixture can be utilized. This mixture goes to the shell side of the pre-heater to heat the incoming reaction mixture. After pre-heating the incoming mixture, it leaves the shell side of preheater at 150 °C and goes to the absorption column from bottom side. Water at 37 °C is showered from the top of the absorption column. The amount of water showered is very critical as it produces the required concentration of the final product. The formalin is removed from the bottom. The temperature of the exiting product is 27 °C. The unreacted reaction mixture i.e., CH$_3$OH, O$_2$, N$_2$ is removed as off gas from top of absorption column at 23 °C.

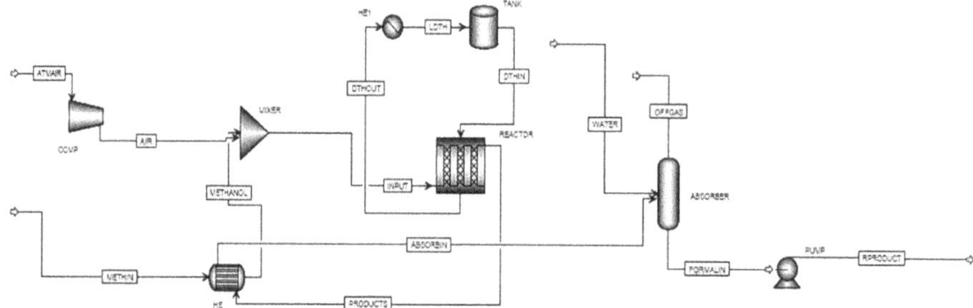

Figure 1. Simulation model of Molybdenum-based formaldehyde process.

3.2. Simulation Model of Silver-Based Formaldehyde Process

Figure 2 presents the ASPEN PLUS simulation of silver-based formaldehyde process. NRTL was selected as the fluid package. The reaction temperature for this process is around 600–650 °C. This process does not incorporate the DTH cycle. The air is passed through a compressor and then mixed with methanol using a mixer. The reacting mixer goes directly into the reactor from the mixer as a pre-heater is not used in this process. The reactor is a fixed bed catalyst type of reactor incorporating the bed of catalyst. The reaction occurs at the catalytic bed and quenching water is used at the bottom of reactor to cool down the product. The mixture of products and unreacted reactants enters the absorption column from bottom where water is showered from top. The unreacted reaction mixture is removed from the top as off gas, air is recycled from this off gas and again passed to the mixer while the product is removed from bottom. The product goes to the distillation column where separation takes place and we get formalin as a bottom product and methanol as a top product, which is then recycled.

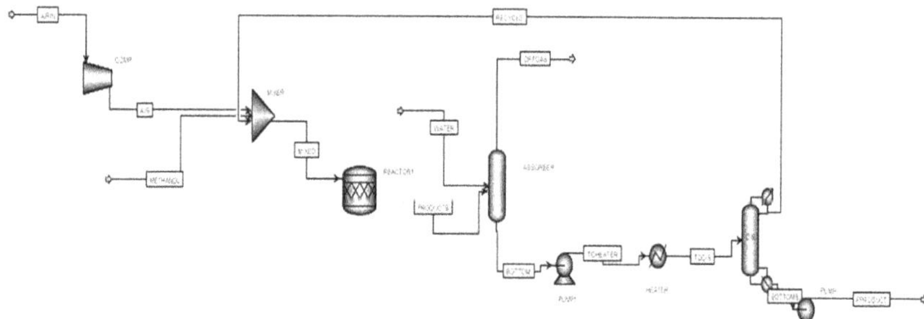

Figure 2. Simulation model of silver-based formaldehyde process.

4. Catalyst Properties

The properties of the molybdenum oxide and silver oxide catalyst used in the production process are shown in Table 1. It can be observed that the conversion of the molybdenum oxide-based plant is 99% and that of the silver oxide-based plant is 85%. The life of the molybdenum oxide catalyst is 12–18 months and that of the silver oxide-based process is 3–8 months. The porosity of the molybdenum oxide catalyst is 0.7 and that of the silver oxide catalyst is 0.5.

Table 1. Properties of catalyst.

Properties	Molybdenum Oxide	Silver Oxide
Methanol Conversion	99%	85%
Utilization time	12–18 months	3–8 months
Regeneration	Difficult	Convenient
Porosity	0.7	0.5

The life of the silver catalyst is highly dependent upon the operating conditions of the formaldehyde reactor. Silver catalysts in the formaldehyde plant are typically used for a period of a few months to a year depending on the reaction temperature and pressure [22]. Generally, sintering occurs in the silver catalyst due to reaction temperatures, which results in high pressure drop over the bed, which decreases the performance of catalyst. It is observed that the water ballast process prolongs the life of catalyst by introducing water with its associated high heat capacity that equally distributes the heat over the catalyst bed resulting in minimizing coke formation and sintering [23,24]. Our objective is the performance comparison of both processes based on the plant size, installation and utility cost, and material and energy balance.

5. Results and Discussion

The results in this section include the materials and energy balance of both plants, costing, payback period, and comparison on the basis of size for the two catalytic processes. Section 5.3 presents the discussions on the results.

5.1. Material and Energy Balance

The materials and energy balance sheets that were obtained from Aspen Plus® are given in Sections 5.1 and 5.2.

5.1.1. Material Balance of Molybdenum Oxide

The material balance of Molybdenum Oxide catalyzed process is shown in Table 2. The table shows the flow rates and densities of all of the components of the process. The methanol in inlet stream is 46.25 kmol/h. The flow rate of oxygen in the inlet air stream is 27.56 kmol/h. The formaldehyde in the product stream is 45.7875 kmol/h.

5.1.2. Energy Balance of Molybdenum Oxide

The energy balance of Molybdenum oxide process is shown in Table 3. The table shows the temperatures, pressures, and enthalpies of all the components of the process. The pressure varies from 1 bar to 1.8 bar during the process and temperature varies from 25 to 246.69 °C during the process. The input air has a temperature of 25 °C and the temperature of DTH going out of condenser is 246.69 °C. The enthalpy in reactor is −18,685.22 cal/mol. The highest pressure during the process is inside the absorption column, which is 1.8 bar.

5.1.3. Material Balance of Silver Oxide

The materials and energy balance of the silver oxide catalyzed process obtained from aspen plus is shown in Table 4. The table shows the material balance for silver oxide plant. The molar flow rate of oxygen in the air stream is 14.28 kmol/h. The molar flow rate of methanol in the inlet stream is 49.5 kmol/h. The formaldehyde in the product stream is 46.728 kmol/h.

Table 2. Material balance of Molybdenum-based process.

Stream ID	A2	AIR	DTHIN	DTHOUT	GH	INPUTAIR	LIQUID	METHANOL	METHIN	OFFGAS	PRODOUT	REACIN	TOTANK	WATER
From	ABSORBER	MIXER	HEATEX1	CONDENSR	EX2	COMP		MIXER	EX2		HEATEX1	REACTOR1	TANK	ABSORBER
To	EX2	COMP	TANK	HEATEX1	HEATEX1		ABSORBER	EX2		ABSORBER	REACTOR1	MIXER	CONDENSR	
Phase	MIXED	VAPOR	LIQUID	LIQUID	VAPOR	VAPOR	LIQUID	MIXED	LIQUID	VAPOR	VAPOR	VAPOR	LIQUID	LIQUID
METHA-01 (kmol/h)	0.4625	0	0	0	0.4625	0	0.4625	46.25	46.25	3.71×10^{-234}	0.4625	46.25	0	0
HYDRO-01 (kmol/h)	0	0	0	0	0	0	0	0	0	0	0	0	0	0
WATER (kmol/h)	45.787	0	0	0	45.787	0	129.787	0	0	2.79×10^{-7}	45.787	0	0	84.0
OXYGE-01 (kmol/h)	4.662	27.556	0	0	4.662	27.556	3.15×10^{-6}	0	0	4.662	4.662	27.556	0	0
FORMA-01 (kmol/h)	45.787	0	0	0	45.787	0	45.549	0	0	0.237	45.787	0	0	0
NITRO-01 (kmol/h)	103.664	103.664	0	0	103.664	103.664	4.25×10^{-5}	0	0	103.664	103.664	103.664	0	0
DIPHE-01 (kmol/h)	0	0	8.50	8.50	0	0	0	0	0	0	0	0	8.50	0
DIPHE-02 (kmol/h)	0	0	8.50	8.50	0	0	0	0	0	0	0	0	8.50	0
Total Flow (kmol/h)	200.364	131.220	17.0	17.0	200.364	131.220	175.80	46.25	46.25	108.564	200.364	177.470	17.0	84.0
Pressure (bar)	1.7	1.7	1	1	1.7	1	1.5	1	1	1.5	1.7	1.5	1	1.8
Vapor Frac	0.883	1	0	0	1	1	0	0.9143976	0	1	1	1	0	0
Liquid Frac	0.116	0	1	1	0	0	1	0.0856024	1	0	0	0	1	1
Solid Frac	0	0	0	0	0	0	0	0	0	0	0	0	0	0
Density (mol/cc)	6.86×10^{-5}	5.59×10^{-5}	6.38×10^{-3}	5.30×10^{-3}	4.83×10^{-5}	4.03×10^{-5}	0.0415	3.90×10^{-5}	0.0234	1.05×10^{-4}	3.57×10^{-5}	5.43×10^{-5}	6.38×10^{-3}	0.0551
Average MW	26.290	28.850	162.211	162.211	26.290	28.850	21.164	32.042	32.042	28.189	26.290	29.682	162.211	18.015
Liq Vol 60F (L/min)	140.703	117.131	43.664	43.664	140.703	117.131	69.122	31.091	31.091	96.851	140.703	148.222	43.664	25.270

Table 3. Energy balance of Molybdenum-based process.

Stream ID	A2	AIR	DTHIN	DTHOUT	GH	INPUTAIR	LIQUID	METHANOL	METHIN	OFFGAS	PRODOUT	REACIN	TOTANK	WATER
From	ABSORBER	MIXER	HEATEX1	CONDENSR	EX2	COMP		MIXER	EX2		HEATEX1	REACTOR1	TANK	ABSORBER
To	EX2	COMP	TANK	HEATEX1	HEATEX1		ABSORBER	EX2		ABSORBER	REACTOR1	MIXER	CONDENSR	
Phase	MIXED	VAPOR	LIQUID	LIQUID	VAPOR	VAPOR	LIQUID	MIXED	LIQUID	VAPOR	VAPOR	VAPOR	LIQUID	LIQUID
Temperature (°C)	63.995	92.459	40	246.69	150	25	31.936	64.20	60	45.263	300	58.928	40	25
Pressure (bar)	1.7	1.7	1	1	1.7	1	1.5	1	1	1.5	1.7	1.5	1	1.8
Enthalpy (cal/mol)	−21,715.29	471.044	13,700.96	28,142.45	−19,910.51	-1.87×10^{-13}	−60,235.33	−48,290.74	−56,109.41	−950.7649	−18,685.22	−12,236.58	13,700.96	−68,262.2

Table 4. Material balance of silver-based process.

Stream ID	AIR	ERT	FORMALIN	LIQUID	METHANOL	OFFGAS	OUT1	REACIN	RECYCLE	TOABSRBR	WATER	WATER1	WATER2
From	MIXER	EX		HEATER	MIXER		DIS	REACTOR1	MIXER	ABSORBER	ABSORBER	EX	
To		REACTOR1	DIS	ABSORBER		ABSORBER	HEATER	MIXER	DIS	EX			EX
Phase	VAPOR	VAPOR	LIQUID	LIQUID	VAPOR	VAPOR	VAPOR	MIXED	LIQUID	VAPOR	LIQUID	LIQUID	MIXED
METHA-01 (kmol/h)	0	6.307	5.996	6.307	49.50	8.36×10^{-222}	6.307	49.810	0.310	6.307	0	0	0
HYDRO-01 (kmol/h)	0	14.943	2.17×10^{-8}	2.54×10^{-8}	0	14.943	2.54×10^{-8}	3.66×10^{-9}	3.66×10^{-9}	14.943	0	0	0
WATER (kmol/h)	0	35.617	110.56	117.617	0	3.73×10^{-6}	117.617	7.057	7.057	35.617	82.0	65.0	65.0
OXYGE-01 (kmol/h)	14.28	0	0	0	0	0	0	14.28	0	0	0	0	0
FORMA-01 (kmol/h)	0	46.728	42.850	46.075	0	0.652	46.075	3.225	3.225	46.728	0	0	0
NITRO-01 (kmol/h)	53.72	53.72	1.53×10^{-5}	1.74×10^{-5}	0	53.719	1.74×10^{-5}	53.72	2.09×10^{-6}	53.72	0	0	0
Total Flow (kmol/h)	68	157.315	159.407	170.0	49.50	69.31554	170.0	128.092	10.592	157.315	82.0	65.0	65.0
Pressure (bar)	1.8	1.8	0.5	1.5	1.1	1.5	1	1.2	0.5	1.8	1.8	1	1
Vapor Frac	1	1	0	0	1	1	1	0.933	0	1	0	0	0.967
Liquid Frac	0	0	1	1	0	0	0	0.066	1	0	1	1	0.032
Solid Frac	0	0	0	0	0	0	0	0	0	0	0	0	0
Density (mol/cc)	7.26×10^{-5}	2.48×10^{-5}	0.039	0.037	3.20×10^{-5}	9.63×10^{-5}	3.22×10^{-5}	4.65×10^{-5}	0.039	5.51×10^{-5}	0.055	0.055	3.33×10^{-5}
Average MW	28.850	24.039	21.771	21.791	32.042	22.427	21.791	29.524	22.083	24.039	18.015	18.015	18.015
Liq Vol 60F (L/min)	60.698	106.782	65.294	69.733	33.276	61.717	69.733	98.414	4.439	106.782	24.668	19.554	19.554

5.1.4. Energy Balance of Silver Oxide

The energy balance of the Molybdenum oxide process is shown in Table 5. The table shows the temperatures, pressures, and enthalpies of all the components of the process. The pressure varies from 0.5 bar to 1.8 bar during the process and temperature varies from 25 °C to 600 °C during the process. The inlet gas is at 25 °C, whereas the reactor has a temperature of 600 °C. The enthalpy in the reactor is −19,619.55 cal/mol. The highest pressure during the process is in the absorption column, which is 1.8 bar.

5.2. Catalyst Performance

The comparison of both catalytic processes' performance is given in Table 6. It can be observed that the size of the Molybdenum Oxide plant is larger due to the greater amount of air requirement, whereas the size of silver oxide plant is lesser.

5.3. Comparsion of Molybdenum- and Silver-Based Processes

It can be observed from the results that the amount of methanol used was slightly different, but the amount of air varied significantly in the two processes. In the molybdenum oxide plant, the methanol was utilized at a rate of 46.25 kmol/h whereas it was utilized at a rate of 49.9 kmol/h in the silver-based plant. The amount of oxygen consumed in the silver-based plant was 27.56 kmol/h whereas the amount of oxygen required was at a rate of 15.12 kmol/h. Therefore, the lesser requirement oxygen in the silver-based plant lowers the utilities cost as well as decreasing the size of the equipment used. The capital cost in the case of silver is greater due to presence of an additional distillation column, however the size of overall plant is smaller though. The payback period of the silver-based plant is 2.8 years while it is 3.5 years for the molybdenum-based plant. The lesser payback period in the silver-based plant is due to a lower utility cost in the case of the silver-based process. The regeneration of silver catalyst is also possible. Therefore, from our results we can see that the silver plant has lower utility cost, shorter payback period, lower amount of oxygen required, and compact size as compared to molybdenum. Figures 3 and 4 show the comparison of the outcomes of the two processes for the formaldehyde formation i.e., Molybdenum-based and silver-based processes.

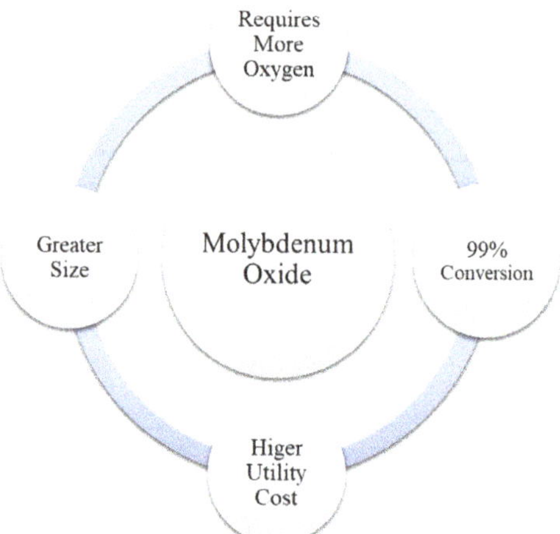

Figure 3. Molybdenum-based process.

Table 5. Energy balance of silver-based process.

Stream ID	AIR	ERT	FORMALIN	LIQUID	METHANOL	OFFGAS	OUT1	REACIN	RECYCLE	TOABSRBR	WATER	WATER1	WATER2
From	MIXER	EX		HEATER	MIXER		DIS	REACTOR1	MIXER	ABSORBER	ABSORBER	EX	
To		REACTOR1	DIS	ABSORBER		ABSORBER	HEATER	MIXER	DIS	EX			EX
Phase	VAPOR	VAPOR	LIQUID	LIQUID	VAPOR	VAPOR	VAPOR	MIXED	LIQUID	VAPOR	LIQUID	LIQUID	MIXED
Temperature (°C)	25	600	28.643	69.411	140	85.748	100	59.374	16.298	120	25	25	99.649
Pressure (bar)	1.8	1.8	0.5	1.5	1.1	1.5	1	1.2	0.5	1.8	1.8	1	1
Enthalpy (cal/mol)	-1.87×10^{-13}	−19,619.55	−59,483.02	−58,554.92	−46,668.7	−1078.707	−50,085.2	−22,892.57	−58,746	−24,078.11	−68,262.2	−68,262.2	−57,471.4

Table 6. Catalyst performance parameters.

Parameters	Molybdenum Oxide	Silver Oxide
Size	Larger	Compact
Capital Cost	$6.2 M	$7.1 M
Utility cost	$2.3 M/Year	$1.65 M/Year
Payback period	3.5 years	2.8 years

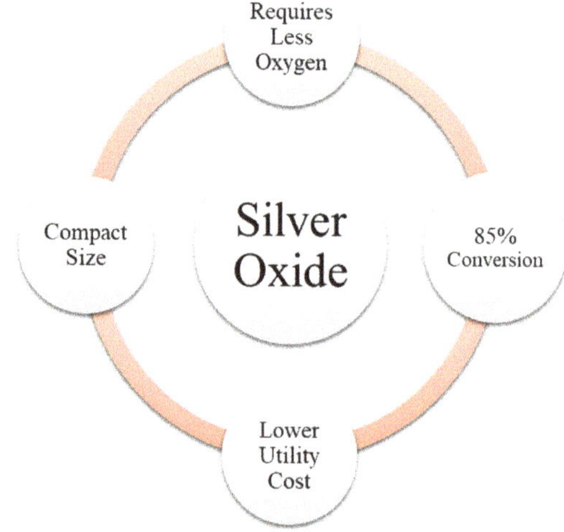

Figure 4. Silver oxide-based process.

6. Conclusions

The performance of two catalysts was studied simultaneously and the results were compared. The molybdenum-based plant has a larger plant size, high utility cost, but low cost of installation. Moreover, it has higher conversion of methanol to product i.e., 99%. The silver-based plant requires half the amount of air as compared to the molybdenum-based process, so it has compact plant size, the installation cost is high, but utility cost is low. However, the conversion is less in this process i.e., 75–85% as compared to the molybdenum-based process. Based on the comparative study and calculations, the silver-based process for formaldehyde production is better on an industrial scale.

Author Contributions: Conceptualization, S.R.N.; Data curation, K.S.; Formal analysis, S.A.A.T.; Funding acquisition, M.B.K.N.; Investigation, M.J.; Methodology, Y.M.; Project administration, S.R.N.; Software, S.A.A.T.; Supervision, S.R.N.; Validation, M.T.M.; Writing—Original draft, S.A.A.T., S.R.N.; Writing—review & editing, F.U., U.S. All authors have read and agreed to the published version of the manuscript.

Funding: This research received no external funding.

Acknowledgments: The authors would like to acknowledge the National University of Sciences and Technology (NUST), Pakistan.

Conflicts of Interest: The authors declare no conflicts of interest.

References

1. Webster, D.E.; Rouse, I.M. Production of Formaldehyde. U.S. Patant 4208353A, 17 June 1980.
2. Takashi, E.; Tamechika, Y.; Saburo, Y. Process for Preparing Formalin by Oxidation of Methanol. U.S. Patant 2908715A, 13 October 1959.
3. Meath, W.B. Production of Formaldehyde. U.S. Patant 2462413A, 22 February 1949.
4. Hall, J.L. Method for the Production of Formaldehyde. U.S. Patant 2384028A, 4 September 1945.

5. McClellan, W.R.; Stiles, A.B. Conversion of Methanol to Formaldehyde. U.S. Patant 3987107A, 19 October 1976.

6. Hoene, D.J. Formaldehyde Process. U.S. Patant 4343954A, 10 August 1982.

7. Hader, R.N.; Wallace, R.D.; McKinney, R.W. Formaldehyde from Methanol. *Ind. Eng. Chem.* **1952**, *44*, 1508–1518. [CrossRef]

8. Salthammer, T.; Mentese, S.; Marutzky, R. Formaldehyde in the Indoor Environment. *Chem. Rev.* **2010**, *110*, 2536–2572. [CrossRef] [PubMed]

9. Tang, X.; Bai, Y.; Duong, A.; Smith, M.T.; Li, L.; Zhang, L. Formaldehyde in China: Production, consumption, exposure levels, and health effects. *Environ. Int.* **2009**, *35*, 1210–1224. [CrossRef] [PubMed]

10. Abdollahi, M.; Hosseini, A. Formaldehyde. In *Encyclopedia of Toxicology*, 3rd ed.; Wexler, P., Ed.; Academic Press: Oxford, UK, 2014; pp. 653–656.

11. Dunky, M. Urea–formaldehyde (UF) adhesive resins for wood. *Int. J. Adhes. Adhes.* **1998**, *18*, 95–107. [CrossRef]

12. Dunky, M. Adhesives Based on Formaldehyde Condensation Resins. *Macromol. Symp.* **2004**, *217*, 417–430. [CrossRef]

13. Critical Review and Exergy Analysis of Formaldehyde Production Processes: Reviews in Chemical Engineering. 1 December 2014. Available online: https://www.degruyter.com/view/j/revce.ahead-of-print/revce-2014-0022/revce-2014-0022.xml (accessed on 26 April 2019).

14. Cheng, W.-H. Methanol and Formaldehyde Oxidation Study over Molybdenum Oxide. *J. Catal.* **1996**, *158*, 477–485. [CrossRef]

15. Adkins, H.; Peterson, W.R. The Oxidation of Methanol with Air Over Iron, Molybdenum, And Iron-Molybdenum Oxides. *J. Am. Chem. Soc.* **1931**, *53*, 1512–1520. [CrossRef]

16. Wachs, I.E.; Madix, R.J. The oxidation of methanol on a silver (110) catalyst. *Surf. Sci.* **1978**, *76*, 531–558. [CrossRef]

17. Brown, M.J.; Parkyns, N.D. Progress in the partial oxidation of methane to methanol and formaldehyde. *Catal. Today* **1991**, *8*, 305–335. [CrossRef]

18. Lefferts, L.; van Ommen, J.G.; Ross, J.R.H. The oxidative dehydrogenation of methanol to formaldehyde over silver catalysts in relation to the oxygen-silver interaction. *Appl. Catal.* **1986**, *23*, 385–402. [CrossRef]

19. Yang, T.-J.; Lunsford, J.H. Partial oxidation of methanol to formaldehyde over molybdenum oxide on silica. *J. Catal.* **1987**, *103*, 55–64. [CrossRef]

20. Qian, M.; Liauw, M.A.; Emig, G. Formaldehyde synthesis from methanol over silver catalysts. *Appl. Catal. Gen.* **2003**, *238*, 211–222. [CrossRef]

21. Waterhouse, G.I.N.; Bowmaker, G.A.; Metson, J.B. Influence of catalyst morphology on the performance of electrolytic silver catalysts for the partial oxidation of methanol to formaldehyde. *Appl. Catal. Gen.* **2004**, *266*, 257–273. [CrossRef]

22. Millar, G.J.; Collins, M. Industrial production of formaldehyde using polycrystalline silver catalyst. *Ind. Eng. Chem. Res.* **2017**, *56*, 9247–9265. [CrossRef]

23. Brenk, M. Silver Catalyst for Formaldehyde Preparation. U.S. Patent 8,471,071, 25 June 2013.

24. Heim, L.E.; Konnerth, H.; Prechtl, M.H. Future perspectives for formaldehyde: Pathways for reductive synthesis and energy storage. *Green Chem.* **2017**, *19*, 2347–2355. [CrossRef]

 *processes* 

Article

Simulation and Experimental Study of a Single Fixed-Bed Model of Nitrogen Gas Generator Working by Pressure Swing Adsorption

Pham Van Chinh [1], Nguyen Tuan Hieu [1], Vu Dinh Tien [2], Tan-Y Nguyen [3,*], Hoang Nam Nguyen [4], Ngo Thi Anh [5] and Do Van Thom [6,*]

[1] Institute of Technology—General Department of Defense Industry, Ha Noi 100000, Vietnam; Pvchinhhc95@gmail.com (P.V.C.); hieuhtip@gmail.com (N.T.H.)

[2] School of Chemical Engineering, Hanoi University of Science and Technology, Hanoi 100000, Vietnam; tien.vudinh@hust.edu.vn

[3] Faculty of Mechanical, Electrical, Electronic and Automotive Engineering, Nguyen Tat Thanh University, Ho Chi Minh City 700000, Vietnam

[4] Modeling Evolutionary Algorithms Simulation and Artificial Intelligence, Faculty of Electrical & Electronics Engineering, Ton Duc Thang University, Ho Chi Minh City 700000, Vietnam; nguyenhoangnam@tdtu.edu.vn

[5] KTHH09, School of Chemical Engineering, Hanoi University of Science and Technology, Hanoi 100000, Vietnam; anh.nt174442@sis.hust.edu.vn

[6] Department of Mechanics, Le Quy Don Technical University, Hanoi 100000, Vietnam

* Correspondence: nty@ntt.edu.vn (T.-Y.N.); thom.dovan.mta@gmail.com (D.V.T.)

Received: 15 August 2019; Accepted: 16 September 2019; Published: 25 September 2019

Abstract: Nitrogen is an inert gas available in the air and is widely used in industry and food storage technology. Commonly, it is separated by air refrigerant liquefaction and fractional distillation techniques based on different boiling temperatures of components in the mixed air. Currently, selective adsorption techniques by molecular sieve materials are studied and applied to separate gases based on their molecular size. In this paper, we simulate and investigate the effect parameters in a single fixed-bed model of a nitrogen gas generator using carbon molecular sieves, following pressure swing adsorption. This study aims to identify the effect of changing parameters so as to select the optimal working conditions of a single fixed-bed model, used as a basis for equipment optimization. This equipment was designed, manufactured, and installed at the Institute of Technology, General Department of Defense Industry, Vietnam to investigate, simulate, and optimize the industrial scale-up.

Keywords: pressure swing adsorption (PSA); carbon molecular sieve (CMS); adsorption; nitrogen; nitrogen generator

---

1. Introduction

In the literature, many studies on molecular sieve adsorption materials, the pressure swing adsorption process, and nitrogen gas generators have been published adopting diverse perspectives. Currently, many manufacturers all over the world create nitrogen gas generators to supply the open market. However, the results of simulations and experimental investigations into the technological parameters of a single fixed-bed device are very limited. Thus, most of these parameters are still known only by manufacturers due to the copyright of molecular sieve adsorption materials and equipment, without clarification via experiment [1].

In [2], a single-bed N_2 gas generator using pressure swing adsorption (PSA) was simulated and investigated to present the material equilibrium equation describing the change in concentration of substance adsorbed in the gas phase and solid phase over time upon changing the height of the bed. The thermal equilibrium equation described the change in temperature over time upon changing the height of the bed, as well as the capacity of heat adsorption, desorption, and transfer through the column wall. The Langmuir adsorption isotherm equation was predicted, whereas the momentum equation described the changing pressure and drop pressure according to the height of the bed and its head and boundary conditions. These equations could be solved using analytical, finite element, and numerical methods or commercial software such as Pascal, Fortran, Matlab, or Visual Basic when determining parameters such as porosity (ε), axial diffusion coefficient (DL), and velocity (u) in the column. However, verifying these models is very difficult because parameters such as concentration and temperature vary smoothly according to the height of the column, and their determination needs a high-precision measuring device. This article was not intended to solve these problems.

In the study conducted by Ashkan et al. [2], the development model was applied to N_2PSA systems, generating good results when compared to experimental and simulated models. The effects of flow, loading rate, cycle time, and column length on purity and product recall were investigated. The results showed that the N_2 purity decreased as the period of the cycle and the feed flow increased. However, upon increasing the feeding rate and bed length, the purity of the product also increased. All of the above conclusions were inversely proportional to the amount of recovery. It was observed that, during a defined cycle time, the effects of feed and discharge flow on purity and recovery were larger than the other parameters. The effects of drop pressure and non-isothermal conditions did not show a significant change.

Furthermore, in [2], there was no equation regarding time-varying pressure upon changing the height of the bed, which can be transformed via the material equilibrium equation due to a change in concentration in terms of it being regarded as an ideal gas. At the same time, the experiments determined some influencing factors such as feeding rate, purging rate, time, and height of the bed with regard to the purity of N_2. However, they did not study the effects of process parameters such as pressure, flow, and concentration of the bed or the influence of pressure, flow, drop pressure, pressurization time, adsorption time, and amount of O_2 gas absorbed at different pressures to find the optimal working regime of the bed. While the findings of other studies [1–14] proposed different mathematical models to simulate and examine the work cycle of the equipment, the use of a single bed as the basis for research cycles and equipment was not mentioned [15–19].

In Vietnam, many companies are willing to use this equipment for small- and medium-scale production, where it is not convenient to transport liquefied N_2. Therefore, the Institute of Technology, General Department of Defense Industry, Vietnam researched, designed, and manufactured an N_2 gas generator from open air using a CMS-240 carbon molecular sieve, implementing a pressure swing adsorption cycle to investigate, simulate, and optimize the industrial scale-up.

In this work, we solve the problems of building a single fixed-bed experimental model to study the unresolved problems in Reference [2], whereby we develop a model of pressure change over time according to the bed height, thus making it easier to determine the parameters according to the height of the bed with selected high-accuracy pressure sensors. This work investigates influencing factors such as the amount of O_2 gas adsorbed and the optimal working mode of the bed. This study is important to determine the mutual influence of technological parameters and to optimize the best working mode. Hence, the ultimate goal is to optimize the N_2 gas generator, allowing an industrial scale-up of the used equipment. The results of this study were used to verify other published theoretical and simulation studies using this equipment. Simulation and experimental studies of a single fixed-bed model to produce N_2 gas using a CMS-240 adsorption material and the PSA cycle were carried out. In particular, the authors built and simulated pressure as a function of the height of the column over time to find the optimal working mode of the bed in case of instability.

This paper is divided into five main sections. A brief introduction is given in Section 1. The model and theoretical basis are presented in Section 2. Calculations and simulations are presented in Section 3. Experimental results and discussions are given in Section 4. Finally, Section 5 concludes some of the remarkable results of this work.

2. The Model and Theoretical Basis

In this section, there are many symbols used to represent the mathematical, physical and chemical quantities, all their meanings can be seen detail in the Abbreviations.

2.1. The Model

Carbon molecular sieves (CMS) are basically a type of activated carbon, but the area of pore size distribution is very narrow, so it can selectively adsorb according to its molecular size (sieve, see Figure 1). The majority of carbon molecular sieves on the market today are made of anthracite coal with a tightly controlled activation process. The capillary structure can be modified by subsequent heat treatment process including cracking of hydrocarbons in the micropore and partial gasification under strict control conditions. Thus, the carbon molecular sieve obtained has an effective pore diameter ranging from 0.4 to 0.9 nm, but the porosity and adsorption capacity will be lower than that of conventional activated carbon. The largest application on an industrial scale is air separation. Surprisingly, surface oxidation by oxygen adsorption does not affect the efficiency of the separation process. It is also used in the process of hydrogen purification or the cleaning of gas mixtures [15–18].

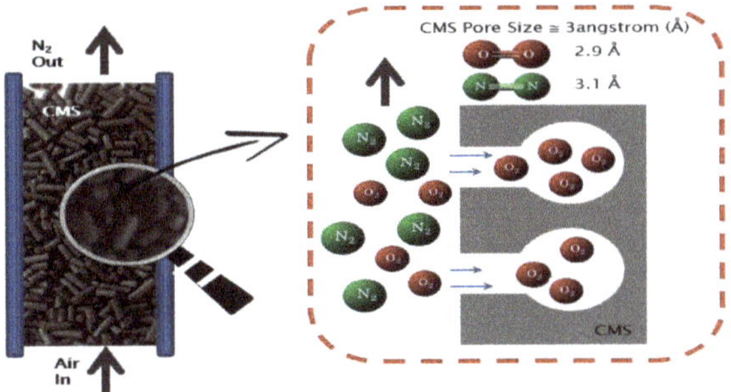

Figure 1. Selective adsorption mechanism of CMS material.

In this study, carbon molecular sieves (CMS-240) is selected to study the adsorption process in a column of N_2 gas generator working according to PSA pressure change cycle [15–18]. Material CMS-240 means that it can produce 240 m^3/h.1ton CMS (N_2 99.5%) at standard conditions. The capacity of the column is 14 L/minute N_2 99.5% at standard conditions.

The construction of a single fixed bed experimental model is shown in Figure 2, including the following main equipment:

- F1-primary filter; C1—piston compressor; T01—compressed air tank; D1—column of silica gel to remove water steam; F2—secondary filter;
- B1-adsorption column (D × H = 102 × 950 mm) contains 3.5 kg CMS-240; T02-N_2 gas product tank;
- FM01—mass flow sensor to measure gas feed; FM02—mass flow sensor to measure N_2 gas product with the purpose of measuring the change of flow over time and determine the amount of adsorption is an important parameter showing the adsorption capacity of the material.
- CT02—nitrogen gas concentration sensor to measure nitrogen gas concentration output with the purpose of measuring the change in gas concentration N_2 at the output of the column over time and pressurization time, adsorption time, discharge pressure time, and desorption time.
- TT01, TT02—temperature sensors to measure the input/output temperature of a single fixed bed model to determine the temperature difference through the column.
- PT-01/06: six pressure sensors installed according to the bed height (10 cm/1 sensor), and the purpose is to measure the pressure change according to the height of the bed and over time. These sensors are deeply inserted into the center of the bed to measure the different pressure between absorbent layers (fitting at the edge of the column will be no difference in pressure).

- The V1 and V7 control valves (on/off; normal close), according to the settings and measurement parameters, are transmitted to S7-1200 PLC computer and recorded on the computer with control monitoring by SCADA interface programmed by WinCC V14.

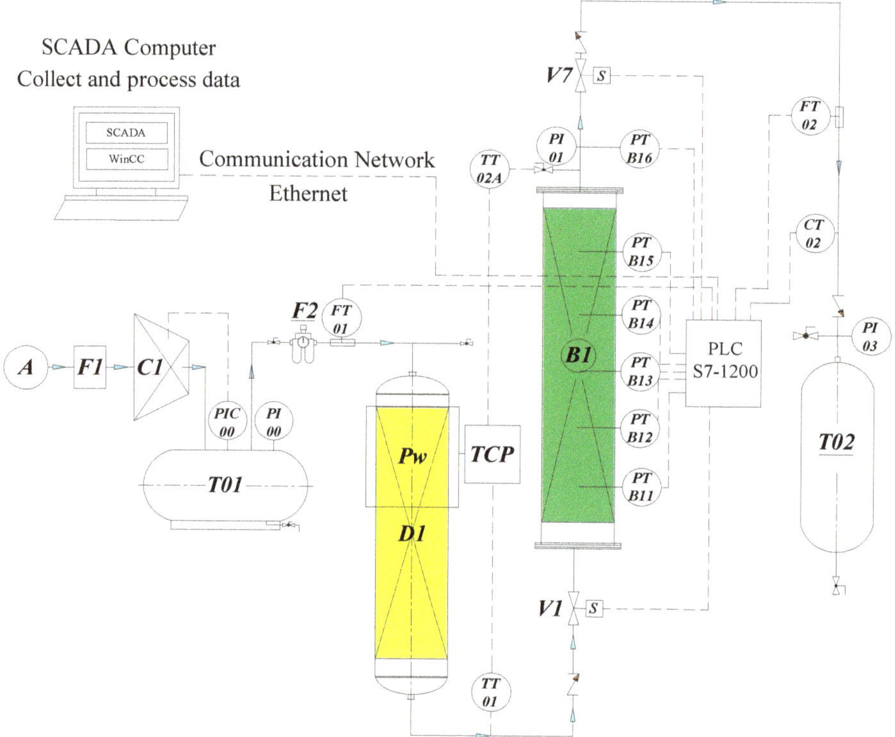

Figure 2. A single bed model.

The real model of this study is shown in Figure 3. To operate this equipment as a single fixed bed model, herein, we use two valves V1 and V7, and the other valves are completely closed.

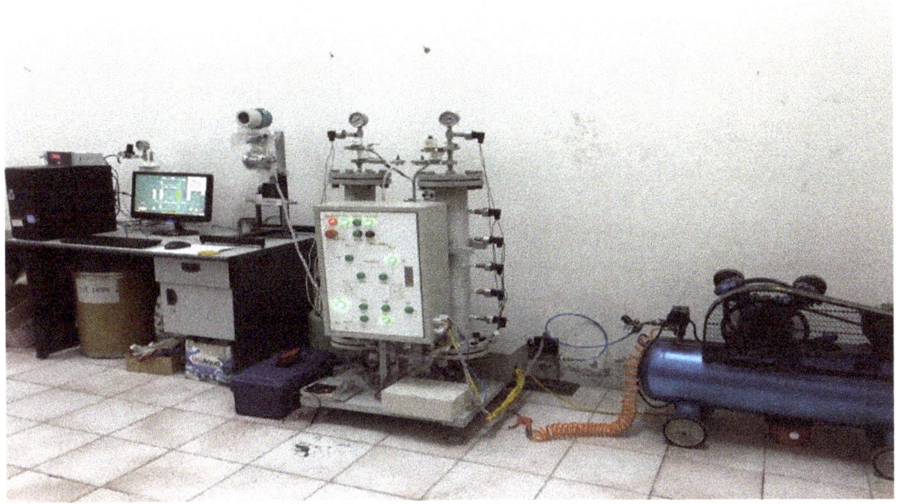

Figure 3. The experimental equipment system.

2.2. The Theoretical Basis

Research on materials and calculations, design and manufacture of laboratory equipment model, and the studied mathematical can be seen in [1–14], which are based on the balance of materials, energy, and momentum equations to find equations to change pressure over time and the height of the bed by analytic transformations. However, to study the rule of changing the technological parameters of a single bed in the equipment, it is necessary to implement the experimental plan as follows:

+ Survey 1 column with different pressures to achieve the highest concentration of N_2 gas with hypothetical pressure, adsorption and desorption time to determine the real-time parameters and best working conditions of the column.
+ Determining the optimal working mode of the column.

Now we have some governing equations as follows

- First, the material equilibrium equation describing the change of the adsorbent concentration (O_2) changes over time and according to the height of the column is expressed as:

$$\frac{\partial C_i}{\partial t} - D_L \frac{\partial^2 C_i}{\partial z^2} + \frac{\partial (C_i u)}{\partial z} + \rho_p \left(\frac{1-\varepsilon}{\varepsilon} \right) \frac{\overline{\partial q_i}}{\partial t} = 0 \tag{1}$$

Equation (1) can be completely solved by Matlab to simulate the change of the concentration according to the height of the column when determining the porosity parameters (ε), axial diffusion coefficient (D_L) and speed apparent (u) by analysis, calculation, and experiment. But testing by experiment is difficult because it is hard to install a device that measures the exact concentration according to the height of the column.

For ideal gas ($C_i = y_i P/RT$), Equation (1) is transformed into following forms:

$$-D_L \frac{\partial^2 C_i}{\partial z^2} + y_i \frac{\partial u}{\partial z} + u \left(\frac{\partial y_i}{\partial z} + y_i \left(\frac{1}{p} \frac{\partial P}{\partial z} - \frac{\partial T}{\partial z} \right) \right) \frac{\partial y_i}{\partial z} + \frac{\partial y_i}{\partial t}$$
$$+ y_i \left(\frac{1}{p} \frac{\partial P}{\partial z} - \frac{1}{T} \frac{\partial T}{\partial z} \right) + \left(\frac{\rho_p RT}{P} \right) \left(\frac{1-\varepsilon}{\varepsilon} \right) \frac{\overline{\partial q_i}}{\partial t} = 0 \tag{2}$$

$$\frac{\partial P}{\partial t} + P\frac{\partial u}{\partial z} + u\frac{\partial P}{\partial t} + PT\left(\frac{\partial}{\partial t}\left(\frac{1}{T}\right) - u\frac{\partial}{\partial z}\left(\frac{1}{T}\right)\right) - 2D_L R\frac{\partial P}{\partial z}\frac{\partial}{\partial z}\left(\frac{1}{T}\right) + \rho_p RT\left(\frac{1-\varepsilon}{\varepsilon}\right)\sum_{i=1}^{n}\frac{\overline{\partial q_i}}{\partial t} = 0 \qquad (3)$$

Equations (2) and (3) describe the change of concentration, pressure, and temperature over time and the height of the column. Solving these equations is very complex; it is often considered that the process is isothermal (for the PSA cycle) and isometric (for the TSA cycle).

In this work, the studied column working under the PSA cycle is considered as an isothermal adsorption process. In order to simulate and verify this model, we must build an equation that describes the change of pressure over time and the height of the column. Respond to the empirical model developed to measure the pressure according to the height of the column.

Results of the analytic changes from changes in concentration to pressure changes as follows

For adsorption process we have:

$$\frac{\partial p_i}{\partial t}\left(1 + \frac{1-\varepsilon}{\varepsilon}K\right) = -\varepsilon.u.\frac{\partial p_i}{\partial z} + \varepsilon.D_L.\frac{\partial^2 p_i}{\partial z^2} \qquad (4)$$

For desorption process we have:

$$\frac{\partial p_{des}}{\partial t}\left(1 + \frac{1-\varepsilon}{\varepsilon}K\right) = -u.\frac{\partial p_{des}}{\partial z} + D_{des}.\frac{\partial^2 p_{des}}{\partial z^2} \qquad (5)$$

Equations (4) and (5) describe the change of pressure over time and according to the height of the column, which depend on the adsorption constant, porosity, axial diffusion coefficient, and speed. These parameters are determined by analysis, calculation, and experiment. This equation will be solved by Matlab. Equation (5) is one case of Equation (4) when porosity $\varepsilon = 1$ in the case of desorption.

- Thermal equilibrium equation is expressed as follows:

$$-K_i\frac{\partial^2 T}{\partial z^2} + \varepsilon\rho_\varepsilon C_{p,g}\left(u\frac{\partial T}{\partial z} + T.\frac{\partial u}{\partial z}\right) + \left(\varepsilon_t\rho_g C_{p,g} + \rho_b C_{p,s}\right)\frac{\partial T}{\partial t} - \rho_b\sum_{i=1}^{n}\left(\frac{\overline{\partial q_i}}{\partial t}(\Delta H_i)\right) + \frac{2h_i}{R_{Bi}}(T - T_w) = 0 \quad (6)$$

Loss of heat through the column wall is defined as

$$\rho_w C_{p,w} A_w\frac{\partial T_{i,w}}{\partial t} = 2\pi R_{Bi}h_i(T - T_w) - 2\pi R_{Bo}h_o(T_w - T_{atm}) \qquad (7)$$

where $A_w = \pi(R^2_{B,o} - R^2_{B,i})$.

The amount of heat expected during the adsorption process is very small, the heat generated during the adsorption process and the steaming process completely passes through the column wall. Herein, the temperature of the column will change insignificantly over time and along the height of the column, so the process is considered as an isothermal state. Then, Equations (6) and (7) are only theoretical and not experimental.

- The momentum equilibrium equation according to Eugrun's is expressed as follows

$$-\frac{dP}{dZ} = a.\mu.u + b.r.u.|u|; a = \frac{150(1-\varepsilon)^2}{4R_p^2\varepsilon^2}; b = \frac{1.75(1-\varepsilon)}{2R_p\varepsilon} \qquad (8)$$

Equation (8) describes the change of pressure according to the height of the column, in addition, it depends not only on parameters such as particle size and porosity, but also it depends on the velocity of the gas flow through the column. In the theoretical calculation, it is usually taken at a constant speed however, in practice, this experimental speed can change up to 20% caused by O_2 being adsorbed. So, the error between theory data and experiment data can reach 20%.

- The Langmuir–Freundlich adsorption equilibrium equation is predicted as

$$q_i^* = \frac{B_i q_{mi} P_i^{ni}}{1 + \sum\limits_{j=1}^{n} B_j P_j^{ni}}$$

(9)

in which

$$q_{mi} = K_1 + K_2 \cdot T; \; B_i = K_3 \exp\left(\frac{K_4}{T}\right); \; n_i = \frac{k_5 + k_6}{T} \frac{\overline{\partial q_i}}{\partial z} = \omega_i\left(q_i^* - \overline{q_i}\right); \; \omega_i = \frac{15 D_{ei}}{r_c^2} = CP_r^5(1 + B_i P_i)^2$$

Equation (9) describes the adsorption load depending on the equilibrium constant, the equilibrium adsorption load and the pressure, temperature. These parameters can be determined experimentally and calculated in the case of isothermal.

In this work, the simulation of the change of pressure according to the height of the column and with the Equations (4) and (5) are established. We must determine the parameters of the model such as porosity, axial diffusion coefficient, and speed to find the solution of the above equations with the initial conditions and given boundary conditions.

For the adsorption process, Equation (4) has some fundamental conditions as follows:

+ Initial condition:

The partial pressure at $(z = 0, t = 0)$ is $p(z, t = 0) = 0$.

+ Boundary conditions: According to Danckwerts standards, we have:

The partial pressure at all times in input position and output of adsorption column are: at $z = z$;

$$p(z = 0, t) = \begin{cases} p^d + \frac{D_L}{u_c} \frac{\partial p}{\partial z}\Big|_{z=0,t}; & 0 \le t \le t^c \\ + \frac{D_L}{u_c} \frac{\partial p}{\partial z}\Big|_{z=0,t}; & t \ge t^c \end{cases}$$

where p^d is the partial pressure of the adsorbed material in the adsorption column at $z = L$;

$$\frac{\partial p}{\partial t}\Big|_{x=L} = 0$$

where t^c is the duration of the adsorption cycle.

For the desorption process, Equation (5) has some fundamental conditions as follows:

+ Initial condition:

The partial pressure at $(z = 0, t = 0)$ is:

$$p(z, t = 0) = p_{des}^o$$

+ Boundary conditions: According to Danckwerts standards, we have:

The partial pressure at all times in input position and output of adsorption column are: at $z = z$;

$$p(z = 0, t) = \begin{cases} p_{des}^d + \frac{D_L}{u_{des}} \frac{\partial p_{des}}{\partial z}\Big|_{z=0,t}; & 0 \le t \le t^c \\ p_{des}^c; & t \ge t^c \end{cases}$$

where p^d is the partial pressure of the adsorbed material in the adsorption column at $x = L$;

$$\left.\frac{\partial p_{\text{des}}}{\partial t}\right|_{z=L} = 0$$

where t^c is the duration of the adsorption cycle.

In this study, we will not simulate the adsorption process. Because when the simulation time is set, this process is very similar to the adsorption process.

3. Calculation and Simulation Results

Calculation to determines the porosity of the column [16–19] includes some steps as follows

+ Determining bulk density $\rho_b = 0.676$ g/cm^3, particle density $\rho_p = 0.78$ g/cm^3 and solid density $\rho_s = 2.174$ g/cm^3 by means of volume, mass weighing, and pycnometer.

The porosity of the column is determined in the following Table 1.

Table 1. Results of calculation porosity of a single fixed bed.

1	Porosity outside or between particles	ε_i	m^3/m^3	$\varepsilon_i = \frac{V_X}{V_T} = 1 - \frac{\rho_b}{\rho_p}$	0.1326
2	Porosity in the grain	ε_p	m^3/m^3	$\varepsilon_p = \frac{V_p}{V_T-V_X} = 1 - \frac{\rho_b}{\rho_s}$	0.556
3	Total porosity	ε_t	m^3/m^3	$\varepsilon_t = \frac{V_p+V_X}{V_T}$ $\varepsilon_t = \varepsilon_i + \varepsilon_p(1 - \varepsilon_i)$	0.615

+ Determining the air velocity through the column [16–19] can be expressed as:

$$u_c = \frac{V_{kk}}{\varepsilon_t \frac{\pi D^2}{4}} \tag{10}$$

When the total pressure reaches the critical pressure, the free step of the gas molecule is approximately the capillary size. Diffusion follows both the Knusen diffusion and molecular diffusion mechanisms:

+ Diffuse in transition zone [16–19] is:

$$\frac{1}{D_L} = \frac{1}{D_{AB}} + \frac{1}{D_k} \tag{11}$$

+ Molecular diffusion [16–19] is:

$$D_{AB} = 0.00158 \frac{T^{3/2}\left(\frac{1}{M_A} + \frac{1}{M_B}\right)^{1/2}}{P \cdot \sigma_{AB}^2 \Omega_{AB}} \tag{12}$$

+ Flow diffusion (Knusen) [16–19] is:

$$D_K = 9700 \cdot R_p \cdot \left(\frac{T}{M}\right)^{1/2} \tag{13}$$

From (11)–(13) we determine the basic parameters of the models (4) and (5) following pressure and temperature:

From the results from Tables 1 and 2 we can simulate adsorption column according to the change of partial pressure of adsorbed material over time and in column height in case of unstable working column V1, V2 are open, the installation time is 460 s to observe the maximum capacity of the column at a pressure from 1 bar to 8 bar. In this paper we only give results at pressures of 5 bar, 5.5 bar and 8 bar for discussion.

Table 2. Results of calculation parameters of a single fixed bed model at different pressure.

Order	Pressure	P	bar	1	2	3	4	5	6	7	8
1	Gas velocity in the column	u_c	m/s	0.058	0.038	0.029	0.023	0.019	0.016	0.014	0.013
2	Temperature	T	K	298	298	298	298	298	298	298	298
3	Pore radius	R_p	10^{-8} cm	2	2	2	2	2	2	2	2
4	Molecular weight	$M(O_2)$	kg/kmol	32	32	32	32	32	32	32	32
5	Flow diffusion coefficient	D_k	10^{-6} cm^2/s	592	592	592	592	592	592	592	592
6	Molecular diffusion coefficient	D_{AB}	10^{-6} cm^2/s	8.6	5.8	4.3	3.5	2.9	2.5	2.2	1.92
7	Axial diffusion coefficient	D_L	10^{-6} cm^2/s	8.5	5.7	4.29	3.44	2.87	2.46	2.16	1.92
8	Equilibrium constant O_2	K	-	9.25	9.25	9.25	9.25	9.25	9.25	9.25	9.25

+   Simulation (4):

By simulating results at pressure of 5 bar by Matlab with initial conditions and boundary conditions above, we have the partial pressure of O_2 initially of 1 bar (corresponding to 20% of the mole of the initial 5 bar gas mixture) with t = 460 s:

Figure 4 shows that the partial pressure of the adsorbent (O_2) decreases over time and according to height of the column corresponding to the gradual reduction of the adsorbed concentration (O_2). This means that the concentration of N_2 increases gradually at the output of the column.

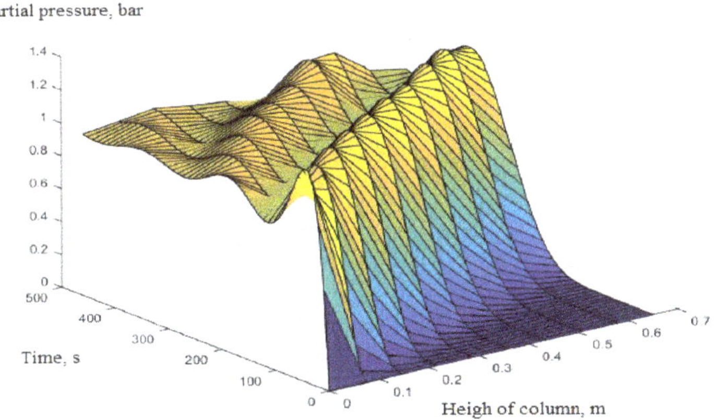

Figure 4. The result of partial pressure (O_2) following over time and height of column at 5 bar.

The following Figures 5 and 6 are the 2D sections of Figure 4 at the time t = 60 s and at the height of column h = 0.65 m.

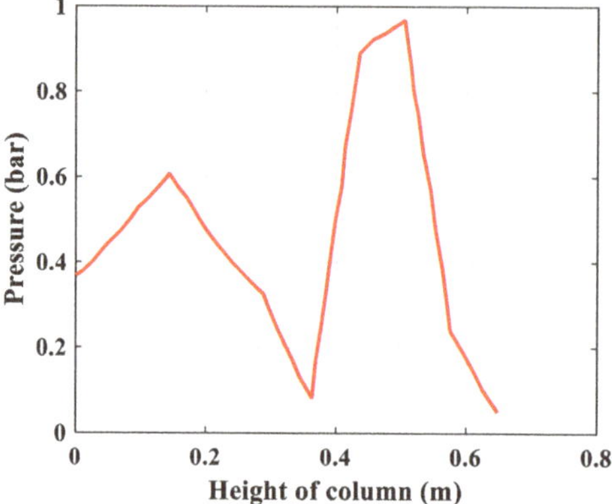

Figure 5. The result of partial pressure (O$_2$) following height of column at 5 bar.

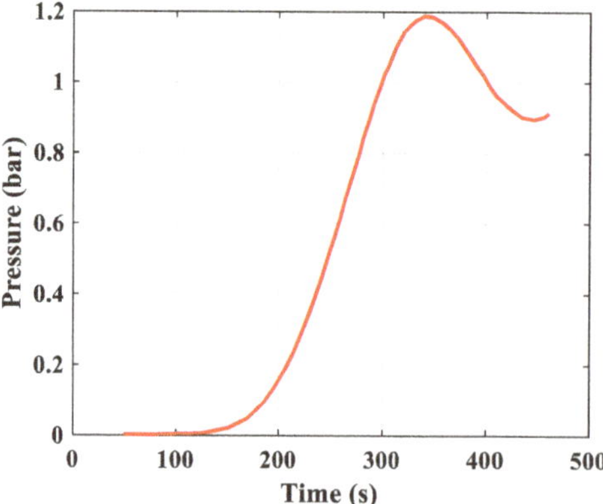

Figure 6. The result of partial pressure (O$_2$) following over time at 5 bar.

Figure 5 shows that the partial pressure 0.91 bar of the adsorbent is reduced to the lowest at 0.65 m height (the end of the column), and the corresponding pressure loss through the column is approximately 0.1 bar.

Figure 6 shows that the partial pressure starts to increase only after about 30 s. It means that at that time the column is saturated, the adsorption capacity decreases. Pressure at 460 s time decreases by approximately 0.1 bar.

Next, by simulation results at maximum pressure of 5.5 barby MATLAB with initial conditions and boundary conditions above, we have the partial pressure of O$_2$ initially of 1.1 bar (corresponding to 20% of the mole of the initial 8 bar gas mixture) with t = 460 s.

The following Figures 8 and 9 are the 2D sections of Figure 7 at the time t = 60 s and at the height of column h = 0.65 m.

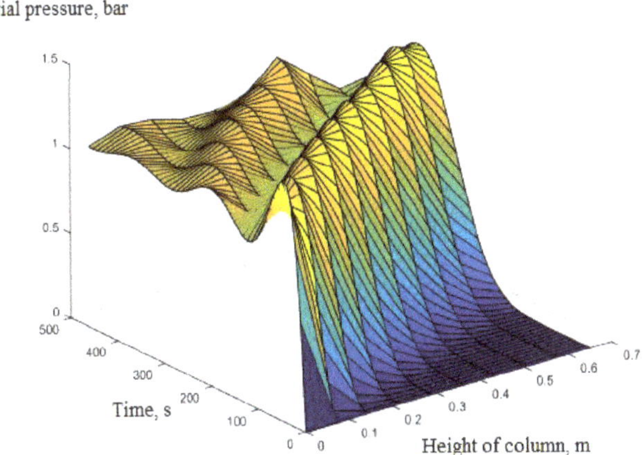

Figure 7. The result of partial pressure (O_2) following over time and height of column at 5.5 bar.

Similar to the case of 5 bar, but the partial pressure decreases with time and the column height is faster, Figure 7 almost likes Figure 4.

Figure 8 shows that the partial pressure of the adsorbent (O_2) decreases signifficantly at the output. This may be the optimal working point of the model (4). Drop pressure through column is 0.148 bar.

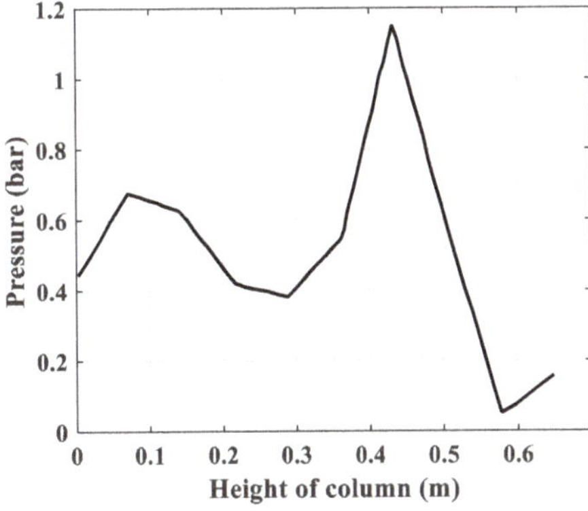

Figure 8. The result of partial pressure (O_2) following height of column at 5.5 bar.

Figure 9 shows that partial pressure starts to increase after about 25 s. It means that when the column is saturated, the adsorption capacity decreases. The pressure at 460 s has increased by approximately 0 bar. That means the partial pressure (O_2) at the output of the column is (O_2) air feed. This confirms again that the pressure of 5.5 bar is the optimal value. Adsorption time is about 25 s.

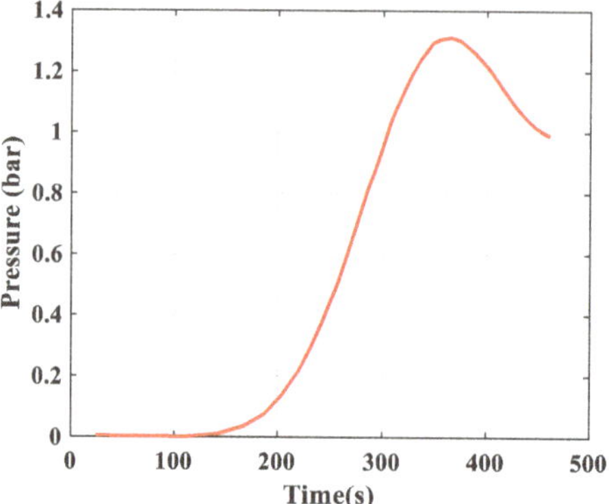

Figure 9. The result of partial pressure (O$_2$) following over time at 5.5 bar.

For simulation results (by Matlab) at maximum pressure of 8 bar with initial conditions and boundary conditions, the partial pressure of initial O$_2$ gas is 1.6 bar (corresponding to 20% of the mole of the initial 8 bar gas mixture) at t = 460 s.

Similar to the case of 5 bar, but the partial pressure decreases with time and the column height is faster, Figure 10 indicates a larger slope.

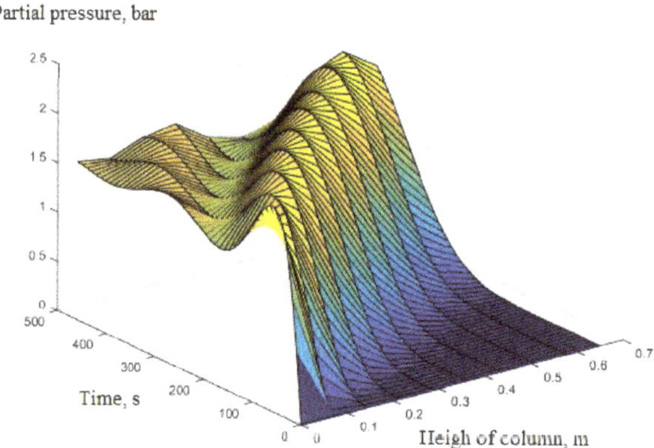

Figure 10. The result of partial pressure (O$_2$) following over time and height of column at 8 bar.

The following Figures 11 and 12 are the 2D sections of Figure 10 at the time t = 60 s and at the height of column h = 0.65 m.

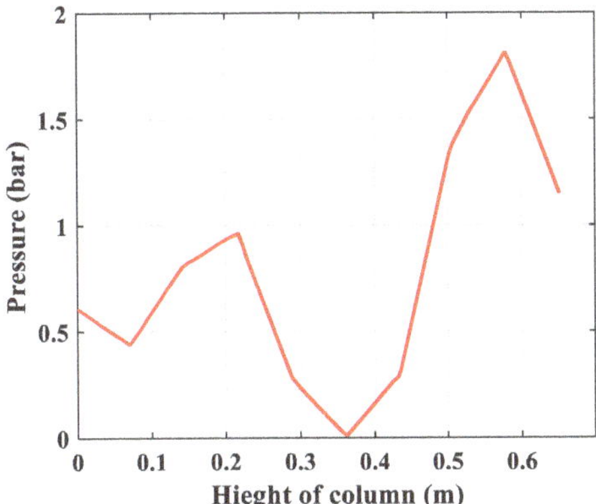

Figure 11. The result of partial pressure (O_2) following height of column at 8 bar.

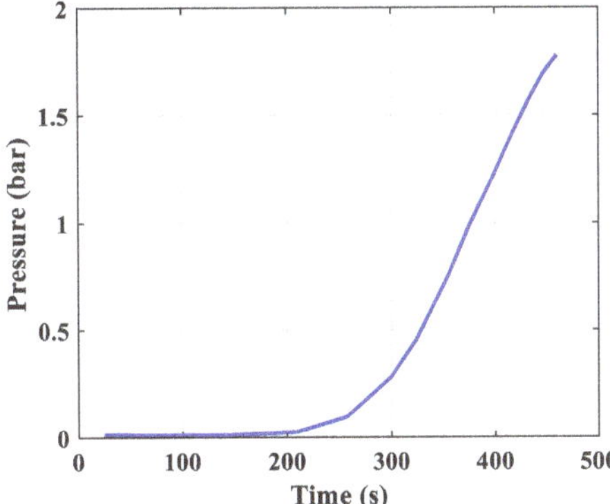

Figure 12. The result of partial pressure (O_2) following over time at 8 bar.

Figure 11 shows that the partial pressure 1.4 bar of the adsorbent is reduced to the lowest at 0.36 m of height (the middle of the column), and the corresponding pressure loss through the column is approximately 0.2 bar. At the end of the column, the partial pressure (O_2) is increased by desorption.

Figure 12 shows that partial pressure starts to increase after about 25 s. It also means that when the column is saturated, the adsorption capacity decreases. The pressure has increased by approximately 0.2 bar at the time t = 460 s. That means the partial pressure (O_2) at the output of the column includes (O_2) air feed and (O_2) desorption.

Drop pressure adsorption and desorption processes are expressed as follow

$$\frac{\Delta P_{hp}}{H} = \frac{150 \cdot \mu (1 - \varepsilon_t)^2}{(d_p \psi)^2 \varepsilon_t^3} v_{hp} + \frac{1.75 \cdot (1 - \varepsilon_t) \rho_g}{\varepsilon_t^3 (d_p \psi)} v_{hp}^2 \tag{14}$$

Drop pressure through the particle layer is expressed as

$$\Delta P_{CMS(ta)} = \lambda_h \frac{2 \cdot H_{CMS}}{d_o} \cdot \frac{\rho \cdot v_{ta}^2}{2} \tag{15}$$

Total drop pressure through a bed is defined as

$$\Delta P_{T(hp)} = \lambda_h \frac{2 \cdot H_{CMS}}{d_o} \cdot \frac{\rho \cdot v_{ta}^2}{2} + \left[\frac{150 \cdot \mu (1 - \varepsilon_t)^2}{(d_p \psi)^2 \varepsilon_t^3} v_{hp} + \frac{1.75 \cdot (1 - \varepsilon_t) \rho_g}{\varepsilon_t^3 (d_p \psi)} v_{hp}^2 \right] \cdot H \tag{16}$$

The calculated results of drop pressure at 5 bar, 5.5 bar, and 8 bar are, respectively, 0.2 bar, 0.16 bar, and 0.14 bar. This calculation result has a slight difference compared with the simulation. Both of these results are very good data to reference with the experimental results given below.

4. Experiment Results and Discussions

To verify the calculation and simulation results above, we conduct the experiment set up according to Figure 2 (the data Table 3 below). The investigation of a single fixed bed from 1 bar to 8 bar is carried out and we assume an adsorption time of 60 s and a desorption time of 400 s to observe the real adsorption and desorption process on real-time graphs and determine the adsorption and desorption time.

Table 3. Set up parameters for the experimental process from 1 bar to 8 bar.

Time, Pressure Set Up	T = 460 s; P = 1 to 8 Bar	
Valve status	$T_1 = 60$ s	$T_2 = 400$ s
V_1	ON	OFF
V_7	ON	ON

According to the principle of technological parameters in the operation such as temperature, pressure, flow, and gas concentration N_2 we can observe the rule when surveying the column at 1 bar pressure. The data resolution is drawn on the graph with higher quality.

+ Temperature: the temperature is considered constant
+ Pressure: The pressure changes over time and according to the height of the column. The experimental results of the pressure changes over time and the pressure distribution of the column height (with 6 sensors CB-1.1 to CB-1.6) are clearly seen in Figure 13 when surveying one column according to setup mode in Table 3 at 1 bar pressure. At the higher pressures, the lower resolution makes it more difficult to see.

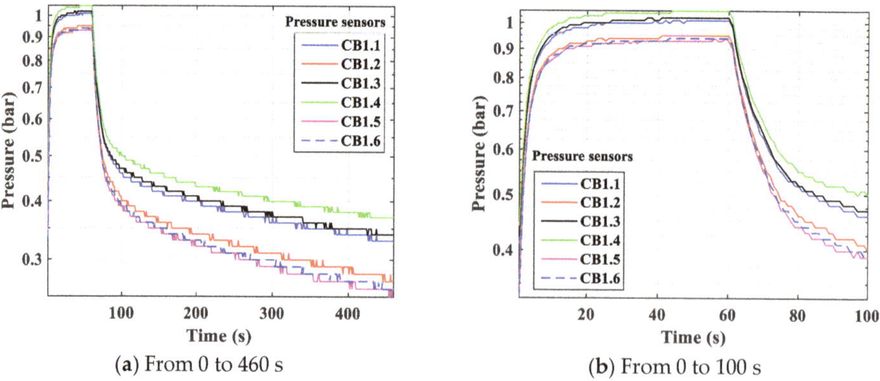

Figure 13. Pressure over time and heigh of a single fixed bed (see Figure 2).

From Figure 13, we see the rule of pressure change in a column over time an adsorption cycle set. One column adsorption cycle includes pressurization time, adsorption time, pressure release time and desorption time. Observing the above graphs shows the form of the pressure line: straight-line pressurization phase with slope coefficient >0 pressure increased rapidly over time, the adsorption phase of the convex curve type slightly increased to constant, straight-line pressure release phase with slope coefficient <0 pressure rapid decrease over time, the desorption phase form concave curve pressure decreases slowly under low pressure and desorption along time. From this figure, we can determine the drop pressure through the column. Figure 13 shows the two-dimensional (2D) section of the change and distribution of the total pressure over time and the height of the column; it is similar to the three-dimensional (3D) simulation of partial pressure presented in Figures 4, 7 and 10, which are processed by MATLAB according to Equation (4).

+   Mass flow input/output: Experimental results of input/output flow stream are obtained by 02 sensors FM1, FM2 (slm) referring to standard conditions. The experimental results at 1 bar pressure are observed in Figure 14 (the blue line is the flow from FM1, the red line is the flow from FM2). The difference between the two lines can determine the adsorption and desorption processes of the column

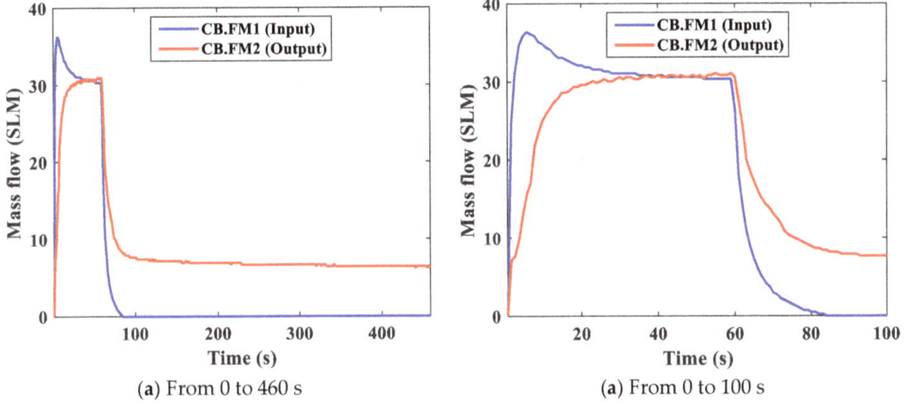

(a) From 0 to 460 s   (a) From 0 to 100 s

Figure 14. Mass flow input/output of a single fixed bed (see Figure 2).

Figure 15, we can see the changing law N_2 gas and O_2 gas concentration at the output of the column over time, here we can determine the pressurization time, adsorption time, pressure release time, desorption time, and the change in concentration according to pressure until saturation. From this image, we can determine the amount of adsorption.

+   Concentration of gas N_2 at the output of the column: Experimental results of concentration N_2 and O_2 obtained by CT-02 sensor are observed in Figure 15 at 1 bar pressure. The highest concentration of N2 reached 82.6%.

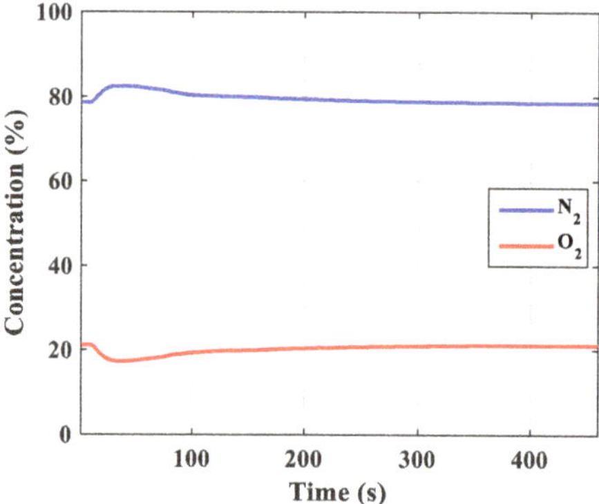

Figure 15. Concentration of gas N_2 at the output of a single fixed bed (see Figure 2).

Figure 15 shows clearly that the concentration of N_2 gas changes over time at the output of the column. The time of hypertension is the time when the concentration does not change, the adsorption time is the time of increasing concentration of N_2 gas, the pressure drop time is constant high concentration-time (this is the time to take reasonable products), and the time of sorption release is the time that N_2 concentration decreases.

Combining Figures 13–15, we can completely determine the parameters of time, amount of adsorbed and pressure loss through the column. The following Table 4 presents experimental results of column survey from 1 bar to 8 bar. However, the 5 bar column has shown saturation.

Table 4. Experimental results from 1 to 5 bar.

Parameter/Experiment	1	2	3	4	5
Pressure, bar	1	2	3	4	5
Number of unstable cycles, n	4	3	3	3	2
Weight of CMS in single bed, kg	3.5	3.5	3.5	3.5	3.5
Pressurization time, s	5	9	9	10	15
Adsorption time, s	27	29	31	33	35
Drop pressure, bar	0.12	0.14	0.16	0.17	0.18
Mass Flow Input FM1, sml	39.65	60.97	84.56	112.16	125.20
Recovery ratio, R	0.62	0.86	0.80	0.77	0.82
Mass Flow Output FM2, sml	24.58	52.43	67.65	86.36	102.66
The best concentration N_2, % (Take the product at atmospheric pressure)	82.60	87.90	88.90	93.40	93.40
O_2 gas flow is calculated adsorption, sml	3.65	5.85	9.40	16.73	18.26
Specific gravity of O_2 at standard conditions, kg/m^3	1.43	1.43	1.43	1.43	1.43
The amount of O_2 gas absorbed per 1 kg CMS-240, kg	0.0007	0.0012	0.0020	0.0038	0.0044

Table 4 is clearly shown in Figures 17–20 below, showing the relationship and interplay between the technological parameters and workability of the column. Survey data to 8 bar demonstrate saturation of the column.

Figures 16–20 are experimental results of one column from 1 bar to 8 bar pressure. We can observe that, at 5 bar pressure, the column reaches saturation state.

Figure 16 presents the flow measurement of input/output streams of the column from 1 bar to 8 bar pressure, and the black line is the amount of adsorbed O_2.

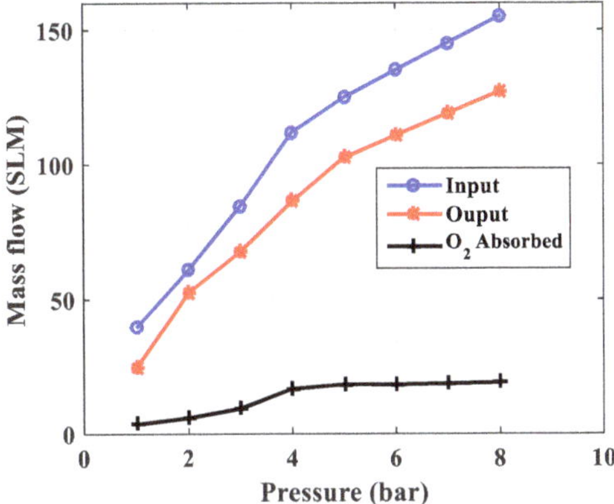

Figure 16. The amount of adsorbed material depends on the adsorption pressure.

Figure 16 shows that when the pressure increases, the inlet/outlet airflow also increases, the amount of adsorbent increases to a certain limit at 5 bar pressure.

Figure 17 shows the experimental results of the boosting time and adsorption time from Figures 13–15 (1 bar to 8bar pressure) according to real changes (when setting the running time is 460 s).

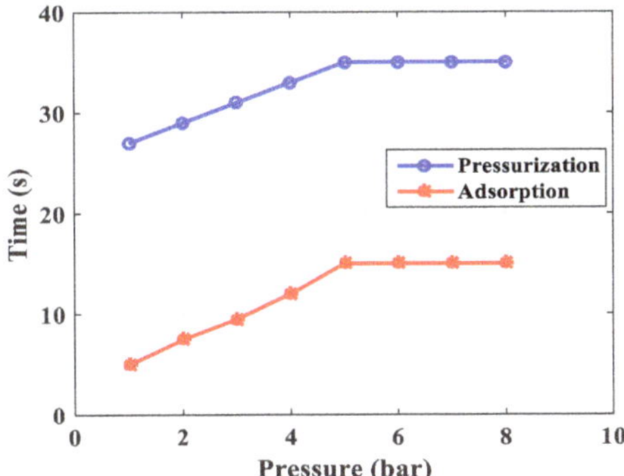

Figure 17. Adsorption time depends on adsorption pressure.

Figure 17 shows that when the pressure increases, pressurization time and adsorption time also increased to the limit at 5 bar pressure.

Figure 18 presents the experimental data of the drop pressure changing over time and according to the height of the column, which is shown in Figure 13. This result is determined by the maximum difference between the sensors CB-1.1 and CB-1.6.

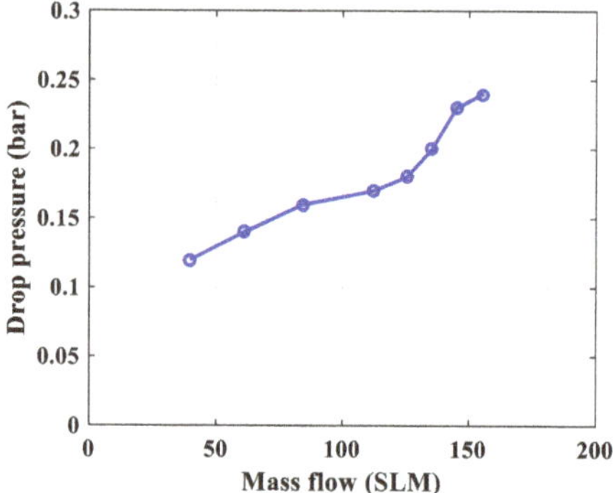

Figure 18. Drop pressure depends on incoming air flow input.

Figure 18 shows that when the mass flow increases, drop pressure also increased.

Figure 19 presents the highest concentration of N_2 gas at the output of the column from 1 bar to 8 bar pressure.

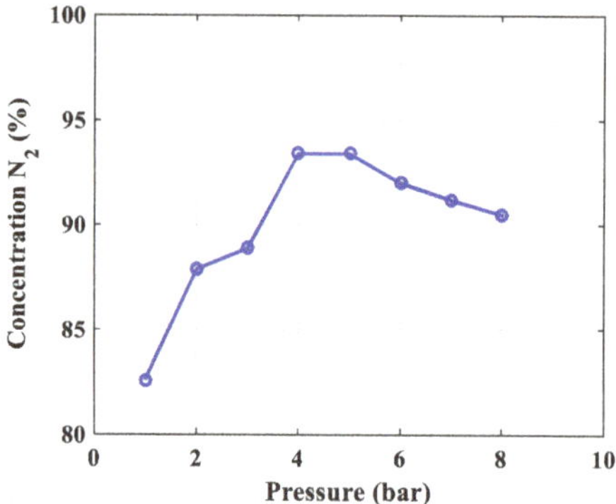

Figure 19. The dependence of N_2 concentration on adsorption pressure.

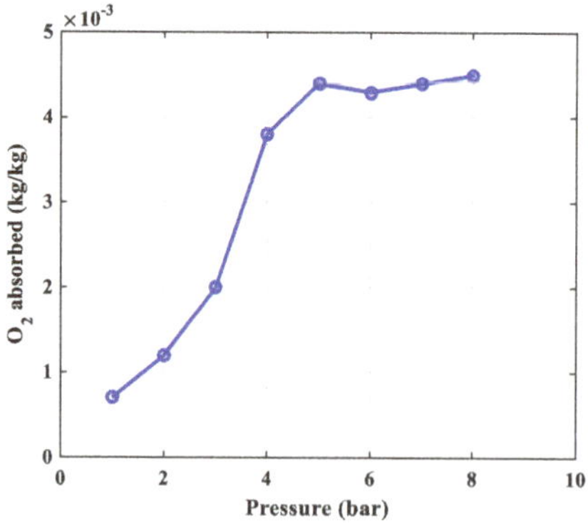

Figure 20. The amount of O2 gas absorbed depends on the pressure.

Figure 19 shows that when the pressure increases, concentration N_2 gas at output also increased to 5 bar, after decreasing due to the saturation.

Figure 20 is the experimental data of absorbed O_2 by determining the difference between input and output flow with sensors FM1 and FM2 from 1 bar to 8 bar pressure.

Figure 20 shows that, when the pressure increases, the amount of O_2 gas absorbed also increased to 5 bar, remaining so as pressure increases further due to the saturation.

From Figures 16–20, we see the optimal working point of the column at a pressure of 5 bar, the concentration N_2 gas product of the column reaches the highest of 93.4%. The maximum amount of adsorbent 18.26 L/minute is equivalent to 0.0044 kg O_2/1 kg CMS-240. The maximum adsorption time of O_2 gas is 35 s. The maximum pressurization is 15 s.

Finally, the compare simulation and experimental results are listed in Table 5.

Table 5. Compare simulation and experimental results.

Order.	Research Parameters	Simulation	Experimental	Conclusion
1	Rules of pressure change, bar	Partial pressure (O_2) changes over time and according to the height of the column.	The change in total pressure over time and the height of the column.	The same rule, but the experiment measured the rule of total pressure and gas concentration N_2 and O_2 at the output of the column. The concentration of O_2 gas decreases in accordance with the height of the column, which can confirm the reliability of the model being established.
2	The optimal working pressure of the column, bar	5.5 bar	5 bar	The optimum pressure is nearly the same, because this error is chosen because the speed in the model is constant, in fact it is changed by 20% because O^2 is adsorbed. In addition, there are errors due to the calculation of porosity and diffusion coefficient. So, this error is acceptable.
3	Drop pressure, bar	0.148 bar	0.18 bar	The drop pressure measured is greater because in the simulation only the material layer is calculated without taking into account the upper and lower filter materials and sieves.
4	Adsorption time, s	25 s	35 s	Experimental adsorption time is greater due to the late flow of air through the empty front and rear cylinders to stabilize the pressure and evenly distribute the gas.

5. Conclusions

Based on the proposed theory and simulation and experimental results, we have some conclusions to highlight as follows. Calculating and simulating adsorption columns is a very difficult task to deal with, however, with a suitable theory and mathematical tools, such as the help of computer software, we can simulate the partial pressure change of the adsorbed substance over time and in the direction height of the column. The partial pressure of the adsorbed material decreases both over time and the height of the column. The optimum working pressure of the column is 5.5 bar and adsorption time is 25 s. This simulation result is nearly accurate and has a margin of about 20% due to the rate of change, which in the simulation we calculated the speed as constant at the time of the column input. Drop pressure in this process is 0.148 bar. From the experimental research results, we understand the adsorption column uses a CMS-240 carbon molecular sieve material to separate N_2 from air working in environmental, isothermal, working regime stability (pressure change over time) can separate N_2 gas to reach a maximum concentration of 93%. Total pressure changes over time and decreases with column height. The optimal working mode at a pressure of 5 bar with adsorption time parameters is determined in Figure 20. Experimental results also show the mutual influence of technological parameters and the adsorption capacity of materials and columns.

The results of simulated and experimental research are very reliable on the basis of practical models and perfect laboratory equipment. The research methods and results are a solid basis for studying two columns and equipment according to the PSA cycle.

Author Contributions: Investigation, P.V.C., Software, H.N.N., V.D.T.; Methodology, P.V.C., H.N.N., N.T.H.; Visualization, V.D.T., T.-Y.N.; Formal Analysis, N.T.A., P.V.C. and T.-Y.N.; Writing—original draft, P.V.C., V.D.T., T.-Y.N.; Writing—review & editing, H.-N.N., D.V.T.

Funding: This research was funded by Institute of Technology—General Department of Defense Industry grant number 100.02-2019.10.

Conflicts of Interest: The authors declare no conflict of interest.

Abbreviations

Symbol	Unit	Meaning
D_L	$[\text{cm}^2 \cdot \text{s}^{-1}]$	Vertical dispersion coefficient (adsorption column)
k_i	$[1 \cdot \text{s}^{-1}]$	Mass transfer factor of the component i
K_i	$[-]$	Equilibrium constant of the component i
Z	$[\text{m}]$	Distance along the column (from the beginning of the column to the end of the column)
ΔP	$[\text{bar}]$	Drop pressure of the column
q_i	$[\text{g} \cdot \text{g}^{-1}]$	The amount of gas absorbed in the adsorbent of the component i
q^*_i	$[\text{g} \cdot \text{g}^{-1}]$	The amount of gas adsorbed in the material adsorbing when equilibrium of the component i
r_p	$[\mu\text{m}]$	Particle radius
ε_t	$\left[\frac{m^3 void}{m^3}\right]$	Total porosity (porosity inside and between particles)
P	$[\text{mmHg, KG/cm}^2]$	Pressure
V	$[\text{m}^3]$	Volume
T	$[\text{K}]$	Absolute temperature
C_i	$[\text{mole/cm}^3]$	The concentration of component i in the gas mixture
t	$[\text{s}]$	Time
U	$[\text{m/s}]$	velocity inside the column
ρ_p	$[\text{g/cm}^3]$	Particle density of adsorbent
K_L	$[\text{J/cm} \cdot \text{s} \cdot \text{K}]$	Thermal conductivity coefficient along the axis
T_w	$[\text{K}]$	The temperature of the column wall
T_{atm}	$[\text{K}]$	Temperature of the surrounding environment
ρ_g	$[\text{g/cm}^3]$	Density of gas
ρ_b	$[\text{g/cm}^3]$	Bulk density of adsorbent
ρ_w	$[\text{g/cm}^3]$	The density of material column wall
C_{pg}	$[\text{J/g} \cdot \text{K}]$	Specific heat capacity of gas
C_{ps}	$[\text{J/g} \cdot \text{K}]$	Specific heat capacity of adsorbent
C_{pw}	$[\text{J/g} \cdot \text{K}]$	Specific heat capacity of material column wall
$-\Delta H$	$[\text{J/mol}]$	Heat effect of adsorption process

h_i	[J/cm^2·K·s]	Internal heating factor
h_o	[J/cm^2·K·s]	External heating factor
R_{Bi}	[cm]	Inner radius of the column
R_{Bo}	[cm]	Outside radius of the column
A_w	[cm^2]	Cross-sectional area of the wall
B	[kPa^{-1}]	Langmuir equation parameters expanded
q_m	[mol/kg]	The equilibrium parameter for the extended Langmuir equation
P_i	[kPa]	Pressure of the component i
K	-	The coefficient of Langmuir equation extends

References

1. Kulkarami, S.J. Pressure Swing Adsorption: A Summary on Investigation in Recent Past. *Int. J. Res. Rev.* **2016**, *3*, 46–49.
2. Ashkan, M.; Masoud, M. Simulation of a Single Bed Pressure Swing Adsorption for Producing Nitrogen. In Proceedings of the International Conference on Chemical, Biological and Environmental Sciences, Bangkok, Thailand, 23–24 December 2011.
3. Smith, A.R.; Klosek, K. A review of air separation technologies and their integration with energy conversion processes. *Fuel Process. Technol.* **2011**, *70*, 115–134. [CrossRef]
4. Carlos Grande, A. Advance in Pressure Swing Adsorption for Gas Separation. *ISRN Chem. Eng.* **2012**, *2012*, 982934.
5. Snehal Patel, V.; Patel, J.M. Separation of High Purity Nitrogen from Air by Pressure Swing Adsorption on Carbon Molecular Sieve. *Int. J. Eng. Res. Technol.* **2014**, *3*, 450–454.
6. Delavar, M.; Nabian, N. An investigation on the Oxygen and Nitrogen separation from air using carbonaceous adsorbents. *J. Eng. Sci. Technol.* **2015**, *10*, 1394–1403.
7. Roy Chowdhury, D.; Sarkar, S.C. Application of Pressure Swing Adsorption Cycle in the quest of production of Oxygen and Nitrogen. *Int. J. Eng. Sci. Innov. Technol.* **2016**, *5*, 64–69.
8. Shafeeyan, M.S.; Daud, W.M.A.W.; Shamiri, A. A review of mathematical modeling of fixed-bed columns for carbon dioxide adsorption. *Chem. Eng. Res. Des.* **2014**, *92*, 961–988. [CrossRef]
9. Zhe, X.U. Mathematically modeling fixed-bed adsorption in aqueous systems. *Appl. Phys. Eng.* **2013**, *3*, 155–176.
10. Ehsan, J.S.; Masoud, M. Pilot-Scale Experiments for Nitrogen Separation from Air by Pressure Swing Adsorption. *S. Afr. J. Chem. Eng.* **2014**, *19*, 42–56.
11. Shokrooi, E.J.; Motlaghian, S.M.A.M. A robust and user friendly sofware (TB-PSA-SS) for numberical simulation of two-bed pressure swing adsorption processes. *Pet. Coal* **2015**, *57*, 13–18.
12. Furtat, I.B. Mathematical model of the process of adsorption. Models and modeling. *ASTU Bull.* **2008**, *42*, 24–30.
13. Akulinin, E.I.; Butler, D.S.; Dvoretsky, S.I.; Simanenkov, A.A. Mathematical Modeling of the Process of Oxygen Enrichment with Air in the Installation of Short-Cycle Adsorption. *TSTU Bull.* **2009**, *2*, 341–355.
14. Akulinin, E.I.; Dvoretsky, D.S.; Dvoretsky, S.I. Investigation of Heat and Mass Transfer Processes for Enriching Air with Oxygen with Hydrogen through Short-Cycle Adsorption Method. *TSTU Bull.* **2016**, *22*, 411–419.
15. Kevin Wood, R.; Liu, Y.A.; Yu, Y. *Design, Simulation and Optimization of Adsorption and Chromatographic Separation: A Hand-on Approach*, 1st ed.; Wiley-VCH Verlag GmbH and Co. KGaA: Hoboken, NJ, USA, 2018.
16. Suzuki, M. *Adsorption Engineering*; University of Tokyo: Tokyo, Japan, 1990.
17. Ruthven, D.M. *Principles of Adsorption and Adsorption Process*; Wiley-VCH Publishers: New York, NY, USA, 1984.
18. Douglas Ruthven, M.; Shamsuzzanman, F.; Kent Knaebel, S. *Pressure Swing Adsorption*; Wiley-VCH Publishers: New York, NY, USA, 1994.
19. Seader, J.D.; Ernest Heney, J.; Keith Roper, D. *Separation Process Principles Chemical and Biochemical Operations*; John Wiley and Sons, Inc.: Hoboken, NJ, USA, 2011.

Article

Production of Butyric Anhydride Using Single Reactive Distillation Column with Internal Material Circulation

Guanghui Chen, Fushuang Jin, Xiaokai Guo, Shuguang Xiang and Shaohui Tao *

College of Chemical Engineering, Qingdao University of Science and Technology, Qingdao 266042, China; guanghui@qust.edu.cn (G.C.); fushuangjin521@163.com (F.J.); guoxiaokai0304@163.com (X.G.); xsg@qust.edu.cn (S.X.)
* Correspondence: tau@qust.edu.cn; Tel.: +86-13869895813

Received: 22 November 2019; Accepted: 16 December 2019; Published: 18 December 2019

Abstract: The traditional two-column reactive distillation (RD) process is used for the production of butyric anhydride, which is synthesized with butyric acid and acetic anhydride via a reversible reaction. In this work, a novel process with a single RD column (SRDC) is designed for the production of butyric anhydride, where the second distillation column for separating excess reactant is removed based on the boiling point profile of the reaction system. Two applications of the proposed SRDC process, namely SRDC with excess butyric acid or acetic anhydride circulating internally, are economically optimized, and the results show that both SRDC processes have a lower total annual cost (TAC) than the traditional two-column process. Furthermore, from the perspective of TAC, the application with an excess feed of butyric acid is better than the application with excess acetic anhydride. The developed technique may also be applied to retrofit other traditional two-column RD processes, where the overhead and bottom products are the lightest and heaviest components of the reaction system, respectively, and no azeotrope is involved in the RD column.

Keywords: butyric anhydride; single reactive distillation column; internal material circulation; dynamic control

---

1. Introduction

Butyric anhydride is an important intermediate product, which is mainly used in the production of cellulose acetate butyrate, butyl butyryl lactate, and spices. The existing production of butyric anhydride is a reversible anhydride exchange reaction of butyric acid and acetic anhydride. The boiling point profile of the reaction systems shows that reactive distillation (RD) can be used to produce butyric anhydride. RD is an intensified process which couples the chemical reaction with distillation separation in one piece of equipment, thus equipment costs are reduced. RD can also improve the conversion and yield of the reversible reaction by removing the product from the reaction system in time; in addition, the concentration of reactants increases and the reaction rate also increases as the products are moved out. In the past decades, the batch RD process has been used to produce butyric anhydride [1], with the purity of butyric anhydride being high. The main disadvantages of the batch process include difficulty of operation, long production period, high labor costs due to the difficulty of automation, and high energy consumption due to the need to steam out the transition components [1].

In recent years, continuous manufacturing processes have been introduced to produce fine chemicals, such as butyric anhydride, due to the numerous advantages over batch-wise manufacturing; these include lower capital and operating costs, improved controllability and product consistency, and lower environmental footprint [2–4].

Gao [1] introduced continuous RD technology for producing butyric anhydride based on the analysis of the physical properties of the reaction system; results revealed that the continuous RD process is feasible for the production of butyric anhydride.

The operation of the continuous RD process can be classified into two types, namely, neat and excess. For neat operation, reactants are fed into the RD column at the same ratio as the stoichiometric of reaction, while the composition of the column overhead and bottom products are not determined in this situation, since strict flow control cannot be guaranteed in practical production; for example, Xu et al. [5] compared the neat and excess operations for synthesizing N-propyl propionate, the dynamic simulation results show that the product purity requirements cannot be ensured during neat operation; Daniel et al. [6] showed the same result with a process combining a continuous stirred tank reactor and an RD column for production of isoamyl acetate on a pilot scale.

Therefore, the excess operation is often adopted in practice; this requires one feed for the reactants according to the reaction stoichiometric, which helps avoid the disadvantage of the neat operation, while an additional distillation column is needed to separate the excess reactant from the product. Chung et al. [7] studied the influence of the excess ratio on the reactant conversion and energy consumption, and the results show that there is an inherent trade-off between the number of reactive trays and energy consumption; Chua et al. [8] compared different choices of excess reactant for producing isopropanol with the RD process.

As a result of the reliability of the excess operation, Li et al. [9] proposed a two-column RD process to produce butyric anhydride with an excess feed of butyric acid, thus the reliability of the process was validated by simulation and experiments.

In this paper, a novel continuous process with a single RD column (SRDC) is proposed to produce butyric anhydride using the excess operation, where the second distillation column in the traditional RD process for separating excess reactant is removed based on the analysis of the boiling point profile of the reaction system. In order to ensure the stable composition of the RD column, an excess reactant is usually needed in the traditional RD process, and for this reason a second distillation column is needed to separate the product from the excess reactant. SRDC completes the separation of the excess reactant and the product in the stripping section or the rectifying section of the RD column. Compared with the traditional RD process, product purification is removed and the total annual cost (TAC) of butyric anhydride production is greatly reduced.

The rest of this paper is organized as follows. Section 2 summarizes the reaction kinetics and thermodynamic data needed for the simulation of the RD process. The traditional two-column RD process for the production of butyric anhydride is simulated in Section 3. Section 4 analyzes the boiling point profile of the reaction system and presents a novel SRDC process with internal material circulation for butyric anhydride production, where two different applications are simulated and compared. Finally the paper is concluded in Section 5.

2. Reaction System

2.1. Synthesis of Butyric Anhydride

The reaction of acetic acid with butyric anhydride is homogeneous; it is also a series of reversible reactions without a catalyst, which is carried out in two steps as shown by Equations (1) and (2). The reaction rates of the two steps are quite different, and the rate of Reaction (1) is much greater than that of the second step. According to Li et al. [10], the most suitable reaction temperature is 343.15 K.

$$CH_3CH_2CH_2COOH + CH_3(CO_3C)CH_3 \Leftrightarrow$$
$$CH_3CH_2CH_2(CO_3C)CH_3 + CH_3COOH \tag{1}$$

$$CH_3CH_2CH_2COOH + CH_3CH_2CH_2(CO_3C)CH_3 \Leftrightarrow$$
$$CH_3CH_2CH_2(CO_3C)CH_2CH_2CH_3 + CH_3COOH \tag{2}$$

2.2. Reaction Kinetics

Li et al. [10] have studied and estimated the kinetics of Reactions (1) and (2) in the temperature range of 323.15 K to 353.15 K, as shown by Equations (3) and (4):

$$-r_{AcAn} = dc_{AcAn}/dt = k_{1+}c_{AcAn}c_{BuAc} - k_{1-}c_{BuAnAc}c_{HAC} \tag{3}$$

$$r_{BuAn} = dc_{BuAn}/dt = k_{2+}c_{BuAnAc}c_{BuAc} - k_{2-}c_{BuAn}c_{HAC}. \tag{4}$$

In Equations (3) and (4), n is the reaction step; + and − represent forward and backward reaction respectively; k_{n+}, k_{n-} are the reaction rate constants for reaction n with unit $dm^3 \cdot mol^{-1} \cdot min^{-1}$, which are calculated by Equations (5)–(8):

$$k_{1+} = 3.29 \times 10^7 e^{(-7115.86/T)} \tag{5}$$

$$k_{1-} = 1.63 \times 10^7 e^{(-6202.74/T)} \tag{6}$$

$$k_{2+} = 7.93 \times 10^4 e^{(-6009.26/T)} \tag{7}$$

$$k_{2-} = 7.45 \times 10^5 e^{(-6594.28/T)}. \tag{8}$$

2.3. Thermodynamics Data

Table 1 shows the information on reactants, products, and the boiling point at a reactive pressure of 20 kPa. There is no binary or ternary azeotropic phenomenon in the system [10].

Table 1. Information of pure components at a reactive pressure on 20 kPa.

Component ID	Component Name	Formula	T_b/K
BuAc	butyric acid	C_3H_7COOH	389.15
AcAn	acetic anhydride	$(CH_3CO)_2O$	364.15
HAC	acetic acid	CH_3COOH	348.15
BuAn	butyric anhydride	$(C_3H_7CO)_2O$	417.15
BuAnAc	Butyric anhydride acetate	$C_3H_7(CO_3C)CH_3$	413.15

As can be seen from Table 1, the reactants of the reaction, namely, acetic anhydride and butyric acid, have boiling points higher than the lighter product acetic acid, and lower than the heavier product butyric anhydride, which is key to the proposed SRDC process with the internal circulation of excess reactant.

Table 2 shows binary interaction parameters estimated using Aspen Plus (Aspen Plus 8.4, Aspen Tech, Bedford, CO, USA, 2014), VLE-HOC is a physical property method. The non-random two liquid (NRTL) activity coefficient model was used to calculate phase equilibrium in this work [11], which is calculated by Equations (9)–(12):

$$\ln \gamma_i = \left(\sum_j \tau_{ji} G_{ji} x_j / \sum_k G_{ki} x_k\right) + \sum_j \left(x_j G_{ij} / \sum_k G_{kj} x_k\right)\left[\tau_{ij} - \left(\sum_k x_k \tau_{kj} G_{kj} / \sum_k G_{kj} x_k\right)\right] \tag{9}$$

$$\tau_{ij} = a_{ij} + b_{ij}/T + c_{ij}/T^2 \tag{10}$$

$$G_{ij} = e^{(-\alpha_{ij}\tau_{ij})} \tag{11}$$

$$\alpha_{ij} = \alpha'_{ij} + \beta'_{ij}T \tag{12}$$

Table 2. Binary interaction parameters.

Component j	AcAn	HAC	HAC
Component i	BuAc	BuAc	AcAn
source	VLE-HOC	VLE-HOC	VLE-HOC
temp. unit	K	K	K
A_{ij}	0	−0.564	0.8146
A_{ji}	0	1.7591	−0.5218
B_{ij}	20.9697	0	−147.1604
B_{ji}	608.8256	0	332.2892
C_{ij}	0.3	0.3	0.3

3. Simulation of Two-Column RD Process for the Production of Butyric Anhydride

According to the above reaction kinetic and thermodynamic data, and the process introduced in the literature [9], the traditional two-column RD process is operated with a 10% excess feed of butyric acid as simulated with Aspen Plus, where the pressure of both columns is 20 kPa, and the residence time of reactive tray is 6 min. Figure 1 shows the process parameters and simulation results of the process streams, where N_F means feed position; N_r, N_{rxn} and N_s represent the number of rectifying, reactive, stripping trays, respectively; and N_T means the total number of column trays. The reactants are fed into the RD column C1; the overhead discharge of C1 is acetic acid, while the bottom discharge is the mixture of excess butyric acid and butyric anhydride, which is fed into distillation column C2, whose overhead discharge is the excess butyric acid to be recycled back to C1, and the bottom discharge is the product butyric anhydride, whose purity is more than 94 mol %, as seen from Figure 1.

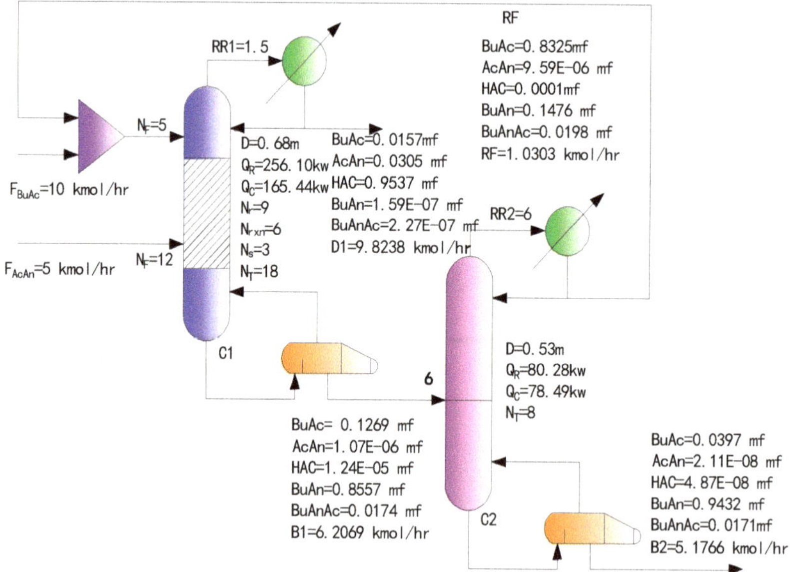

Figure 1. Traditional two-column reactive distillation (RD) process.

4. Single RD Column with Internal Material Circulation

4.1. Principle and Two Applications

Before the reactants are fed, it is supposed that the excess operation has been used to prepare the initial bottom materials of the RD column before start-up, e.g., the mole ratio of acetic anhydride to butyric acid is 1.1:2 for the bottom materials. The neat operation is used for the RD column after the process is running in a steady state. As a result of the different boiling points of the pre-excessed acetic anhydride reactant and the acetic acid, they can be separated easily from in the rectifying section of the column. As shown by Figure 2a, the acetic anhydride is always excessive and circulates in the column, thus the conversion of butyric acid is very promising, similar to the case of the traditional two-column RD process with the excess feed of acetic anhydride. Of course, the condenser duty of the RD column in Figure 2a must be tuned carefully to implement the internal circulation of the excessed acetic anhydride.

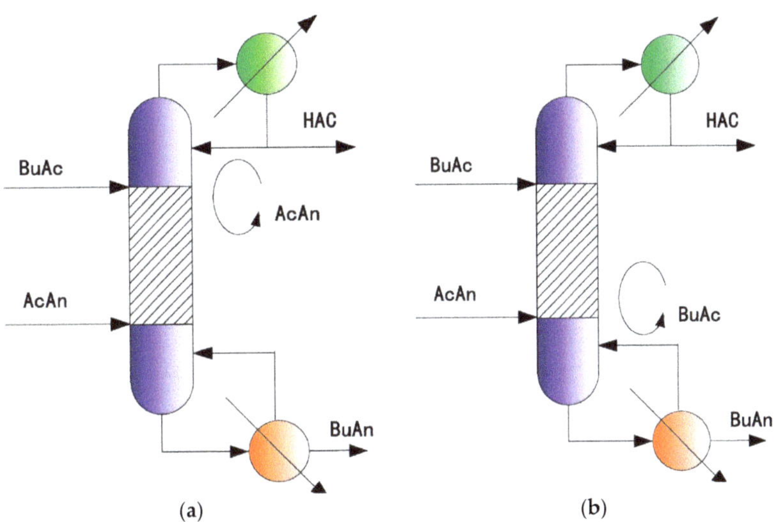

Figure 2. A single RD column (SRDC) with internal circulating excess reactant. (**a**) SRDC with internal circulation of pre-excess AcAn; (**b**) SRDC with internal circulation of pre-excess BuAc.

Similarly, with an excess of butyric acid in the initial bottom material, the internal circulation of butyric acid can also be realized by tuning the reboiler duty of the RD column in Figure 2b carefully, since the difference in boiling points is also significant between butyric acid and butyric anhydride.

The above two applications of the single RD column with excess reactant circulating internally are the bases of the proposed SRDC process.

4.2. Simulation of the SRDC Process

The application of internal circulation of excess reactant depends on tuning the condenser duty or reboiler duty of the SRDC. Therefore, it is important to simulate the processes shown in Figure 2a,b whose corresponding flowsheets in ASPEN are shown with Figure 3a,b. The position of the reactive section is fixed, but the simulation strategy is different depending on the different excess feed. When acetic anhydride is in excess, the rectifying section and the reactive section are separated from the stripping section, the tear flow is supplemented with an excess of 10% acetic anhydride, and the rectifying section is used to separate the excess acetic anhydride and acetic acid. When butyric acid is in excess, the rectifying section is separated from the reactive section and the stripping section, the tear flow is supplemented with an excess of 10% butyric acid, and the stripping section is used for the separation of the excess butyric acid and butyric anhydride.

In Figure 3a, which represents the pre-excess of acetic anhydride, the upper column with the condenser is the only rectifying section of the SRDC, as shown in Figure 2a; the only material discharged from overhead is the acetic acid byproduct, the excess acetic anhydride circulates internally, while the lower column with the reboiler only includes the reactive and stripping sections of the SRDC, whose bottom discharge is the butyric anhydride product.

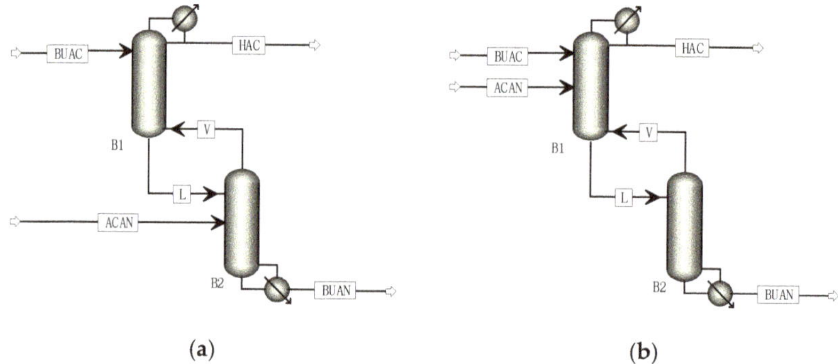

Figure 3. Flowsheets of SRDC with internal circulation of excess reactant. (**a**) SRDC with internal circulation of pre-excess AcAn; (**b**) SRDC with internal circulation of pre-excess BuAc.

In Figure 3b, which represents the pre-excess of butyric acid, the upper column with the condenser only includes the reaction and rectifying sections of the SRDC, as shown in Figure 2b, whose overhead discharge is also the acetic acid byproduct; while the lower column with the reboiler only represents the stripping section of the SRDC, where the pre-excess butyric acid is separated from butyric anhydride and circulates internally; the bottom discharge is also the butyric anhydride product.

4.3. Simulation Results of the Optimized SRDC

On the basis of the flowsheets presented in Figure 3, sensitivity-based optimization was carried out to optimize the proposed two applications of the SRDC process, and the results are shown in the following.

Under the condition of a pre-excess of acetic anhydride, the simulation results of the optimized SRDC process are shown in Figure 4, the SRDC3 is the SRDC with excess acetic anhydride circulating internally.

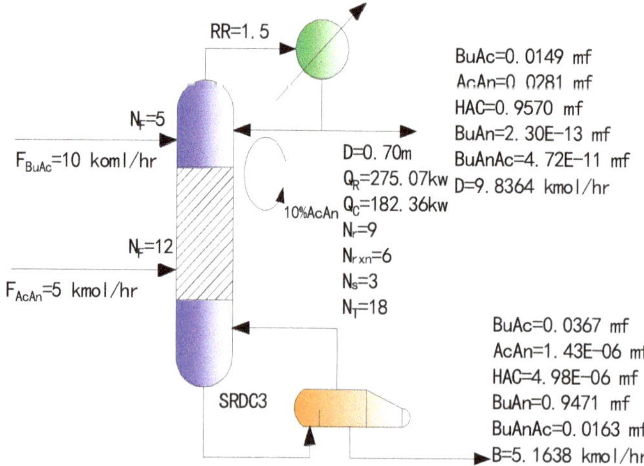

Figure 4. SRDC3 (SRDC with excess acetic anhydride circulating internally) with excess acetic anhydride circulating internally.

As seen from Figure 4, the purity of the butyric anhydride product is higher than in the traditional two-column process. Figure 5 shows the temperature and liquid composition profiles between the SRDC trays.

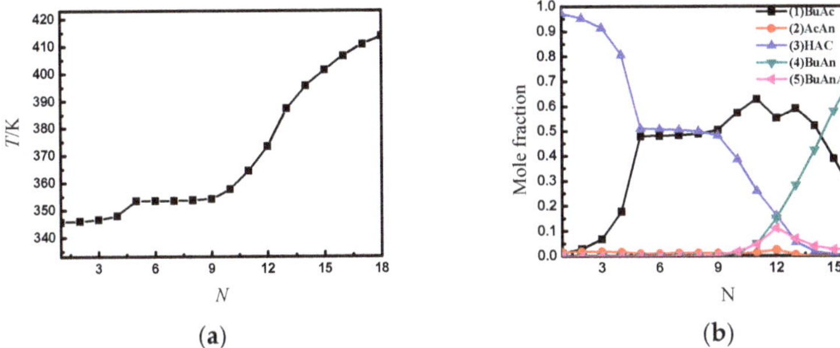

(a) **(b)**

Figure 5. The temperature and liquid composition profile between the SRDC3 trays. (**a**) The temperature profile between SRDC3 trays; (**b**) the liquid composition profile between SRDC3 trays.

Figure 6 shows the simulation results of the optimized SRDC under the condition of excess butyric acid, where the purity of the butyric anhydride product is also higher than that of the two-column process, the SRDC4 is the SRDC with excess butyric acid circulating internally.

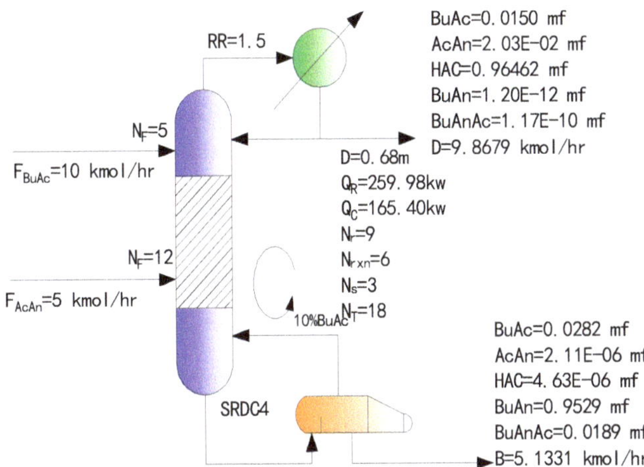

Figure 6. SRDC4 (SRDC with excess butyric acid circulating internally) with excess butyric acid circulating internally.

Figure 7 shows the temperature and liquid composition profiles between the trays of the SRDC4, which are similar to those under the condition of an excess feed of acetic anhydride.

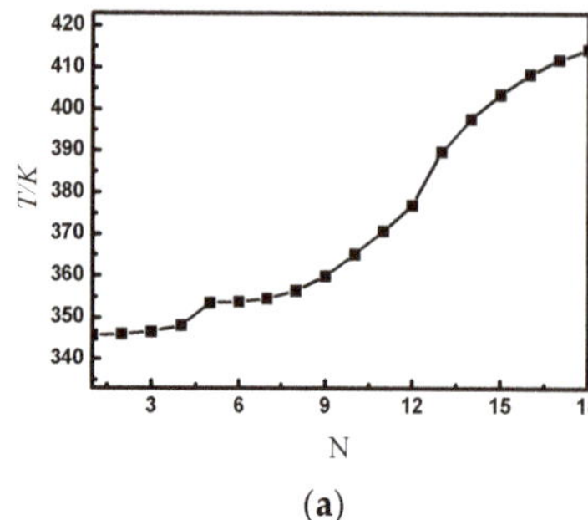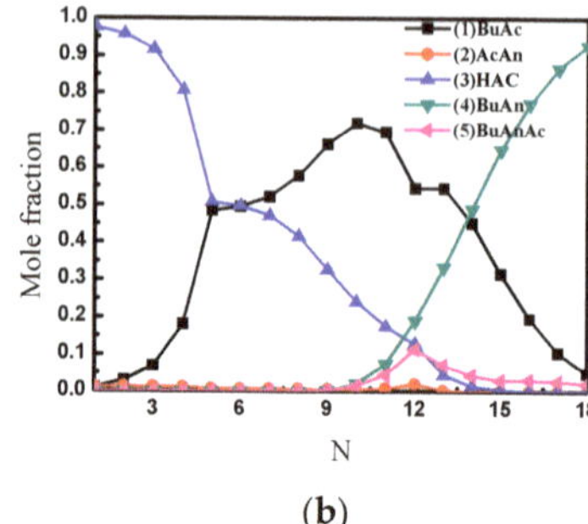

Figure 7. The temperature and liquid composition profile between SRDC4 trays. (**a**) The temperature profile between SRDC4 trays; (**b**) the liquid composition profile between SRDC4 trays.

The results of the three processes are compared in Table 3.

Table 3. Results of the three processes.

Parameter		Traditional Two-Column RD Process		SRDC3	SRDC4
		Column C1	Column C2		
Conversion of BuAc/%		95.37		95.69	96.40
Conversion of AcAn/%		96.40		96.70	97.60
Mole fraction of bottom/%	BuAc		3.97	3.67	2.80
	AcAn		2.11×10^{-6}	1.43×10^{-4}	2.11×10^{-4}
	HAC		4.87×10^{-6}	4.98×10^{-4}	4.63×10^{-4}
	BuAn		94.32	94.71	95.29
	BuAnAc		1.71	1.63	1.89
Mole fraction of top/%	BuAc	1.57		1.49	1.50
	AcAn	3.05		2.81	2.03
	HAC	95.37		95.70	96.46
	BuAn	1.59×10^{-5}		2.3×10^{-11}	1.20×10^{-10}
	BuAnAc	2.27×10^{-5}		4.72×10^{-9}	1.17×10^{-8}
Temperature/K	Bottom	408.24	413.55	414.11	414.38
	Top	345.77	391.50	345.76	345.76

The conversion of reactants and the purity of the product are higher than in the two-column process.

4.4. Economic Comparison

In this section, the traditional two-column process with an excess of butyric acid and the proposed two applications of the SRDC process are compared from the perspective of TAC, the calculation of which is shown in Equation (13) [12], consisting of capital cost and energy cost. The capital cost includes the cost of the column, tray, and heat exchanger, and the energy cost includes the cost of steam and cooling water. As shown in Table 4, the TAC of the three processes are calculated and compared.

$$TAC = capital\ cost / payback\ period + energy\ cost \tag{13}$$

As seen in Table 4, both SRDC processes are more economical than the traditional two-column process. In addition, the excess of butyric acid is more economical than the excess of acetic anhydride for the SRDC process; this condition is the result of the removal of the second distillation column,

Table 4. Economic comparison.

Parameter	Traditional Two-Column RD Process		SRDC3	SRDC4
	Column C1	Column C2		
ID_r (m)	0.79	0.53	0.79	0.68
ID_{rxn} (m)	0.68	-	0.70	0.67
ID_s (m)	0.90	-	0.89	0.95
Total number of trays (N_T)	18	8	18	18
Condenser duty (Q_C) (km)	165.44	78.49	182.36	165.40
Heat transfer area of condenser (A_C) (m^2)	5.49	1.13	5.49	5.49
Reboiler duty (Q_R) (km)	256.10	80.28	275.07	256.98
Heat transfer area of reboiler (A_R) (m^2)	18.10	7.21	23.39	23.84
Shell cost (106 $)	0.096	0.027	0.097	0.090
Heater exchanger (HX) cost (106 $)	0.070	0.034	0.085	0.085
Energy cost (106 $/year)	0.058	0.018	0.062	0.058
Capital (106 $)	0.166	0.061	0.182	0.175
Total annual cost (TAC) (106 $/year)	0.152		0.123	0.116

5. Conclusions

In this paper, a novel process for a single reactive distillation column with excess reactant circulating internally is proposed for the production of butyric anhydride. Two applications with different reactant pre-excesses are proposed, and the corresponding simulation method is also presented. On the basis of the simulation, the proposed SRDC process is compared to the traditional two-column process, and the results show the economical superiority of the SRDC. A more detailed analysis of the proposed process will be presented in the future.

Author Contributions: Conceptualization, S.T.; methodology, S.T.; software, F.J.; validation, G.C. and S.X.; formal analysis, F.J. and X.G.; investigation, G.C. and S.X.; resources, S.T.; data curation, F.J.; writing-original draft preparation, S.T. and F.J.; writing-review and editing, G.C., F.J., X.G., S.X., and S.T.; supervision, S.X.; project administration, G.C., F.J., X.G., S.X. and S.T. All authors have read and agreed to the published version of the manuscript.

Funding: This work is supported by the National Natural Science Foundation of China (No. 21776145). The Innovation and Training Plan of the University Students in Qingdao University of Science and Technology (No. 201810426125).

Conflicts of Interest: The authors declare no conflict of interest. The funders had no role in the design of the study; in the collection, analyses, or interpretation of data; in the writing of the manuscript, or in the decision to publish the results.

Nomenclature

AcAn	Acetic Anhydride
A_C	the heat transfer area of condenser
A_R	the heat transfer area of reboiler
B	bottom rate
BuAc	butyric acid
BuAn	butyric anhydride
BuAnAc	butyric anhydride acetate
D	distillate rate
F	the fresh feed of the reactive distillation column
HAC	acetic acid
HX	heater exchanger

ID_r	the internal diameter of the rectifying section
ID_{rxn}	the internal diameter of the reaction section
ID_s	the internal diameter of the stripping section
k_{n+}	the forward reaction rate constants, $dm^3 \cdot mol^{-1} \cdot min^{-1}$
k_{n-}	the backward reaction rate constants, $dm^3 \cdot mol^{-1} \cdot min^{-1}$
N	the number of trays
N_T	the total number of trays
N_F	the feed stage of the fresh feed
N_r	the number of trays in rectifying section
N_{rxn}	the number of trays in reactive section
N_s	the number of trays in stripping section
N_T	the total number of trays
Q_R	reboiler duty
Q_C	condenser duty
RR	the reflux ratio
RF	the recycled flow of BuAc
RD	reactive distillation
SRDC	single reactive distillation column
TAC	total annual cost
T_b	boiling point

References

1. Gao, F. Development of Continuous Reactive Distillation Process for Production of Butyric Anhydride. Master's Thesis, Qingdao University of Science and Technology, Qingdao, China, 2007.
2. Sahlodin, A.-M.; Barton, P.-I. Optimal Campaign Continuous Manufacturing. *Ind. Eng. Chem. Res.* **2015**, *54*, 11344–11359. [CrossRef]
3. Tian, H.; Zhao, S.; Zheng, H.; Haung, Z. Optimization of coproduction of ethyl acetate and butyl acetate by reactive distillation. *Chin. J. Chem. Eng.* **2015**, *23*, 667–674. [CrossRef]
4. Mallaiah, M.; Kishore, K.-A.; Reddy, G.-V. Catalytic Reactive Distillation for the Esterification Process: Experimental and Simulation. *Chem. Biochem. Eng. Q.* **2017**, *31*, 293–302. [CrossRef]
5. Xu, H.; Ye, Q.; Zhang, H.; Qin, J.-W.; Li, N. Design and control of reactive distillation–recovery distillation flowsheet with a decanter for synthesis of N-propyl propionate. *Chem. Eng. Process.* **2014**, *85*, 38–47. [CrossRef]
6. Gonzalez, D.-R.; Bastidas, P.; Rodriguez, G.; Gil, I. Design alternatives and control performance in the pilot scale production of isoamyl acetate via reactive distillation. *Chem. Eng. Res. Des.* **2017**, *123*, 347–359. [CrossRef]
7. Chung, Y.-H.; Peng, T.-H.; Lee, H.-Y.; Chen, C.-L.; Chien, I.-L. Design and control of reactive distillation system for esterification of levulinic Acid and n-Butanol. *Ind. Eng. Chem. Res.* **2015**, *54*, 3341–3354. [CrossRef]
8. Chua, W.-J.; Rangaiah, G.-P.; Hidajat, K. Design and optimization of isopropanol process based on two alternatives for reactive distillation. *Chem. Eng. Process.* **2017**, *118*, 108–116. [CrossRef]
9. Li, B.-C.; Han, X.-P.; Zhang, W.-L.; Geng, C.-X. Simulation and experimental research on butyric anhydride synthesis by reactive distillation. *J. Chem. Ind. Eng.* **2017**, *45*, 72–78.
10. Li, B.-C.; Xue, X.-X.; Zhang, W.-L.; Geng, C.-X. Study on the kinetics of preparation of butyric anhydride by acylation. *J. Chem. Ind. Eng.* **2017**, *45*, 49–53.
11. Renon, H.; Prausnitz, J.-M. Local compositions in thermodynamic excess functions for liquid mixtures. *AI ChE J.* **1968**, *14*, 135–144.
12. Luyben, W.-L. Design and control of a methanol reactor/column process. *Ind. Eng. Chem. Res.* **2010**, *49*, 6150–6163. [CrossRef]

 *processes*

Article

Assessment of the Total Volume Membrane Charge Density through Mathematical Modeling for Separation of Succinic Acid Aqueous Solutions on Ceramic Nanofiltration Membrane

Agata Marecka-Migacz [1], Piotr Tomasz Mitkowski [1,*], Jerzy Antczak [2], Jacek Różański [1] and Krystyna Prochaska [2]

[1] Division of Chemical Engineering and Equipment, Institute of Chemical Technology and Engineering, Poznan University of Technology, 60-965 Poznań, Poland
[2] Division of Chemical Technology, Institute of Chemical Technology and Engineering, Poznan University of Technology, 60-965 Poznań, Poland
* Correspondence: piotr.mitkowski@put.poznan.pl; Tel.: +48-61-665-3334

Received: 8 July 2019; Accepted: 17 August 2019; Published: 23 August 2019

Abstract: Nanofiltration of aqueous solutions of succinic acid with the addition of sodium hydroxide or magnesium hydroxycarbonate has been investigated experimentally and modeled with the comprehensively described Donnan–Steric partitioning model. The experimental retentions of acid at the same pH varied between 16% and 78%, while the estimated total volume membrane charge densities were in the range of −35.73 and +875.69 mol/m^3. This work presents a novel insight into the modeling of nanofiltration and investigates the relations between the estimated total volume membrane charge densities, ionic strength, and component concentration on the performance of ceramic membrane. In addition, this study takes into consideration other parameters such as pH regulation and viscosities of solutions.

Keywords: nanofiltration; total volume membrane charge density; modeling DSPM model; ceramic membrane; ionic strength

---

1. Introduction

Diminishing resources of fossil fuels, vulnerability of their prices, global warming, and environmental pollution result in a large interest in renewable and unconventional energy sources. Hence, one of these resources is biodiesel, which is currently under rigorous investigation [1]. Biodiesel is composed from renewable biological sources, such as vegetable oils and animal fats. It is biodegradable and nontoxic, and is characterized by low emission profiles and therefore is environmentally benign [2,3], unfortunately significant amounts of waste are generated during its production, which mainly consists of the glycerol phase. According to data presented by BP [4], the global biofuel production increased on the average by 5.13% (+4.1 Mtoe) in 2017 compared to 2014. In 2017, in the Europe region, biodiesel production reached approximately 14.167 million tons of oil equivalent (toe) whereas in North America nearly 38.190 Mtoe. Additionally, global primary energy consumption was projected to grow by 1.5 % per year between 2012 and 2035, whereas the energy from renewable resources, including biofuels, was expected to grow by 6.4% per year [5]. The biodiesel industry produces approximately 600 million tons of crude glycerol [6]. Therefore, due to the increase of glycerol waste, the technologies which allow for its beneficial reuse are of great interest. Many researchers focused on the biotechnological processing of glycerol by microbial bioconversion, delivering additional benefit to the environment in view of lower pollution. Through appropriate relation of microorganism clones, they postulate that it is possible to obtain low molecular weight organic compounds which are economically desirable, such as succinic acid [6–8]. In recent years, the interest in succinic acid production is growing because of the possibility to use it in the production of polymers and biodegradable plastics, surfactants, detergents, electrolytic coatings, and pharmaceutical active agents [9,10]. Generally, the technology of succinic acid biosynthesis ends at the stage of obtaining a post-culture liquid called post-fermentation broth, in which succinic acid produced by bacteria is dissolved [11–14]. Unfortunately, due to complex fermentation process, the main product is contaminated with various metabolites, especially with organic acids: acetic, formic, and others at low concentrations. The post-fermentation broth also contains wastes from the fermentation medium: residual glycerol, mineral salts, and minor amounts of polyols and proteins. Therefore, the success of bioconversion is determined by the process of purification [15–19]. Different traditional methods are used to separate organic compounds from the post-fermentation broth e.g., precipitation, distillation, liquid–liquid extraction, ion-exchange, adsorption, crystallization or esterification, however there are alternatives to the traditional purification methods, such as the membrane techniques, for example nanofiltration (NF) [15,20,21], which is one of the newer membrane pressure methods for separation of liquid mixtures. The NF membrane can be either organic in nature (i.e., polymeric) or inorganic. Inorganic membranes, especially the ceramic ones, are more suitable for use with organic solvents due to their excellent chemical, mechanical, and thermal stability [22]. It is worth to notice that there is a commercially available ceramic membrane with a hydrophobized surface for organic solvent nanofiltration (OSN). In general, transport characteristics are much less investigated compared to polymeric membranes. Additionally, due to the multifaceted nature of interactions between membrane, solvent, and solute, the prediction of transport mechanism in OSN is much more complicated compared to aqueous applications [23]. Nevertheless, mathematical description of NF separations, whether it is organic or an aqueous solution containing organic compounds, is of great interest for many researchers [24–34] and it is far from truly being predictive.

In general, NF combines the removal of uncharged components at nanoscale with charge effects between the solution and the surface of the membrane [24]. Acids possess very low molecular weight (MW) (e.g., the MW of succinic acid is equal to 118.09 g/mol) in comparison to cut-off of ceramic membranes, therefore the so-called sieving effect cannot play the main role. In such case, the retention of acid is very low and process efficiency is unsatisfactory. In order to enhance the separation of acid, the pH value is adjusted to obtain a dissociated form of acids above the pK_a (for succinic acid $pK_{a1} = 4.22$ [35]). When the dissociated form of the acid is separated in NF, the main separation mechanism changes to the electrostatic repulsion. Nevertheless, for a charged compound, both steric hindrance and electrostatic interactions are responsible for efficient separation—i.e., retention rate. Another important parameter in the transport and interpretation of retention is the membrane charge present along the surface of a membrane and also through the pores [32]. A strong charge present at the membrane surface has a crucial effect on the ion retention by the membrane [36] and therefore, the knowledge regarding the electrochemical interactions which occur at the membrane surface should offer the possibility to influence and describe the permeate flux, fouling tendency, retention of components, and cleaning conditions of the NF membranes more comprehensively. Understanding the ion-transport mechanism through a ceramic NF membrane is challenging and essential for further optimization of the NF processes.

The only available method adequate for characterizing the inherent membrane charge is the zeta potential of membrane. In the standard streaming potential measurements, the zeta potential is determined in order to link it to the streaming potential by the classical Helmholtz–Smoluchowski equation [37]. Streaming potential measurement gives the opportunity to obtain reliable information regarding the surface properties of ceramic membranes. Unfortunately, such measurement method requires sample in the flat, powder or eventually in fiber forms which requires destruction of a tubular membrane. Therefore, authors of this study postulate using the mathematical model to determine the total volume membrane charge density through parameter estimation based on the retention experiments.

Modeling of a nanofiltration membrane performance comprises two aspects: flux and rejection predictions [27]. Generally, the NF transport description models should consider the interaction between the charged membrane and ionic solutes [38]. Many charged membrane transport theories have been proposed which account for electrostatic effects as well as diffusive and convective flow to describe the solute separation [39], such as the steric-hindrance pore model (SHP), electric-steric-hindrance pore model (ESHP) [40], Teorell–Mayer–Sievers model [41], frictional model or space-charge model [42]. However, the most popular and widely adopted mathematical model for NF process is based on the extended Nernst–Planck (eN-P) equation [39]. Bowen and co-authors [25,43–45] proposed the Donnan-steric-partitioning model (DSPM) which arose from the eN-P equation. The DSPM has been also used by many authors [32,46–49] with fairy good results. However, prediction of separation or membrane charge is very limited, and modeling is used to describe and correlate rejection results.

In order to predict the separation performance, it is important to evaluate the membrane charge density in well-defined solutions. Therefore, the authors aim to consider each ion and water in the modeling, and therefore to obtain values of the total volume membrane charge densities through mathematical modeling. As a case study, the nanofiltration of aqueous solutions of succinate sodium and magnesium salts in pH equal to 9 and in two different concentrations (3.6 and 36.0 g/L) were studied experimentally and modeled with a comprehensively described DSPM model. Aside from sodium hydroxide, magnesium hydroxycarbonate ($4MgCO_3 \times Mg(OH)_2 \times 5H_2O$) was also selected for pH adjustment of the feeds. It was because the magnesium hydroxycarbonate was used in the fermentation process reported in [50] to keep pH at approximately 8–9 in the bioreactor and at around 8.5 in the post-fermentation broth.

2. Experimental Methods

2.1. Experimental and Operating Conditions

The pilot plant presented in Figure 1 was used in the study of the separation of succinic acid by the NF technique. The exploited system was equipped with two membrane modules, whereas all research was carried out on the only one of them. The used membrane module was equipped with a tubular ceramic mono-channel membrane (TiO_2 active layer, TAMI Industries, Nyons, France) with a cut-off of 450 Da, support mean porosity 38.5 mm^2, membrane mean porosity 3 μm, open porosity 30–40%, external diameter 10 mm, channel diameter and filtration area per 0.6 m tube equal to 0.0125 m^2 [51]. The transmembrane pressure (TMP) was set to 1.5 MPa. The process temperature for all experiments was fixed to 300 ± 2.0 K. The system was operated in a continuous mode, which means that the permeate and retentate were recirculated to the feeding tank continuously. The total volume of each investigated solution was equal to 12 L.

After each experiment, the ceramic membrane was cleaned to recover its initial permeance according to the procedure described in details elsewhere [51]. All cleaning steps were carried out at a flow system, the deionized water was filtered for 60 min, and the resulting flux was compared to the initial water flux. The decline in water flux after cleaning was not observed.

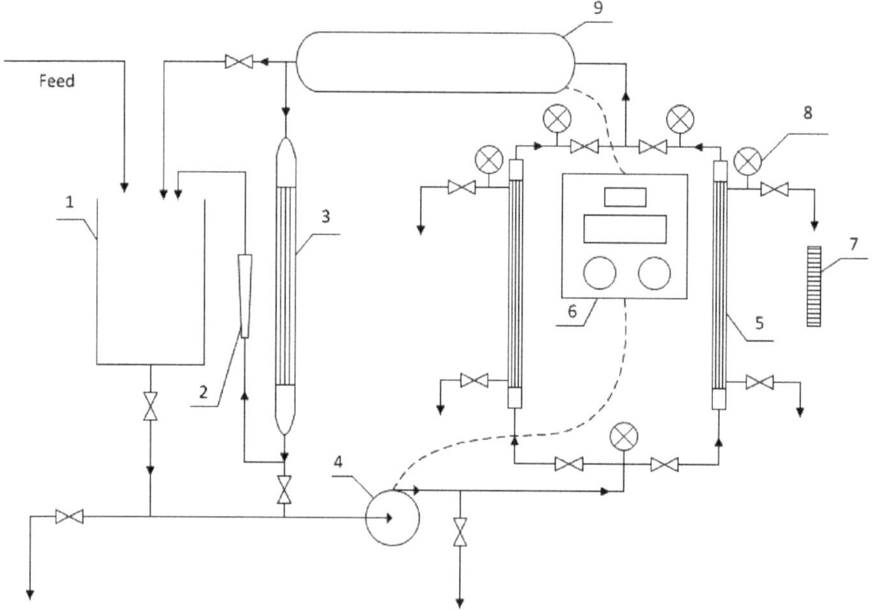

Figure 1. Pilot installation of the NF process: 1—feed tank; 2—rotameter; 3—heat exchanger; 4—pump; 5—membrane module; 6—temperature and flow rate controller; 7—permeate measuring cylinder; 8—pressure gauge; 9—heater.

2.2. Materials

Three one-main-component model solutions were tested in the reported study. All of them were prepared by dissolving succinic acid, either 3.6 or 36.0 g/L, in deionised water with an electrolytic conductivity not exceeding 3 μS/cm. The pH of each of the solutions was adjusted by addition of sodium hydroxide in model solutions 1 (MS1) and 2 (MS2), and addition of magnesium hydroxycarbonate in model solution 3 (MS3), with precise control of added amounts of each substance. The pH of all model solutions was equal to 9 ± 0.7. The compositions and concentrations of all model solutions were presented in Table 1. Analytical grade succinic acid was purchased from Sigma-Aldrich (pure p.a.) while sodium hydroxide and magnesium hydroxycarbonate from Avantor Performance Materials Poland S.A. (pure p.a).

Table 1. Compositions of investigated solutions

Model Solution	Concentration of Succinic Acid (g/L)	Compound Used to Regulate pH	Amount of Added Compound Which Regulated pH (g/L)
MS1	3.6	NaOH	2.49
MS2	36.0	NaOH	24.26
MS3	3.6	$4MgCO_3 \times Mg(OH)_2 \times 5H_2O$	3.85

2.3. Analysis

The concentration of sodium ions was measured based on microwave induced plasma with optical emission spectrometry (MIP-OES) using a PLASMAQUANT 100 (Carl Zeiss, Jena, Germany), which included a microwave energy generator and resonator delivered by Plazmatronika, Wroclaw. The resonant cavity was combined with a microwave energy generator operating with frequency equal to 2.45 GHz. The generator has two cooling systems: aqueous (magnetron cooling) and pneumatic (external cavity wall cooling). More details regarding the MIP-OES used in that analysis was published elsewhere [52]. The analysis was performed using solutions with a concentration equal to 0.5000 g/L NaOH (microgranules, analytically pure, POCH, Gliwice, Poland), 0.500 g/L $4MgCO_3 \times Mg(OH)_2 \times 5H_2O$, 3.6, and 36.0 g/L $C_4H_6O_4$ (purum p.a., >99.0%, Fluka Analytical, Sigma-Aldrich), which were appropriately diluted in order to prepare the analytical curve in the range of 0.5–30.0 mg/L. Standard solutions and samples were introduced to the excitation source (argon-helium plasma) with a rate of 1.5×10^{-3} L/min delivered by peristaltic pump (SPETEC, Erding, Germany) and concentric atomizer (Meinhard Glass Products, Golden, CO, USA) equipped with cyclone cloud chamber (EPOND, Vevey, Switzerland).

The contents of succinic acid and succinates in feed solutions, retentate and permeate obtained during the NF processes were determined by high-performance liquid chromatography method using HP Agilent 1100 Series system (Germany). The apparatus was equipped with an autosampler, interface (HP 35900), RI Detector (HP 1047A), pump (HP 1050), and column Rezex ROA-OrganicAcid H+ (8%), Phenomenex®. The carrier phase was a 2.5 mM H_2SO_4 solution with the flow of 0.5×10^{-3} L/min. The column temperature at the input to the detector was equal to 50 °C and pressure of 5.6 MPa. Before measurement, all samples were acidified to pH $\leq$ 2 by addition of 25% H_2SO_4 in an amount of 100×10^{-6} L to 1000×10^{-6} L of investigated solution. The pH of solutions was measured by laboratory pH-meter CP-505 (Elmetron, Poland) equipped with EPS-1 electrode.

The obtained concentrations of components were used to calculate retentions (R) according to Equation (1):

$$R = \left(1 - \frac{C_{p,i}}{C_{f,i}}\right) \cdot 100\% \tag{1}$$

All rheological studies were conducted using a rotational stress rheometer Physica MCR 501 (Anton Paar) with a double gap concentric cylinder measuring geometry (DG26.7, cup diameter 27.59/23.83 mm, bob diameter 26.66/24.66 mm). The experiments were conducted in the scanning shear rate measurements with increasing shear rates in order to obtain flow curves of succinic acid solutions in the range from 10 to 500 s^{-1}.

3. Modeling of Nanofiltration

3.1. Theory of the Utilized Model

In order to describe the ion transport through the NF ceramic membrane, the comprehensively described DSPM (ddDSPM) was proposed. The ddDSPM is based on the eN-P equation, extensively utilized by Bowen and co-authors [25,30,33,44]. Description of solute fluxes in the DSPM considers convection, diffusion, and electromigration mechanism. Convection occurs due to the applied pressure difference over the membrane, diffusion due to the concentration gradient across the membrane and finally charge effects related to electrostatic repulsion between the charged membrane and a charged organic compound [53]. The proposed ddDSPM explicitly takes all ions, solutes, and solvent into account. The whole ddDSPM model used in this work consists of Equations (2)-(19) set. Equation (2) describes the solvent velocity (V), which depends on membrane properties such as pore size (r_p), porosity (A_k) and thickness of active layer (Δx), properties of separated solutions—i.e., osmotic pressures ($\Delta \pi$) and feed viscosity (η)—and process parameter such as transmembrane pressure (ΔP). Difference of osmotic pressures defined by Equation (3) is calculated according to Equations (6) and (7) which are based on feed ($x_{f,i}$) and permeate ($x_{p,i}$) molar fractions (Equations (4) and (5)). Equation (8) defines ratio of solute ($r_{s,i}$) to pore radius, which is used along with Equation (9) to compute diffusive ($K_{d,i}$) and convective ($K_{c,i}$) hindrance factors for each component i present in the mixture. The main equations of the ddDSPM model describe the gradient of individual ion concentration ($c_{m,i}$) expressed by Equation (12) and electric potential gradient (ψ) across the membrane active layer thickness presented by Equation (13). The equations described above are solved under the condition that the membrane poses effective membrane charge density X_d, which is present in Equation (14), and that the separated mixture is electroneutral (Equation (15)). The presented model is constructed under the assumption of the Donnan exclusion mechanism on the feed-membrane interface which is expressed in the form of Equation (16). The components retentions are expressed by Equation (17). In order to solve the model equations set discussed above, there is a need to define a set of boundary conditions representing component concentrations at the membrane feed ($c_{m(0^+),i}$) and permeate ($c_{p,i}$) sides; the boundary conditions are presented by Equations (18) and (19). In summary, the ddDSPM consists of $17 + 14NC$ variables (listed in Table 2) in $7 + 11NC$ equations which were provided with appropriate descriptions in Table 3.

$$V = \frac{r_p^2(\Delta P - \Delta \pi)A_k}{8\eta_s \Delta x} \tag{2}$$

$$\Delta \pi = \pi_{feed} - \pi_{permeate} \tag{3}$$

$$x_{f,i} = \frac{C_{f,i}}{\sum\limits_{i=1}^{NoComp} C_{f,i} + C_{f,H_2O}} \tag{4}$$

$$x_{p,i} = \frac{C_{p,i}}{\sum\limits_{i=1}^{NoCom} C_{p,i} + \left(\tilde{V}_w\right)^{-1}} \tag{5}$$

$$\pi_{feed} = \frac{RT}{\tilde{V}_w} \sum\limits_{i=1}^{NoComp} x_{f,i} \tag{6}$$

$$\pi_{permeate} = \frac{RT}{\tilde{V}_w} \sum\limits_{i=1}^{NoComp} x_{p,i} \tag{7}$$

$$\lambda_i = \frac{r_{s,i}}{r_p} \tag{8}$$

$$\phi_i = (1 - \lambda_i)^2 \tag{9}$$

$$K_{d,i} = 1 - 2.3\lambda_i + 1.154\lambda_i^2 + 0.224\lambda_i^3 \tag{10}$$

$$K_{c,i} = (2 - \phi_i)\left(1 + 0.054\lambda_i - 0.988\lambda_i^2 + 0.441\lambda_i^3\right) \tag{11}$$

$$\frac{dc_{m,i}}{dx} = \frac{V}{K_{d,i}D_i}\left(K_{c,i}c_{m,i} - C_{p,i}\right) - \frac{F}{RT}z_i c_{m,i}\frac{d\psi}{dx} \tag{12}$$

$$\frac{\psi_{x=0} - \psi_{x=\delta}}{\delta} = \frac{d\psi}{dx}\bigg|_{x=0} = \frac{\sum\limits_{i=1}^{NoComp} \left(\frac{z_i V}{D_i}\left(K_{c,i}c_{m(0^+),i} - C_{p,i}\right)\right)}{\frac{F}{RT}\sum\limits_{i=1}^{NoComp}\left(z_i^2 c_{m,i}\right)} \tag{13}$$

$$\sum\limits_{i=1}^{NoComp} c_{m(0^+),i}z_i = -X_d \tag{14}$$

$$\sum\limits_{i=1}^{NoComp} C_{p,i}z_i = 0 \tag{15}$$

$$c_{m(0^+),i} = C_{f,i}\phi_i \exp\left(-\frac{z_i F}{RT}\psi_D\right) \tag{16}$$

$$R_i = 1 - \frac{C_{p,i}}{C_{f,i}} \tag{17}$$

and boundary conditions

$$x = 0^+ \rightarrow c_{m,i} = c_{m(0^+),i} \tag{18}$$

$$x = \Delta x \rightarrow c_{m,i} = C_{p,i} \tag{19}$$

Table 2. Variables in the ddDSPM model (*NC*—number of separated components)

Differential variables		Unit	Number
Concentration of ion in the membrane	$c_{m,i}$	mol/m^3	*NC*
Algebraic and implicit variables		**Unit**	**Number**
Potential gradient inside the membrane pore	ψ	V	1
Ratio of solute to pore radius	λ_i	–	*NC*
Steric term	ϕ_i	–	*NC*
Hindrance factor for diffusion	$K_{d,i}$	–	*NC*
Hindrance factor for convection	$K_{c,i}$	–	*NC*
Ion concentration in the permeate	$C_{p,i}$	mol/m^3	*NC*
Retention coefficient	R_i	%	*NC*
Solvent velocity (volume flux)	V	m^3/(m^2·s)	1
Donnan potential	ψ_D	V	1
Osmotic pressure difference	$\Delta\pi$	Pa	1
Osmotic pressure on the feed side	π_{eed}	Pa	1
Osmotic pressure on the permeate side	$\pi_{permeate}$	Pa	1
Molar fraction on the feed side	$x_{f,i}$	mol/mol	*NC*
Molar fraction on the permeate side	$x_{p,i}$	mol/mol	*NC*
Ion concentration in the membrane in the surface directly contacting with the feed	$c_{m(0+),i}$	mol/m^3	*NC*
Parameters and known variable		**Unit**	**Number**
Effective membrane charge density	X_d	mol/m^3	1
Pore radii	r_p	m	1
Ion radii	$r_{s,i}$	m	*NC*
Transmembrane pressure	ΔP	Pa	1
Ideal gas constant	R	J/(mol·K)	1
Faraday constant	F	C/mol	1
Temperature	T	K	1
Solvent viscosity	η_s	Pa·s	1
Thickness of membrane active layer	Δx	m	1
Molar volume of water	$\tilde{V}_w$	m^3/mol	1
Diffusion coefficient of ion	D_i	m^2/s	*NC*
Mean membrane porosity	A_k	%	1
Charge of individual ion	z_i	–	*NC*
Ion concentration in the feed	$C_{f,i}$	mol/m^3	*NC*
Water molar concentration in feed	$C_{f,H2O}$	mol/m^3	1
Total number of variables: 17 + 14*NC*			

Although, the eN-P equations (Equations (12) and (13)) have been commonly used for the calculation of ion rejection by RO and NF membranes, they have rarely been applied to organic solutes [31]. Moreover, the ddDSPM model presented above in this work was not only used for process simulation but also for the estimation of parameter X_d which is present in Equation (14), i.e., the total volume membrane charge density across the membrane active layer. Thus, the estimated X_d is evaluated under the assumption of the constant surface charge and constant surface potential at the interface of pore entrance. In detail, the constant surface charge means that the pore wall surface charge density is identical to the free surface charge density, e.g., measured on particles. Whereas constant surface potential means that the pore wall surface potential at the pore entrance is identical to the free surface potential, which is related by some researchers [54] to the Donnan potential at the feed–membrane interface and therefore it is equal to the zeta potential ζ. Additionally, since X_d is present in electroneutrality condition (Equation (14)), it actually combines all electrochemical interactions in close membrane neighborhood, i.e., those between solutes, solvents, and membrane material. Therefore, naming X_d as the total volume charge density across the membrane active layer is justified.

Table 3. List of equations in the ddDSPM model (*NC*—number of separated components).

Description of Equations	Equations	Number of Equations
Solvent velocity (volume flux) based on Hagen–Poiseuille-type relationship	(2)	1
Osmotic pressure difference across the membrane	(3)	1
Component molar fraction in feed	(4)	*NC*
Component molar fraction in permeate	(5)	*NC*
Osmotic pressure at the feed side	(6)	1
Osmotic pressure at the permeate side	(7)	1
Ratio of the solute radii to the pore radii	(8)	*NC*
Steric partitioning coefficient	(9)	*NC*
Hindrance factor for diffusion	(10)	*NC*
Hindrance factor for convection	(11)	*NC*
Concentration gradient inside the membrane pore	(12)	*NC*
Potential gradient inside the membrane pore	(13)	1
Electroneutrality conditions in the membrane	(14)	1
Electroneutrality conditions in the permeate	(15)	1
Donnan steric partitioning	(16)	*NC*
Retention coefficient	(17)	*NC*
Boundary condition at the membrane feed side	(18)	*NC*
Boundary condition at the membrane active layer thickness	(19)	*NC*
Total number of equations: $7 + 9NC + 2NC$ (boundary conditions)		

3.2. Determination of Total Volume Membrane Charge Density Values in Nanofiltration

The degree of freedom (DOF) of the presented model is equal to $10 + 3NC$, where *NC* stands for number of solutes present in the mixture. In order to solve the derived ddDSPM, the DOF must be equal to zero, therefore values of all parameters and known variables need to be provided. As already mentioned, during the parameter estimation in the ddDSPM, each ion existing in the solution is considered, even those originating from the sodium hydroxide or magnesium hydroxycarbonate, which were used for regulation of the pH of separated solutions. Values of diffusion coefficient D_i, ions charge z_i and radius of ions $r_{i,s}$ used in all calculations were presented in Table 4. Due to lack of the data, the Stokes–Einstein Equation (20) was used to determine the ionic radius of the succinate anion.

$$r_{i,s} = \frac{k_B T}{D_i 6\pi\eta} \tag{20}$$

In Figure 2, the values of viscosities of investigated solutions were presented in comparison to pure water viscosities. It is important to notice that the difference between viscosities of model solutions and water varied between 4% and 24%.

The modeling in this study is considering each ion presented in the system, even ions originating from solutions used to set the desired values of pH. Such detailed approach is innovative in modeling of NF processes. Until now, researchers dealing with modeling with the DSPM model, did not consider ions originating from solutions used for regulating pH or at least had not shown it explicitly. The solutes dissociate in aqueous solutions, then they deliver specific ionic forms to the separated feed. Authors are convinced that the presence of additional ions (such as Na^+, OH^-, Mg^{2+}, or CO_3^{2-}) may influence the total volume membrane charge density. It was also assumed in the ddDSPM, that the concentrations of the components in the feed are constant (i.e., steady state model), transmembrane pressure for the entire duration of the process is constant, pores are straight cylindrical in shape and of length equal to the effective membrane layer thickness. Due to the crossflow velocity and achieved Reynolds numbers within the membrane module in experiments equal to 2.3 m/s and 19,293, respectively, it was also assumed that concentration polarization effect and fouling phenomena are negligible. Additionally, as it is well known that the NF ceramic TiO_2 membrane has a support layer (Al_2O_3), the influence of that layer was neglected in the view of ratio of ions radii to support layer pore radii and the assumption that support layer is uncharged (neutral).

Table 4. Characteristics of all ions present in the model solutions

Type of Ion	$r_{i,s} \cdot 10^{10}$ m	$D_i \cdot 10^9$ m^2/s	z_i
$C_4H_4O_4^{2-}$	2.52	0.99 [55]	−2
H^+	0.01 [56]	11.81 [55]	+1
Na^+	1.02 [56]	1.33 [57]	+1
OH^-	1.33 [58]	6.70 [55]	−1
Mg^{2+}	0.72 [58]	0.71 [59]	+2
CO_3^{2-}	1.78 [58]	0.96 [59]	−2

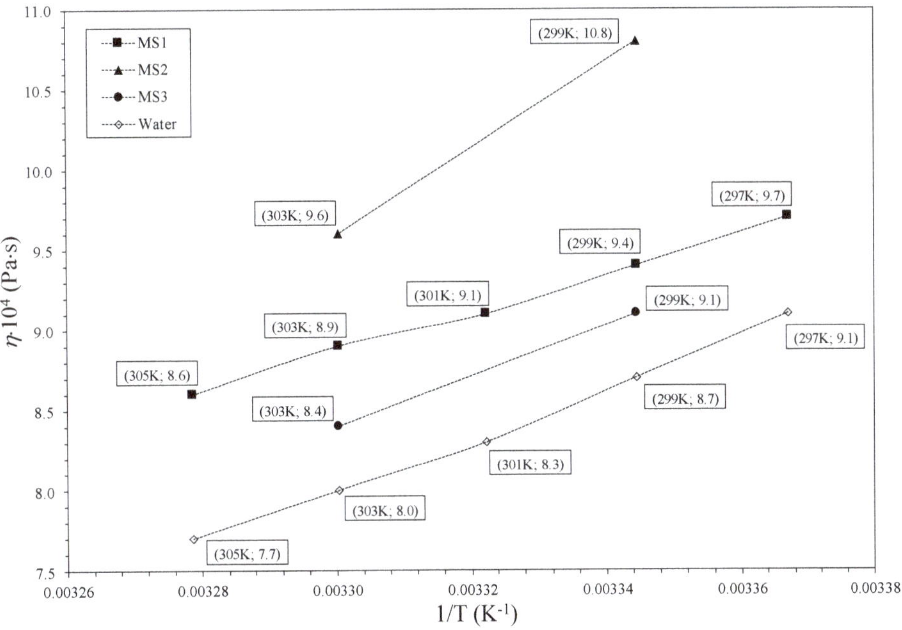

Figure 2. Experimental viscosity values of model solutions (MS1, MS2, MS3) and reference water viscosity in relation to temperature. Data in labels are ordered according to temperature expressed in Kelvin and value of viscosity.

The parameter estimations were conducted in the gPROMS software, which employs a rigorous optimization-based approach for model validation by offering parameter estimation capabilities, i.e., fitting model parameters to experimental data. Parameter estimation in gPROMS is based on the maximum likelihood formulation which provides simultaneous estimation of parameters in the physical model of the process [60]. Assuming independent, normally distributed measurement errors, with zero means and standard deviations, that maximum likelihood goal can be achieved through the objective function presented by Equation (21) [60]. In cases discussed in this study, the parameters estimation problems had the following values of parameters following Equation (21): $NE = 3$, $NV = 1$, $NM = 1$, $N = 3$.

$$\Phi = \frac{N}{2}\ln(2\pi) + \frac{1}{2}\min_{X_d}\left\{\sum_{i=1}^{NE}\sum_{j=1}^{NV_i}\sum_{k=1}^{NM_{ji}}\left[\ln\left(\sigma_{ijk}^2\right) + \frac{\left(c_{ijk,mes} - c_{ijk}\right)^2}{\sigma_{ijk}^2}\right]\right\} \tag{21}$$

4. Results and Discussion

Generally, it is well known that the following key parameters primarily affect the rejection of organic solutes during NF separations and are related to:

- Solute parameters: molecular size, acid dissociation constant, hydrophobicity/hydrophilicity, and diffusion coefficient,
- Membrane properties: cut-off, pore size, surface charge,
- Feed composition: pH, ionic strength, hardness, and the presence of organic matter.

It has to be clearly stated that the negatively hydrophilic solutes can be rejected by electrostatic repulsion through negatively charged membrane surfaces. Electrostatic interactions between charged solutes and a porous membrane have been frequently reported to be an important rejection mechanism. Ions such as Na^+, K^+, Ca^{2+}, and Mg^{2+} in feed water reduced the negative zeta potential of a membrane.

The experimentally obtained retention rates in the NF process for all model solutions were presented in Figure 3. The obtained retentions in process time slightly changed only initially and reached constant values after 40 min for whole processes which lasted up to 240 min. It has to be noticed, that two model solutions containing 3.6 g/L of succinic acid have displayed totally different values of retention. When sodium hydroxide was used as a pH regulator, the retention rate varied between 67% and 77% (see MS1 in Figure 3). When magnesium hydroxycarbonate was used, the retention decreased from significantly reaching a range of 20–22% (see MS3 in Figure 3). On the other hand, when the concentration of succinic acid increased from 3.6 g/L (MS1) to 36 g/L (MS2), the retention decreased form 67–77% to 3–16%. When retention profiles of MS1 and MS2 were analyzed separately, the initial increase of retention rate could be related to the fouling caused by concentration polarization, internal pore blocking or cake formation as it was described in [61]. However, since the MS2 has 10 times higher concentration than MS1, then it should be expected that retention rate would be kept at the same level, or that it should increase if there fouling would occur; but this was not the case. Therefore, it can be assumed that there is no typical fouling but a type of electrokinetic saturation of surface charges, the mathematical description of which is unknown to authors and requires further detailed studies.

The concentrations of sodium in the feed and permeate have changed significantly in the end of MS2 separation experiments, which is not the case in MS1 (Figure 4). That observation indicates that there is a mechanism of ion binding by the pore wall at higher concentration of sodium. Based on the obtained results, the ion binding for lower concentrations of succinic acid and pH-regulators (MS1 and MS3) can be excluded.

4.1. Impact of Dynamic Viscosity on Modeled Permeate Flux

The obtained dynamic viscosity values for all model solutions were presented in relation to the temperature in Figure 2. In modeling, it is generally assumed that viscosities of aqueous solutions with low solute concentrations are equal to water viscosity. For the MS1 at 303 K, the obtained viscosity value differed by approximately 10% in comparison with reference water viscosity. Such difference can influence the modeling results, which is highlighted by comparison of volumetric fluxes calculated with experimentally obtained viscosities and with viscosity being equal to water at certain temperatures which was presented in Figure 5. In general, the experimental viscosities in calculations of volumetric fluxes for investigated solutions resulted in a decrease of calculated volumetric fluxes from 7.5% for MS1 and up to 66.4% for MS3 to those calculated with pure water viscosities. It has to be stated that including the experimentally obtained viscosities resulted in better description of the experimental volume flux for MS2. However, for MS1 and MS3 cases the discrepancies between utilization of water and experimental viscosities were rather far from the experimental, but still of the same order. Generally, for the investigated solutions, the use of experimental viscosities with combination of the Hagen–Poiseuille equation resulted in an underestimation of the calculated volume flux.

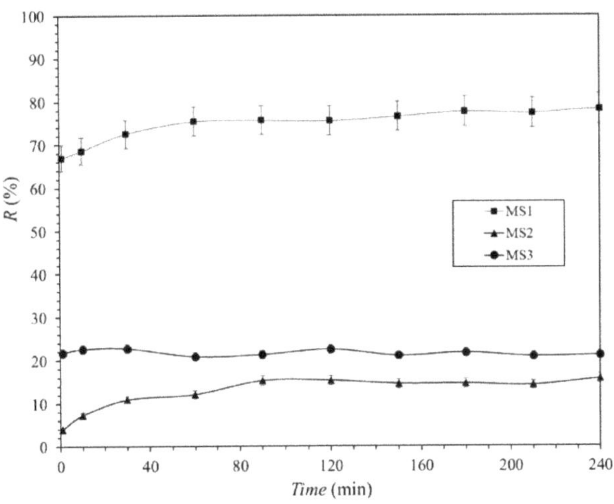

Figure 3. Experimental retention rates achieved for model solutions: MS1 (3.6 g/L succinic acid at pH = 9.7 regulated with granulate NaOH), MS2 (36.0 g/L succinic acid at pH = 8.8 regulated with granulate NaOH) and MS3 (3.6 g/L succinic acid at pH = 8.7 regulated with $4MgCO_3 \times Mg(OH)_2 \times 5H_2O$) in relation to duration of the NF process.

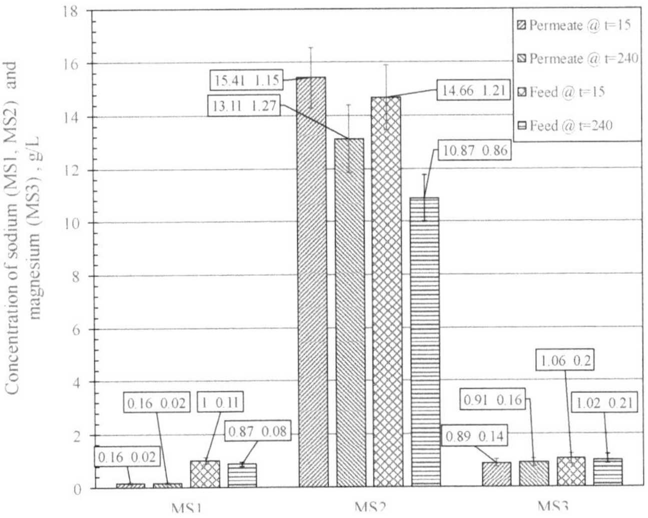

Figure 4. Concentration of sodium and magnesium cations in permeate and feed sides after 15 and 240 min of separation.

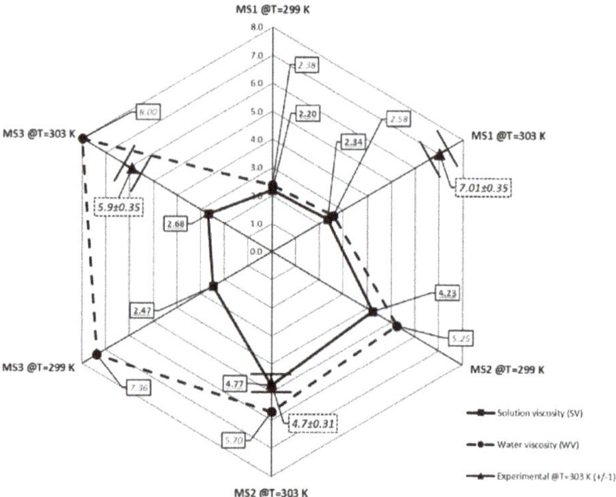

Figure 5. Comparison of calculated volume fluxes of model solutions (MS1, MS2, MS3) (10^{-5} m^3/(m^2·s)) with experimental viscosities (SV) and pure water viscosities (WV) at different temperatures.

4.2. Comparison of the ddDSPM Model with the Standard Approach

The standard approach considers only concentrations of solutes and ions coming from dissolved components, in the presented cases from succinic acid. The ddDSPM model takes into account all solutes, ions, and solvent, which include succinic acid, pH regulating solutions and water. In the standard approach, the model consists of 29 equations and 46 variables ($NC = 2$), while in the ddDSPM there are 62 equations with 88 variables ($NC = 5$), which were solved with the aim to estimate X_d. Additionally, estimation of X_d was also conducted under direct consideration of values of experimental fluxes (V_{exp}) instead of volume fluxes calculated with the Hagen–Poiseuille equation. All estimation results were compared in Figure 6. In the standard approach (i.e., DSPM model, ions coming from pH regulator not included), X_d is changing between −35.59 and +278.09 mol/m^3, while in the detailed approach (i.e., ddDSPM model, ions coming from pH regulator included) the changes ranged between −35.73 and +875.69 mol/m^3. The use of the water viscosity in the ddDSPM resulted in X_d values ranging between −34.98 and +939.67 mol/m^3 and use of experimental values of volume fluxes (V_{exp}) in ddDSPM (i.e., ddDSPM model, ions coming from pH regulator included) resulted in X_d values ranging between −19.57 and +871.74 mol/m^3. It is important to highlight that the retention obtained with the models overlaid with the experimental values, regardless of which model was used.

Based on the obtained results, it is difficult to postulate which modeling approach is the best, since it is impossible to experimentally obtain the overall membrane charge density X_d. However, it is evident that the obtained values of X_d for higher concentration of solutes (MS2) differ significantly depending on the applied modeling approach. At first glance, it seems that there is no clear reason to use the ddDSPM model for diluted solution such as MS1; however, when feed contains a higher number of components or polyanions like in MS3, the ddDSPM would be recommended. The presented results clearly show that use of experimentally obtained volumetric flux in comparison to the ddDSPM with experimentally obtained viscosities does not significantly influence the X_d when computed volumetric flux is in good comparison to the experimental one (MS2). Otherwise, use of experimental volumetric flux in estimation of X_d is recommended.

4.3. Variation of the Overall Volume Charge Densities in Relation to Used pH Regulator

Comparison of the estimated values of the total volume membrane charge densities with standard and detailed models was presented in Figure 6 along with ionic strength and obtained experimental rejections for each investigated solution. The ionic strength I for each model solution was calculated according to Equation (22).

$$I = \frac{1}{2}\sum_{i=1}^{NC}\left(c_i \cdot z_i^2\right) \qquad (22)$$

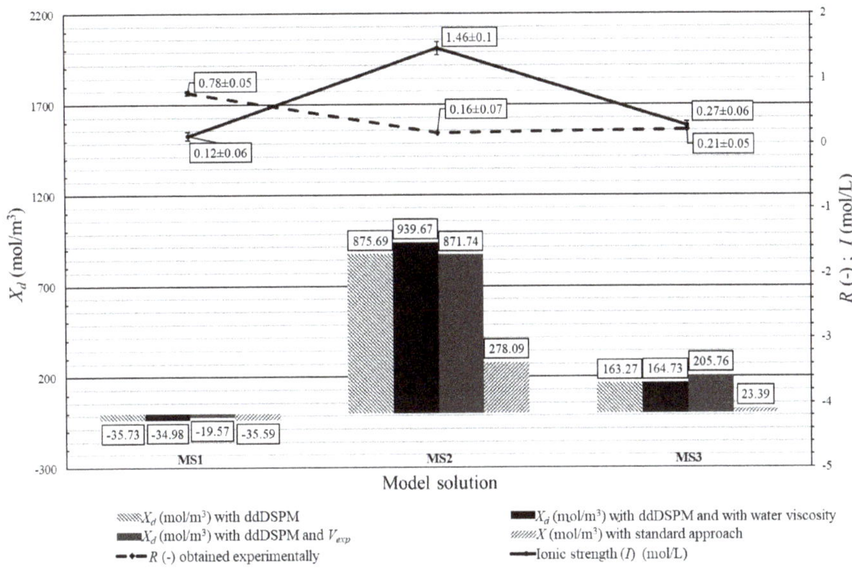

Figure 6. Comparison of estimated total volume membrane charge densities (X_d) with different modeling approaches with relation to ionic strength (I) and experimentally obtained retentions (R) (ddDSPM—detailed described Donnan–Steric Partitioning Model; DSPM—Donnan–Steric Partitioning Model with standard approach; V_{exp}—experimental volume flux).

As presented in Figure 6, the estimated values of X_d for each model solutions were clearly different, but these values were adequate to retention rates obtained in the NF processes. In general, the higher ionic strength of the separated solutions resulted in a higher value of the total volumetric charge of the membrane. It should be noted that X_d values for all model solutions were values obtained for the experiments which reached the steady state and these values should not be confused with the fixed membrane charge because the X_d relates all electrokinetic phenomena present in the vicinity of membrane surface and not only those related to the membrane surface as it is in the case of fixed membrane charge [62–64]. The achieved X_d value for MS1 (3.6 g/L succinic acid of pH = 9.7 regulated with granulate NaOH) equal to −35.73 mol/m^3 may indicate the presence of strong electrostatic repulsion between succinate anions present in aqueous solution and membrane surface functional groups, hence the X_d reached a negative value and retention achieved 78%. For two other cases, the X_d values were positive. Such behavior could be related to the amphoteric characteristics of TiO$_2$ active membrane layer, although it is generally explained by the charge change at isoelectric point at specific pH [46].

$$-Ti-OH+H_3O^+ \rightarrow -Ti-OH_2^+ +H_2O \quad \text{at pH} < IEP \qquad (23)$$

$$-Ti-OH+OH^- \rightarrow -Ti-O^- +H_2O \quad \text{at pH} > IEP \qquad (24)$$

In the studied cases, when pH was equal to approximately 9, it was expected that the active layer would possess negatively charged groups. The succinate anion has the lowest diffusion coefficient among other ions present in the system (Table 4), therefore succinate anions should grind and penetrate the membrane as the last anions. Although such strongly positive X_d value obtained for MS2 (36.0 g/dm^3 succinic acid at pH = 8.8 regulated with granulate NaOH) may reflect high concentration of sodium cations, their selective adsorption on membrane surface, which results in formation of additional surface layer and appearance of the electrostatic attraction of the succinic anions, resulted in very low retention (16%). In cases of basic salts, it is very difficult to determine which hypothetical mechanisms might influence the separation and could result in such charge values. Although the estimated value for the MS3 (3.6 g/dm^3 of succinic acid with pH = 8.7 regulated with 4MgCO$_3 \times$ Mg(OH)$_2 \times$ 5H$_2$O) was equal to +163.27 mol/m^3, which suggests the selective adsorption of magnesium cations or other cations on the membrane active layer surface, also within the pores.

In general, the increase of the membrane charge densities with pH increase might be caused by the selective adsorption of ions at the membrane-separated mixture interface and additional adsorption in membranes' pores in active layer [65]. In this work, authors consider the total volume membrane charge density as the sum of the fixed membrane charge density and the number of adsorbed ions in the whole active membrane volume and its close vicinity at the feed side and membrane pores. The possible mechanism for the formation of membrane charge assumes that ions are partitioned from the bulk solution into the membrane pore under the influence of the Donnan potential. Among the partitioned ions in the membrane pores, either cations or anions are adsorbed selectively by the pore walls. Next, the adsorbed ions are bound on the pore wall and provide the electric charge to the membrane, which is specific to the investigated cases. In view of all that was stated above, the values of total volume membrane charge densities X_d will always be different depending on the type of solution subjected to the NF process, although the pH values are the same. Despite that pH for all studied solutions was equal to approximately 9, the X_d values varied between −35.73 and +875.69 mol/m^3 (Figure 6). Therefore, it can be assumed that, the mechanism of selective ion adsorption plays a significant role despite similar pH values, the schematic explanation of which was presented in Figure 7. The overall volume membrane charge density in MS1 is negative because not all negative membrane surface group are associated with cations present in the solution, therefore succinic ions are repulsed by the membrane interface. The MS3 contains a similar amount of added magnesium hydroxide as sodium hydroxide in MS1. Since the magnesium ion possesses two positive charges, it can interact with two negative membrane sites and more negative membrane sites can be associated. Therefore, in MS3, the succinic ions can permeate more easily than in case of MS1. The lowest retention observed for MS2 could be related to the highest overall volume membrane charge due to high concentration of dissociated succinic acid and the highest amount of added sodium hydroxide, which in that case could compensate all present negative membrane charges.

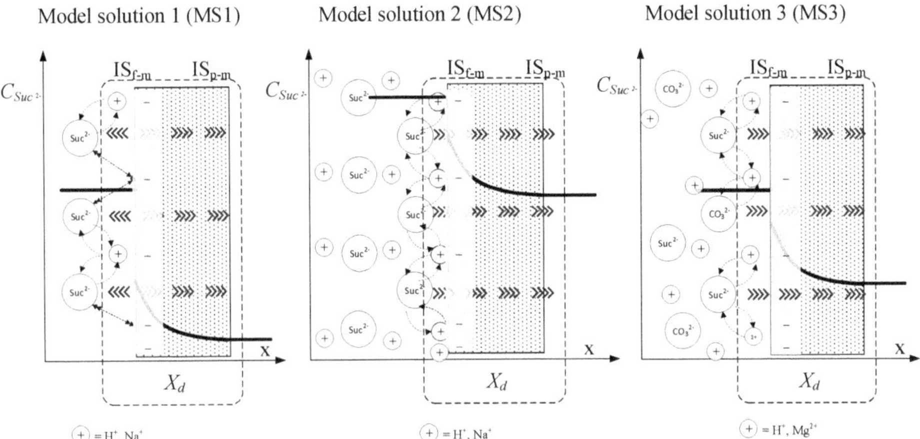

Figure 7. Schematic representation of separation mechanisms in nanofiltration of aqueous solutions of succinic acid with different pH regulators (for details about MS1, MS2, and MS3 see Table 4, X_d—total volume membrane charge density; $C_{suc2}{}^-$—molar concentration of succinate anion; IS*f-m*—feed-membrane interface; IS*f-m*—permeate-membrane interface).

5. Conclusions

The main goal of the performed computer-aided simulations was to estimate the membrane surface charge densities with the use of the comprehensively described Donnan–Steric partitioning model, which is derived from the extended Nernst–Planck equation with Donnan partitioning assumption. The obtained total volume membrane charge densities X_d are consistent with experimental values of retention. Values such as charge density of the membrane are very important for explanation of the mechanism of ions transport across the membrane, defining retention, and describing influences on electrostatic repulsion between ions and membrane.

The presented work can be summarized in the following points: (1) separation of organic acids, such as succinic acid, on the ceramic membrane is affected by the acid concentration and the type of used pH regulator; (2) experiments clearly show that increase of the concentration of components can decrease the retention of separated components (MS1 and MS2); (3) use of a different pH regulator can dramatically change the performance of the separation (MS1 and MS3); (4) in case of low succinic acid concentration, the membrane functional group (-TiO$_2$) dissociation has the greatest impact on the charge formation, whereas in case of higher succinic acid concentrations, it is postulated that the charge formation is determined by adsorption of specific ions on the membrane active layer; (5) defining separation performance at certain pH can be totally different in the vicinity of other components or impurities; (6) detailed description of experiments is required in order to compute X_d, even reporting the amount of added pH regulator; (7) including all components in mathematical modeling can allow for better understanding of nanofiltration separation.

Based on the obtained results and authors knowledge, it is impossible to state which modeling approach is the best, however it is clear to recommend the comprehensively described Donnan–Steric partitioning model to analyze the separation of multicomponent mixtures as it gives reasonable results and takes into account all components present in the mixture.

Author Contributions: Conceptualization, A.M.-M., P.T.M. and K.P.; Methodology, K.P., P.T.M., and J.R.; Modeling and simulation, A.M.-M.; Validation, A.M.-M. and P.T.M.; Experimental investigation, A.M.-M. and J.A.; Resources, K.P. and J.R.; Writing—original draft preparation, A.M.-M.; Writing—review and editing, A.M.-M., P.T.M., K.P., and J.R.; Supervision, P.T.M.

Funding: This research was supported by the Ministry of Science and Higher Education in Poland subsidy for Poznan University of Technology, Faculty of Chemical Technology (grant nos. 03/32/SBAD/0902 and 03/32/SBAD/0901).

Conflicts of Interest: The authors declare no conflict of interest.

Nomenclature

A_k	mean membrane porosity, %
$C_{f,H2O}$	water molar concentration in feed, mol/m^3
$C_{f,i}$	ion concentration in the feed, mol/m^3
$C_{p,i}$	ion concentration in the permeate, mol/m^3
D_i	diffusion coefficient of component or ion, m^2/s
F	Faraday constant, C/mol
I	ionic strength, mol/dm^3
$IS_{f\text{-}m}$	feed-membrane interface
$IS_{f\text{-}m}$	permeate-membrane interface
$K_{c,I}$	hindrance factor for convection
$K_{d,I}$	hindrance factor for diffusion
MS	model solution, numbering and details according the Table 1
N	total number of measurements taken during all the experiments
NC	number of components
NE	number of experiments performed
NM_{ij}	number of measurements of the j-th variable in the i-th experiment
NV_i	number of variables measured in the i-th experiment
R	ideal gas constant, J/(mol × K)
R_i	retention coefficient of component i
T	temperature, K
V	solvent velocity (volume flux), m^3/(m^2 × s)
$\tilde{V}_w$	molar volume of water, m^3/mol
X_d	effective membrane charge density, mol/m^3
c_{ijk}	k-th predicted value of variable j in experiment i
$c_{ijk,mes}$	k-th measured value of variable j in experiment i
$c_{m(0+),i}$	ion concentration in the membrane in the surface directly contacting with the feed, mol/m^3
$c_{m,i}$	concentration of ion in the membrane, mol/m^3
k_B	Boltzmann constant (1.38×10^{-23} J/K)
r_p	pore radii, m
$r_{s,i}$	ion radii, m
$x_{f,i}$	molar fraction on the feed side, mol/mol
$x_{p,I}$	molar fraction on the permeate side, mol/mol
z_i	charge of individual ion
Greek Letters	
θ	set of model parameters to be estimated
Ψ	potential gradient inside membrane pore, V
Ψ_D	Donnan potential, V
$\Delta\pi$	osmotic pressure difference, Pa
ΔP	transmembrane pressure, Pa
Δx	thickness of membrane active layer, m
η	viscosity, Pa × s
ηs	solvent viscosity, Pa × s
λi	ratio of solute to pore radius
$\pi_{permeate}$	osmotic pressure on the permeate side, Pa
π_{feed}	osmotic pressure on the feed side, Pa
σ^2_{ijk}	variance of the k-th measurement of variable j in experiment i
Φ_i	steric term

References

1. Lisiecki, P.; Chrzanowski, Ł.; Szulc, A.; Ławniczak, Ł.; Białas, W.; Dziadas, M.; Owsianiak, M.; Staniewski, J.; Cyplik, P.; Marecik, R.; et al. Biodegradation of diesel/biodiesel blends in saturated sand microcosms. *Fuel* **2014**, *116*, 321–327. [CrossRef]

2. Chrzanowski, Ł.; Dziadas, M.; Ławniczak, Ł.; Cyplik, P.; Białas, W.; Szulc, A.; Lisiecki, P.; Jeleń, H. Biodegradation of rhamnolipids in liquid cultures: Effect of biosurfactant dissipation on diesel fuel/B20 blend biodegradation efficiency and bacterial community composition. *Bioresour. Technol.* **2012**, *111*, 328–335. [CrossRef] [PubMed]

3. Ma, F.; Hanna, M.A. Biodiesel production: A review. *Bioresour. Technol.* **1999**, *70*, 1–15. [CrossRef]

4. BP Statistical Review of World Energy. Available online: https://www.bp.com/content/dam/bp/business-sites/en/global/corporate/pdfs/energy-economics/statistical-review/bp-stats-review-2018-full-report.pdf (accessed on 22 January 2019).

5. Lalman, J.A.; Shewa, W.A.; Gallagher, J.; Ravella, S. Biofuels Production from Renewable Feedstocks. In *Quality Living Through Chemurgy and Green Chemistry*; Springer: Berlin/Heidelberg, Germany, 2016; pp. 193–220. ISBN 9783662537022.

6. Gao, C.; Yang, X.; Wang, H.; Rivero, C.P.; Li, C.; Cui, Z.; Qi, Q.; Lin, C.S.K. Robust succinic acid production from crude glycerol using engineered Yarrowia lipolytica. *Biotechnol. Biofuels* **2016**, *9*, 179. [CrossRef] [PubMed]

7. Antczak, J.; Regiec, J.; Prochaska, K. Separation and concentration of succinic adic from multicomponent aqueous solutions by nanofiltration technique. *Pol. J. Chem. Technol.* **2014**, *16*, 1–4. [CrossRef]

8. Becker, J.; Lange, A.; Fabarius, J.; Wittmann, C. Top value platform chemicals: Bio-based production of organic acids. *Curr. Opin. Biotechnol.* **2015**, *36*, 168–175. [CrossRef] [PubMed]

9. Piotrowska, E.; Szewczyk, K.W.; Jaworska, M.M.; Konieczna-Mordas, E. Biotechnologiczne wytwarzanie kwasu bursztynowego. *Inż. Ap. Chem.* **2012**, *51*, 171–173.

10. Vinoth Kumar, R.; Pakshirajan, K.; Pugazhenthi, G. Malic and succinic acid: potential C4 platform chemicals for polymer and biodegradable plastic production. In *Platform Chemical Biorefinery: Future Green Chemistry*; Elsevier: Amsterdam, The Netherlands, 2016; pp. 159–179. ISBN 978-0-12-802980-0.

11. Van Heerden, C.D.; Nicol, W. Continuous succinic acid fermentation by Actinobacillus succinogenes. *Biochem. Eng. J.* **2013**, *73*, 5–11. [CrossRef]

12. Carvalho, M.; Matos, M.; Roca, C.; Reis, M.A.M. Succinic acid production from glycerol by Actinobacillus succinogenes using dimethylsulfoxide as electron acceptor. *New Biotechnol.* **2014**, *31*, 133–139. [CrossRef]

13. Bradfield, M.F.A.; Nicol, W. Continuous succinic acid production by Actinobacillus succinogenes in a biofilm reactor: Steady-state metabolic flux variation. *Biochem. Eng. J.* **2014**, *85*, 1–7. [CrossRef]

14. Pateraki, C.; Patsalou, M.; Vlysidis, A.; Kopsahelis, N.; Webb, C.; Koutinas, A.A.; Koutinas, M. Actinobacillus succinogenes: Advances on succinic acid production and prospects for development of integrated biorefineries. *Biochem. Eng. J.* **2016**, *112*, 285–303. [CrossRef]

15. Staszak, K.; Woźniak, M.; Sottek, M.; Karaś, Z.; Prochaska, K. Removal of fumaric acid from simulated and real fermentation broth. *J. Chem. Technol. Biotechnol.* **2015**, *90*, 432–440. [CrossRef]

16. Erickson, B.; Nelson; Winters, P. Perspective on opportunities in industrial biotechnology in renewable chemicals. *Biotechnol. J.* **2012**, *7*, 176–185. [CrossRef] [PubMed]

17. López-Garzón, C.S.; Straathof, A.J.J. Recovery of carboxylic acids produced by fermentation. *Biotechnol. Adv.* **2014**, *32*, 873–904. [CrossRef] [PubMed]

18. Efe, Ç.; Pieterse, M.; van der Wielen, L.A.M.; Straathof, A.J.J. Separation of succinic acid from its salts on a high-silica zeolite bed. *Chem. Eng. Process. Process Intensif.* **2011**, *50*, 1143–1151. [CrossRef]

19. Pérez-Cisneros, E.S.; Mena-Espino, X.; Rodríguez-López, V.; Sales-Cruz, M.; Viveros-García, T.; Lobo-Oehmichen, R. An integrated reactive distillation process for biodiesel production. *Comput. Chem. Eng.* **2016**, *91*, 233–246. [CrossRef]

20. Zaman, N.K.; Rohani, R.; Abdul Shukor, M.H.; Mohamad, A.W. Purification of High Value Succinic Acid from Biomass Fermentation Broth via Nanofiltration. *Indian J. Sci. Technol.* **2016**, *9*, 95244. [CrossRef]

21. Zaman, N.K.; Rohani, R.; Mohammad, A.W.; Isloor, A.M.; Jahim, J.M. Investigation of succinic acid recovery from aqueous solution and fermentation broth using polyimide nanofiltration membrane. *J. Environ. Chem. Eng.* **2017**. [CrossRef]

22. Wang, Y.; Wang, X.; Liu, Y.; Ou, S.; Tan, Y.; Tang, S. Refining of biodiesel by ceramic membrane separation. *Fuel Process. Technol.* **2009**, *90*, 422–427. [CrossRef]

23. Blumenschein, S.; Böcking, A.; Kätzel, U.; Postel, S.; Wessling, M. Rejection modeling of ceramic membranes in organic solvent nanofiltration. *J. Memb. Sci.* **2016**, *510*, 191–200. [CrossRef]

24. Hilal, N.; Al-Zoubi, H.; Darwish, N.A.; Mohamma, A.W.; Abu Arabi, M. A comprehensive review of nanofiltration membranes: Treatment, pretreatment, modelling, and atomic force microscopy. *Desalination* **2004**, *170*, 281–308. [CrossRef]

25. Bowen, W.R.; Welfoot, J.S. Predictive modelling of nanofiltration: Membrane specification and process optimisation. *Desalination* **2002**, *147*, 197–203. [CrossRef]

26. Mitkowski, P.T.; Buchaly, C.; Kreis, P.; Jonsson, G.; Górak, A.; Gani, R. Computer aided design, analysis and experimental investigation of membrane assisted batch reaction-separation systems. *Comput. Chem. Eng.* **2009**, *33*, 551–574. [CrossRef]

27. Van der Bruggen, B.; Mänttäri, M.; Nyström, M. Drawbacks of applying nanofiltration and how to avoid them: a review. *Sep. Purif. Technol.* **2008**, *63*, 251–263. [CrossRef]

28. Vandezande, P.; Gevers, L.E.M.; Vankelecom, I.F.J. Solvent resistant nanofiltration: Separating on a molecular level. *Chem. Soc. Rev.* **2008**, *37*, 365–405. [CrossRef] [PubMed]

29. Szymczyk, A.; Fatin-Rouge, N.; Fievet, P.; Ramseyer, C.; Vidonne, A. Identification of dielectric effects in nanofiltration of metallic salts. *J. Memb. Sci.* **2007**, *287*, 102–110. [CrossRef]

30. Szymczyk, A.; Fievet, P. Investigating transport properties of nanofiltration membranes by means of a steric, electric and dielectric exclusion model. *J. Memb. Sci.* **2005**, *252*, 77–88. [CrossRef]

31. Bellona, C.; Drewes, J.E.; Xu, P.; Amy, G. Factors affecting the rejection of organic solutes during NF/RO treatment—A literature review. *Water Res.* **2004**, *38*, 2795–2809. [CrossRef]

32. Teixeira, M.; Rosa, M.; Nystrom, M. The role of membrane charge on nanofiltration performance. *J. Memb. Sci.* **2005**, *265*, 160–166. [CrossRef]

33. Broniarz-Press, L.; Mitkowski, P.T.; Szaferski, W.; Marecka, A. Modelowanie procesu odzysku fumaranu diamonu metodą nanofiltracji. *Inż. Ap. Chem.* **2014**, *53*, 223–224.

34. Mitkowski, P.T.; Broniarz-Press, L.; Marecka, A. Knowledge and model-based framework for nanofiltration modelling. In Proceedings of the 4th European Young Engineers Conference 2015, Warsaw, Poland, 27–29 April 2015.

35. Perrin, D.D.; Dempsey, B.; Serjeant, E.P. *pKa Prediction for Organic Acids and Bases*; Springer: Dordrecht, The Netherlands, 1981; ISBN 978-94-009-5885-2.

36. Puhlfürß, P.; Voigt, A.; Weber, R.; Morbé, M. Microporous TiO_2 membranes with a cut off <500 Da. *J. Memb. Sci.* **2000**, *174*, 123–133.

37. Zhang, Q.; Jing, W.; Fan, Y.; Xu, N. An improved Parks equation for prediction of surface charge properties of composite ceramic membranes. *J. Memb. Sci.* **2008**, *318*, 100–106. [CrossRef]

38. Kong, F.; Yang, H.; Wang, X.; Xie, Y.F. Assessment of the hindered transport model in predicting the rejection of trace organic compounds by nanofiltration. *J. Memb. Sci.* **2016**, *498*, 57–66. [CrossRef]

39. Ang, W.L.; Mohammad, A.W. Mathematical modeling of membrane operations for water treatment. In *Advances in Membrane Technologies for Water Treatment*; Elsevier: Amsterdam, The Netherlands, 2015; pp. 379–407. ISBN 9781782421214.

40. Marchetti, P.; Jimenez Solomon, M.F.; Szekely, G.; Livingston, A.G. Molecular separation with organic solvent nanofiltration: A critical review. *Chem. Rev.* **2014**, *114*, 10735–10806. [CrossRef] [PubMed]

41. Darvishmanesh, S.; Buekenhoudt, A.; Degrève, J.; Van der Bruggen, B. General model for prediction of solvent permeation through organic and inorganic solvent resistant nanofiltration membranes. *J. Memb. Sci.* **2009**, *334*, 43–49. [CrossRef]

42. Tanimura, S.; Nakao, S.I.; Kimura, S. Transport equation for a membrane based on a frictional model. *J. Memb. Sci.* **1993**, *84*, 79–91. [CrossRef]

43. Bowen, W.R.; Mukhtar, H. Characterisation and prediction of separation performance of nanofiltration membranes. *J. Memb. Sci.* **1996**, *112*, 263–274. [CrossRef]

44. Bowen, W.R.; Mohammad, A.W. Diafiltration by nanofiltration: Prediction and optimization. *AIChE J.* **1998**, *44*, 1799–1812. [CrossRef]

45. Bowen, W.R.; Mohammad, A.W. A theoretical basis for specifying nanofiltration membranes-dye/salt/water streams. *Desalination* **1998**, *117*, 257–264. [CrossRef]

46. Van Gestel, T.; Vandecasteele, C.; Buekenhoudt, A.; Dotremont, C.; Luyten, J.; Leysen, R.; Van der Bruggen, B.; Maes, G. Salt retention in nanofiltration with multilayer ceramic TiO_2 membranes. *J. Memb. Sci.* **2002**, *209*, 379–389. [CrossRef]

47. Bernata, X.; Fortuny, A.; Stüber, F.; Bengoa, C.; Fabregat, A.; Font, J. Recovery of iron (III) from aqueous streams by ultrafiltration. *Desalination* **2008**, *221*, 413–418. [CrossRef]

48. Izadpanah, A.A.; Javidnia, A. The ability of a nanofiltration membrane to remove hardness and ions from diluted seawater. *Water* **2012**, *4*, 283–294. [CrossRef]

49. Ramadan, Y.; Pátzay, G.; Szabó, G.T. Transport of NaCl, $MgSO_4$, $MgCl_2$ and Na_2SO_4 across DL type nanofiltration membrane. *Period. Polytech. Chem. Eng.* **2010**, *54*, 81. [CrossRef]

50. Prochaska, K.; Antczak, J.; Regel-Rosocka, M.; Szczygiełda, M. Removal of succinic acid from fermentation broth by multistage process (membrane separation and reactive extraction). *Sep. Purif. Technol.* **2018**, *192*, 360–368. [CrossRef]

51. Antczak, J.; Szczygiełda, M.; Prochaska, K. Nanofiltration separation of succinic acid from post-fermentation broth: Impact of process conditions and fouling analysis. *J. Ind. Eng. Chem.* **2019**, *77*, 253–261. [CrossRef]

52. Matusiewicz, H.; Ślachciński, M.; Hidalgo, M.; Canals, A. Evaluation of various nebulizers for use in microwave induced plasma optical emission spectrometry. *J. Anal. At. Spectrom.* **2007**, *22*, 1174. [CrossRef]

53. Braeken, L.; Bettens, B.; Boussu, K.; Van der Meeren, P.; Cocquyt, J.; Vermant, J.; Van der Bruggen, B. Transport mechanisms of dissolved organic compounds in aqueous solution during nanofiltration. *J. Memb. Sci.* **2006**, *279*, 311–319. [CrossRef]

54. Labbez, C.; Fievet, P.; Thomas, F.; Szymczyk, A.; Vidonne, A.; Foissy, A.; Pagetti, P. Evaluation of the "DSPM" model on a titania membrane: Measurements of charged and uncharged solute retention, electrokinetic charge, pore size, and water permeability. *J. Colloid Interface Sci.* **2003**, *262*, 200–211. [CrossRef]

55. Ilie, O.; van Loosdrecht, M.C.M.; Picioreanu, C. Mathematical modelling of tooth demineralisation and pH profiles in dental plaque. *J. Theor. Biol.* **2012**, *309*, 159–175. [CrossRef]

56. Periodic Table of Elements: Hydrogen—H. Available online: http://environmentalchemistry.com/yogi/periodic/H.html (accessed on May 22, 2018).

57. Lide, D.R. *CRC Handbook of Chemistry and Physics*, 84th ed.; CRC Press: Boca Raton, FL, USA, 2003.

58. Atomic and Ionic Radii—Wired Chemist. Available online: http://www.wiredchemist.com/chemistry/data/atomic-and-ionic-radii (accessed on 22 May 2018).

59. Li, Y.-H.; Gregory, S. Diffusion of ions in sea water and in deep-sea sediments. *Geochem. Cosmochim. Acta* **1974**, *38*, 703–714.

60. Process Systems Enterprise Limited gPROMS. *Model Builder Documentation 2014. Release 4.0.0-April 2014*; Process Systems Enterprise: London, UK, 2014 20 April.

61. Aydiner, C.; Kaya, Y.; Gönder, Z.B.; Vergili, I. Evaluation of membrane fouling and flux decline related with mass transport in nanofiltration of tartrazine solution. *J. Chem. Technol. Biotechnol.* **2010**, *85*, 1229–1240. [CrossRef]

62. Kamo, N.; Kobatake, Y. Fixed charge density effective to membrane phenomena—Part III. Transference number of small ions. *Kolloid-Z. Z. Polym.* **1971**, *249*, 1069–1076. [CrossRef]

63. Bandini, S.; Bruni, L. Transport Phenomena in Nanofiltration Membranes. In *Comprehensive Membrane Science and Engineering*; Elsevier: Amsterdam, The Netherlands, 2010; Volume 2, pp. 67–89. ISBN 9780080932507.

64. Hagmeyer, G.; Gimbel, R. Modelling the rejection of nanofiltration membranes using zeta potential measurements. *Sep. Purif. Technol.* **1999**, *15*, 19–30. [CrossRef]

65. Takagi, R.; Larbot, A.; Cot, L.; Nakagaki, M. Effect of Al_2O_3 support on electrical properties of TiO_2/Al_2O_3 membrane formed by sol–gel method. *J. Memb. Sci.* **2000**, *177*, 33–40. [CrossRef]

Article

Modeling and Economic Optimization of the Membrane Module for Ultrafiltration of Protein Solution Using a Genetic Algorithm

Tuan-Anh Nguyen [1],* and Shiro Yoshikawa [2]

1   Faculty of Chemical Engineering, Ho Chi Minh City University of Technology, VNU-HCM, 268 Ly Thuong Kiet, Ho Chi Minh City 70000, Vietnam
2   Department of Chemical Science and Engineering, Tokyo Institute of Technology, 2-12-1, Ookayama, Meguro-ku, Tokyo 152-8550, Japan; syoshika@chemeng.titech.ac.jp
*   Correspondence: anh.nguyen@hcmut.edu.vn

Received: 8 November 2019; Accepted: 16 December 2019; Published: 18 December 2019

Abstract: The performance of cross-flow ultrafiltration is greatly influenced by permeate flux behavior, which depends on many factors, including solution properties, membrane characteristics, and operating conditions. Currently, most research focuses on improving membrane performance, both in terms of permeability and selectivity. Only a few studies have paid attention to how the membrane module is configured and operated. In this study, the geometric design and operating conditions of a membrane module are considered as multivariable optimization variables. The objective function is the annual cost. The cost consists of a capital investment depending on the plant scale and an operating expense associated with energy consumption. In the optimization problem, the channel dimensions (width × length × height), and operating conditions (the inlet pressure and recirculation flow rate) were considered as decision variables. The operating configuration of the membrane plant is assumed to be feed and bleed mode, and a model including the pressure drop is introduced. The model is used to simulate the membrane plant and calculate the membrane area and energy usage, which are directly related to the total cost. The genetic algorithm is used for the optimization. The effect of individual parameters on the total cost is discussed.

Keywords: modeling; optimization; ultrafiltration; membrane module; cross-flow; protein solution

---

1. Introduction

The performance of cross-flow ultrafiltration is greatly influenced by the behavior of permeate flux. The flux declination depends on many factors, including solution properties, membrane characteristics, and operating conditions. Currently, most research has focused on improving membrane performance in terms of permeability and selectivity [1,2]. Only a few studies have paid attention to how the membrane module is configured and operated [3].

The selection of a specific membrane module geometry for a specific application is affected by several factors: the fabrication method, energy consumption, and the possibility of fouling [4]. Manufacturers usually suggest the geometric design of the membrane module from the fabrication aspect [4]. There is little evidence that energy efficiency is primarily considered in their approach. Presently, with an increase in energy costs and a decrease in membrane price, energy efficiency should be given greater emphasis in the design of membrane module geometry. Therefore, a module design methodology which includes energy consideration is necessary.

In addition, the determination of operating conditions is normally derived from experience, obtained from the handbook or by recommendation from the membrane manufacturer. However, the performance of the membrane system, which is governed by the permeate flux equation, varies considerably between different situations and applications. In this manner, a general methodology for the design and operation scheme should be studied for any specific application.

In our earlier work [5], a simple combined model, which simultaneously considers pore blockage and cake filtration, was proposed and proved its potential for estimation of flux decline in cross-flow ultrafiltration of protein solution. In our other study [6], the steady-state permeate flux was predicted and correlated to operating conditions from the model. The methodology can be generalized for any particular process to determine the steady state operation equations of the membrane for protein separation. From the operation equation, an optimal design for any certain application can be obtained. This design is dependent on various factors, such as plant capacity, desired recovery, and the energy cost at the specific location. The optimal design will give additional information for supporting the assessment of existing membrane modules and the fabrication and operation decisions of the new membrane system.

However, there are few studies on the optimization of membrane processes and cost estimation. For example, in the study of Wiley, Fell, and Fane [4], membrane module design for brackish water desalination was optimized. However, the configuration is single-pass operation, and the cost only included the membrane cost and energy cost. In the study of Sethi and Wiesner [7], a cost model for the removal of natural organic matter was developed, but the optimization was not conducted. In membrane technology, the feed and bleed configuration, which merges the batch and the single-pass operations, is widely used for continuous full-scale operation [8,9]. In our previous study [10], a simulation and optimization model of a feed and bleed membrane system was investigated. The model was solved by a simple discretization method, and some discontinuous points of the objective function appeared. Moreover, the fixed cost was assumed independently of the system size. Therefore, in this study, the optimization of a membrane module for protein ultrafiltration is considered, in which the geometric design and operating conditions are variables. A system of ordinary differential equations is developed and numerically solved to improve the accuracy of the model. The objective function is the annual cost. The cost consists of various types of capital investments depending on the plant scale and an operating expense associated with energy consumption. The core facilities of the membrane plant are classified into pumps, valves and pipes, instruments and controls, vessels and frames, and miscellaneous. The cost of each category is individually correlated to plant scale, specifically the membrane area. A model incorporating the pressure loss is introduced to simulate the membrane plant and calculate the membrane area and energy usage, which are directly related to the total cost. In the optimization problem, the channel dimensions (width × length × height) and operating conditions (the inlet pressure and recirculation flow rate) were considered as decision variables.

Among various optimization techniques, evolutionary algorithms which are a broad class of population-based metaheuristic optimization techniques inspired by the evolution of species in nature, have been effectively utilized in different fields to solve numerous optimization problems [11]. Evolutionary algorithms, including the genetic algorithm, have some advantages such as being derivative-free and having durable robustness, flexibility, and the ability to discover the global optimum. Therefore, the genetic algorithm is used in this study to find the most cost-effective design and operation of the membrane module.

2. Process Configuration and Model Calculation

In this section, the configuration of the plant operation is introduced. The system of ordinary differential equations for modeling the membrane plant is developed. The results from the simulation are essential for the estimation of the membrane area and energy usage, which are two significant factors affecting the total cost of the plant.

2.1. Membrane Plant Configuration

The operational configuration assumed in developing the model for simulation of membrane plant is presented in this section. The plant's operation is supposed to be a continuous feed and bleed mode, as shown schematically in Figure 1. A summary of the notations is presented in Table 1. There are two important pumps in this configuration: the feed pump supplies the appropriate transmembrane pressure, while the recirculation pump supports the required cross-flow rate through the modules. At startup, the feed pump is used to fill the recirculation/module loop, and then the recirculation pump is started. After the system is stabilized, a small fraction of the flow is continually withdrawn as concentrate at a flow rate (R).

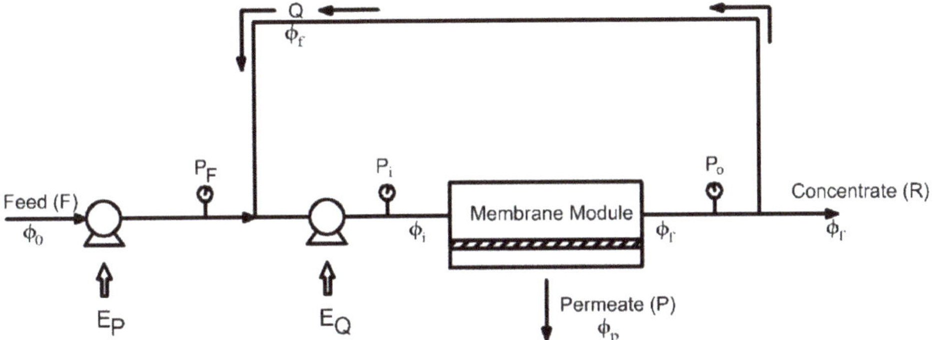

Figure 1. Schematic configuration of feed-and-bleed mode membrane system.

Table 1. Summary of system configuration notations.

Notation	Name and Units
F	Feed flow rate (m³/h)
R	Retentate (concentrate) flow rate (m³/h)
Q	Recirculation flow rate (m³/h)
P	Permeation flow rate (m³/h)
P_F	Pressure at outlet of feed pump (kPa)
P_i	Pressure at the inlet of membrane module (kPa)
P_o	Pressure at the inlet of membrane module (kPa)
E_P	Energy supplied by feed pump (kW)
E_Q	Energy supplied by recirculation pump (kW)
ϕ_0	initial concentration of protein solution (m³/m³)
ϕ_i	inlet concentration of protein solution (m³/m³)
ϕ_f	final concentration of protein solution (m³/m³)
ϕ_P	protein concentration of permeate flux (m³/m³)

2.2. Modeling of Membrane Modules

In ultrafiltration, the pressure is also a significant factor affecting the flux behavior; therefore, the friction loss should be taken into consideration. The model proposed in this study is the improvement of our previous work [10]. In [10], rather than using incremental length step, the channel is subdivided into a number of segments by the protein concentration from the inlet (ϕ_i) to the outlet (ϕ_f). The concentration of the protein solution is expressed as a volume fraction. For each control volume, the transport properties, the permeate flux equation in [6], and friction loss formulae are applied. Then, the length of each increment and the pressure of the next segment can be obtained by mass balance. However, the results show that there are some discontinuous points due to discretization. Therefore, in this study, the chain rule of differentiation [12] is applied and a system of ordinary differential equations was developed. The derivatives are taken with respect to the concentration instead of the length. The mathematical form is discussed in detail as follows.

The overall mass balance and component material balance give:

$$F = R + P$$
$$F\phi_0 = R\phi_f + P\phi_p \tag{1}$$

Material balance around the mixing junction of the feed and recirculation stream provides:

$$F\phi_0 + Q\phi_f = (F + Q)\phi_i \tag{2}$$

In Karasu's study [13], the author reconstituted various solutions with different protein concentrations from the whey protein concentrated powder (ALACENTM, acquired from NZMP Ltd.) and measured their rheological properties. Therefore, the correlation equations from [13] can be used to estimate the properties of protein solution in the simulation.

The correlation equation for the viscosity of protein solution is:

$$\mu_m = 8.94 \times 10^{-4} \exp(13.5482\phi_m) \tag{3}$$

in which μ_m, ϕ_m are the viscosity and concentration of the mixture (solution), respectively.

The density of protein solution (ρ_m) can be approximated from the density of water (1000 kg/m^3) and the density of dry protein powder (1360 kg/m^3)

$$\rho_m = 1000(1 - \phi_m) + 1360\phi_m \tag{4}$$

The permeate flux through the membrane is (obtained from [6])

$$\frac{J}{u} = 3.66 \times 10^{-7} \left(\frac{P}{\rho_m u^2}\right)^{0.27} \left(\frac{\rho_m u d_h}{\mu_m}\right)^{0.52} \tag{5}$$

in which P is the transmembrane pressure, u is the cross-flow velocity, and d_h is the hydraulic diameter of the flow channel.

From the shear-induced diffusion or surface transport model, the permeate flux depends linearly on the membrane shear rate [14]. The increase of module height will decrease the shear rate at the membrane surface. In the experiment of this study, the height of the module channel remained constant. Therefore, in order to extend the application of the permeate flux equation to other systems which has different module heights, the correction factor d_{h0}/d_h should be introduced as follows.

$$\frac{J}{u} - 3.66 \times 10^{-7} \left(\frac{P}{\rho_m u^2}\right)^{0.27} \left(\frac{\rho_m u d_h}{\mu_m}\right)^{0.52} \left(\frac{d_{h0}}{d_h}\right) \tag{6}$$

where d_{h0} is the value of hydraulic diameter of the membrane module used in the study [5].

The equation for permeate flux can be rewritten as:

$$J = 5.124 \times 10^{-9} \left(\frac{P}{\rho_m u^2}\right)^{0.27} \left(\frac{\rho_m u d_h}{\mu_m}\right)^{0.52} \left(\frac{u}{d_h}\right) \tag{7}$$

or in terms of shear rate at the membrane surface $\dot{\gamma} = \frac{6u}{h} = \frac{12u}{d_h}$, where h is the channel height:

$$J = 4.27 \times 10^{-10} \left(\frac{P}{\rho_m u^2}\right)^{0.27} \left(\frac{\rho_m u d_h}{\mu_m}\right)^{0.52} \dot{\gamma} \tag{8}$$

The flow rate and cross-flow velocity at the entrance of the membrane module

$$Flow_i = F + Q \tag{9}$$

$$u_i = \frac{F+Q}{A_{cross}} = \frac{F+Q}{h \times w} \tag{10}$$

where $w \times h$ is the cross-sectional area of the flow channel, w is the width, and h is the height of the membrane module.

The flow rate and velocity change along the length of the membrane module are determined by the mass balance with the assumption that there is no protein in the permeation, as:

$$Flow \times \phi_m = const \tag{11}$$

$$d(Flow \times \phi_m) = 0 \tag{12}$$

After applying the product rule and then rearranging, we obtain

$$\frac{d(Flow)}{d\phi_m} = -\frac{Flow}{\phi_m} \tag{13}$$

$$u = \frac{Flow}{h \times w} \tag{14}$$

The mass balance for the element control volume along the length of the membrane:

$$d(Flow) = -J \times d(wz) = -J\,w\,dz \tag{15}$$

Incorporating with Equation (13), the differential equation for the length can be obtained:

$$dz = -\frac{d(Flow)}{w\,J} = \frac{Flow \times d\phi_m}{\phi_m\,w\,J} \tag{16}$$

The pressure loss is calculated by the Darcy–Weisbach equation in differential form as follows

$$dP = -4f\rho_m \frac{dz}{d_h}\frac{u^2}{2} \tag{17}$$

in which f is the friction factor

$f = \frac{24}{Re}$ for laminar flow in a rectangular channel (Re < 2000) [15].

$f = \frac{0.079}{Re^{0.25}}$ for turbulent flow (Re > 2000) (smooth pipes, Blasius correlation [16]).

Incorporating with Equation (16), the differential equation for the pressure can be obtained:

$$\frac{dP}{d\phi_m} = -4f\rho_m \frac{1}{d_h}\frac{u^2}{2}\frac{Flow}{\phi_m}\frac{1}{w\,J} \tag{18}$$

Finally, the system of ordinary equations for the membrane module system was developed as follows.

$$\left\{ \begin{array}{l} \dfrac{d(Flow)}{d\phi_m} = -\dfrac{Flow}{\phi_m} \\[4pt] \dfrac{dz}{d\phi_m} = \dfrac{Flow}{\phi_m\,w\,J} \\[4pt] \dfrac{dP}{d\phi_m} = -4f\rho_m\dfrac{1}{d_h}\dfrac{u^2}{2}\dfrac{Flow}{\phi_m}\dfrac{1}{w\,J} \\[4pt] \mu_m = 8.94 \times 10^{-4}\exp(13.5482\phi_m) \\[4pt] \rho_m = 1000(1-\phi_m) + 1360\phi_m \\[4pt] J = 5.124 \times 10^{-9}\left(\dfrac{P}{\rho_m u^2}\right)^{0.27}\left(\dfrac{\rho_m u d_h}{\mu_m}\right)^{0.52}\left(\dfrac{\mu}{d_h}\right) \\[4pt] u = \dfrac{Flow}{h \times w} \\[4pt] f = \dfrac{24}{Re} \text{ if } Re < 2000 \\[4pt] f = \dfrac{0.079}{Re^{0.25}} \text{ if } Re > 2000 \end{array} \right. \tag{19}$$

In the interval $[\phi_i, \phi_f]$ and the initial condition

$$\begin{cases} Flow(\phi_i) = F + Q \\ z(\phi_i) = 0 \\ P(\phi_i) = P_i \end{cases} \tag{20}$$

The system of the ordinary equations can be solved by the Runge–Kutta 4th order method [17]. The solution is the flow rate, the length, and the pressure at the final concentration. After the total length was determined, the membrane area was obtained:

$$A_{membrane} = w \times L \tag{21}$$

The total energy generated by the pumps:

$$\begin{aligned} E_{pumps} &= E_P + E_Q \\ &= F \times P_F + \{F \times (P_i - P_F) + Q \times (P_i - P_o)\} \\ &= F \times P_i + Q \times \Delta P_{drop} \end{aligned} \tag{22}$$

In this equation,

E_P, E_Q are the power supplied by the feed pump and recirculation pump, respectively
P_F, P_i, P_o are the pressure at the outlet of the feed pump, at the inlet and outlet of the membrane module, respectively
ΔP_{drop} is the pressure drop in the membrane module

For desirable feed flow rate, recirculation flow rate, and initial and final concentration, the total membrane area and energy usage are calculated from the simulation model with the governing equation of permeate flux. The total membrane area and energy usage are the important factors in the estimation of the total cost of the filtration plant, and the correlation to the total cost is discussed in the next section.

2.3. Cost Estimation

2.3.1. Operating Cost

The operating cost takes into account the energy expense of the pumps and the cost of membrane replacement. The annual energy expense of the pumps is calculated as

$$C_{energy} = \frac{E_{pumps}}{\eta} \times 8000 \times \frac{3600}{1000} \times energy\ cost \quad [\$/year] \tag{23}$$

E_{pumps} is the power supplied of the two pumps, feed pump and recirculation pump, and is evaluated by Equation (22). The assumption used in the calculation of annual energy expense is 8000 h/year operation (24 h/day and 333 days/year). η is the efficiency of the pumps, which is set to be 0.7. Energy cost is the electricity price as \$/kWh, and is supposed to be 0.08, which is the price for the industrial sector in the United States [18].

The membrane replacement cost is calculated as

$$C_{membrane} = A_{membrane} \times c_{membrane} \times \left(\frac{A}{P}\right) \quad [\$/year] \tag{24}$$

where $C_{membrane}$ [\$/m^2/year] is the membrane replacement cost calculated per year and $c_{membrane}$ [\$] is the membrane price per unit area. $\left(\frac{A}{P}\right)$ is the amortization factor, which presents the time value of money [19] and is calculated as a function of interest rate i and the membrane life.

The membrane price is usually about 200 \$/m^2 [9], and membrane life is 12–18 months. Therefore, the membrane replacement cost per year is roughly estimated as 200 \$/m^2/year for the interest of $I = 8\%$.

2.3.2. Capital Cost

It is empirically observed that all capital cost components, such as construction cost, pump cost, and equipment cost, are correlated with the plant capacity, size of the main equipment, or other design parameters, in the power-law form [20]:

$$\text{cost} = k(\text{size})^n \tag{25}$$

Instead of directly relating the capital cost of the membrane to capacity, Sethi and Wiesner [7] proposed an approach for predicting capital costs of membrane processes. In this approach, the capital cost is divided into several major categories and the capital cost for each category is developed to improve the accuracy. Membrane system costs generally consist of the costs of pumps and other manufactured equipment. The correlation of each category to the plant capacity is discussed in detail as follows.

Pump Capital Cost

The capital cost for pumps can be estimated using the power-law relation as shown in Equation (25), where the relevant size parameter is the power of the pump:

$$C^*_{\text{pump}} = k_{\text{p}}(\text{power of pump})^{n_{\text{p}}} \tag{26}$$

The cost correlation for general-purpose-single and two-stage-single-suction centrifugal pumps discussed in Perry, et al. [21] was used in this study to calculate the pump capital cost. This relation is as follows:

$$C^*_{\text{pump}} = I \times f_1 \times f_2 \times L \times 81.27 \times (Q \times P)^{0.4} \tag{27}$$

in which

I: a cost index ratio for updating the cost to the recent year
f_1: an adjust factor for pump construction material
f_2: an adjust factor for suction pressure range
L: a factor used to incorporate labor costs
Q: flow capacity of the pump [m^3/h]
P: pressure outlet of the pump [kPa]

The cost index, I in Equation (27), can be referred to as the chemical engineering (CE) index and obtained from "Chemical Engineering" magazine. The values of CE index of several years were also tabulated in [22]. The cost index I calculated in this study is 2.4 to update the pump cost.

Usually, 40% of the cost is required for labor to install manufactured equipment [23], thus $L = 1.4$. The factors f_1 and f_2 can be found in [21]. For the application in food industries, the material was set to stainless steel, indicating that $f_1 = 1.5$. Because the operating pressure of most UF and MF applications are below 10 bar (1 MPa), $f_2 = 1.0$ was used in this study.

In the system configuration used in this study, the size factor ($Q \times P$ in Equation (27)) of the two pumps (feed pump and recirculation pump) is calculated as

$$\text{pump size} = (F + Q) \times P_{\text{i}} \tag{28}$$

Capital Cost of Other Equipment

In membrane applications, the membrane area required is the most significant factor which determines the size of the plant [24]. Therefore, the membrane area should be the size parameter used in power-law correlation expression for the estimation of the capital cost of various components.

Sethi [25] segregated nonmembrane equipment and facilities, excluding the pumps which have already been evaluated in Pump Capital Cost, into four main categories: (1) pipes and valves; (2) instruments and controls; (3) tanks and frames; and (4) miscellaneous. The author investigated different sources of data and proposed the calibrated cost correlation for various membrane system components as the following equations:

1. Pipes and valves

$$C^*{}_{\text{PV}} = 6000(A_{\text{membrane}})^{0.42} \quad [\$] \tag{29}$$

2. Instruments and controls

$$C^*{}_{\text{IC}} = 1500(A_{\text{membrane}})^{0.66} \quad [\$] \tag{30}$$

3. Tanks and frames

$$C^*{}_{\text{TF}} = 3100(A_{\text{membrane}})^{0.53} \quad [\$] \tag{31}$$

4. Miscellaneous

$$C^*{}_{\text{MI}} = 8000(A_{\text{membrane}})^{0.57} \quad [\$] \tag{32}$$

Annual Capital Cost

The annual capital cost will be obtained from the capital cost based on the amortization factor, as

$$C_{\text{capital cost}} = C^*{}_{\text{capital}} \times \left(\frac{A}{P}\right) \quad [\$/\text{year}] \tag{33}$$

For the plant design year of 20 years and the interest rate 8%, the amortization factor will be about 0.1.

3. Formulizations of the Problem

3.1. Fix Parameters and Design Variables

The optimization problem is the geometric design and operating condition of the membrane system. The operation is under steady-state conditions. In the design problem, some variables should meet the requirement of the process. These are inlet variables and usually are feed flow Γ, inlet concentration ϕ_0, and outlet concentration ϕ_f. The protein is assumed to be entirely rejected by the membrane ($\phi_p = 0$).

The decision variables investigated in the optimization problem were: channel geometry (width × length × height), the inlet pressure (P_i), and recirculation flow rate (Q).

3.2. Objective Function

The objective function consists of both capital cost and operating cost expressed per year:

$$\text{minimum } f(\mathbf{x}) = \text{annual total cost} = \text{annual capital cost} + \text{annual operating cost} \tag{34}$$

The capital cost includes the cost of membranes, pumps, pipes, and valves, instruments and controls, tanks and frames, miscellaneous (buildings, storage, etc . . .).

The operating cost includes membrane replacements and pumping energy cost.

3.3. Constraints

The pressure drop should ensure that the pressure at the outlet point will be positive. This constraint is easily set by defining that the objective function is infinity if the outlet pressure is negative.

The bounded constraint also affects the optimization results. Based on typical available data, the physically meaningful ranges, the manufacturing problem, operating properties, etc., the decision variables are usually bounded on a finite range:

$$\mathbf{x}_l \leq \mathbf{x} \leq \mathbf{x}_u \tag{35}$$

where $\mathbf{x}$ is the vector of decision variables $\mathbf{x} = (P_i, Q, w, h)$, subscript l and u indicate the lower and upper bound.

The system parameters and variables are summarized in Table 2.

Table 2. System parameters and variables.

Parameters	Value
Feed flow rate (m^3/h)	0.02–200
Inlet pressure (kPa)	200–1000
Recirculation flow rate (m^3/h)	0–50
Initial solid fraction (m^3/m^3)	0.1
Final solid fraction (m^3/m^3)	0.4
Plant design year (year)	20
Interest rate (%)	8
Energy price ($/kWh)	0.08
Efficiency of pumps (%)	70
Operating temperature (°C)	25
Module height (mm)	0–100
Module width (m)	0–30

3.4. Optimization by Genetic Algorithm

Genetic algorithms (GA), a class of evolutionary algorithms (EA), are adaptive heuristic search algorithms inspired by the evolutionary ideas of natural selection and genetics. The genetic algorithms start with arbitrarily picked parent chromosomes from the search space, each representing a candidate solution, to form an initial population. The fitness of each individual is indicated by the value of the objective function. The population of individual solutions is repeatedly modified in an analogous way to processes happening in nature, such as selection, reproduction by cross-over, and mutation. Over successive generations, the population evolves, and the optimal solution might be achieved by the principles of "survival of the fittest". More detailed discussions of the genetic algorithm can be found elsewhere, such as in [26].

In this study, the parameters of GA, population size, crossover probability, mutation probability values, were set to be, 100, 1.0, and 0.30, respectively. The selection was based on roulette wheel selection with elitism, which means that the best individual always survives to the next generation. The number of generations was 500. Due to the minimization of the annual total cost, the fitness function was defined as:

$$fitness = \begin{cases} 0, & \text{if the annual total cost} > 5 \times 10^5 \\ 5 \times 10^5 - \text{annual total cost}, & \text{otherwise} \end{cases} \tag{36}$$

4. Results and Discussion

4.1. Effect of Recirculation Flow Rate on the Total Cost

The relationship between the recirculation flow rate and the annual total cost is shown in Figure 2. Other design variables are held constant, inlet pressure TMP = 500 kPa, module height h = 5 mm, module width w = 1 m.

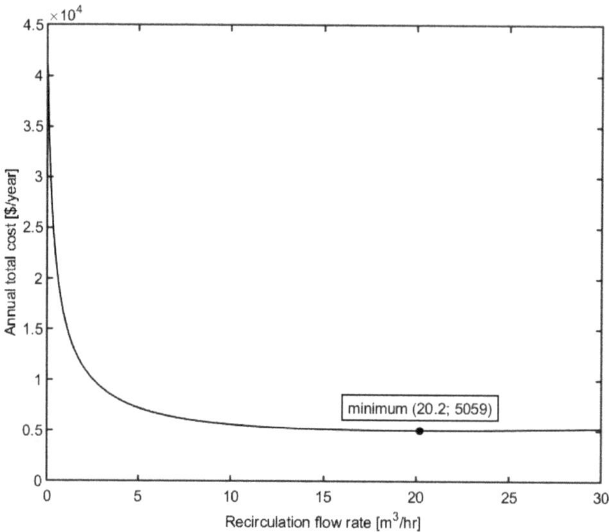

Figure 2. Effect of recirculation flow rate on the annual total cost of the plant.

As illustrated in the figure, the increase in the recirculation flow rate initially reduces the total cost of the membrane plant. However, further increasing the recirculation flow rate causes the total cost to increase. This might be explained by considering the permeate flux and power consumption. The rise in flow rate leads to an increase in the Reynolds number and the permeate flux, as demonstrated in the Equation (7). Consequently, the required membrane area can be reduced, which results in a decrease in capital investment and membrane replacement cost. Hence, the annual total cost of the plant is minimized. However, from the Equations (17) and (22), it can be obtained that the pressure drop increases rapidly in the turbulent flow regime and faster than the increase in permeate flux (the exponents of the velocity dependency are 1.75 and 1, respectively). The pressure loss is directly related to the power consumption of the pumps or the energy cost. Therefore, as the recirculation flow rate increases, the annual total cost decreases and reaches a minimum value, and then rises again with a further increase in flow rate.

4.2. Effect of Module Inlet Pressure Operation

Figure 3 demonstrates the relation between the operating pressure at the inlet and the annual total cost. Other design variables are held constant, recirculation flow rate Q = 35 m³/h, module height h = 15 mm, module width w = 1 m.

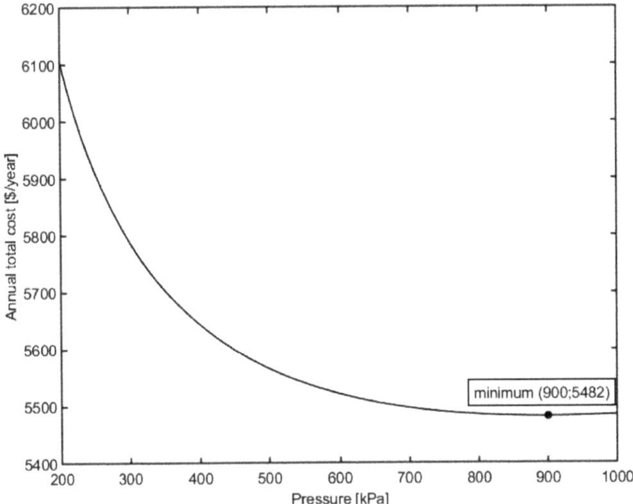

Figure 3. Effect of inlet pressure on the total cost of the plant.

As illustrated in the figure, the increase in the inlet pressure initially minimizes the annual total cost of the membrane plant. However, a further increase in pressure causes the annual total cost to increase. The explanation is as follows. The permeate flux benefits from the increase in pressure operation with the order of dependence of 0.26, as obtained from Equation (7). Hence, the required membrane area can be reduced, which results in a decrease in the capital investment and membrane replacement cost. Therefore, the annual total cost of the membrane plant is minimized. However, from Equations (22) and (27), it can be obtained that both the power consumption and the initial investment for the pumps rise much faster than the increase of permeate flux (the exponents of the pressure dependency are 1 and 0.4, respectively). Therefore, as the inlet operating pressure increases, the annual total cost decreases and reaches a minimum value, and then rises again with a further increase in operating pressure.

4.3. Effect of Module Height

Figure 4 shows the effect of module height on the total cost. Other design variables are kept constant, inlet pressure TMP = 500 kPa, recirculation flow rate Q = 45 m^3/h, module width w = 1 m.

From the figure, it is shown that the decrease in the height of the membrane module will decrease the total cost of the membrane plant. However, a further decrease in module height makes the total cost also increase. The reason is similar to the effect of the recirculation flow rate. The decrease in module height will increase the cross-flow velocity and thus increase the permeate flux. Therefore, the membrane area decreases, which leads to a decrease in the membrane replacement cost. Consequently, the annual total cost of the membrane plant will decrease. However, high velocity affects the pressure loss and hence, energy consumption. Therefore, the annual total cost initially decreases and reaches a minimum value, then rises again with a further decrease in module height.

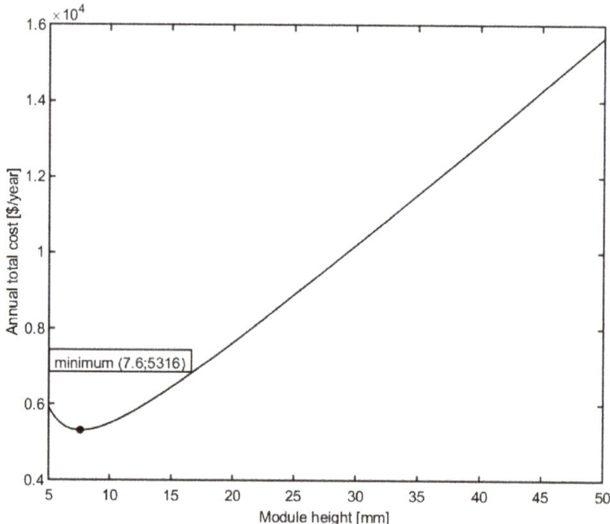

Figure 4. Effect of module height on the total cost of the plant.

4.4. Effect of Module Width

Figure 5 shows the effect of module height on the total cost. Other design variables are kept constant, inlet pressure TMP = 500 kPa, recirculation flow rate Q = 35 m^3/h, module height h = 10 mm.

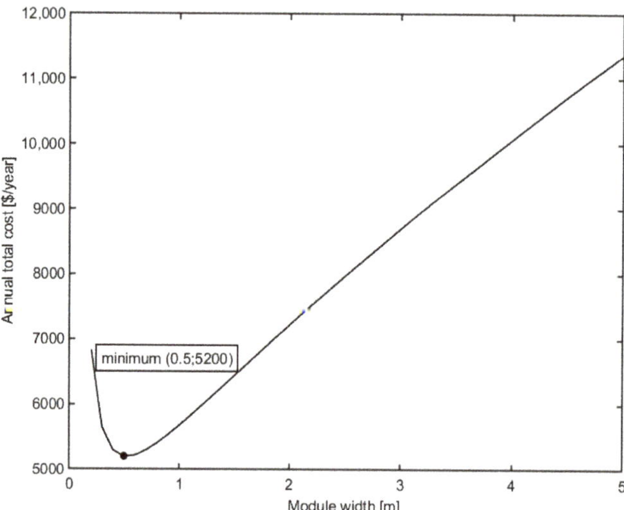

Figure 5. Effect of module width on the total cost of plant.

Similar to the effect of the recirculation flow rate, it is shown that the decrease in the height of the membrane module will decrease the total cost of the membrane plant. However, a further decrease in module height makes the total cost also increase. The decrease in module width will increase the cross-flow velocity and shear rate, thus increase the permeate flux. Therefore, the membrane area decreases, which leads to a decrease of membrane replacement cost, and the annual total cost of the membrane plant will decrease. However, high velocity affects the pressure loss and hence, energy consumption. Therefore, the annual total cost initially decreases and reaches a minimum value, then rises again with a further decrease in module height.

4.5. Optimum Design

The optimum design of various feed flow rate designs was done in order to illustrate the method. The lower bound for the membrane width is 0.1 m and the lower bound for the module height is 0.5 mm. The results of the design are shown in Table 3.

Table 3. Optimum design of the membrane module.

Feed [m³/h]	ϕ_0 [-]	ϕ_f [-]	Pressure [kPa]	Recirculation [m³/h]	Width [m]	Height [mm]	Total Cost [$/yr]
0.02	0.1	0.4	523	2.8	0.1	5.0	1.29×10^3
0.2	0.1	0.4	1000	4.9	0.1	8.9	4.30×10^3
2	0.1	0.4	1000	0.2	0.1	6.9	1.18×10^4
20	0.1	0.4	987	0.2	1.3	5.0	5.60×10^4
200	0.1	0.4	1000	0.8	11.1	5.0	3.65×10^5

From Table 3, it can be obtained that the operating pressure of the optimum design plant is at a high value when the feed flow rate is not so low. When operating at high pressure, the permeate flux increases, resulting a reduction in the membrane area. However, if the operating pressure is too high, it will affect the mechanical stability of the system in addition to the pump cost expression. Therefore, the plant should operate at the maximum allowance of pressure.

Figure 6 shows the total cost per unit flow rate design. The cost per unit flow rate design decreases when the plant size becomes larger. It reflects the economics of scale.

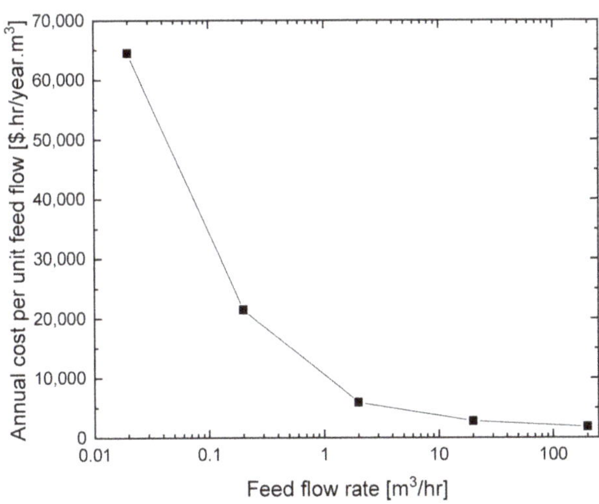

Figure 6. The behavior of cost per unit flow rate design in optimum condition with plant capacity.

From the results, it can be shown that the membrane plant should be designed as long in channel length but short in channel width to achieve a high cross-flow velocity. The recirculation flow rate is also operated at a moderate value to maintain relatively high cross-flow velocity, but to not consume too much energy for the pump. The channel height should be configured at several millimeters.

From the results it can also be obtained that at different requirements of the process, such as the desired feed flow rate, the membrane plant configuration and operation condition will significantly change. The tendency is difficult to predict. Therefore, the design requirement and permeate flux expression strongly affect the design, selection, and operation strategy in membrane separation processes. There is no general rule, and for a particular system, the permeate flux and the correlation with operating conditions and membrane geometry should be investigated. The design and operating strategy strongly depend on this correlation.

5. Conclusions

In this study, a simulation of a membrane plant was discussed. A system of ordinary differential equations was developed to accurately simulate the membrane module. The system of equations reflects the change of flow properties along the membrane and incorporates the pressure drop. The steady-state permeate flux presented in [6] is assumed as the governing equation for the filtration process. An economic model including the capital cost as well as the operating cost was used to evaluate the performance of the membrane plant. The effect of operating and design parameters, such as recirculation flow rate, inlet pressure, and membrane module geometry, on the economic viewpoint of membrane application was investigated. All the parameters have a critical point which minimizes the annual cost while the other parameters are fixed. The search for an optimum design using genetic algorithm suggests that the system should be operated at high pressure and short channel width. Other parameters should be further investigated so that the cross-flow velocity is maintained at a high value, but does not consume too much energy. It also suggests that the design, selection, and operating strategy strongly depend on the requirement of the process and the permeate flux and its correlation to system parameters. The procedure presented in this study might be extended for the general design of a particular process in which the governing equations are known.

Author Contributions: Conceptualization, T.-A.N.; Methodology, T.-A.N.; Supervision, S.Y.; Writing—original draft, T.-A.N.; Writing—review & editing, S.Y. All authors have read and agreed to the published version of the manuscript.

Funding: This research received no external funding.

Acknowledgments: The authors would like to thank Japan International Cooperation Agency (JICA) and Tokyo Institute of Technology for the kind support during the preparation of the manuscript.

Conflicts of Interest: The authors declare no conflict of interest.

References

1. Mores, P.L.; Arias, A.M.; Scenna, N.J.; Caballero, J.A.; Mussati, S.F.; Mussati, M.C. Membrane-Based Processes: Optimization of Hydrogen Separation by Minimization of Power, Membrane Area, and Cost. *Processes* **2018**, *6*, 221. [CrossRef]
2. Ramírez-Santos, Á.A.; Bozorg, M.; Addis, B.; Piccialli, V.; Castel, C.; Favre, E. Optimization of multistage membrane gas separation processes. Example of application to CO_2 capture from blast furnace gas. *J. Membr. Sci.* **2018**, *566*, 346–366. [CrossRef]
3. Ohs, B.; Lohaus, J.; Wessling, M. Optimization of membrane based nitrogen removal from natural gas. *J. Membr. Sci.* **2016**, *498*, 291–301. [CrossRef]
4. Wiley, D.E.; Fell, C.J.D.; Fane, A.G. Optimisation of membrane module design for brackish water desalination. *Desalination* **1985**, *52*, 249–265. [CrossRef]
5. Nguyen, T.-A.; Yoshikawa, S.; Karasu, K.; Ookawara, S. A simple combination model for filtrate flux in cross-flow ultrafiltration of protein suspension. *J. Membr. Sci.* **2012**, *403–404*, 84–93. [CrossRef]
6. Nguyen, T.-A.; Yoshikawa, S.; Ookawara, S. Steady State Permeate Flux Estimation in Cross-Flow Ultrafiltration of Protein Solution. *Sep. Sci. Technol.* **2014**, *49*, 1469–1478. [CrossRef]
7. Sethi, S.; Wiesner, M.R. Performance and Cost Modeling of Ultrafiltration. *J. Environ. Eng.* **1995**, *121*, 874–883. [CrossRef]
8. Zeman, L.J.; Zydney, A.L. *Microfiltration and Ultrafiltration: Principles and Applications*; Taylor & Francis: New York, NY, USA, 1996.
9. Cheryan, M. *Ultrafiltration and Microfiltration Handbook*; Taylor & Francis: New York, NY, USA, 1998.

10. Nguyen, T.-A.; Yoshikawa, S.; Ookawara, S. Effect of operating conditions in membrane module performance. *Asean Eng. J. Part B* **2015**, *4*, 4–13.

11. Sumathi, S.; Hamsapriya, T.; Surekha, P. *Evolutionary Intelligence: An Introduction to Theory and Applications with Matlab*; Springer: Berlin/Heidelberg, Germany, 2008.

12. Apostol, T.M. *Mathematical Analysis*; Addison-Wesley: Boston, MA, USA, 1974.

13. Karasu, K. A study on permeation phenomena in cross-flow ultrafiltration producing a compressible cake layer. Ph.D. Thesis, Tokyo Institute of Technology, Tokyo, Japan, 2010.

14. Porter, M.C. Concentration Polarization with Membrane Ultrafiltration. *Prod. R&D* **1972**, *11*, 234–248. [CrossRef]

15. White, F.M. *Fluid Mechanics*; McGraw Hill: New York, NY, USA, 2011.

16. Pritchard, P.J. *Fox and McDonald's Introduction to Fluid Mechanics*, 8th ed.; John Wiley & Sons: Hoboken, NJ, USA, 2010.

17. Buzzi-Ferraris, G.; Manenti, F. *Differential and Differential-Algebraic Systems for the Chemical Engineer: Solving Numerical Problems*; Wiley: Hoboken, NJ, USA, 2015.

18. U.S. Energy Information Administration. Average Price of Electricity to Ultimate Customers: Total by End-Use Sector. In *Electric Power Monthly*; Table 5.6.A.; U.S. Department of Energy: Washington, DC, USA, 2019.

19. Park, C.S. *Fundamentals of Engineering Economics: International Edition*; Pearson Education Limited: London, UK, 2013.

20. Guthrie, K.M. *Process Plant Estimating, Evaluation, and Control*; Craftsman Book Company of America: Carlsbad, CA, USA, 1974.

21. Perry, R.H.; Chilton, C.H.; Perry, J.H. *Chemical Engineers' Handbook*, 5th ed.; McGraw-Hill: New York, NY, USA, 1973.

22. Green, D.W.; Perry, R.H. *Perry's Chemical Engineers' Handbook*, 8th ed.; McGraw-Hill Education: New York, NY, USA, 2007.

23. Holland, F.A.; Wilkinson, J.K. Process Economics. In *Perry's Chemical Engineers' Handbook*, 7th ed.; Perry, R.H., Green, D.W., Eds.; McGraw-Hill: New York, NY, USA, 1997.

24. Mir, L.; Michaels, S.L.; Goel, V.; Kaiser, R. Crossflow Microfiltration: Applications, Design, and Cost. In *Membrane Handbook*; Ho, W.S.W., Sirkar, K.K., Eds.; Springer: Boston, MA, USA, 1992; pp. 571–594.

25. Sethi, S. Transient Permeate Flux Analysis, Cost Estimation, and Design Optimization in Crossflow Membrane Filtration. Ph.D. Thesis, Rice University, Houston, TX, USA, 1997.

26. Haupt, R.L.; Haupt, S.E. *Practical Genetic Algorithms*; Wiley: Hoboken, NJ, USA, 2004.

Article

Sustainable Synthesis Processes for Carbon Dots through Response Surface Methodology and Artificial Neural Network

Musa Yahaya Pudza [1,*], Zurina Zainal Abidin [1,*], Suraya Abdul Rashid [1], Faizah Md Yasin [1], Ahmad Shukri Muhammad Noor [2] and Mohammed A. Issa [1]

[1] Department of Chemical and Environmental Engineering, Faculty of Engineering, Universiti Putra Malaysia, Serdang 43400, Selangor, Malaysia; suraya_ar@upm.edu.my (S.A.R.); fmy@upm.edu.my (F.M.Y.); academicfinest@gmail.com (M.A.I.)

[2] Department of Computer and Communication System Engineering, Faculty of Engineering, Universiti Putra Malaysia, Serdang 43400, Selangor, Malaysia; ashukri@upm.edu.my

* Correspondence: pudzamusa@gmail.com (M.Y.P.); zurina@upm.edu.my (Z.Z.A.)

Received: 30 July 2019; Accepted: 30 August 2019; Published: 5 October 2019

Abstract: Nowadays, to ensure sustainability of smart materials, it is imperative to eliminate or reduce carbon footprint related to nano material production. The concept of design of experiment to provide an optimal synthesis process, with a desired yield, is indispensable. It is the researcher's goal to get optimum value for experiments that requires multiple runs and multiple inputs. Herein, is a reliable approach of utilizing design of experiment (DOE) for response surface methodology (RSM). Thus, to optimize a facile and effective synthesis process for fluorescent carbon dots (CDs) derived from tapioca that is in line with green chemistry principles for sustainable synthesis. The predictions for fluorescent CDs synthesis from RSM were in excellent agreement with the artificial neural network (ANN) model prediction by the Levenberg–Marquardt back propagation (LMBP) algorithm. Considering R^2, root mean square error (RMSE) and mean absolute error (MAE) have all revealed a positive hidden layer size. The best hidden layer of neurons were discovered at point 4-8, to confirm the validity of carbon dots, characterization of surface morphology and particles sizes of CDs were conducted with favorable confirmations of the unique characteristics and attributes of synthesized CDs by hydrothermal route.

Keywords: tapioca; response surface methodology; artificial neural network; carbon dots; hydrothermal; photoluminescence; organic

---

1. Introduction

The process of optimizing synthesis parameters (factors) to provide high quality organic carbon dots (CDs) represents a complex process. It is a similitude of a search in the dark. Several researchers have reported a low value of photoluminescence, long hours synthesis, and high volume of resources used in the process of CDs [1,2]. Attempting to synthesize CDs needs optimization by an appropriate mathematical model, which is embedded with the task to optimize the synthesis process in terms of quality criteria and prediction with less error. This is necessary due to the influence of factors that may or may not affect the quality of yield. So far, the synthesis of sustainable organic CDs with high performance yield is still being researched. In the past, synthesis processes that involved multiple factors were conducted by varying a single factor while others were kept constant, known as one variable at a time (OVAT), but this method is time consuming. It became imperative to formulate multivariate statistics that substantially reduce the numbers of experiments [3].

Nwobi-Okoye and Ochieze [4] made a study based on the comparison of Response Surface Methodology (RSM) and Artificial Neural Network (ANN) data to validate the prediction of aluminum alloy A356/cow horn particulate composites hardness. The study confirmed that ANN with R^2 of 0.9921 exhibited better accuracy than the RSM with R^2 of 0.9583 in predicting the hardness values of the composites [4]. The ANN model is generally based upon artificial intelligence (machine learning), under which a predefined set of data is being trained [5,6], validated, and tested for prediction purposes. Due to this constraint, it is worthy to note that the values predicted by ANN will not often be the best predicted values, but will be within the range of the experimental study [7,8].

RSM is a technique that has been widely applied for defining the interactions between various process parameters and responses with the various desired criteria and taking note of the significance of these process parameters on the desired responses [9]. However, RSM is reported not to be desirable to optimize the non-linear system study that possesses minimal difference in parts, processing boundaries, or investigated data sets because it affects the overall properties of material [9]. The prediction of RSM is based on a first or second order polynomial equation, hence, it is inadequate enough to capture non-linear behavior and can give a non-reliable estimation of photoluminescent quantum yield of fluorescent carbon dots of organic origin. Consequently, the application of the artificial neural network (ANN) can be employed to checkmate and surmount concerns of using a lone RSM in predicting the non-linear system. The concept of ANN is an independent method that uses a model that effectively handles nonlinearity in responses that concerns the synthesis and photoluminescent quantum yield of carbon dots.

Formulating an ANN model which accepts a small data set of experimental runs while supplying a useful output in the synthesis of an advanced nanoparticle (CDs) is studied here. From the studies conducted, there are no results published by using the Levenberg–Marquardt back propagation (LMBP) algorithm in the synthesis of fluorescent carbon dots from tapioca powder. With this in view, the LMBP training algorithm was built and deployed in the current study to predict the photoluminescent quantum yield of the synthesized carbon dots from tapioca powder.

In this report, sample data were acquired from design of experiment for response surface methodology (RSM). The training and predictions by an artificial neural network (ANN) were carried out by different multi inputs, and multi output ANN, developed using the Levenberg–Marquardt back propagation (LMBP) algorithm to predict the fluorescent properties of carbon dots synthesized in the study.

2. Mathematical Models and Analytical Methods

2.1. Response Surface Method and Mathematical Model

The design software utilized was Design-Expert Version 11.0.5. Central composite design (CCD) was adopted for the analysis of effects. These consist of four (4) independent variables: Temperature (X1), Dosage (X2), Time (X3), and W/Ace/NaOH ratio (X4), as shown in Table 1 below. A total sum of 30 experimental runs were carried out, photoluminescent quantum yield (PLQY) was the response considered and expressed as the dependent variable shown in Table 2. The inputs variables were expressed individually as a function of independent variables. A second-order polynomial equation was used to express PLQY (Y1) as an independent variable. Given in the equation below:

$$Y = \beta_0 + \sum_{i=1}^{4} \beta_i X_i + \sum_{i=1}^{4} \beta_{ii} X_i^2 + \sum_{i<j}^{4} \beta_{ij} X_i X_j \tag{1}$$

where, Y represents the response variable, β_0 is a constant, β_i, β_{ii}, and β_{ij} are the linear, quadratic, and cross-product coefficients, respectively. Xi and Xj are the respective levels of the independent variables. Three dimensional (3D) surface response plots were generated by varying two variables at a time within the experimental range and holding the other two constant at the central point. Furthermore, a test of statistical significance was based on the total error criteria with a confidence level of 95.0%. Below is the response surface methodology (RSM) design summary.

Table 1. Independent variables used in the Response Surface Methodology (RSM) design.

Factor	Name	Units	Low Actual	High Actual	Low Coded	High Coded	Mean	Std. Dev.
A (X_1)	Temp	°C	75.00	175.00	−1.000	1.000	125.0	40.825
B (X_2)	Dosage	g	0.100	0.50	−1.000	1.000	0.30	0.163
C (X_3)	Time	min	45.00	105.00	−1.000	1.000	75.00	24.495
D (X_4)	W/Ace/NaOH	mL	8.00	40.00	−1.000	1.000	24.00	13.064

Table 2. Design of experiment for response surface methodology report.

Std Order	Factor-A Temperature (°C)	Factor-B Dossage (gram)	Factor-C Time (min)	Factor-D Solvent (mL) ($H_2O/C_3H_6O/NaOH$)	Exp. Actual Value (PLQY)	Pred. Value	Res. Value
1	75	0.10	45	8.00	14.67	14.41	0.26
2	175	0.10	45	8.00	21.05	20.89	0.15
3	75	0.50	45	8.00	22.80	22.35	0.45
4	175	0.50	45	8.00	19.96	19.13	0.83
5	75	0.10	105	8.00	14.00	13.01	0.99
6	175	0.10	105	8.00	25.27	26.13	−0.86
7	75	0.50	105	8.00	20.15	21.52	−1.36
8	175	0.50	105	8.00	24.87	24.94	−0.06
9	75	0.10	45	40.00	24.82	24.39	0.42
10	175	0.10	45	40.00	20.99	19.88	1.11
11	170	0.1	100	12.00	27.75	27.38	0.37
12	175	0.50	45	40.00	12.53	13.16	−0.63
13	75	0.10	105	40.00	17.90	18.99	−1.09
14	175	0.10	105	40.00	21.04	21.13	−0.09
15	75	0.50	105	40.00	22.75	22.55	0.21
16	175	0.50	105	40.00	14.46	14.98	−0.51
17	54	0.30	75	24.00	24.28	24.57	−0.30
18	195	0.30	75	24.00	23.89	23.80	0.08
19	125	0.02	75	24.00	24.49	25.08	−0.59
20	125	0.58	75	24.00	26.73	26.35	0.38
21	125	0.30	32	24.00	18.53	20.74	−2.22
22	125	0.30	117	24.00	23.04	21.03	2.01
23	125	0.30	75	1.37	16.74	16.98	−0.24
24	125	0.30	75	46.63	17.02	16.99	0.03
25	125	0.30	75	24.00	23.53	23.58	−0.05
26	125	0.30	75	24.00	24.53	23.58	0.94
27	125	0.30	75	24.00	22.89	23.58	−0.69
28	125	0.30	75	24.00	22.53	23.58	−1.06
29	125	0.30	75	24.00	23.93	23.58	0.34
30	125	0.30	75	24.00	24.53	23.58	0.94

Predicted Value = Pred. value, Res. Value = Residual value. Study Type: Response Surface, Runs: 30, Initial Design: Central Composite, Design Model: Quadratic.

2.2. Artificial Neural Network Mathematical Model and Method

Response surface methodology data along with the experimental data was collected from sample formulation, totaling 30 samples. These sample data were used in the training of overall data. Note; the regular data collected where normalized to a range between 0 and 1 using Equation (2) below.

$$X_{norm} = \frac{X - X_{min.}}{X_{max-}X_{min}} \tag{2}$$

The X_{norm} is the normalized value, X is the variable, X_{min} and X_{max} are the minimum and maximum values among the data set.

The normalization is necessary to execute the sigmoid transfer function effectively. The network model was programmed by codes of multilayer perceptron (MLP) along with the training algorithm of back propagation (BP), which consists of an input layer (the input variables of temperature, time, dosage, and solvent ratio), a hidden layer and an output layer (photoluminescent quantum yield) which was the response generated from the experimental values. These three node layers are neurons that utilizes non-linear activation function as given in Figure 1 below.

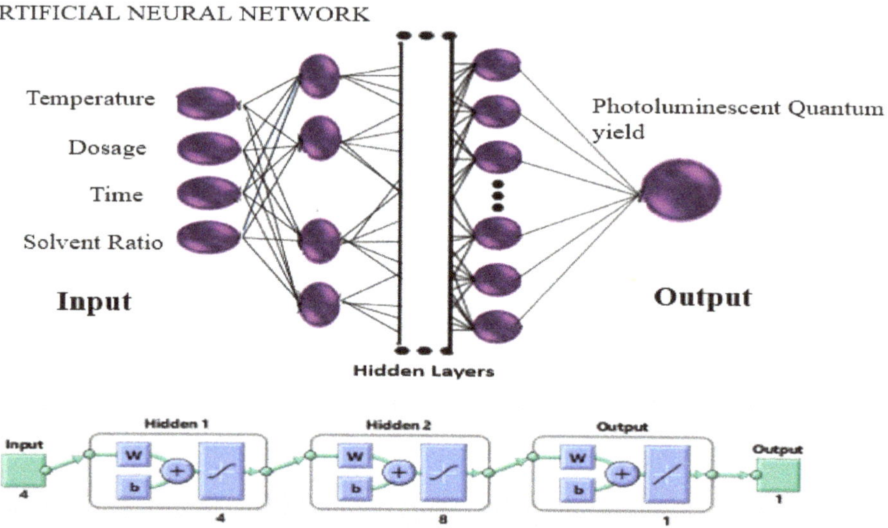

Figure 1. Architecture of multilayer perceptron neural network.

Data tests have shown that a pair of hidden layer resulted in a high performance value. Thus, after multiple iterations for the best data set performance, artificial neural network topologies were selected based on the log-sigmoid transfer function (Equation (3)), linear transfer function in the output layer (Equation (4)), and best performance criteria of coefficient of determination (R^2) at Equation (5), mean absolute error (*MAE*) at Equation (6), and root mean square error (*RMSE*) at Equation (7).

$$Logsig\,(n) = \frac{1}{1 + exp(-n)} \tag{3}$$

$$Pureline\,(n) = n \tag{4}$$

$$R^2 = 1 - \sum_{i=1}^{n}\left[\frac{\left(y^i_{pred} - y^i_{targ}\right)^2}{\left(y_{avg,targ} - y^i_{targ}\right)^2}\right] \tag{5}$$

$$MAE = \frac{1}{N}\sum_{i=1}^{N}|y^i_{pred} - y^i_{targ}| \tag{6}$$

$$RMSE = \sqrt{\sum_{i=1}^{N}\frac{\left(y^i_{Pred} - y^i_{targ}\right)^2}{N}} \tag{7}$$

where, n is the number of experimental data, y^i_{Pred} is the predicted value and y^i_{targ} is real value obtained from experimental data, $y_{avg,targ}$ is the average experimental value. However, the value of R^2 is the amount of reduction in the variability of the response by using a repressor variable in the model. R^2 close to 1 is desirable and the root mean square error (*RMSE*) is required to be negligibly infinitesimal [10].

The process of developing an artificial intelligence model for the prediction and optimization of fluorescent carbon dots followed the flow chart in Figure 2. The chart demonstrates the procedural flow involved in the formulation of the artificial neural network for photoluminescent quantum yield prediction and optimization for the synthesized fluorescent carbon dots (see Section 2.3 for synthesis approach).

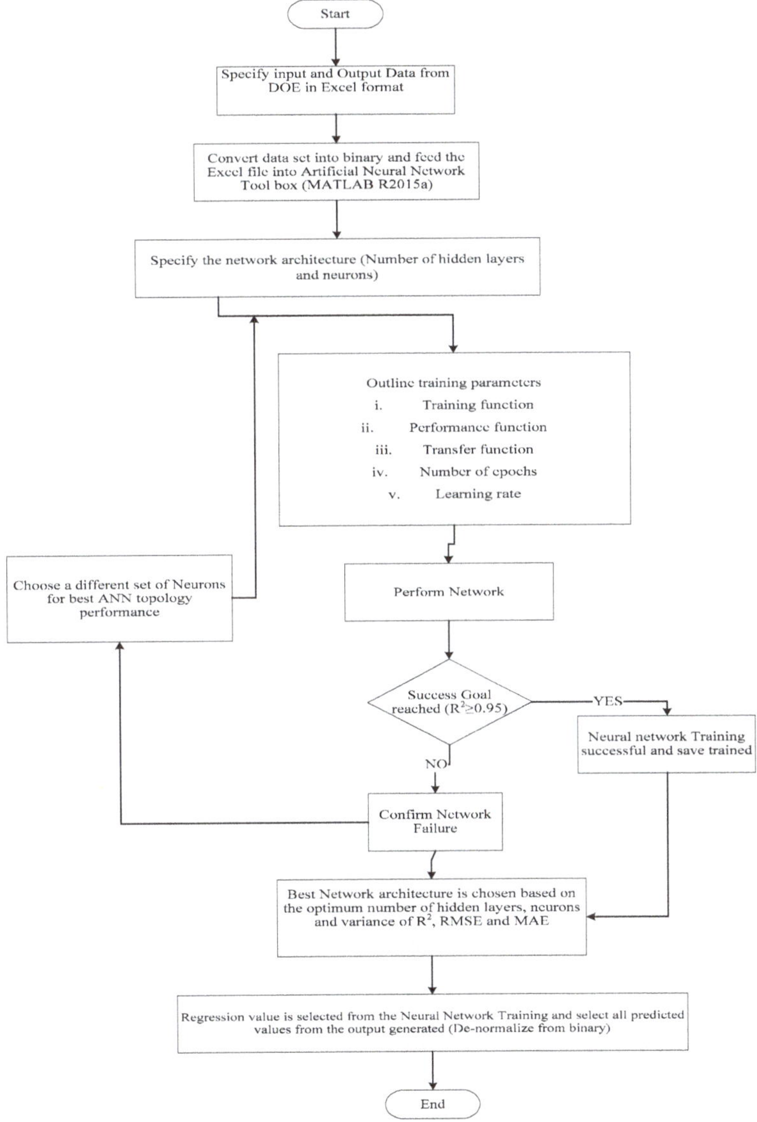

Figure 2. Artificial neural network model flow chart.

A multi input and a single output artificial neural network model was developed by utilizing the Levenberg–Marquardt back propagation (*LMBP*) algorithm to effectively predict the photoluminescent quantum yield of synthesized fluorescent carbon dots [11,12].

2.3. Synthesis of Carbon Dots (CDs)

An environmental suitable technique for producing carbon dots, (hydrothermal synthesis process), were adopted from the response surface methodology analysis. The best photoluminescent quantum yield data were used here for the report of the response. A small quantity, (0.1 g), of tapioca flour was mixed in 12 mL prepared solvent ratio (deionized water + sodium hydroxide + acetone), see Figure 3 for mechanism flow. This mixture was placed in a stainless steel hydrothermal reactor in a convection oven at a temperature of 170 °C for a period of 1 h 40 min. This study has successfully reduced the needed temperature and time needed for synthesizing CDs [13–15].

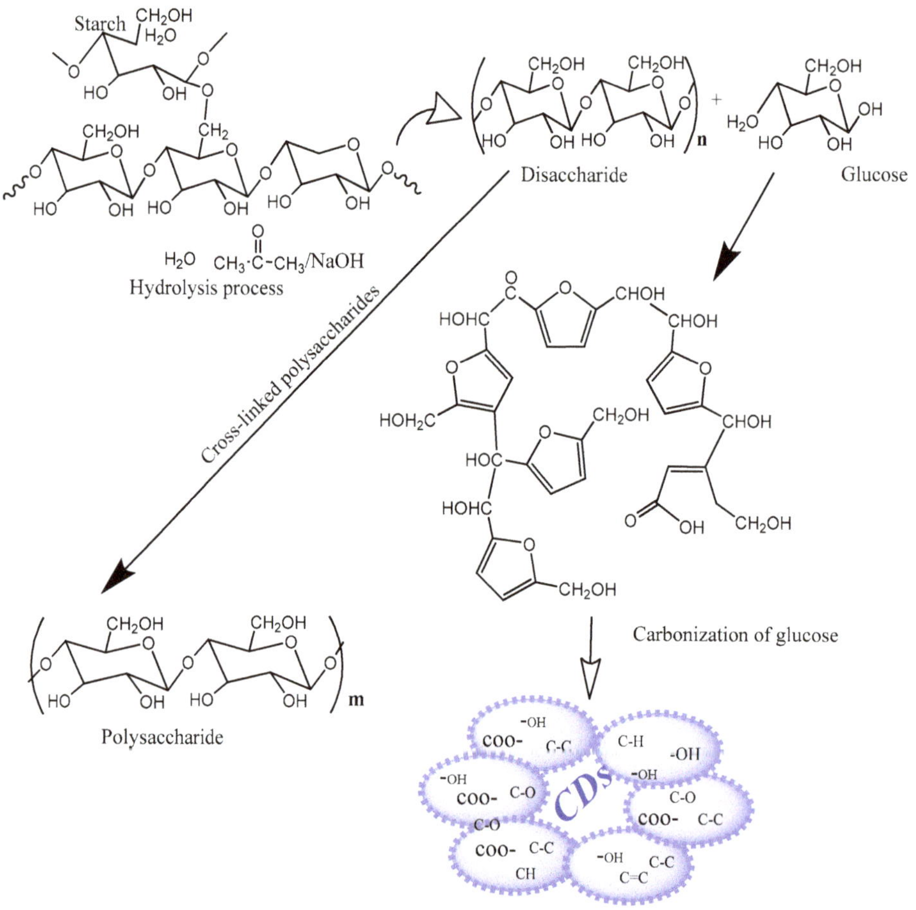

Figure 3. Mechanism for synthesis of carbon dots.

The mixture was centrifuged for 20 min at 3000 rpm. For clarity (no substantial suspended solids detected). The photoluminescent quantum yield was thus calculated by;

$$Q = Q_R \left(\frac{GRAD}{GRAD_R} \right) \left(\frac{e^2}{e_R^2} \right) \tag{8}$$

3. Result and Discussion

3.1. Response Surface Methodology

The design of experiment data (ref: Table 1) was used as independent variables inputs. The best outputs of photoluminescent quantum yield (*PLQY*) of carbon dots, the predicted and actual experimental values of photoluminescent quantum yield, is reported in Table 2 below.

It shows a positive model with an R^2 value of 0.956 as revealed on the fits statistics (Table 6), a high experimental value of PLQY of 27.75%, and a predicted value of 27.38% with a residual value of 0.37%. The experimental data was then used to calculate the coefficient of polynomial equation for the response yield with the inputs data of temperature, time, dosage, and solvent ratio; by adopting Equation (1).

The results of design of experiment was computed by central composite design (CCD) as stated earlier. Photoluminescent quantum yield of the predicted and experimental yield was given by a model equation as in Equation (1) and represented by the expression below.

A = temperature, B = dosages, C = time, D = solvent ($H_2O/C_3H_6O/NaOH$)

Now, let;

$A = X_1$, $B = X_2$, $C = X_3$, $D = X_4$ (refer to RSM Table 2)

i   Final equation in terms of coded (predicted) factors (full model): Also see Table 3 below for R^2 values and lack of fit for the polynomial regression equation.

$$\text{Photoluminescent quantum yield (Response)} = 23.63 - 0.1732\,X_1 + 0.03498\,X_2 + 0.1905\,X_3 - 0.0802\,X_4 - 2.38\,X_1\,X_2 + 1.61\,X_1\,X_3 - 2.68\,X_1\,X_4 + 0.1894\,X_2\,X_3 - 1.30\,X_2\,X_4 - 0.9353\,X_3\,X_4 + 0.2905\,X^2{}_1 + 1.07\,X^2{}_2 - 1.38\,X^2{}_3 - 3.36\,X^2{}_4.$$

ii   Final equation in terms of actual factors (full model): Also see Table 4 below for R^2 values and lack of fit for the polynomial regression equation.

$$\text{Photoluminescent quantum yield (Response)} = -3.3822 + 0.03866\,X_1 + 22.7576\,X_2 + 0.1385\,X_3 + 1.3133\,X_4 - 0.2379\,X_1\,X_2 + 0.0010\,X_1\,X_3 - 0.0033\,X_1\,X_4 + 0.0315\,X_2\,X_3 - 0.4069\,X_2\,X_4 - 0.0019\,X_3\,X_4 + 0.0001\,X^2{}_1 + 26.8733\,X^2{}_2 - 0.0015\,X^2{}_3 - 0.0131\,X^2{}_4.$$

Table 3. The fitted quadratic model in terms of coded variables.

Response	2nd Order Polynomial Equation	Regression (*p*-Value)	R^2	R^2 (Adjusted)	Lack of Fit
PLQY	$23.63 - 0.1732\,X_1 + 0.03498\,X_2 + 0.1905\,X_3 - 0.0802\,X_4 - 2.38\,X_1\,X_2 + 1.61\,X_1\,X_3 - 2.68\,X_1\,X_4 + 0.1894\,X_2\,X_3 - 1.30\,X_2\,X_4 - 0.9353\,X_3\,X_4 + 0.2905\,X^2{}_1 + 1.07\,X^2{}_2 - 1.38\,X^2{}_3 - 3.36\,X^2{}_4.$	0.0001	**0.9563**	**0.9155**	0.1685

Table 4. The fitted quadratic model in terms of actual variables.

Response	2nd Order Polynomial Equation	Regression (*p*-Value)	R^2	R^2 (Adjusted)	Lack of Fit
PLQY	$-3.3822 + 0.03866\,X_1 + 22.7576\,X_2 + 0.1385\,X_3 + 1.3133\,X_4 - 0.2379\,X_1\,X_2 + 0.0010\,X_1\,X_3 - 0.0033\,X_1\,X_4 + 0.0315\,X_2\,X_3 - 0.4069\,X_2\,X_4 - 0.0019\,X_3\,X_4 + 0.0001\,X^2{}_1 + 26.8733\,X^2{}_2 - 0.0015\,X^2{}_3 - 0.0131\,X^2{}_4.$	0.0001	**0.9563**	**0.9155**	0.1685

Analysis of Variance and Model Statistical Report

The data set for the response surface methodology as generated by the software made the model to fit in to quadratic significance, and the analysis of variance (ANOVA) for statistical significance of the quadratic model is computed on Table 5 below;

Table 5. ANOVA for quadratic model.

Source	Sum of Squares	df	Mean Square	F-Value	p-Value	
Model	446.88	14	31.92	23.45	<0.0001	significant
A-Temperature	0.5324	1	0.5324	0.3912	0.5411	
B-Dosage	2.14	1	2.14	1.57	0.2289	
C-Time	0.6458	1	0.6458	0.4745	0.5014	
D-W/Ace/NaOH	0.1132	1	0.1132	0.0832	0.7770	
AB	83.27	1	83.27	61.18	<0.0001	
AC	38.33	1	38.33	28.16	<0.0001	
AD	101.98	1	101.98	74.93	<0.0001	
BC	0.5274	1	0.5274	0.3875	0.5430	
BD	24.01	1	24.01	17.64	0.0008	
CD	12.38	1	12.38	9.10	0.0087	
A^2	0.7806	1	0.7806	0.5735	0.4606	
B^2	10.48	1	10.48	7.70	0.0142	
C^2	17.76	1	17.76	13.05	0.0026	
D^2	106.01	1	106.01	77.89	<0.0001	
Residual	20.42	15	1.36			
Lack of Fit	16.94	10	1.69	2.44	0.1685	Not significant
Pure Error	3.47	5	0.6947			
Cor Total	467.30	29				

F-value of 23.45 implies the model is significant and the subsequent values for the parameters show their degree of effects on the response of photoluminescent quantum yield for fluorescent carbon dots. Herewith, in Table 5, the most effective single–multiple parameter is the solvent ratio (D^2) with an *F*-value of 77.89. While, interactive most effective parameters are temperature and solvent ratio (AD) with *F*-value 74.93. The least effective single parameter is temperature (A^2) with an *F*-value of 0.5735 and the least effective interactive parameters are dosage and time (BC) with an *F*-value of 0.3875 [16,17].

Model *p*-values less than 0.0500 indicate the model terms are significant [18]. In this case AB, AC, AD, BD, CD, B^2, C^2, D^2 are significant model terms. Values greater than 0.1000 indicate the model terms are not significant, hence, individual lone factors are independent and are non-effective. Favorable interactive effects were observed between temperature and dosage (AB), temperature and time (AC), temperature and solvent (AD), dosage and solvent ratio (BD), and time and solvent ratio (CD), while the individual factor effects were observed with dosage (D^2), time (C^2), and solvent ratio (D^2). Non favorable interactive effects were observed with dosage and time (BC), and the multiple factor of non-effect is temperature (A^2). If there are many insignificant model terms (not counting those required to support hierarchy), model reduction may be considered to improve the model, however, in this study the non-effects are few and infinitesimal. The lack of fit *F*-value of 2.44 implies the lack of fit is not significant relative to the pure error. Non-significant lack of fit is an excellent requirement, since it is needed for the model to fit [19].

The experimental R^2 of 0.9563 in Table 6, shows a significant response [20]. The predicted R^2 of 0.7689 is in reasonable agreement with the adjusted R^2 of 0.9155; i.e., the difference is less than 0.2 [21]. Adequate precision, measures the signal to noise ratio, thus, ratio greater than 4 is desirable. Ratio of 17.519 indicates an adequate signal.

Table 6. Fit statistics summary.

Std. Dev.	1.17	R^2	0.9563
Mean	21.40	Adjusted R^2	0.9155
C.V. %	5.45	Predicted R^2	0.7689
		Adequate Precision	17.5186

This model is significant to navigate the design space. It is necessary for a model to comply with the following;

i   A significant model: Large *F*-value with $p < 0.05$.

ii   Insignificant lack-of-fit: *F*-value with $p > 0.10$.

iii   Adequate precision >4 [18–20].

Furthermore, from Figure 4 below, the three dimensional (3D) plots have shown interactive responses to the favorable yield of photoluminescent quantum yield at 27.75% and the intercept value of 23.63 on the 2nd order polynomial equation of actual values is suitable.

In Figure 4A the interactive behavior of temperature (A) at 170 °C and dosage (B) at 0.1 g has an effective interaction with an *F*-value of 61.18 and a *p*-value less than 0.0001 with a value of −0.2379 in the 2nd order polynomial equation of actual factors. Figure 4B is an interactive effect of temperature (A) and time (C) at 1 h 40 min. It records a linear effect on the response value of photoluminescent quantum yield with a favorable *p*-value that is less than 0.001 and a coefficient value of 1.6143 on the polynomial equation. The interactive effect of temperature (A) and solvent ratio W/Ace/NaOH (D) at 12 mL have the best effect on the response yield of photoluminescent quantum yield with an *f*-value of 74.93 and a *p*-value less than 0.0001, coefficient of −2.68 as seen in the 2nd order polynomial equation with a linear response as shown in Figure 4C.

Figure 4D shows a non-effective interaction between dosage (B) and time (C) on the response of photoluminescent quantum yield. *p*-value at 0.5430 and *f*-value of 0.3875 all fall short of the requirement of a *p*-value <0.100 and an *f*-value that is extremely low. The interactive effect of dosage (B) and solvent ratio W/Ace/NaOH (D) at *p*-value 0.0008 as shown in the 3D plot on Figure 4A,C,E,F is the most important factor in the sustainability of environmental resources. Environmental resources management is an essential factor to be considered in the synthesis of products, with high emphasis on minimal resource requirement [18].

3.2. Photoluminescent Quantum Yield

The photoluminescent quantum yield (PLQY) of the CDs was ascertained by Equation (8) as in Section 2.3. Using quinine sulfate added to H_2SO_4 to form 0.1 M solution, optical density of 0.00, 0.02, 0.04, 0.06, 0.08, 0.1 were obtained; at absorption wavelength of 340 nm and dilution was made from the synthesized carbon dots solution. The procedure, is an established process of calculating quantum yields of photoluminescent substances. Quinine sulphate as reference quantum yield was held at 54.6 [22–26].

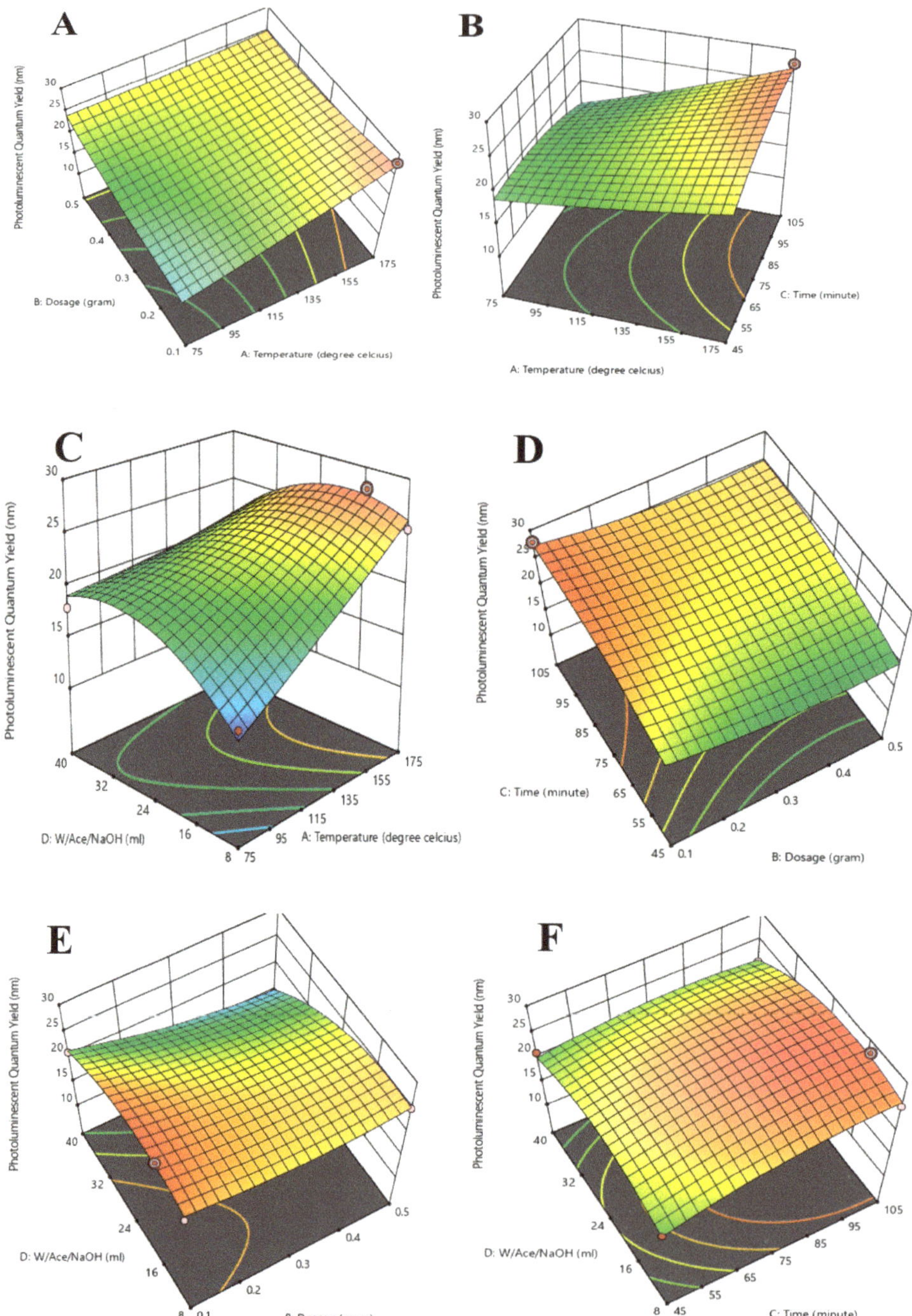

Figure 4. Experimental factors for 3D surface plots of tapioca powder conversion to fluorescent carbon dots.

3.3. Evaluation Performance between Artificial Neural Network (ANN) and Response Surface Methodology (RSM) on the Yield of Photoluminescent Quantum Yield

Artificial neural Network (ANN) is a system that mimics the naturally inspired computational model. Thus, it emulates the workings of the human brain to take in certain connections among information inputs and yield outputs through trained data [27–30].

From Table 7 below, it shows the relationship between the response surface methodology and the artificial neural network performance of the trained data (see Figure 5). The best output for photoluminescent quantum yield was obtained at No. 11 at actual experimental value of 27.75%, RSM predicted value of 27.38%, and ANN predicted value of 26.25%. The training of the data set was conducted by Matlab R2015a (8.5.0.197613), utilizing Lavenberg–Marquardt algorithm (LMA). The LMA is based on the training neural network through iteration and reiteration of data set weight and bias values as shown on Table 8 [31–33].

Table 7. Response surface methodology and artificial neural network.

Std Order	Factor-A Temperature (°C)	Factor-B Dossage (gram)	Factor-C Time (min)	Factor-D Solvent (mL) ($H_2O/C_3H_6O/NaOH$)	Exp. Actual Value (PLQY%)	RSM Pred. Value	ANN. Pred. Value
1	75	0.10	45	8.00	14.67	14.41	12.46
2	175	0.10	45	8.00	21.05	20.89	21.31
3	75	0.50	45	8.00	22.80	22.35	22.74
4	175	0.50	45	8.00	19.96	19.13	19.93
5	75	0.10	105	8.00	14.00	13.01	14.76
6	175	0.10	105	8.00	25.27	26.13	24.84
7	75	0.50	105	8.00	20.15	21.52	19.99
8	175	0.50	105	8.00	24.87	24.94	24.96
9	75	0.10	45	40.00	24.82	24.39	24.51
10	175	0.10	45	40.00	20.99	19.88	21.10
11	170	0.1	100	12.00	27.75	27.38	26.25
12	175	0.50	45	40.00	12.53	13.16	15.80
13	75	0.10	105	40.00	17.90	18.99	17.56
14	175	0.10	105	40.00	21.04	21.13	21.07
15	75	0.50	105	40.00	22.75	22.55	23.82
16	175	0.50	105	40.00	14.46	14.98	18.69
17	54	0.30	75	24.00	24.28	24.57	23.45
18	195	0.30	75	24.00	23.89	23.80	24.89
19	125	0.02	75	24.00	24.49	25.08	24.09
20	125	0.58	75	24.00	26.73	26.35	25.17
21	125	0.30	32	24.00	18.53	20.74	16.69
22	125	0.30	117	24.00	23.04	21.03	23.65
23	125	0.30	75	1.37	16.74	16.98	15.06
24	125	0.30	75	46.63	17.02	16.99	17.38
25	125	0.30	75	24.00	23.53	23.58	23.72
26	125	0.30	75	24.00	24.53	23.58	23.72
27	125	0.30	75	24.00	22.89	23.58	23.72
28	125	0.30	75	24.00	22.53	23.58	23.72
29	125	0.30	75	24.00	23.93	23.58	23.72
30	125	0.30	75	24.00	24.53	23.58	23.72

Predicted Value = Pred. value, Study Type: Response Surface, Runs: 30, Initial Design: Central Composite, Design Model: Quadratic.

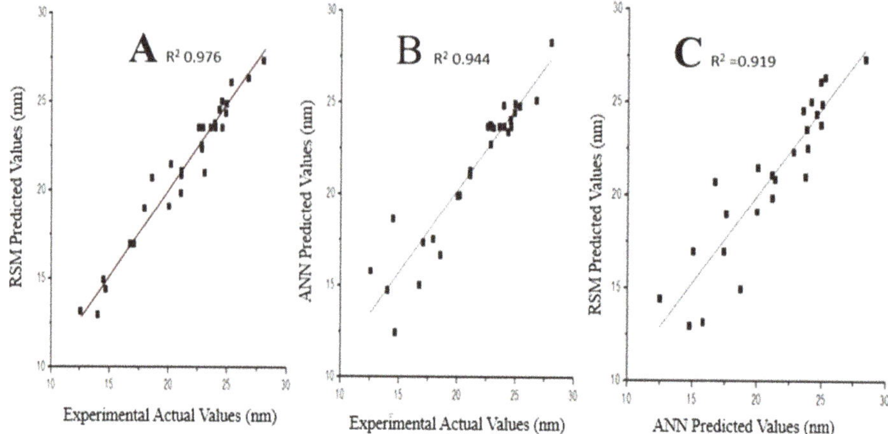

Figure 5. Parity plots; (**A**) Response Surface Methodology (RSM) predicted against experimental actual values. (**B**) Artificial Neural Network (ANN) predicted against experimental actual values. (**C**) RSM predicted values against ANN predicted values.

Table 8. Artificial neural network (ANN) optimum values for hidden layer sizes and corresponding transfer functions ('tansig' and 'logsig') for minimum error fittings for train, validation, and test of optimized RSM data.

Hidden Neurons	Train			Validation			Test			All
	R^2	RMSE	MAE	R^2	RMSE	MAE	R^2	RMSE	MAE	R^2
4-8 *	0.99	0.09	0.06	0.99	0.07	0.04	0.99	0.08	0.06	0.99
8-4	0.99	0.03	0.01	0.95	0.10	0.09	0.95	0.11	0.09	0.94
8-16	0.99	0.04	0.03	0.78	0.17	0.13	0.96	0.12	0.09	0.93
9-19	0.99	0.02	0.01	0.93	0.13	0.08	0.99	0.11	0.09	0.96
11-4	0.99	0.04	0.03	0.85	0.11	0.09	0.95	0.14	0.11	0.96
11-7	0.99	0.02	0.01	0.97	0.07	0.06	0.82	0.12	0.09	0.97
11-9	0.99	0.02	0.01	0.99	0.11	0.09	0.83	0.15	0.12	0.96
13-9	0.95	0.07	0.05	0.92	0.14	0.13	0.94	0.12	0.09	0.93
13-13	0.99	0.03	0.02	0.76	0.11	0.09	0.93	0.16	0.13	0.95
17-10	0.99	0.02	0.01	0.90	0.12	0.13	0.99	0.06	0.05	0.96
17-18	0.99	0.02	0.01	0.91	0.11	0.08	0.94	0.12	0.09	0.97
19-4	0.99	0.03	0.02	0.52	0.12	0.09	0.97	0.12	0.09	0.96
19-6	0.98	0.07	0.05	0.83	0.12	0.11	0.99	0.02	0.02	0.96

* 4-8 is the chosen model based on its comparative high R^2 value.

As shown in Figure 5 below, the R^2 value is a good revelation of the compatibility of each data set to each other.

The RSM values shows a high R^2 value of 0.9563 than the ANN R^2 of 0.944 with a negligible residual value of 0.0123 between the RSM and ANN. Hence, the ANN is a very good method of validating RSM data set [21]. The process of validating the Levenberg–Marquardt back propagation model for the response yield of photoluminescent quantum yield adopted in this study were done by deploying different adjustable topologies (see Table 8) in training of the network performance. From the 13 topologies deployed hidden layers between 4 and 20; the best hidden layer configuration with high coefficient of determination and low training error was gained at 4–8, as evident in Figure 6.

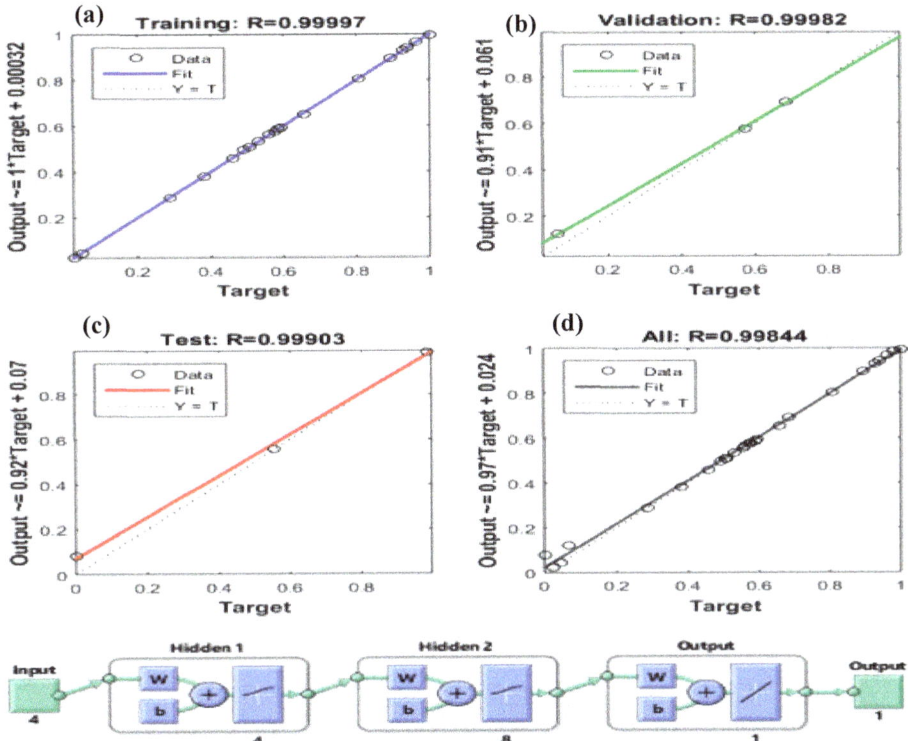

Figure 6. Artificial neural network model coefficient of determination and error relationship of data set of fluorescent carbon dots. (**a**) Trained R^2 output, (**b**) Validated R^2 output, (**c**) Test R^2 output, and (**d**) Overall R^2 output of data set.

Figure 6 below shows a high value of $R^2 \geq 95$ for the output parameter of photoluminescent quantum yield which complies with the expected results from training of data set by the Levenberg–Marquardt algorithm [17,29].

3.4. Characterization and Properties of Carbon Dots

The need to analyze carbon dots for their characteristic attributes is very essential. These can be done by determining the particle sizes and morphological patterns using high resolution devices [34]. The atomic force microscopy (AFM), high resolution transmission electron microscopy (HrTEM), and field emission scanning emission microscopic (FESEM) techniques have been utilized for this purposes.

3.4.1. Atomic Force Microscopy (AFM) and High Resolution Transmission Electron Microscopy (HrTEM) of Carbon Dots (CDs)

CDs particle size distribution and morphology have been investigated (Figure 7). Figure 7a shows the three dimensional (3D) plot of the morphological pattern of carbon dots while Figure 7b depicts 62 counts of CDs with a mean height and diameter of 4.054 nm and 44.032 nm, respectively, which is a confirmation of the nano-dimension of CDs. Figure 7c represents the histogram of the 62 carbon dots count and Figure 7d is the HrTEM of carbon dots at less than 10 nm and a lattice spacing at 0.24 nm. HrTEM analysis for the carbon dots were investigated to determine the actual size and shape of the carbon dots. The image clearly depicts the synthesized CDs as well dispersed in water with a spherical petal shape and fine size distribution of about 3.0–5.0 nm in diameter shown in Figure 7.

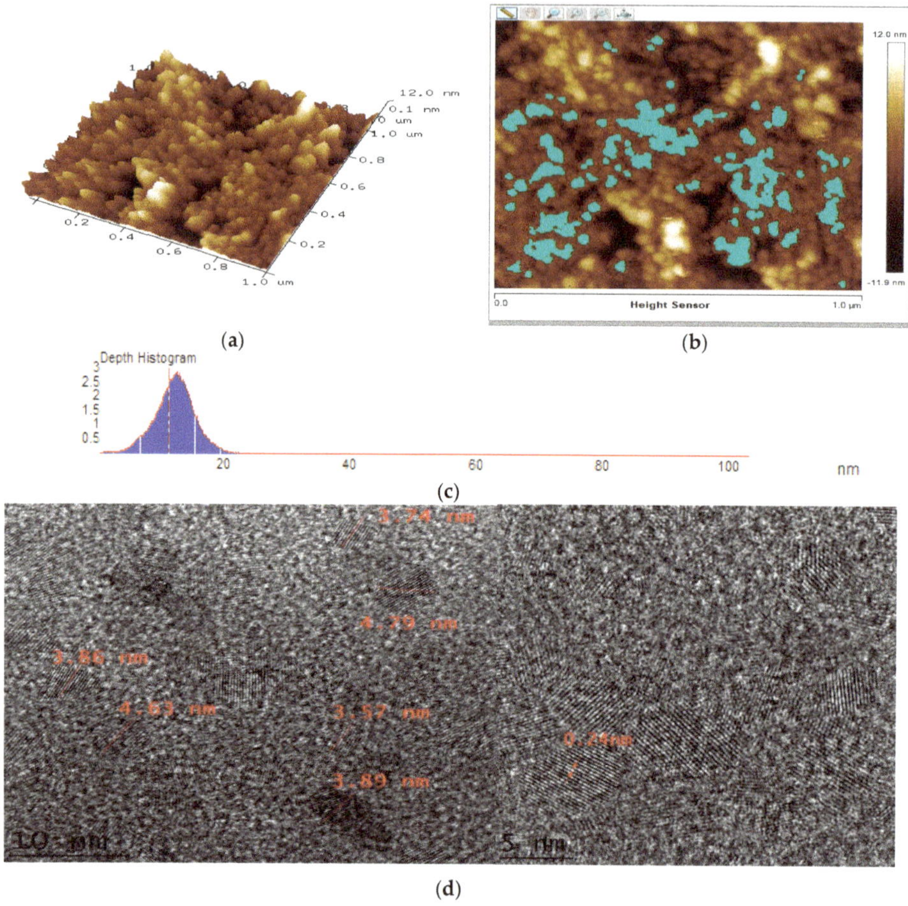

Figure 7. Atomic force microscopy (AFM) and high resolution transmission electron microscopy (HrTEM) of carbon dots (CDs). (**a**) 3D CDs particles, (**b**) 2D CDs particles, (**c**) CDs particle distribution, and (**d**) CDs particle sizes and lattice space.

The HrTEM provides a validation of the nano dimensions present in the synthesized CDs via hydrothermal route which is in agreement with semiconductors synthesized at the nano scale [35,36].

The CDs lattice spacing of 0.24 nm renders it suitable in membrane filtration application. More so, the sizes of CDs are below 5 nm, which means there are numerous surface sites for adsorption application purposes in wastewater treatment.

In addition, Table 9 shows a detailed presentation of the size distribution of CDs obtained by measuring the heights, areas, and diameters of 62-single carbon dots observed under atomic force microscopy.

Table 9. Analysis of the atomic force microscopy of CDs.

Parameter	Mean	Minimum	Maximum
Total Count	62	62	62
Height	4.054 (nm)	2.174 (nm)	8.486 (nm)
Area	2086.516 (nm^2)	381.470 (nm^2)	18,005.371 (nm^2)
Diameter	44.032 (nm)	22.039 (nm)	151.411 (nm)

As presented on Table 9. The CDs mean diameter of 44.034 nm and mean area of 2086.516 nm^2 possess the attributes of a suitable adsorbent for adsorbing environmental pollutants [13].

3.4.2. Field Emission Scanning Electron Microscopy (FESEM) and EDx of Tapioca-Derived Carbon Dots

From the conducted FESEM analysis, it is found that the actual shape of the CDs is in the form of a flower shaped petals, spherical in nature, as can be seen in Figure 8a,b. The EDX study determined the elemental compositions of carbon dots, and the results shows presence of C (31.64%), 0 (55.84%), Na (10.99%), Si (1.45%), and K (0.079%) as in Figure 8c. The Na signal in EDX spectrum is due to the sodium hydroxide constituent of carbon dots synthesis and silicon is as result of glass substrate for drying carbon dots during the sample preparation for FESEM.

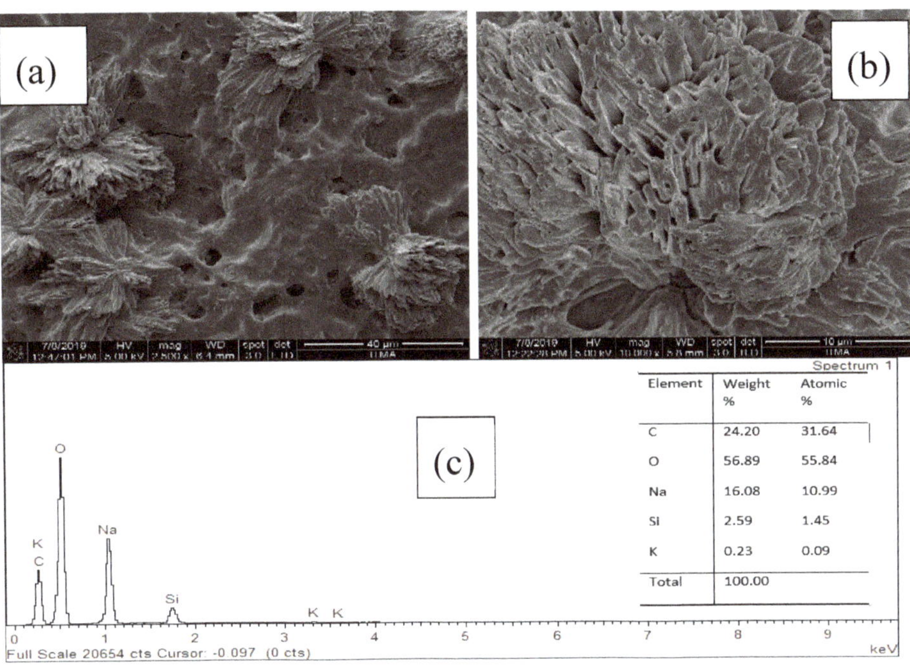

Element	Weight %	Atomic %
C	24.20	31.64
O	56.89	55.84
Na	16.08	10.99
Si	2.59	1.45
K	0.23	0.09
Total	100.00	

Figure 8. FESEM at (**a**) 40 μm (**b**) 10 μm and (**c**) EDx of tapioca-derived carbon dots.

3.4.3. Properties of Carbon Dots (CDs)

The use of UV lamp to assess the quality of fluorescent carbon dots have been applied as a source to obtain a blue/green color in the near visible region of color band group. The UV irradiations absorbed by the carbon dots and excited through absorbing the energy leading to an electron excited state. The molecules of carbon dots with extended Pi-electron provides the basics for the fluorescence emission of carbon dots. The tapioca-derived carbon dot is a wavelength dependent photoluminescent ionic solution in the visible range with a surface abundant with hydroxyl and carboxylic/carboxyl moieties [13].

CDs indicated a strong optical absorption in the UV region (230–340 nm) with a tail extending to the visible range as presented in Figure 9.

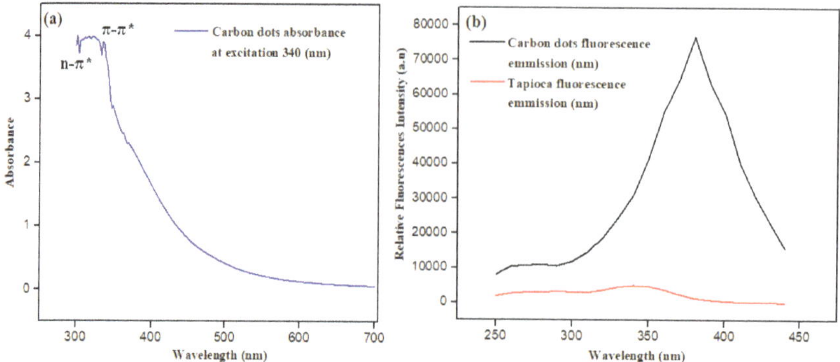

Figure 9. Optical properties of CDs. (**a**) UV-visible absorption and emission spectra of as prepared CDs dispersed in water. (**b**) Fluorescence emission spectra of carbon dots and reference material (Tapioca).

Absorption shoulders in the spectrum may be due to the π-π^* (pi to pi star transition) of C=C bonds or n-π^* (n to pi star transition) of C=O [37]. The uniqueness of CDs is the photoluminescence emitted by it. Based on past study, it shows the dependency of intensity and wavelength emission towards excitation wavelength [38]. This is due to the different size of particles and surface chemistry and different emissive traps on CDs surface that can be related by the synthesis method.

The wavelength dependence behavior makes CDs possible to be applied in multi-color imaging applications. It has been suggested that there are separate emissions by CDs core and surface states whereby size, surface, and defects are responsible for the emission properties [13]. The color of CDs most of the time is related to the surface groups which compares to particles size and normally CDs show strong photoluminescence from blue to green wavelength. In terms of chemical properties, different synthesis methods of CDs lead to different chemical structure, such as polymer chains, oxygen based and amino-based groups [39].

The main challenge with carbon dots is the agglomeration of the particles due to strong particle interactions. It can be postulated that the agglomeration of particles over time is the very reason that these CDs emit green fluorescence since it had been left for 3 months. The resultant change in colour due to agglomeration of carbon dot particle can be observed in Figure 10 (A and B) [40].

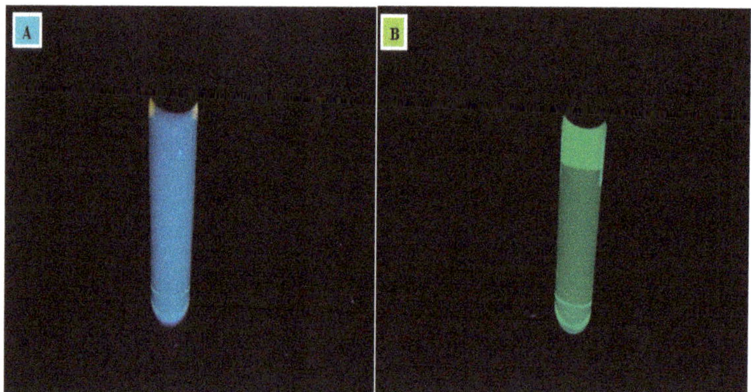

Figure 10. Tapioca powder as source of fluorescent carbon dots. (**A**) Carbon dots after synthesis emitting blue fluorescence. (**B**) Carbon dots emitting green fluorescence after 3 months of synthesis.

It is well-known that there is a relationship between emission wavelength of quantum dots and particle size, i.e., the smaller the particle size is, the shorter the emission wavelength [41]. It is reasonable to speculate that this law is also applicable to carbon dots. Strong green photoluminescence offers unique superiority, because most of the current carbon dots emit blue fluorescence under UV irradiation. This kind of carbon dots had been synthesized [42].

4. Conclusions

The concept of synthesizing CDs (at zero dimension) on an industrial scale requires automation by first being able to predict the possible outcomes based upon the intended experimental factors. Thus, this study has applied tapioca as the material for optimization to achieve the set objective of large scale synthesis by means of prediction through a reliable approach of utilizing design of experiment (DoE) for a response surface methodology (RSM), to optimize a facile and effective synthesis process of fluorescent carbon dots from tapioca powder (starch) via hydrothermal synthesis route. The prediction for optimized fluorescent carbon dots synthesis from RSM is in excellent agreement with artificial neural network prediction by the Levenberg–Marquardt back propagation (LMBP) algorithm in terms of R^2, root mean square error and mean absolute error. Positive hidden layer sizes have resulted in the ANN prediction of PLQY of fluorescent carbon dots at 26.25% and RSM predicted value of 27.38% at R^2 values of 0.94 and 0.95, respectively. The best parameters values for the synthesis of carbon dots were at 170 °C for 1 h 40 min with solvent ratio of 12 mL and dosage 0.1 g. These optimization and prediction process have produced sustainable, efficient, and reliable fluorescent carbon dots, which is energy saving in a manageable time, along with a decreased dosage with optimum quality output.

Moreso, to confirm the validity of carbon dots, characterization of surface morphology and particles size carbon dots were conducted with favorable confirmations of the unique characteristics and attributes of synthesized carbon dots by hydrothermal route.

Author Contributions: M.Y.P., As the first author; made the Study conception and design (wrote the ANN codes), Acquisition of data and Drafting of manuscript. Z.Z.A., As the corresponding author; contributed in the Study conception and design (RSM analysis), Critical revision major scientific advisor through clinical experience. S.A.R., As a co-author; served as scientific advisor, critically reviewed the study proposal. F.M.Y., As a co-author; contributed in the Analysis and interpretation of data as well as critically reviewed the study proposal. A.S.M.N., As a co-author; contributed in the Analysis and interpretation of data. M.A.I., As a co-author; contributed in the interpretation of data.

Funding: The authors would like to thank Universiti Putra Malaysia, Malaysia as this reported research work is funded by the UPM under the GP-IPS/2017/9556800 grant.

Data Availability Statement: The [ANN and RSM] data used to support the findings of this study may be released upon application to Zurina Zainal Abidin, who can be contacted at [zurina@upm.edu.my]. This is an ongoing research.

Conflicts of Interest: The authors declare that there is no conflict of interest whatsoever regarding the publication of this paper.

References

1. Zhou, J.; Sheng, Z.; Han, H.; Zou, M.; Li, C. Facile synthesis of fluorescent carbon dots using watermelon peel as a carbon source. *Mater. Lett.* **2012**, *66*, 222–224. [CrossRef]

2. Das, R.; Bandyopadhyay, R.; Pramanik, P. Carbon quantum dots from natural resource: A review. *Mater. Today Chem.* **2018**, *8*, 96–109. [CrossRef]

3. Witek-Krowiak, A.; Chojnacka, K.; Podstawczyk, D.; Dawiec, A.; Pokomeda, K. Application of response surface methodology and artificial neural network methods in modelling and optimization of biosorption process. *Bioresour. Technol.* **2014**, *160*, 150–160. [CrossRef] [PubMed]

4. Nwobi-Okoye, C.C.; Ochieze, B.Q. Age hardening process modeling and optimization of aluminum alloy A356/Cow horn particulate composite for brake drum application using RSM, ANN and simulated annealing. *Def. Technol.* **2018**, *14*, 336–345. [CrossRef]

5. Huang, D.; Zhao, W.; Tang, Y.; Huang, S.; Cao, W. Matching algorithm of missile tail flame based on back-propagation neural network. In Proceedings of the Fourth Seminar on Novel Optoelectronic Detection Technology and Application, Nanjing, China, 24–26 October 2017; p. 1069702.

6. Lee, J.H.; Delbruck, T.; Pfeiffer, M. Training Deep Spiking Neural Networks Using Backpropagation. *Front. Mol. Neurosci.* **2016**, *10*, 508. [CrossRef] [PubMed]

7. Whittington, J.C.R.; Bogacz, R. An Approximation of the Error Backpropagation Algorithm in a Predictive Coding Network with Local Hebbian Synaptic Plasticity. *Neural Comput.* **2017**, *29*, 1229–1262. [CrossRef] [PubMed]

8. Wisesty, U.N.; Warastri, R.S.; Puspitasari, S.Y. Leukemia and colon tumor detection based on microarray data classification using momentum backpropagation and genetic algorithm as a feature selection method. *J. Phys. Conf. Ser.* **2018**, *971*, 012018. [CrossRef]

9. Chamoli, S. ANN and RSM approach for modeling and optimization of designing parameters for a V down perforated baffle roughened rectangular channel. *Alex. Eng. J.* **2015**, *54*, 429–446. [CrossRef]

10. Baş, D.; Boyacı, I.H.; Boyaci, I.H. Modeling and optimization II: Comparison of estimation capabilities of response surface methodology with artificial neural networks in a biochemical reaction. *J. Food Eng.* **2007**, *78*, 846–854. [CrossRef]

11. Nazari, A.; Riahi, S. Prediction split tensile strength and water permeability of high strength concrete containing TiO_2 nanoparticles by artificial neural network and genetic programming. *Compos. Part B Eng.* **2011**, *42*, 473–488. [CrossRef]

12. Nazari, A.; Riahi, S. Artificial neural networks to prediction total specific pore volume of geopolymers produced from waste ashes. *Neural Comput. Appl.* **2013**, *22*, 719–729. [CrossRef]

13. Arumugam, N.; Kim, J. Synthesis of carbon quantum dots from Broccoli and their ability to detect silver ions. *Mater. Lett.* **2018**, *219*, 37–40. [CrossRef]

14. Diao, H.; Li, T.; Zhang, R.; Kang, Y.; Liu, W.; Cui, Y.; Wei, S.; Wang, N.; Li, L.; Wang, H.; et al. Facile and green synthesis of fluorescent carbon dots with tunable emission for sensors and cells imaging. *Spectrochim. Acta Part A Mol. Biomol. Spectrosc.* **2018**, *200*, 226–234. [CrossRef] [PubMed]

15. Shahla, A.F.; Masoud, S.N.; Davood, G. Hydrothermal green synthesis of magnetic Fe_3O_4-carbon dots by lemon and grape fruit extracts and as a photoluminescence sensor for detecting of *E. coli* bacteria. *Spectrochim. Acta Part A Mol. Biomol. Spectrosc.* **2018**, *203*, 481–493.

16. Fermoso, J.; Gil, M.V.; Arias, B.; Plaza, M.; Pevida, C.; Pis, J.; Rubiera, F. Application of response surface methodology to assess the combined effect of operating variables on high-pressure coal gasification for H2-rich gas production. *Int. J. Hydrog. Energy* **2010**, *35*, 1191–1204. [CrossRef]

17. Lee, H.V.; Yunus, R.; Juan, J.C.; Taufiq-Yap, Y. Process optimization design for jatropha-based biodiesel production using response surface methodology. *Fuel Process. Technol.* **2011**, *92*, 2420–2428. [CrossRef]

18. Kefasa, H.M.; Robiah, Y.; Umer, R.; Yun, T.Y. Modified sulfonation method for converting carbonized glucose into solid acid catalyst for the esterification of palm fatty acid distillate. *Fuel* **2018**, *229*, 68–78. [CrossRef]

19. Rashid, U.; Anwar, F.; Ashraf, M.; Saleem, M.; Yusup, S. Application of response surface methodology for optimizing transesterification of Moringa oliefera oil: Biodiesel production. *Energy Convers Manag.* **2011**, *52*, 3034–3042. [CrossRef]

20. Joglekar, A.; May, A. Product excellence through design of experiments. *Cereal Foods World* **1987**, *32*, 857–868.

21. Betiku, E.; Okunsolawo, S.S.; Ajala, S.O.; Odedele, O.S. Performance evaluation of artificial neural network coupled with generic algorithm and response surface methodology in modeling and optimization of biodiesel production process parameters from shea tree (*Vitellaria paradoxa*) nut butter. *Renew. Energy* **2015**, *76*, 408–417. [CrossRef]

22. Da Silva Souza, D.R.; Larissa, D.C.; Joao, P.M.; Fabiano, V.P. Luminescent carbon dots obtained from cellulose. *Mater. Chem. Phys.* **2018**, *203*, 148–155. [CrossRef]

23. Horiba Scientific. *A Guide to Recording Fluorescence Quantum Yield*; Middlesex HA7 IBQ; Horiba UK Limited: Northampton, UK, 2018.

24. Ashby, S.P.; Thomas, J.A.; Coxon, P.R.; Bilton, M.; Brydson, R.; Pennycook, T.J.; Chao, Y. The effect of alkyl chain length on the level of capping of silicon nanoparticles produced by a one-pot synthesis route based on the chemical reduction of micelle. *J. Nanoparticle Res.* **2013**, *15*, 1425–1428. [CrossRef]

25. Sahu, S.; Behera, B.; Maiti, K.; Mohapatra, S. Simple one-step synthesis of highly luminescent carbon dots from orange juice: Application as excellent bioimaging agents. *Chem. Commun.* **2012**, *48*, 8835–8837. [CrossRef] [PubMed]

26. Jhonsi, M.A.; Thulasi, S. A novel fluorescent carbon dots derived from tamarind. *Chem. Phys. Lett.* **2016**, *661*, 179–184. [CrossRef]

27. Yadav, A.M.; Chaurasia, R.C.; Suresh, N.; Gajbhiye, P. Application of artificial neural networks and response surface methodology approaches for the prediction of oil agglomeration process. *Fuel* **2018**, *220*, 826–836. [CrossRef]

28. Ahmad, H.; Ali, A.N.; Ahmad, J.J.; Reza, K.J.; Mansour, B.; Hasan, B.; Fardin, G.P. Application of response surface methodology and artificial neuralnetwork modeling to assess non-thermal plasma efficiency insimultaneous removal of BTEX from waste gases: Effect of operatingparameters and prediction performance. *Process Saf. Environ. Prot.* **2018**, *119*, 261–270.

29. Shailendra, S.S.; Shraddha, S.; Rathindra, M.B. Preparation of Drug Eluting Natural Composite Scaffold Using Response Surface Methodology and Artificial Neural Network Approach. *Tissue Eng. Regen. Med.* **2018**, *15*, 1–13.

30. Vatankhah, E.; Semnani, D.; Prabhakaran, M.P.; Tadayon, M.; Razavi, S.; Ramakrishna, S. Artificial neural network for modeling the elastic modulus of electrospun polycaprolactone/gelatin scaffolds. *Acta Biomater.* **2014**, *10*, 709–721. [CrossRef]

31. Mukherjee, I.; Routroy, S. Comparing the performance of neural networks developed by using Levenberg–Marquardt and Quasi-Newton with the gradient descent algorithm for modelling a multiple response grinding process. *Expert Syst. Appl.* **2012**, *39*, 2397–2407. [CrossRef]

32. Wilamowski, B.M.; Yu, H. Improved Computation for Levenberg–Marquardt Training. *IEEE Trans. Neural Netw.* **2010**, *21*, 930–937. [CrossRef] [PubMed]

33. Varol, T.; Çanakçi, A.; Ozsahin, S. Prediction of effect of reinforcement content, flake size and flake time on the density and hardness of flake AA2024-SiC nanocomposites using neural networks. *J. Alloy. Compd.* **2018**, *739*, 1005–1014. [CrossRef]

34. Saud, P.S.; Pant, B.; Alam, A.-M.; Ghouri, Z.K.; Park, M.; Kim, H.-Y. Carbon quantum dots anchored TiO2 nanofibers: Effective photocatalyst for waste water treatment. *Ceram. Int.* **2015**, *41*, 11953–11959. [CrossRef]

35. Venkatesham, T.; Tanner, A.N.; Denis, O.D.; Ümit, Ö.; Indika, U.A. Ge1−xSnx alloy quantum dots with composition- tunable energy gaps and near-infrared photoluminescence. *Nanoscale* **2018**, *10*, 20296–20306.

36. Siddique, A.B.; Pramanick, A.K.; Chatterjee, S.; Ray, M. Amorphous Carbon Dots and their Remarkable Ability to Detect 2,4,6-Trinitrophenol. *Sci. Rep.* **2018**, *8*, 9770. [CrossRef] [PubMed]

37. Li, L.; Li, L.; Chen, C.; Cui, F. Green synthesis of nitrogen-doped carbon dots from ginkgo fruits and the application in cell imaging. *Inorg. Chem. Commun.* **2017**, *86*, 227–231. [CrossRef]

38. Zhao, C.; Jiao, Y.; Hu, F.; Yang, Y. Green synthesis of carbon dots from pork and application as nanosensors for uric acid detection. *Spectrochim. Acta Part A Mol. Biomol. Spectrosc.* **2018**, *190*, 360–367. [CrossRef] [PubMed]

39. Barman, M.K.; Patra, A. Current status and prospects on chemical structure driven photoluminescence behaviour of carbon dots. *J. Photochem. Photobiol. C Photochem. Rev.* **2018**, *37*, 1–22. [CrossRef]

40. Wanekaya, A.K. Applications of nanoscale carbon-based materials in heavy metal sensing and detection. *Analyst* **2011**, *136*, 4383–4391. [CrossRef]

41. Liu, H.; Ye, T.; Mao, C. Fluorescent Carbon Nanoparticles Derived from Candle Soot. *Angew. Chem. Int. Ed.* **2007**, *46*, 6473–6475. [CrossRef]

42. Zhang, Y.; Gao, Z.; Zhang, W.; Wang, W.; Chang, J.; Kai, J. Fluorescent carbon dots as nanoprobe for determination of lidocaine hydrochloride. *Sens. Actuators B Chem.* **2018**, *262*, 928–937. [CrossRef]

Article

Process of Natural Gas Explosion in Linked Vessels with Three Structures Obtained Using Numerical Simulation

Qiuhong Wang [1,*], Yilin Sun [1], Xin Li [1], Chi-Min Shu [2], Zhirong Wang [3,*], Juncheng Jiang [3], Mingguang Zhang [3] and Fangming Cheng [1]

[1]   College of Safety Science and Engineering, Xi'an University of Science and Technology, Xi'an 710054, China; sunlynlin@163.com (Y.S.); lixinpersonal@outlook.com (X.L.); chengfm@xust.edu.cn (F.C.)

[2]   Process Safety and Disaster Prevention Laboratory, National Yunlin University of Science and Technology, Douliou, Yunlin 64002, Taiwan; shucm@yuntech.edu.tw

[3]   Jiangsu Key Laboratory of Hazardous Chemicals Safety and Control, College of Safety Science and Engineering, Nanjing Tech University, Nanjing 210009, China; jcjiang@njtech.edu.cn (J.J.); mingguang_zhang@njtech.edu.cn (M.Z.)

*   Correspondence: wangqh@xust.edu.cn (Q.W.); wangzhirong@njtech.edu.cn (Z.W.)

Received: 9 December 2019; Accepted: 30 December 2019; Published: 2 January 2020

Abstract: Combinations of spherical vessels and pipes are frequently employed in industries. Scholars have primarily studied gas explosions in closed vessels and pipes. However, knowledge of combined spherical vessel and pipe systems is limited. Therefore, a flame acceleration simulator was implemented with computational fluid dynamics software and was employed to conduct natural gas explosions in three structures, including a single spherical vessel, a single spherical vessel with a pipe connected to it, and a big spherical vessel connected to a small spherical vessel with a pipe. These simulations reflected physical experiments conducted by at Nanjing Tech University. By changing the sizes of vessels, lengths of pipes, and ignition positions in linked vessels, we obtained relevant laws for the time, pressure, temperature, and concentrations of combustion products. Moreover, the processes of natural gas explosions in different structures were obtained from simulation results. Simulation results agreed strongly with corresponding experimental data, validating the reliability of simulation.

Keywords: combination system; flame acceleration simulator (FLACS); pipe length; ignition position

---

1. Introduction

Flammable gases are employed extensively in the production of petrochemicals, and regularly stored and transported using linked vessels. A typical linked system comprises closed vessels connected with pipelines [1]. However, widespread fires and explosions frequently occur in such structures because the flames and shock waves that result from local gas explosions can propagate through the pipeline [2–4]. Although the prevention and mitigation techniques for fires and explosions have been continuously improving, the number of accidents has not decreased in recent years. For instance, an explosion and a fire accident caused by a gas leak considerably damaged two nearby buildings, on 23 December 2008, in a coal gasification plant in Hunan province in China [5]. Furthermore, an explosion occurred in 2013 in Dalian Bay, China, due to an operation that violated safety protocols and ignited flammable gases in a vessel, which caused the death of two individuals and severely injured two others. The explosion that occurred in natural gas pipes in 2017 in Guizhou Province, China, caused the death of eight individuals and injured 35 individuals. Therefore, conducting further studies on gas explosions occurring in linked vessels is of substantial importance for preventing and mitigating the damage caused by gas explosions.

Research on the structure of linked vessels has been conducted in past decades. Generally, compared with methane explosions occurring in vented single vessels, gas explosions in linked vessel systems constitute a complex process. This complexity can lead to high explosion strength in linked vessels [6]. Furthermore, numerous influential factors, including venting sizes and positions, ignition positions, and pipe lengths, could affect this process, among which ignition positions or obstacles could considerably change flame propagation, accelerate the speed of fire, and rapidly increase explosion pressures [7–12]. The venting size and position also play a prominent role in gas explosions [13–18]. Studies have demonstrated that the pressure piling exists in linked vessels, and the explosion strength is principally affected by the pipe length to volume ratio [19–25].

Currently, along with the development of computer techniques, numerical simulations have become more advanced. Numerical simulation has numerous advantages over physical testing; numerically simulated experiments are cost-effective, user-friendly, and compatible with a limitless number of experimental devices. Some scholars have examined gas explosions by using numerical simulations. With experimental data collected from the literature and the computational fluid dynamics (CFD) software AutoReaGas, Maremonti et al. [24] simulated gas explosions in two linked vessels. They not only demonstrated that turbulence induced in the second vessel was a major factor influencing violence of the explosion but also verified the validity of their CFD code. Deng et al. [25] conducted an explosion experiment with a CH_4 and CO mixture in a 20-L nearly-spherical tank and then used flame acceleration simulator (FLACS) software to mimic the gas explosion of the experiment. They compared their simulation results with experimental data to prove the reliability of the simulation. Ferrara et al. [26] modeled gas explosions vented through ducts by using a two-dimensional (2D) axisymmetric CFD model based on the unsteady Reynolds-averaged Navier–Stokes approach. Simulation results evidenced that the severity of ducted explosions is mainly influenced by vigorous secondary explosions occurring in the duct. Valeria et al. [27] used a validated large-eddy simulation model to study the mechanism underlying vented gas explosions in the presence of obstacles. Methane–air mixtures with different composition ratios, variously shaped obstacles, and area block ratios were investigated. The influences of the combustion rate and venting rate on both the number and intensity of overpressure peaks were observed.

In this study, to systematically analyze how the vessel size, vessel structure, ignition position, and length of connection pipes influence natural gas explosions, the strongly validated N–S solver tool of FLACS [28], which has been developed continuously for more than 40 years for predicting the consequences of gas explosions [29,30], was employed. This tool includes a three-dimensional CFD code that solves Favre-averaged transport equations for mass, momentum, enthalpy, turbulent kinetic energy, rate of dissipation of turbulent kinetic energy, mass-fraction of fuel, and mixture-fraction on a structured Cartesian grid using a finite volume method. The Reynolds-averaged Navier-Stokes equations are closed by invoking the ideal gas equation of state and the standard k–ε model for turbulence. Furthermore, one of the key features that distinguishes FLACS from most commercial CFD codes is the use of the distributed porosity concept for representing complex geometries on relatively coarse computational meshes [28]. With this approach, large objects and walls are represented on-grid, whereas smaller objects are represented sub-grid. The pre-processor Porcalc reads the grid and geometry files and assigns volume and area porosities to each rectangular grid cell. In the simulations, the porosity field represents the local congestion and confinement, and this allows sub-grid objects to contribute with flow resistance (drag), turbulence generation and flame folding in the simulations.

Simulated-pressure data were compared with the experimental results received from Nanjing Tech University, where the explosion experiment with 10 vol% methane was conducted [13]. In addition, results revealed the distributions of temperature and concentrations of gas products occurring during gas explosions. This study mainly aimed to contribute to this research field by exploring the characteristics of gas explosion in linked vessels.

2. Simulation Objects

2.1. Experimental Apparatus and Geometric Model

The special-designed experimental device and its size and structure are displayed in Figure 1. This experimental system consisted of two different spherical vessels, where one is big and the other is small comparatively, and they connected each other with pipes having a varied length [13].

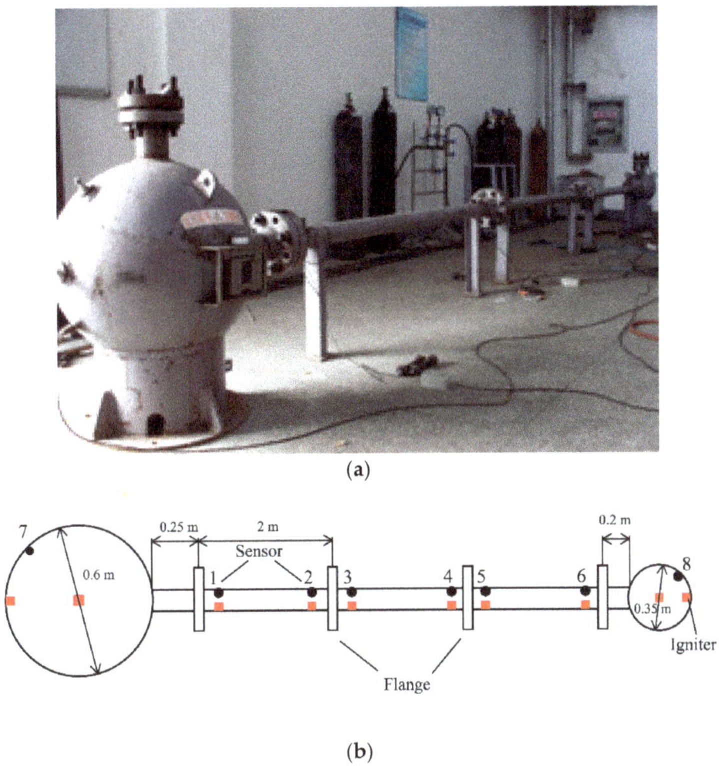

Figure 1. Picture and schematic diagram of the experimental device for the gas explosion. (**a**) Photo of the experimental device; (**b**) scheme structural diagram of linked vessels.

Two spherical vessels were fabricated from steel with a design pressure of 20 MPa and their internal diameters were 0.6 and 0.35 m (0.113 m^3 and 0.022 m^3 in volume, respectively). Seamless steel pipes, with an inner diameter and a thickness of 0.06 and 0.015 m, respectively, were connected using flanges and bolts. The pipes were divided into three parts (each part was 2 m in length), so that they could be combined into pipes with varied lengths of 2.45, 4.45, and 6.45 m between the big and small spherical vessels. Suitable monitor points were placed as dictated by the experiment. The entire explosion reactor with three different structures and the mesh are given in Figure 2.

Although only pressure data could be obtained from the experiment, the data regarding the temperature and explosion products were determined through numerical simulation. It provided further analysis and prediction dealing with unidentified explosion processes in linked vessels.

Based on the physical experiments, this study conducted natural gas explosions in three systems, which were a single spherical vessel, a single spherical vessel connected with a pipe, and a big spherical vessel connected to a small spherical vessel, individually. Meanwhile, multiple influence factors, including different structures, sizes of the vessels, lengths of the connection pipes, and ignition positions were investigated in detail.

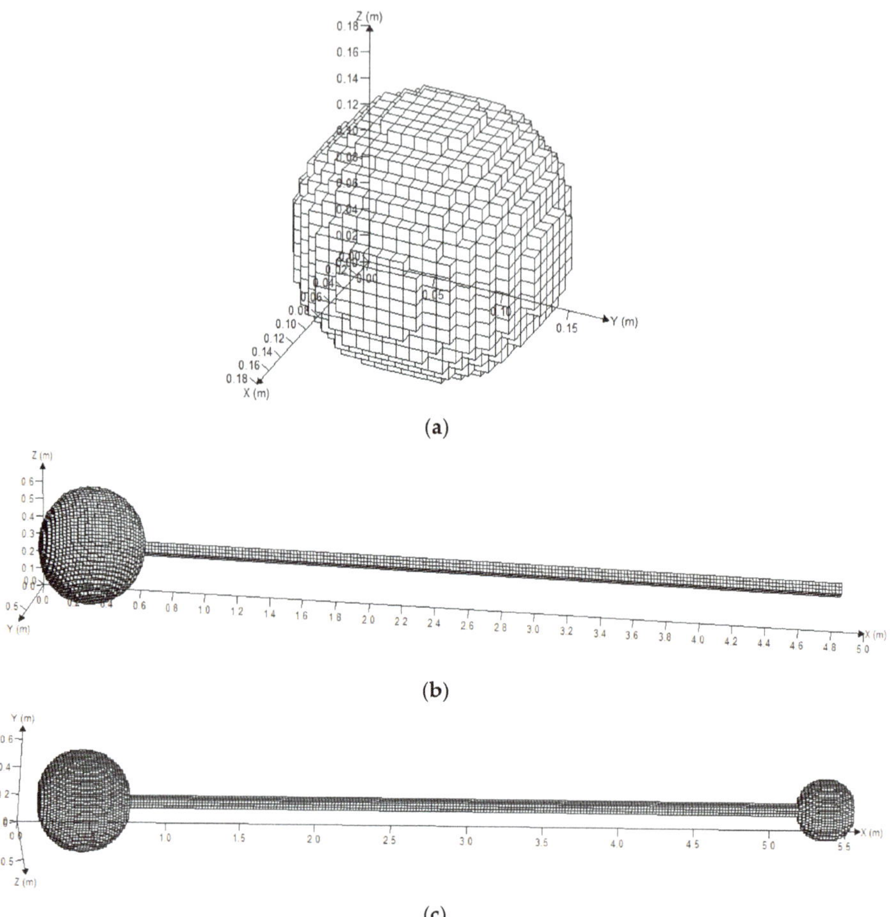

(**a**)

(**b**)

(**c**)

Figure 2. Grid division on the model of the three structures. (**a**) Grid division on the model of the spherical vessel; (**b**) grid division on the model of the single sphere with a pipe; (**c**) grid division on the model of the big vessel connected with the small vessel.

2.2. Boundary Conditions and Initial Conditions

According to the experimental conditions and the FLACS User's Manual [28], the initial conditions were set as standard pressure and temperature and the boundary conditions were set as Euler equations (inviscid flow equations), which is suitable for most explosion simulations. In FLACS, Euler equations are discretized for a boundary element. Namely, the momentum and continuity equations are solved on the boundary in the case of outflow. The concentration of methane used in the experiment was 10 vol%, and the vessel was filled entirely with the gas mixture. The parameter for the relative fuel–oxygen concentration employed in FLACS is denoted by equivalence ratio (ER), which is defined in Equation (1):

$$ER = \frac{\left(m_{fuel}/m_{oxygen}\right)_{actual}}{\left(m_{fuel}/m_{oxygen}\right)_{stoichiometric}} = \frac{\left(V_{fuel}/V_{oxygen}\right)_{actual}}{\left(V_{fuel}/V_{oxygen}\right)_{stoichiometric}} \tag{1}$$

where m is the gas mass and V is the gas volume. Accordingly, the ER of methane–oxygen was 1.058. In addition, the default choice of heat switch is close in FLACS, as large-scale explosions are not much influenced. Activating radiation model will let walls and objects have background temperature, and gas heat loss from radiation will calculate as well as radiation from hot objects around. This is useful for small-scale confined explosions with important heat effects and is suitable to apply in this study. Furthermore, based on the experiment results in small closed vessels, Luo et al. [31] confirmed that adding a radiation model is an effective way to improve the precision of calculated results. They found that the simulation results without adding the radiation model had a large error with an average error of 10.05%, while the simulation results with adding the radiation model had a small error with an average error of 1.88%. Therefore, adding a radiation model to the simulation of gas explosions was able to produce a sound agreement with experimental results. As a result, to examine a realistic situation, this study included a radiation model to improve the accuracy of the results.

3. Simulation Results with Different Structures

3.1. Simulation Results in the Single Spherical Vessel

3.1.1. Influence of Vessel Size on the Explosion Pressure

The center of the large sphere was at (0.35, 0.35, 0.35); the sensor mounted on the sidewall was at (0.075, 0.35, 0.35); the center of small sphere was at (0.19, 0.19, 0.19); and the sensor mounted on the sidewall was at (0.025, 0.19, 0.19). The simulation was performed following the methane explosion experiment at 10 vol% in two single vessels. The results of the simulation are plotted in Figure 3.

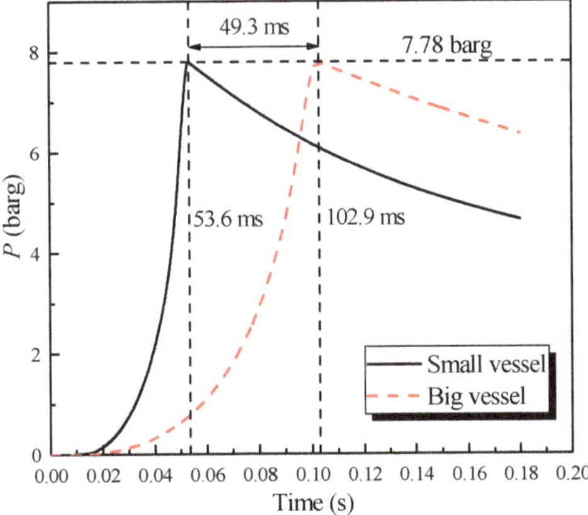

Figure 3. Natural gas explosion pressure varying with time in two single vessels.

The results revealed that the pressure plots at the center of the spheres and the sensor on the sidewall could be superposed. As shown in Figure 3, a nearly consistent maximum explosion pressure can be observed in the big/small vessel, which were 7.78 and 7.79 barg, respectively. However, the maximum explosion pressure in two vessels peaked at 102.9 and 53.6 ms, respectively, and there was a considerable difference of 49.3 ms, where the amplitude can be attained to 48%. Therefore, the rate of pressure rise in the big vessel was lower than that of the small vessel. As a result, without changing the initial conditions of the scenario, the peak pressure was same, but by contrast, the rate of the pressure rise was only relevant to the vessel size. Additionally, this finding was consistent with the "explosion cubic law" [32].

3.1.2. Simulation Results of the Explosion Parameters in the Single Vessel

The distributions on the X-Y section of explosion parameters containing pressure, temperature, and explosion products at the 10 vol% CH_4 mixture in the big vessel are presented in Figure 4.

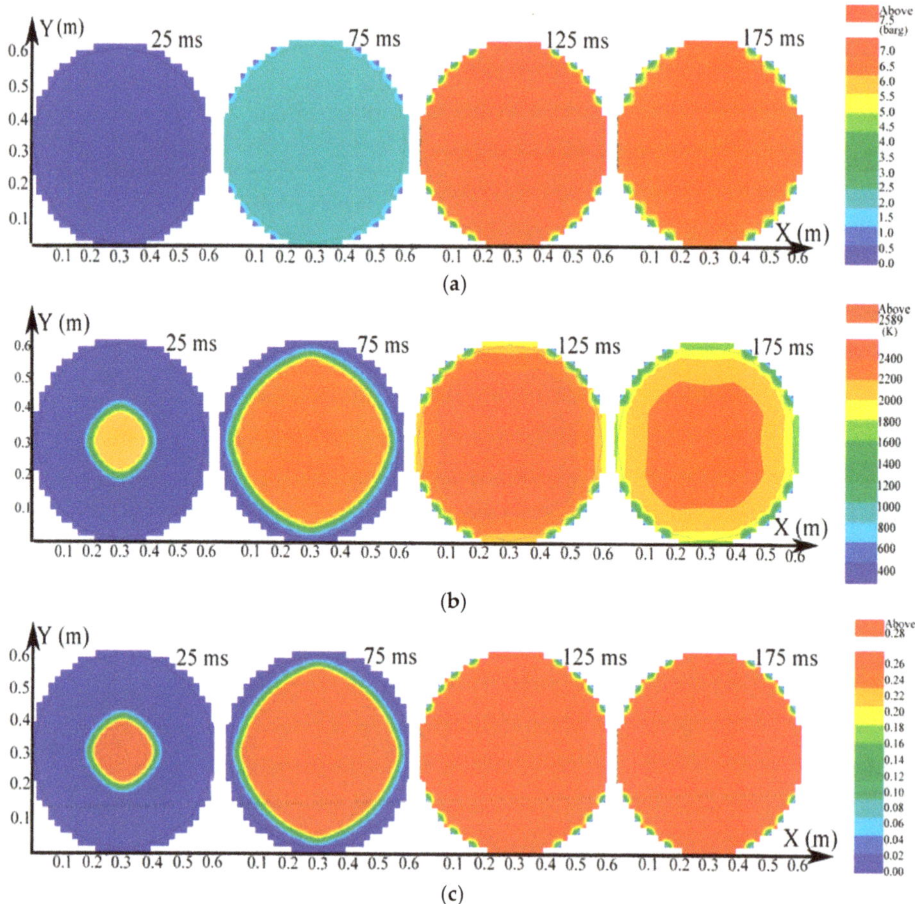

Figure 4. Contours of the natural gas explosion parameters containing pressure, temperature, and concentration of products in the larger vessel. (**a**) Distribution of pressure; (**b**) distribution of temperature; (**c**) distribution of the concentration of products.

As evidenced in Figure 4a, the pressure in the larger vessel was practically uniform at any specific time. The explosion wave expanded into a spherical wave and the temperature was high near the igniter but gradually decreased until it approached the leading edge of the combustion wave in Figure 4b. Here, the combustion wave reached the sidewall at 125 ms. The temperature in the big vessel began to decrease when the explosion ended. As indicated in Figure 4c, when more natural gas participated in the reaction, the distribution of the explosion products spread further until the gas was consumed. The laws of the distribution of pressure, temperature, and explosion products in the small vessel were the same as those in the big vessel.

The changing temperatures and concentrations of the explosion products in the vessels are plotted in Figure 5. When the explosion occurred, the temperature at the center of the vessel enhanced rapidly, reaching 2161 K at 6 ms. Then the rate of the temperature rise dropped, and the temperature at the center of the small vessel reached 2580 K at 53 ms, whereas in the big vessel it reached 2581 K at 101 ms. The center and sidewall reached the highest temperature simultaneously, but the energy was lost when the combustion wave contacted the sidewall. The highest temperature near the sidewall was lower than that at the center of the sphere. After it was ignited at the center of the sphere, the concentration of products at the center escalated promptly, reaching its highest value (27.7%) at 12 ms in Figure 5b. The theoretical concentration of products reached 28.35% based on the methane combustion chemical equation. Accordingly, the simulation results fitted the theoretical values.

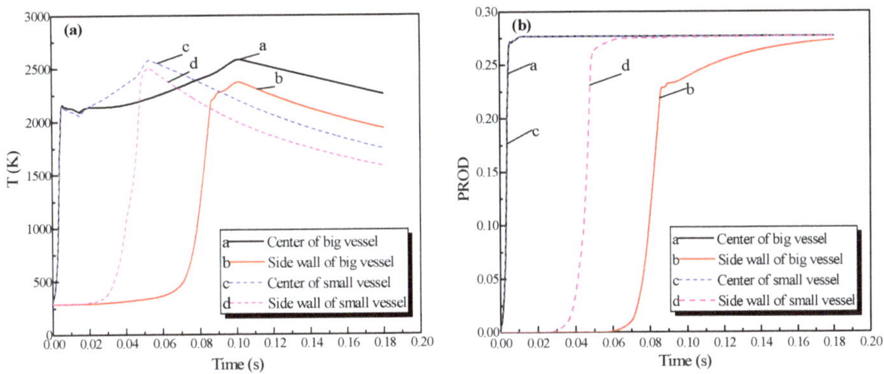

Figure 5. Profiles of the natural gas explosion parameters in the single vessel. (**a**) Profiles of explosion temperatures against time in the vessels; (**b**) concentration profiles against time for the explosion products in the vessels.

3.2. Simulation Results in for Single Vessel with Pipes Connected to IT

3.2.1. Simulation Results of the Explosion Parameters in the Single Vessel with Pipes Connected to It

The big sphere with the 4.25-m pipe was adopted as an example; the igniter was located at the center of the sphere. The distributions of the pressure, temperature, and concentration of products of the methane explosion are shown in Figure 6. The plots of the explosion temperature and the concentration of products are delineated in Figure 7.

Figure 6a shows that the peak pressure in the single vessel with a pipe connected to it was lower than that of the single vessel. As provided by Figure 6b, because of the oscillation sparked by the blast, the gas in the pipe was carried to the vessel, and a jet flow formed. Moreover, from Figure 6c, when the ignition source was located at the center of the vessel, the combustion wave spread 2.3 m in the pipe when it spread from the center to the sidewall in the vessel. Therefore, compared with the vessels alone, the pipes significantly accelerated the spread of the explosion.

In virtue of the oscillation of the explosion, the plots of temperature displayed a fluctuation at the center of the vessel (in Figure 7a). Moreover, the energy was lost when the combustion wave contacted the sidewall. The highest temperature near the sidewall was low and the fluctuations were small comparatively. The plots in Figure 7b reveal that as the explosion wave spread, the entire vessel was filled with combustion products, which reflected that the variation of products was not be affected by the oscillation.

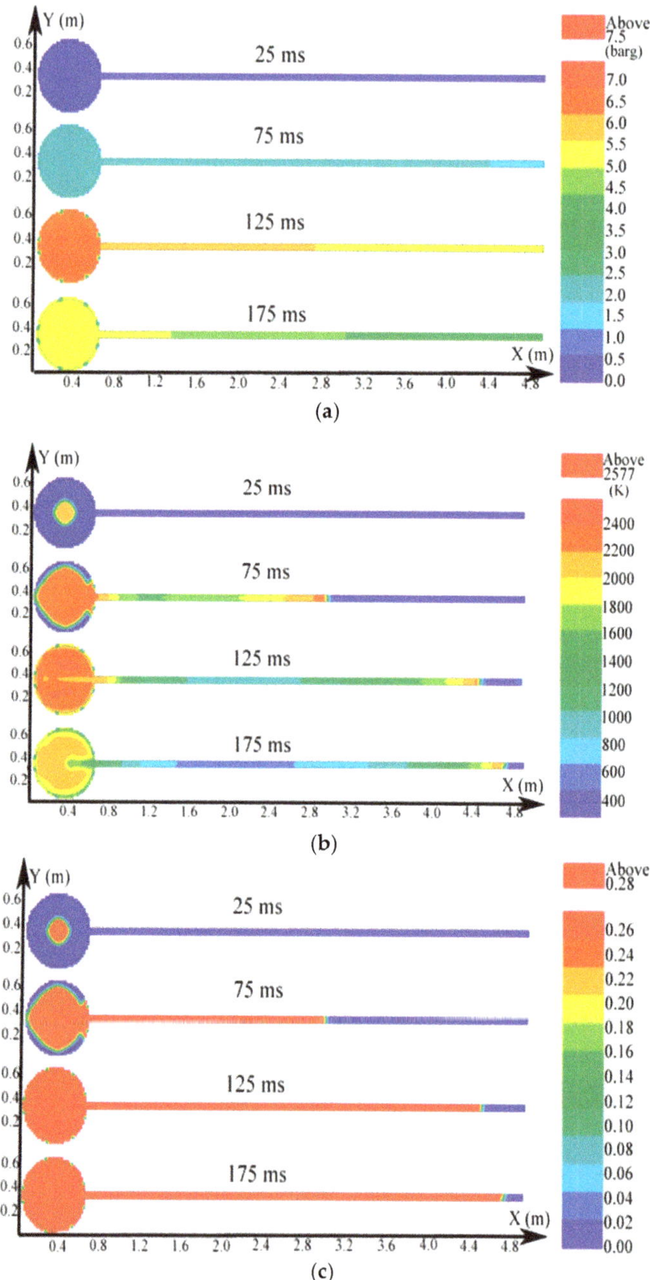

Figure 6. Contours of the natural gas explosion parameters (pressure, temperature, and concentration of products) in the single vessel with a pipe connected to it; (**a**) distribution of pressure. (**b**) distribution of temperature; (**c**) distribution of the concentration of products.

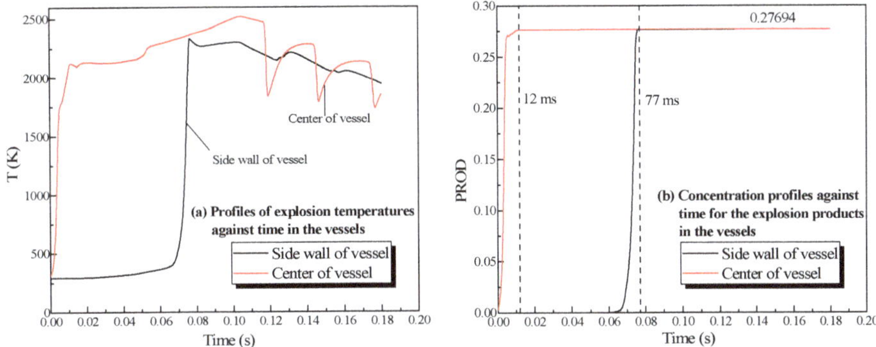

Figure 7. Profiles of the natural gas explosion parameters (temperature and concentration of products) in the single vessel with a pipe connected to it.

3.2.2. Influence of Changing the Ignition Position

The big vessel with the 4.25-m pipe was adopted as an example, the gas was ignited at the sidewall of the big vessel, the center of the sphere, location 1, and location 2 (in Figure 2). The plots of the explosion pressure as it changed with time are illustrated in Figure 8.

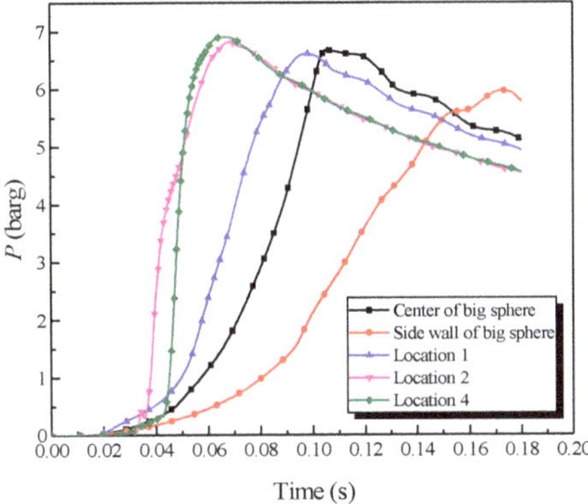

Figure 8. Natural gas explosion pressure versus time by changing the ignition positions.

As revealed in Figure 8, changing the ignition position substantially influenced the explosion pressure in the big vessel. The peak pressure locations in the big vessel with different ignition positions could be sorted from highest to lowest as follows: location 4, location 2, the center, location 1, and the sidewall, orderly. By contrast, the maximum rate of pressure rise could be sorted as follows: location 4, location 2, location 1, the center, and the sidewall, orderly. The combustion wave contacted the sidewall during the explosion when the ignition source was located at the sidewall. Due to the wall effect, the abundant energy was consumed, which resulted in the explosion intensity to be weaker than at the other ignition positions. Accordingly, the explosion pressure and the rate of pressure rise were diminished. Locations 1, 2, and 4 were located within the pipe, where the explosion occurred with a tighter constraint and the pipes accelerated the explosion wave. Location 4 was at the end of the pipe, at which the acceleration distance was the longest. The explosion intensity was most vigorous when the combustion wave spread to the big vessel. Therefore, the explosion pressure and the rate of pressure rise were the largest when the ignition source was at location 4.

3.2.3. Influence of the Length of the Connection Pipes

The ignition source was located at the center of the vessel. The explosion pressure in different lengths of pipe is shown in Figure 9.

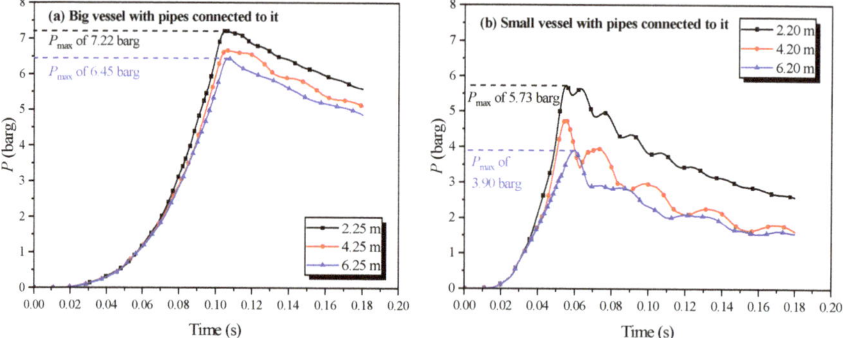

Figure 9. Natural gas explosion pressure versus time by changing the length of the pipe connected to the single vessel.

Figure 9 shows that the size effect [33] of each pipe length had a noticeable influence on the explosion intensity in the vessel. The peak explosion pressure lessened as increasing length of the connection pipe, with the influence being the greatest in the small vessel. In any single vessel with a connected pipe, the energy from the explosion was consumed by the wall. More energy was lost when the lengths of the connected pipes increased on account of the greater area of the inner wall. Consequently, the peak explosion pressure decreased with the increase in the lengths of the connected pipes.

3.3. Simulation Results of the Big Vessel Connected to the Small Vessel

3.3.1. Simulation Results of the Explosion Parameters in the Big Vessel Connected to the Small Vessel

Regarding the simulation of the big vessel with the 4.45-m pipe and the small vessel, the ignition source was located at the center of the big vessel. Changes in the distribution of the pressure, temperature, and concentration of products are presented in Figure 10. The plots of the temperature and the concentration profiles of the explosion products are depicted in Figure 11.

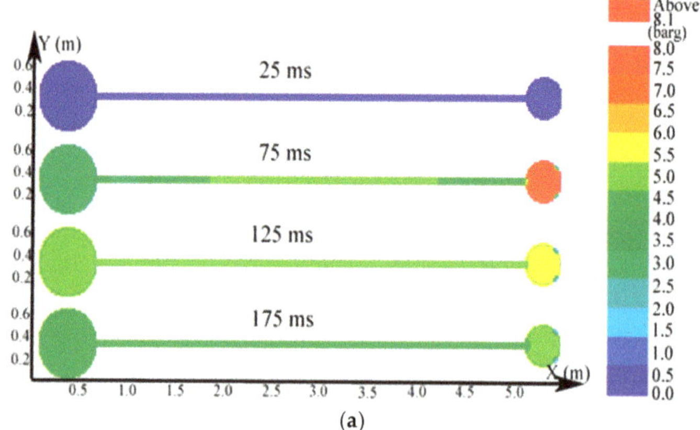

Figure 10. *Cont.*

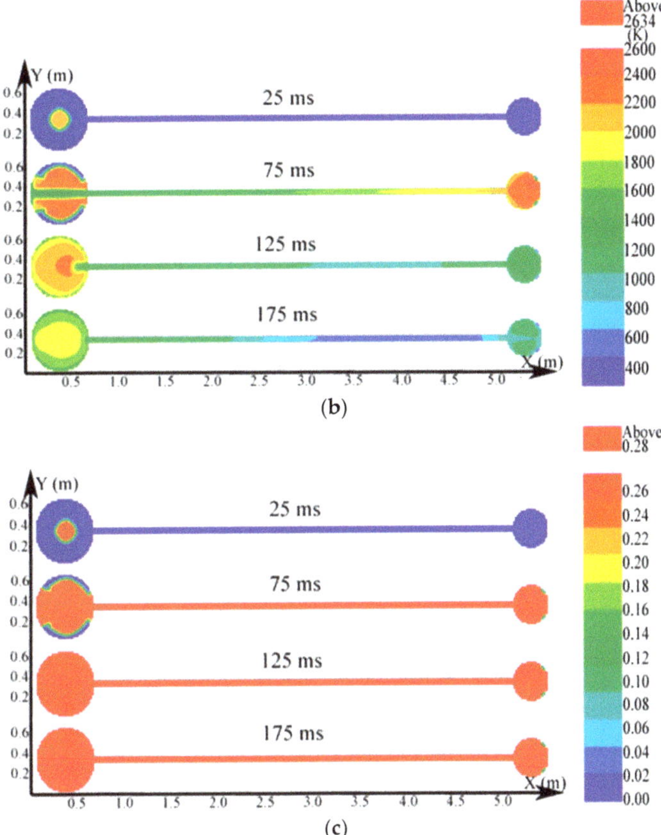

Figure 10. Contours of the natural gas explosion parameters in the big vessel connected to the small vessel. (**a**) Distribution of pressure; (**b**) distribution of temperature; (**c**) distribution of the concentration of products.

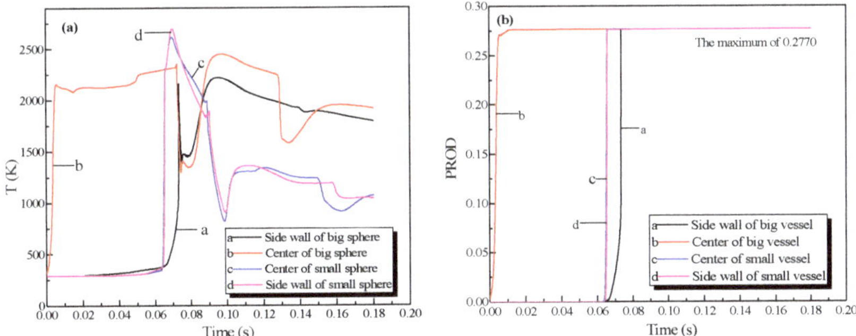

Figure 11. Profiles of the natural gas explosion parameters (temperature and concentration of products versus time) in the big vessel connected to the small vessel. (**a**) Profiles of explosion temperatures against time in the vessels; (**b**) concentration profiles against time for the explosion products in the vessels.

As illustrated in Figure 10a, the peak pressure was attained in the small vessel at 75 ms and was considerably higher than that in the big vessel. The temperature in the pipe was lower than that in the sphere because energy was absorbed by the wall in the long pipe in Figure 12b. Gas at a lower temperature was brought into the sphere as a result of the oscillation originated from the explosion wave. Here, the low-temperature jet cuts the temperature distribution in the sphere to a symmetrical structure.

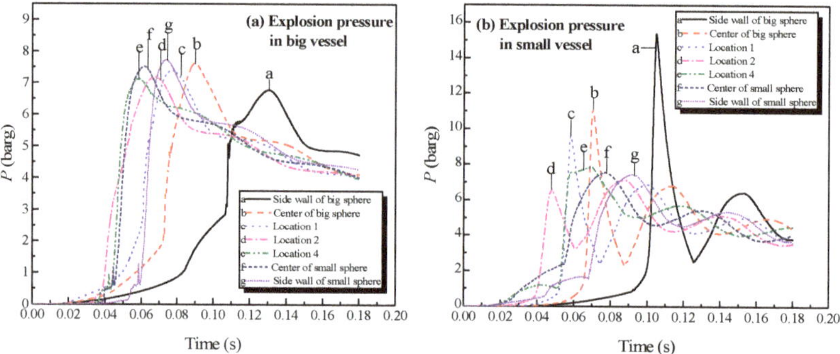

Figure 12. Natural gas explosion pressure versus time in the large and small vessels at different ignition positions.

In Figure 11a, the explosion wave reached its highest temperature (2700 K) at 70 ms in the small vessel. By contrast, the temperature was 2400 K in the big vessel at the same time. Because of the oscillation of the combustion wave, the temperature at the center of the big vessel plummeted to 1300 K at 74 ms after the combustion wave spread into the small vessel, which then formed a second peak.

Figure 11b shows that the concentration of products enhances rapidly at 62 ms. Namely, the combustion wave spread into the small vessel through the connected pipe at that moment. Furthermore, the combustion wave spread to the sidewall of the big vessel at 65 ms. This result again demonstrated the fact that the pipes accelerated the combustion wave.

3.3.2. How Changing the Ignition Position Influenced the Gas Explosion

Considering the big vessel with the 4.45-m pipe and small vessel as an example, the ignition positions were the sidewall of the big vessel, the center of the big vessel, location 1, location 2, location 4, location 6, the center of the small vessel, and the sidewall of the small vessel, as depicted in Figure 2. A methane mixture with a concentration of 10 vol% was intentionally selected in the simulations. The plots of the explosion pressure in the large and small vessels are shown in Figure 12.

As exhibited in Figure 12a, the peak pressure was attained in the big vessel when the ignition was positioned at the sidewall of the small vessel. The minimum pressure value was obtained when the ignition was positioned at the sidewall of the big vessel. Figure 12b shows that the peak pressure decreased when the ignition was moved from the sidewall in the big vessel to location 2, then the peak pressure increased slightly when the ignition was moved from location 2 to the sidewall in the small vessel. Therefore, the pressure piling [21,22] originated from the oscillation was the weakest at location 2.

3.3.3. How Changing the Length of the Connection Pipe Influenced the Gas Explosion

In the experiment, the vessel that contained the ignition source was defined as the initiating vessel, and the other was defined as the secondary vessel. The explosion results were determined variably with the change in the length of the pipe, as drawn in Figure 13.

Figure 13a,b exhibited that the length of the pipe affected the pressure in the initiating vessel. The peak pressure in the initiating vessel was reached when the length of the connecting pipe was 4.45 m. As the length increased, the peak pressure in the initiating vessel increased, yet decreased when the pipe was longer than 4.45 m. The acceleration from the pipes on the combustion wave reinforced the explosion intensity, but as the length of the pipe increased, more energy was dissipated by the cause of the wall. When the length of the connecting pipe was excessive, a more momentous influence on the acceleration from the wall than that from the pipes could be demonstrated. Comparing Figure 13a together with Figure 13c,b together with Figure 13d revealed that the peak pressure in the initiating vessel was lower than that in the secondary vessel. Moreover, when the initiating vessel was the big vessel, the peak pressure in the second vessel was considerably higher.

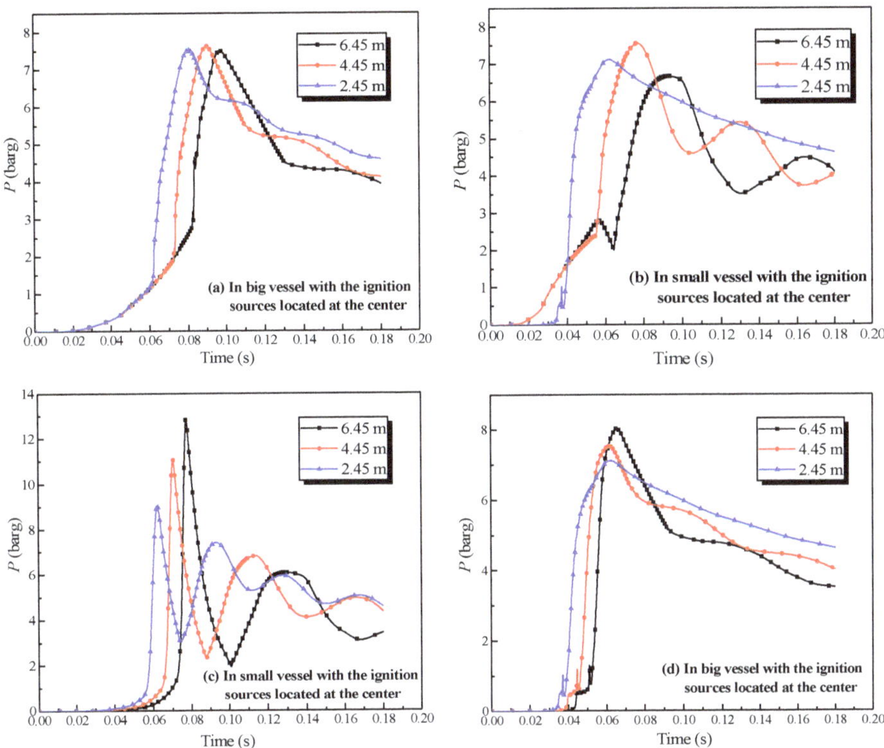

Figure 13. Natural gas explosion pressure-time curves in the vessels with changing the length of the pipe.

4. Error Analysis

The simulation results and experimental data were compared for two different structures, which were the big vessel connected to the small vessel with the 4.45-m pipe and the 6.45-m pipe. The peak pressure values at different ignition positions from the simulation and experiment are shown in Figure 14.

In Figure 14a, "A1" to "A7" correspond to the adopted ignition positions: the sidewall of the big vessel, the center of the big vessel, location 1, location 2, location 4, the center of small vessel, and the sidewall of the small vessel. Similarly, in Figure 14b, "B1" to "B8" represent the adopted ignition positions: the sidewall of the big vessel, the center of the big vessel, location 1, location 2, location 4, location 6, the center of the small vessel, and the sidewall of the small vessel.

Figure 14 shows that the variation that depended with the ignition positions of the peak pressure in the simulation and experiment were analogous. From Figure 14a, the average error between the simulation and experimental data is 7.7% in the big vessel with the 4.45-m pipe connected to it. The average error is shown in Figure 14b is 8.04%. Overall, 70% of the simulation data were marginally larger than those of the experimental data in all the compared points, in which the average difference was 6.97%. Generally, an experiment was controlled by more factors in comparison with a simulation. Even though the radiation model was employed, with the aim of improving simulated accuracy, some degree of discrepancy still could be observed. The increase of the temperature in the wall of the pipes was limited in the physical experiment. A considerable temperature difference existed between the combustion products and the wall of the pipe. The energy was dissipated on account of thermal convection, thermal conduction, and thermal radiation. As a result, the simulation data had higher values than the experimental data [34]. Consequently, on the one hand, the simulation effectively reproduced the development of the gas explosion in linked vessel. On the other hand, it correctly predicted the variations of pressure, temperature, and the changing concentration of products with different factors.

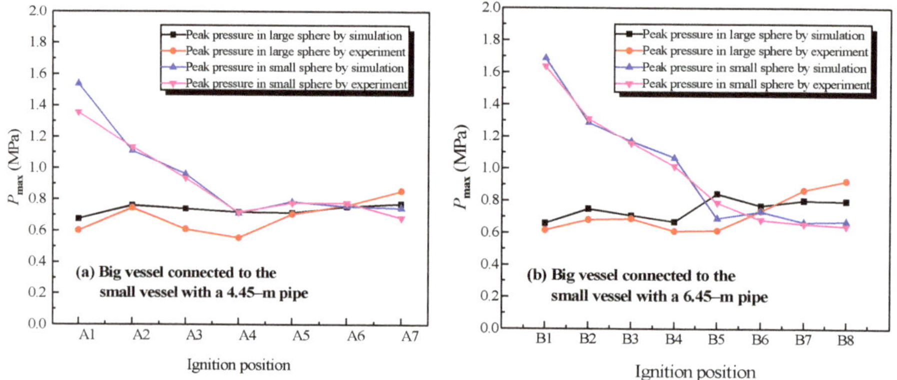

Figure 14. Comparison between the simulation data and experimental data of the natural gas explosion pressure versus time in the linked vessels.

5. Conclusions

In this study, natural gas explosions were conducted, employing the simulation tool of FLACS in three systems, including a single vessel, a single vessel with a pipe connected to it, and a big vessel connected to a small vessel with a pipe. The main conclusions of this study are as follows:

- When the explosions occurred in the same initial conditions, the peak explosion pressure maintained unchanged. However, the maximum rate of pressure rise decreased when the volume of the closed vessels increased, which is following the explosion cubic law.
- The length of the connecting pipe profoundly affected the explosion intensity in the single vessel with a pipe connected to it and in the big vessel connected to the small vessel. Furthermore, an oscillation of the explosion wave developed in the linked vessels.
- The ignition position affected the explosion intensity in the linked system extremely. The explosion intensity was weakest when the ignition source was positioned at the center of the system.
- When the explosions occurred in the linked system, the explosion intensities in the secondary vessels were stronger than those in the initiating vessels.
- Using FLACS software for the simulation of the small–scale linked system, the average difference attained 6.97%. Accordingly, the simulation results agreed reasonably well with the experimental results and the actual explosion conditions were reflected in the FLACS simulation to an acceptable extent.

Author Contributions: All authors contributed equally to the research presented in this paper and to the preparation of the final manuscript. Conceptualization, Q.W.; Formal analysis, Y.S. and X.L.; Methodology, Z.W. and M.Z.; Supervision, J.J.; Software, F.C.; Writing—original draft, Y.S. and X.L.; Writing—review, Q.W. and C.-M.S. All authors have read and agreed to the published version of the manuscript.

Funding: The study was financially supported by the National Key R &D Program of China (2016YFC0800100); the National Nature Science Foundation of China (51504190); and the Natural Science Basic Research Plan in Shaanxi Province of China (2017JM5068).

Conflicts of Interest: The authors declare no conflict of interest.

References

1. Kundu, S.K. Understanding and eliminating pressure fluctuations in an extended chlor-alkali plant due to the size detail of seal pots: A design correlation. *Process Saf. Environ. Prot.* **2010**, *88*, 91–96. [CrossRef]

2. Ni, T.W.; Bi, T.T.; Yang, Z.G. Failure analysis on abnormal perforation of super large diameter buried gas pipeline nearby metro. *Eng. Fail. Anal.* **2019**, *103*, 32–43. [CrossRef]

3. Chen, C.H.; Sheen, Y.N.; Wang, H.Y. Case analysis of catastrophic underground pipeline gas explosion in Taiwan. *Eng. Fail. Anal.* **2016**, *65*, 39–47. [CrossRef]

4. Yin, W.; Fu, G.; Yang, C.; Jiang, Z.; Zhu, K.; Gao, Y. Fatal gas explosion accidents on Chinese coal mines and the characteristics of unsafe behaviors: 2000–2014. *Saf. Sci.* **2017**, *92*, 173–179. [CrossRef]

5. Zhao, X.; Chunming, J.; Yanping, W.; Jiwu, Y.; Zheng, W. Simulation study on an explosion accident in China. *Process Saf. Prog.* **2014**, *33*, 56–63. [CrossRef]

6. Wang, Z.R.; Pan, M.Y.; Jiang, J.C. Experimental investigation of gas explosion in single vessel and connected vessels. *J. Loss Prev. Process Ind.* **2013**, *26*, 1094–1099. [CrossRef]

7. Zhang, K.; Wang, Z.; Chen, Z.; Jiang, F.; Wang, S. Influential factors of vented explosion position on maximum explosion overpressure of methane-air mixture explosion in single spherical container and linked vessels. *Process Saf. Prog.* **2018**, *37*, 248–255. [CrossRef]

8. Kindracki, J.; Kobiera, A.; Rarata, G.; Wolanski, P. Influence of ignition position and obstacles on explosion development in methane-air mixture in closed vessels. *J. Loss Prev. Process Ind.* **2007**, *20*, 551–561. [CrossRef]

9. Zuo, Q.; Wang, Z.; Zhen, Y.; Zhang, S.; Cui, Y.; Jiang, J. The Effect of an Obstacle on Methane-Air Explosions in a Spherical Vessel Connected to a Pipeline. *Process Saf. Prog.* **2016**, *36*, 1–7. [CrossRef]

10. Zhang, K.; Wang, Z.; Ni, L.; Cui, Y.; Zhen, Y.; Cui, Y. Effect of one obstacle on methane–air explosion in linked vessels. *Process Saf. Environ. Prot.* **2016**, *105*, 217–223. [CrossRef]

11. Zhang, K.; Wang, Z.; Gong, J.; Liu, M.; Dou, Z.; Jiang, J. Experimental study of effects of ignition position, initial pressure and pipe length on H_2-air explosion in linked vessels. *J. Loss Prev. Process Ind.* **2017**, *50*, 295–300. [CrossRef]

12. Na'inna, A.M.; Phylaktou, H.N.; Andrews, G.E. Explosion flame acceleration over obstacles: Effects of separation distance for a range of scales. *Process Saf. Environ. Prot.* **2017**, *107*, 309–316. [CrossRef]

13. Cui, Y.; Wang, Z.; Jiang, J.; Liu, X. Size effect on explosion intensity of methane-air mixture in spherical vessels and pipes. *Procedia Eng.* **2012**, *45*, 483–488. [CrossRef]

14. Zhang, K.; Wang, Z.; Yan, C.; Cui, Y.; Dou, Z.; Jiang, J. Effect of size on methane-air mixture explosions and explosion suppression in spherical vessels connected with pipes. *J. Loss Prev. Process Ind.* **2017**, *49*, 785–790. [CrossRef]

15. Zhang, B.; Bai, C.; Xiu, G.; Liu, Q.; Gong, G. Explosion and flame characteristics of methane/air mixtures in a large-scale vessel. *Process Saf. Prog.* **2014**, *33*, 362–368. [CrossRef]

16. Yu, M.; Wan, S.; Zheng, K.; Guo, P.; Chu, T.; Wang, C. Effect of side venting areas on the methane/air explosion characteristics in a pipeline. *J. Loss Prev. Process Ind.* **2018**, *54*, 123–130. [CrossRef]

17. Wan, S.; Yu, M.; Zheng, K.; Wang, C.; Yuan, Z.; Yang, X. Effect of side vent size on a methane/air explosion in an end-vented duct containing an obstacle. *Exp. Therm. Fluid Sci.* **2019**, *101*, 141–150. [CrossRef]

18. Cui, Y.Y.; Wang, Z.R.; Ma, L.S.; Zhen, Y.Y.; Sun, W. Influential factors of gas explosion venting in linked vessels. *J. Loss Prev. Process Ind.* **2017**, *46*, 108–114. [CrossRef]

19. Di Benedetto, A.; Salzano, E. CFD simulation of pressure piling. *J. Loss Prev. Process Ind.* **2010**, *23*, 498–506. [CrossRef]

20. Di Benedetto, A.; Salzano, E.; Russo, G. Predicting pressure piling by semi-empirical correlations. *Fire Saf. J.* **2005**, *40*, 282–298. [CrossRef]

21. Phylaktou, H.; Andrews, G.E. Gas explosions in linked vessels. *J. Loss Prev. Process Ind.* **1993**, *6*, 15–19. [CrossRef]

22. Zhang, Q.; Jiang, J.; You, M.; Yu, Y.; Cui, Y. Experimental study on gas explosion and venting process in interconnected vessels. *J. Loss Prev. Process Ind.* **2013**, *26*, 1230–1237. [CrossRef]

23. Singh, J. Gas explosions in inter-connected vessels: Pressure piling. *Process Saf. Environ. Prot. Trans. Inst. Chem. Eng. Part B* **1994**, *72*, 220–228.

24. Maremonti, M.; Russo, G.; Salzano, E.; Tufano, V. Numerical simulation of gas explosions in linked vessels. *J. Loss Prev. Process Ind.* **1999**, *12*, 189–194. [CrossRef]

25. Deng, J.; Cheng, F.; Song, Y.; Luo, Z.; Zhang, Y. Experimental and simulation studies on the influence of carbon monoxide on explosion characteristics of methane. *J. Loss Prev. Process Ind.* **2015**, *36*, 45–53. [CrossRef]

26. Ferrara, G.; Willacy, S.K.; Phylaktou, H.N.; Andrews, G.E.; Di Benedetto, A.; Salzano, E.; Russo, G. Venting of gas explosion through relief ducts: Interaction between internal and external explosions. *J. Hazard. Mater.* **2008**, *155*, 358–368. [CrossRef]

27. Di Sarli, V.; Di Benedetto, A.; Russo, G. Using Large Eddy Simulation for understanding vented gas explosions in the presence of obstacles. *J. Hazard. Mater.* **2009**, *169*, 435–442. [CrossRef]

28. GexCon, A.S. *FLACS V10.3 User's Manual*; GexCon AS: Bergen, Norway, 2014.

29. Bleyer, A.; Taveau, J.; Djebaïli-Chaumeix, N.; Paillard, C.E.; Bentaïb, A. Comparison between FLACS explosion simulations and experiments conducted in a PWR Steam Generator casemate scale down with hydrogen gradients. *Nucl. Eng. Des.* **2012**, *245*, 189–196. [CrossRef]

30. Middha, P.; Hansen, O.R.; Grune, J.; Kotchourko, A. CFD calculations of gas leak dispersion and subsequent gas explosions: Validation against ignited impinging hydrogen jet experiments. *J. Hazard. Mater.* **2010**, *179*, 84–94. [CrossRef]

31. Luo, Z.M.; Zhang, Q.; Hua, W.; Cheng, F.M.; Tao, W.; Deng, J. Numerical simulation of gas explosion in confined space with FLACS. *J. China Coal Soc.* **2013**, *38*, 1381–1387. (In Chinese)

32. Bartknecht, W. *Dust Explosions: Course, Prevention, Protection*; Springer Science & Business Media: Berlin, Germany, 2012; ISBN 3642739458.

33. Ivanov, A.G. Dynamic fracture and scale effects (survey). *J. Appl. Mech. Tech. Phys.* **1994**, *35*, 430–442. [CrossRef]

34. Wang, Z.R.; Jiang, J.C.; Zhou, C. Experimental investigation of gas explosion characteristic in linked vessels. *Explos Shock Wave* **2011**, *31*, 69–74. (In Chinese)

MDPI

St. Alban-Anlage 66

4052 Basel

Switzerland

Tel. +41 61 683 77 34

Fax +41 61 302 89 18

www.mdpi.com

Processes Editorial Office

E-mail: processes@mdpi.com

www.mdpi.com/journal/processes